METHODS IN CELL BIOLOGY

VOLUME 28
Dictyostelium discoideum:
Molecular Approaches to Cell Biology

Series Editor

LESLIE WILSON

Department of Biological Sciences
University of California, Santa Barbara
Santa Barbara, California

METHODS IN CELL BIOLOGY

Prepared under the Auspices of the American Society for Cell Biology

VOLUME 28
Dictyostelium discoideum:
Molecular Approaches to Cell Biology

Edited by

JAMES A. SPUDICH

DEPARTMENT OF CELL BIOLOGY
STANFORD UNIVERSITY SCHOOL OF MEDICINE
STANFORD, CALIFORNIA

1987

ACADEMIC PRESS, INC.
Harcourt Brace Jovanovich, Publishers

Orlando   San Diego   New York   Austin
Boston   London   Sydney   Tokyo   Toronto

ACADEMIC PRESS, INC.
Orlando, Florida 32887

United Kingdom Edition published by
ACADEMIC PRESS INC. (LONDON) LTD.
24–28 Oval Road, London NW1 7DX

LIBRARY OF CONGRESS CATALOG CARD NUMBER: 64-14220

ISBN 0–12–564128–1 (alk. paper)

PRINTED IN THE UNITED STATES OF AMERICA

87 88 89 90   9 8 7 6 5 4 3 2 1

8/8/87

CONTENTS

v

4. *Molecular Biology in Dictyostelium: Tools and Applications*
   W. Nellen, S. Datta, C. Reymond, A. Sivertsen,
   S. Mann, T. Crowley, and R. A. Firtel

PART II. VESICULAR TRAFFIC IN CELLS

5. *Purification and Characterization of Dictyostelium
   discoideum Plasma Membranes*
   Catherine M. Goodloe-Holland and Elizabeth J. Luna

6. *Endocytosis and Recognition Mechanisms in
   Dictyostelium discoideum*
   Günter Vogel

PART III. CELL MOTILITY

11. Isolation of the Actin Cytoskeleton from Amoeboid Cells of Dictyostelium
Annamma Spudich

12. Preparation of Deuterated Actin from Dictyostelium discoideum
Deborah B. Stone, Paul M. G. Curmi, and Robert A. Mendelson

13. Amino-Terminal Processing of Dictyostelium discoideum Actin
Peter A. Rubenstein, K. L. Redman, L. R. Solomon, and D. J. Martin

PART IV. CHEMOTAXIS

18. *Identification of Chemoattractant-Elicited Increases in Protein Phosphorylation*
Catherine Berlot

19. *Agar-Overlay Immunofluorescence: High-Resolution Studies of Cytoskeletal Components and Their Changes during Chemotaxis*
Yoshio Fukui, Shigehiko Yumura, and Toshiko K. Yumura

PART V. CELL ADHESION AND CELL–CELL RECOGNITION

20. *Cell Adhesion: Its Quantification, Assay of the Molecules Involved, and Selection of Defective Mutants in Dictyostelium and Polysphondylium*
Salvatore Bozzaro, Rainer Merkl, and Günther Gerisch

21. *Discoidins I and II: Endogenous Lectins Involved in
    Cell–Substratum Adhesion and Spore Coat Formation*
    S. H. Barondes, D. N. W. Cooper, and W. R. Springer

PART VI. DEVELOPMENT AND PATTERN FORMATION

22. *Methods for Manipulating and Investigating Developmental
    Timing in Dictyostelium discoideum*
    David R. Soll

27. *Homologous Recombination in Dictyostelium as a Tool for the Study of Developmental Genes*
    Arturo De Lozanne

Appendix. *Codon Frequency in Dictyostelium discoideum*
    Hans M. Warrick

PREFACE

The central point of this book is to present *Dictyostelium* as a valuable eukaryotic organism for those interested in molecular studies that require a combined biochemical, structural, and genetic approach. The book is not meant to be a comprehensive compilation of all methods involving *Dictyostelium,* but instead is a selective set of chapters that demonstrates the utility of the organism for molecular approaches to interesting cell biological problems.

This book developed out of multiple discussions with Les Wilson over the years concerning the need to have a focused investigation on one eukaryotic organism for the study of problems in cell biology. The field of biochemistry has advanced rapidly in large part by focusing on *Escherichia coli* and certain bacteriophages in the past three decades. Since then, more complex eukaryotic biological problems, such as membrane shuttling, cell motility, chemotaxis, cell–cell recognition, and development and pattern formation, have been studied in a wide variety of cell types. This diversity is very important for many reasons, including the fact that no one eukaryotic organism has emerged that is as suitable as *E. coli* has been to the biochemists. However, there are many benefits that derive from a large number of scientists studying the same organism from multiple points of view. In this way, an extensive biochemical, structural, and genetic foundation is laid which all researchers can build upon and utilize for their own studies. Among eukaryotes, one good candidate for intensive study is *Dictyostelium discoideum.* Not only does this cell behave as a typical eukaryote, but a large physiological and biochemical foundation has already been laid for this organism. Furthermore, recent development of vectors for the transformation of *Dictyostelium* opens up this organism to the modern world of molecular genetics.

A number of criteria are essential in choosing an organism to focus on for molecular studies. First and foremost the organism must be easily manipulated biochemically. Only then can one hope to elucidate the structure and function of proteins that represent even minor components of the cells. Also critical is that conventional methods of genetics exist or can be developed. Cloning strategies must also be possible. This book presents methods currently available in these and other important areas, as they are being applied to *Dictyostelium discoideum.* The book is divided into two major sections. The first deals with general principles and the second presents many examples of applications of these principles to modern cell biological problems.

This book is intended for two categories of investigators. One category consists of those who already work on *Dictyostelium* and wish to have a description of a variety of molecular techniques currently available. The other consists of those who are searching for an appropriate organism in which to study the cell biological problem of their particular interest. In 1971, I spent the better part of a year searching for the most appropriate organism for my own research interests. *Dictyostelium* was originally introduced to me by Bill Loomis, who has generously provided me with materials and advice for the past 15 years. His books on *Dictyostelium* [Loomis, W. F. (1975). *"Dictyostelium discoideum*: A Developmental System." Academic Press, New York; Loomis, W. F., ed. (1982). "The Development of *Dictyostelium discoideum."* Academic Press, New York.] have been extremely valuable for researchers interested in this organism, and the current volume, being more oriented around methods, is meant to complement those texts. In editing this volume, I have benefited greatly from the advice and criticism that were given at various times by Margaret Clarke, Rex Chisholm, Peter Devreotes, Richard Firtel, Bill Loomis, Harold McGee, Anna Spudich, and Les Wilson.

JAMES A. SPUDICH

Part I.   General Principles

This section deals with general principles regarding the use of *Dictyostelium*. The first chapter by Spudich briefly outlines the purpose of the book (see also the Preface) and introduces the organism. Some biochemical considerations of importance are also detailed in that chapter. In the next three chapters, general methods are presented for growth and manipulation of *D. discoideum* and for the use of classical genetics and molecular biological techniques. In Chapter 2, Sussman describes how to cultivate cells using bacteria as a food source as well as axenically (meaning in the absence of bacteria), how to maintain and preserve strains, and what conditions to use for synchronous development (see also Soll, Chap. 22). In Chapter 3, Loomis describes genetic tools for *D. discoideum*. His descriptions include mutagenesis, selection and screening techniques, mutant analysis, and procedures involving genetic exchange (see also this volume, Chapter 27). Molecular biological procedures applied to *D. discoideum* are then presented in Chapter 4 by Nellen *et al.* They describe details of how to transform *D. discoideum* and also provide numerous methods and cautions for nucleic acid isolations, preparation of genomic and cDNA libraries (see also Chisholm, Chap. 25), mapping and sequencing of *Dictyostelium* genes, and analysis of transcripts. Additional information on all these approaches is found in the subsequent chapters (Parts II–VI), which deal with specific applications of these techniques to problems of interest in cell biology.

Chapter 1

Introductory Remarks and Some Biochemical Considerations

JAMES A. SPUDICH

Department of Cell Biology
Stanford University School of Medicine
Stanford, California 94305

I. Introductory Remarks

This book is about molecular approaches to cell biology, and it focuses on *Dictyostelium discoideum* as an excellent eukaryotic organism for the study of cell biological problems. This organism grows happily in a nutritive environment, doubling about every 4–10 hours. Bacteria, which are readily ingested by *Dictyostelium* in large numbers, serve as a favorite food. When this organism is confronted by conditions of starvation, it has the marvelous capacity to become social, and in so doing $\sim 10^5$ cells communicate and cooperate in the formation of a fruiting body that contains a differentiated form of the cells, known as spores, which are very resistant to harsh environmental conditions. This process of differentiation and development can be made to occur in a highly synchronous fashion in the laboratory, the entire process requiring only ~ 24 hours. For investigators interested in vesicular traffic in cells, chemotaxis, cell adhesion, and many other cell biological processes, all of these phenomena are prominent in the life of this organism. Its usefulness in the laboratory is evident from the chapters contained in this book as well as in a variety of other books, such as those by Loomis (1975, 1982), and from a very large number of primary articles in the literature that deal with this organism. In the field in which my colleagues and I work, cell motility and the relationship of dynamic changes in the cytoskeleton to external signals, many of the approaches described in this book are being used to provide important information.

3

One good example of the multifaceted approach afforded by *Dictyostelium* is the study of the role of myosin phosphorylation in cell motility and signal transduction. Experiments by Rahmsdorf *et al.* (1978), Malchow *et al.* (1981), and subsequently by Berlot *et al.* (1985) revealed changes in the state of phosphorylation of myosin which are clearly coupled to the cAMP signaling that leads to chemotaxis of *Dictyostelium*. A transient increase in phosphorylation occurs *in vivo* on both the myosin light chain and on the heavy chain in response to the chemoattractant (Berlot *et al.*, 1985, 1987). These experiments were possible because *Dictyostelium* is easily grown, can be obtained in large amounts for serious biochemical work, synchronously undergoes a chemotactic response to cAMP as a cell suspension, and can be labeled *in vivo* with [^{32}P]orthophosphate. Other biochemical experiments have shown that phosphorylation of the light chain is necessary for *Dictyostelium* myosin to move on actin filaments (Griffith and Spudich, 1987), as measured in an *in vitro* myosin movement assay (Sheetz *et al.*, 1984). The heavy-chain phosphorylation, on the other hand, inhibits myosin thick filament formation (Kuczmarski and Spudich, 1980), and may be important for terminating a chemotactic response by inducing the disassembly of the transiently formed contractile apparatus (Berlot *et al.*, 1987).

This detailed biochemical information is now being complemented by molecular genetic studies. Eventually these studies should include site-directed mutagenesis of the regions of interest on the molecule, coupled with expression of the altered gene in an appropriate host, such as *E. coli*, yeast, or *Dictyostelium*. These expressed, modified proteins could then be analyzed by appropriate assays for functional alterations. The *Dictyostelium* myosin heavy chain has been cloned (De Lozanne *et al.*, 1985, 1987), using a polyclonal antibody to myosin and the λ*gt11* system described by Young and Davis (1983). This extremely large gene (6.3 kb) exists as a single copy in *Dictyostelium* and contains no introns. The gene has been completely sequenced and the amino acid sequence consequently determined (Warrick *et al.*, 1986). The heavy-chain phosphorylation site is about three-fourths of the way down the myosin tail from the myosin head–tail junction (Pagh *et al.*, 1984; E. Kuczmarski, personal communication. A 60,000-Da piece of the myosin containing this site has been expressed in and purified from *Escherichia coli*. Under appropriate conditions, this expressed tail fragment assembles into thick filaments with the same 143 Å periodicity seen in native myosin thick filaments. Furthermore, this fragment serves as a substrate for the heavy-chain kinase that is involved in the chemoattractant-elicited increase in heavy-chain phosphorylation (De Lozanne *et al.*, 1987). An exciting new development (see De Lozanne, Chap. 27) is the discovery of homologous recombination in *Dictyostelium*. This technique has been used to disrupt the native myosin gene and express instead the first half of the myosin molecule.

Thus, using a combination of molecular genetic and biochemical approaches, one can probe the structural features of the myosin molecule that are required for its activity and regulation. These and other approaches make it reasonable to hope that we will eventually understand the role of myosin in the chemotactic response. This is only one example of the power that the study of *Dictyostelium* offers to students of cell biology.

II.   Some Biochemical Considerations

In the chapters that follow, various approaches and techniques are outlined in detail. Biochemical considerations can be found in nearly every chapter. Editing all the chapters revealed an apparent need to emphasize two important biochemical considerations, and I therefore refer to them here.

A.   Inhibition of Proteolysis in Lysates

One essential criterion for choosing a eukaryotic cell for molecular studies is that the molecules of interest can be isolated in their native, intact state, and that the yield is sufficient for biochemical studies. It should be noted that *Dictyostelium* is a phagocytic cell and therefore is rich in lysozomes. Special care should be taken in preparing extracts to protect against protease activity which can alter many proteins of interest. Uyemura *et al.* (1978) dealt with this problem for cells cultured in HL5 medium. They defined buffer conditions for cell lysis which largely eliminate proteolysis. Critical components are 40 mM sodium pyrophosphate and 30% (w/v) sucrose. The sucrose undoubtedly helps to stabilize lysozomes. When sucrose and pyrophosphate are omitted from the lysis buffer, substantial proteolysis is apparent. The addition of sucrose without pyrophosphate stabilizes many of the proteins, but actin is still reduced in amount compared to lysis conditions that include pyrophosphate as well. New potent protease inhibitors are available which may allow more variability in buffer conditions for lysis. Stone *et al.* (Chap. 12, this volume) used the Uyemura *et al.* (1978) procedure to isolate deuterated actin from *Dictyostelium*. They observed that crude extracts of *Dictyostelium* grown on bacteria have greater protease activity than extracts of amoebae cultured in HL5 medium. They found that inclusion of chymostatin, leupeptin, and tosyl lysine chloromethyl ketone (TLCK) effectively inhibited *Dictyostelium* protease activity.

One important obstacle in our earlier work on myosin purification was that we had not found any condition under which we could freeze *Dictyostelium* vegetative cells without consequent cleavage in crude extracts of the ~240-kDa myosin heavy chain to a lower molecular weight species. Thus, in our experience, we have had to purify proteins from freshly prepared cells. The freezing and thawing probably ruptures lysozomes and results in protease release. The availability of new protease inhibitors has prompted us to reexamine the prospect of starting with frozen cells. Vegetative cells grown in HL5 medium were harvested at 8×10^6 cells/ml and sedimented at 700 g for 10 minutes. The cells were resuspended in 10 mM Teola, pH 8, 10 mM sodium pyrophosphate, resedimented, and then resuspended in two volumes of 10 mM Teola, pH 8, 30% sucrose, 12 or 40 mM sodium pyrophosphate, 5 mM DTT, 2 mM EDTA, 30 mg/liter PMSF (from a 6 mg/ml stock solution in ethanol), 5 mg/liter leupeptin (from a 5 mg/ml stock solution in H_2O; Sigma), and 5 mg/liter pepstatin A (from a 5 mg/ml stock solution in methanol; Sigma) for 5 minutes. The cell suspension was then frozen by swirling in a dry ice–ethanol bath and stored at $-20°C$. The cells were then thawed and extracts were examined by SDS–PAGE. Western blotting with myosin antibody was also used to examine the integrity of the 240-kDa heavy chain. The presence of the protease inhibitors effectively eliminated proteolysis in the extracts (K. Niebling and J. A. Spudich, unpublished observations). These results suggest that one can now accumulate and store cells in a frozen state, and, when needed, very large amounts of starting material can be thawed for large-scale protein preparations.

A variety of methods can be used for preparing cell lysates. In our early work on actin and myosin, we used homogenization with a Potter homogenizer (Clarke and Spudich, 1974; Spudich, 1974). While this is one of the more gentle methods of cell lysis, *Dictyostelium* cells are difficult to lyse, and a great deal of energy and patience is required to obtain significant cell breakage by this method. If homogenization is used, breakage is easier in very low ionic strength; salt can be added to increase the ionic strength immediately after lysis. Homogenization is difficult to use if large volumes of cells are involved. Sonication is an effective method of lysing both small and large volumes of cells, and we now routinely use this method (Uyemura *et al.*, 1978; Griffith and Spudich, 1987). The cell suspension should be kept in an ice bath, and short bursts of sonication should be used at the lowest setting possible that results in cell breakage. The breakage can be monitored by light microscopy, where the number of intact cells can be scored in a hemacytometer.

B.   Radioactive Labeling of Proteins *in Vivo*

Since *Dictyostelium* can be grown in a chemically defined medium (Franke and Kessin, 1977), proteins can be purified from this organism labeled with carbon-14, tritium, or sulfur-35 with known specific radiocactivity. That is, the specific radioactivity of the protein will be the same as that of the precursor used in the medium. For example, Simpson and Spudich (1980) isolated radiolabeled actin from cells grown in the Franke and Kessin medium, modified to contain 0.5

mM rather than 2 mM L-methionine, and supplemented with 5 mCi of L-[^{35}S]methionine per liter, at the highest specific radioactivity available (1000 Ci/mmol; Amersham). ^{35}S-Labeled actin prepared from these cells had a specific radioactivity of 3 μCi/mg. This method, therefore, results in significant labeling of a protein of interest. This protein can then be used in biochemical experiments where the specific radioactivity of that protein is known.

Cells can also be labeled with [^{32}P]orthophosphate of known specific radioactivity. Kuczmarski and Spudich (1980) isolated myosin phosphorylated *in vivo* on both the myosin light chain and on the heavy chain by growing log phase cells for four generations in [^{32}P]orthophosphate (20 μCi/ml) in the Franke and Kessin medium, with total phosphate decreased to 0.4 mM. The molar ratio of phosphate per subunit of myosin was then determined by cutting the appropriate band from an SDS gel of the purified myosin, digesting the gel piece with 30% H_2O_2 for 16 hours at 90°C, and assaying in scintillation fluid. Berlot *et al.* (1985) labeled developed *Dictyostelium* cells with [^{32}P]orthophosphate to examine the role of myosin phosphorylation in the chemotactic response of these cells. The labeling procedure is described in detail by C. Berlot in Chapter 18 of this volume.

III. Concluding Remarks

I have described here a few of the biochemical considerations that we have found to be important in our work on *Dictyostelium*. Inhibition of protease activity in crude extracts is of great importance, and buffer conditions and effective protease inhibitors are described. These inhibitors should become common reagents in all laboratories interested in biochemical studies using *Dictyostelium*. In addition to the points made in this chapter, numerous other biochemical considerations and techniques are found throughout the book and in nearly all the chapters. In the next three chapters general methods are presented for growth and manipulation of *D. discoideum* and for the use of classical genetics and molecular biological techniques.

REFERENCES

Berlot, C. H., Spudich, J. A., and Devreotes, P. N. (1985). *Cell* **43**, 307–314.
Berlot, C. H., Devreotes, P. N., and Spudich, J. A. (1987). *J. Biol. Chem.*, in press.
Clarke, M., and Spudich, J. A. (1974). *J. Mol. Biol.* **86**, 209–222.
De Lozanne, A., Lewis, M., Spudich, J. A., and Leinwand, L. A. (1985). *Proc. Natl. Acad. Sci. U.S.A.* **82**, 6807–6810.
De Lozanne, A., Berlot, C., Leinwand, L., and Spudich, J. A. (1987) (submitted).
Franke, J., and Kessin, R. (1977). *Proc. Natl. Acad. Sci. U.S.A.* **74**, 2157–2161.

Griffith, L. M., and Spudich, J. A. (1987). *J. Cell Biol.*, in press.

Kuczmarski, E. R., and Spudich, J. A. (1980). *Proc. Natl. Acad. Sci. U.S.A.* **77,** 7292–7296.

Loomis, W. F. (1975). "Dictyostelium discoideum: A Developmental System." Academic Press, New York.

Loomis, W. F., ed. (1982). "The Development of Dictyostelium discoideum." Academic Press, New York.

Malchow, D., Bohme, R., and Rahmsdorf, H. J. (1981). *Eur. J. Biochem.* **117,** 213–218.

Pagh, K., Maruta, H., Claviez, M., and Gerisch, G. (1984). *EMBO J.* **3,** 3271–3278.

Rahmsdorf, H. J., Malchow, D., and Gerisch, G. (1978). *FEBS Lett.* **88,** 322–326.

Sheetz, M. P., Chasan, R., and Spudich, J. A. (1984). *J. Cell Biol.* **99,** 1867–1871.

Simpson, P. A., and Spudich, J. A. (1980). *Proc. Natl. Acad. Sci. U.S.A.* **77,** 4610–4613.

Spudich, J. A. (1974). *J. Biol. Chem.* **249,** 6013–6020.

Uyemura, D. G., Brown, S. S., and Spudich, J. A. (1978). *J. Biol. Chem.* **253,** 9088–9096.

Warrick, H. M., De Lozanne, A., Leinwand, L., and Spudich, J. A. (1986). *Proc. Natl. Acad. Sci. U.S.A.,* **83,** 9433–9437.

Young, R. A., and Davis, R. W. (1983). *Proc. Natl. Acad. Sci. U.S.A.* **80,** 1194–1198.

Chapter 2

Cultivation and Synchronous Morphogenesis of Dictyostelium under Controlled Experimental Conditions

MAURICE SUSSMAN

Department of Biological Sciences
University of Pittsburgh
Pittsburgh, Pennsylvania 15260

I. Introduction

"Good God! Molecular Fascism!"

Some years ago, I had the rare pleasure of teaching jointly with Prof. Sydney Brenner a section of the physiology course at the Marine Biological Laboratory, Woods Hole, Massachusetts. (That summer, he was searching for the one perfect nematode and found *Caenorhabditis elegans* in a pool of animal feces.) One day, I handed him a Petri dish containing a black 47-mm Millipore filter supported by an absorbent pad saturated with a buffered salt solution. On the filter were ~1000 *Dictyostelium discoideum* aggregates of almost identical size and every last one at that instant assuming the shape of the classical Mexican hat. He looked through a dissecting microscope and in his finest Churchillian manner invoked the name of his holy template in the above quotation.

That is the point of choosing *D. discoideum* as a subject for molecular developmental study. By careful microbiological and physiological practice, an investigator can force a population of *D. discoideum* to proceed synchronously and with invariant timing through a developmental sequence that involves complex morphogenetic movements and cytodifferentiation and ultimately generates an organized multicellular assembly. And in the finest traditions of fascism, mo-

9

lecular or otherwise, he or she can exert total environmental control and can sacrifice the entire population at any stage of the proceedings.

To accomplish this, the population must be genetically and physiologically homogeneous and the environmental conditions reproducible. What follows is a compendium of methods of strain preservation, maintenance, and cultivation, and techniques of cell preparation and incubation to achieve these ends. All have been used in our laboratory through the years and found to be satisfactory. The ones that originated there should be credited to a group of ingenious, hard-working and very nice postdoctoral fellows, graduate students, and technicians, and one admirable coinvestigator Wife.

In his book, K. B. Raper (1984) discusses many of the same techniques that are described herein.

II. Cell Cultivation

Cell populations are cultivated for any of a number of purposes: to maintain strains; to isolate mutants, revertants, or ploidal variants from a parent stock; to harvest large numbers of cells for isolation of DNA, plasma membranes, proteins, and other entities; to obtain physiologically homogeneous populations of vegetative amoebae which can then be made to proceed synchronously through part or all of the morphogenetic sequence leading to fruiting body construction. In order to achieve these ends, the myxamoebae are grown axenically or in association with bacteria, on solid substratum or in liquid suspension, clonally or in mass culture. Detailed protocols that cover these exigencies are given below.

A. Bacterial Stock Plates

Prepare a turbid suspension of *Klebsiella aerogenes* in SM broth (Sussman, 1966) or Bonner's salt solution (Bonner, 1947). Dispense and spread 0.2-ml aliquots on SM plates. Incubate overnight at 37°C. Make 1 week's supply, store in the cold, and use each plate for 1 day only. Loopsful of bacteria harvested from these serve to inoculate the two-membered dictyostelid cultures.

B. Clonal Plates (Sussman, 1951)

Make a faintly turbid ($1-2 \times 10^7$/ml) suspension of spores or amoebae in 20 mM phosphate, pH 6.5, or cold distilled H_2O. If a precise count is needed, use a counting chamber. Dilute serially in 1–2 log steps. Dispense 0.2-ml aliquots of the last dilution containing 10–50 spores or amoebae on SM/5 agar plates, add 2–3 drops of a turbid suspension of *K. aerogenes,* and spread evenly over the entire surface with a glass spreader. Allow to dry with lids off in a sterile hood for a few minutes. If many plates are to be spread (as, for example, in a mutant hunt), mix the last myxamoeboid or spore dilution with the bacterial suspension and dispense measured aliquots. Plaques will appear in 2–3 days at 22°C. The bacterial lawn in dishes slightly more than half-filled with SM/5 agar is thick enough to support respectable growth within the plaques, but the excreted metabolic wastes are not sufficient to interfere with the development of the myxamoebae. Hence, clones can be scored for morphogenetic phenotype unequivocally. If the intent is to isolate mutants or make viable counts, use SM agar. Larger numbers of plaques can be accommodated thereon, since they do not spread as rapidly.

C. Mass Stock Plates

These serve two purposes: (a) to collect large numbers of amoebae, aggregates, or fruits when a high degree of morphogenetic synchrony is not required, and (b) to provide a ready supply of spores for inoculation. Use the same procedure as described above except that the inoculum should be increased to ~10^6 spores or myxamoebae per plate. The resulting bacterial lawn will be cleared evenly by about 36–40 hours, leaving a homogeneous layer of ~10^9 stationary-phase amoebae. Fruit construction is complete by ~72 hours, but we generally wait until 4 days to use the spores for inoculations. Spores of *D. discoideum* on plates stored up to 10 days in the cold retain satisfactory viability and do not exhibit an excessively long lag phase on subculture.

D. Stab Stock Plates

Some mutants do not construct fruiting bodies. To carry these, inoculate SM agar plates with 2–3 drops of bacterial suspension, spread, and dry in a sterile hood. Inoculate with myxamoebae at one edge. Incubate at 22°C for a few days and store in the cold.

E. Growth Plates for Morphogenetic Experiments

Cells harvested from these plates are employed for experiments that require synchronous development on solid substratum or in shaken suspension. The cells must be cultivated in an even lawn of bacteria so that none run out of prey and begin to acquire aggregation competence prematurely. They must be harvested while still in the exponential growth phase and yet at a time when the bacterial supply has diminished to a point where the amoebae can be centrifugally washed free of bacteria easily and with optimal yield. For some years now we have employed a very reliable and convenient system referred to but not heretofore described in detail. The two major innovations are listed here:

1. Relatively large numbers of spores (or myxamoebae) and bacteria are inoculated. Depending on the exact number of the former, the lag and exponential growth phases are reduced to 24–30 hours. The onset of the stationary phase is sharp and predictable to within ±1 hour. This is much more convenient for the planning of experiments than the usual 38- to 44-hour regime, and the developmental synchrony is much better.

2. NZCase agar is employed. It is much richer than SM and the yield of myxamoebae is correspondingly larger. Hence fewer growth plates need be prepared.

Suspend spores or amoebae in SM broth at the desired density. (See Table I for the relation between inoculum density and time of harvest.) To collect the large number of spores needed, tap the inverted stock plate smartly on the bench top to deposit the spores on the lid. Introduce a small volume of broth and suspend the spores therein by scratching the surface with the pipet tip. To 6 ml spore suspension, add four large loopsful of bacteria scraped from the lawn of a bacterial stock plate. Dispense 0.5-ml aliquots on NZCase plates and spread evenly (and assiduously) over the surface. Use a triangular spreader fashioned from a 2.5-mm glass rod. Spread the inoculum while rotating the plate for ~10 seconds until fluid begins to be absorbed by the agar. Using the heel of the spreader, traverse the plate in closely spaced lines. Let plates stay lidless in a sterile hood until

TABLE I

RELATION BETWEEN NUMBERS OF SPORES INOCULATED ON NZCase AGAR PLATES AND TIMES AT WHICH MYXAMOEBAE CAN BE HARVESTED FOR EXPERIMENTS REQUIRING A HIGH ORDER OF DEVELOPMENTAL SYNCHRONY[a]

Incubation time to reach stationary phase (hours)	Inoculum size (spores $\times$ 10^{-6}/plate)		
	D. discoideum		D. mucoroides
	Wild type (DdB)	FR-17	Wild type (S-2)
24	45		66
26	26	100	37
28	15	58	25
30	8.8	38	13
32	5.1	20	7.6

[a] The plates are incubated at 22°C until the bacterial lawns are partially cleared. The data were taken from our laboratory methods book.

almost dry. The yield of amoebae is $1-2 \times 10^9$/plate. The NZCase plates should be poured more than two-thirds full, stored at room temperature, and used for ≤ 1 week.

F. Growth with Bacteria in Shaken Suspension

Two methods are available. In the first (Gerisch, 1959), the bacteria (*Escherichia coli* or *K. aerogenes*) are pregrown for 24 hours to a stationary-phase density of $\sim 10^9$/ml in 1-liter batches of Davis minimal medium in 2.8-liter Fernbach flasks shaken at 37°C. The bacteria are harvested, washed in sterile 40 m*M* pH 6.4 phosphate, and resuspended in 200 ml of the buffer. Aliquots, 20% (v/v) in Erlenmeyer flasks, are seeded with spores or amoebae at a density of 10^5/ml and incubated at 22°C on a rotary shaker or wrist shaker. Spores should be washed to remove the germination autoinhibitor (Ceccarini and Cohen, 1967). The cells grow at the same rate as they do on plates ($T_d = 3.2$ hours). The log–stationary phase transition is sharp and the yield $1-2 \times 10^7$/ml. This system is particularly useful for the preparation of homogeneously labeled amoebae. For example, using Davis minimal medium (without citrate) containing 1 mCi of ^{14}C/μl glucose at a concentration of 0.2%, we typically recover 50–60% of the label in the washed bacteria and 70% of that in the amoebae (Franke and Sussman, 1973).

The second method of liquid cultivation (Sussman, 1961) involves the concurrent growth of amoebae and bacteria in SM broth. A stationary-phase culture of *K. aerogenes* is diluted 1:10 in SM broth, and spores or amoebae are inoculated at a density of 10^5/ml. This ratio ensures that the bacteria will not be prematurely wiped out yet not overgrow the culture. The system is particularly useful for sustained log phase passage, as for example in the isolation of haploid and aneuploid segregants of diploid strains (Sussman and Sussman, 1963).

G. Axenic Cultivation

The axenic growth of *D. discoideum* was first accomplished (Sussman and Sussman, 1967) in a medium (HL5 plus liver concentrate and calf serum) that had originally been devised for the cultivation of amoeboflagellates (Balamuth, 1964) and modified for the growth of *Neigleria gruberi* by Chandler Fulton. The axenic cultivation of *D. discoideum* involved the serial selection of a strain that could barely subsist and then one, Ax-1, that could grow exponentially with a T_d of 10–12 hours to a density of $1-2 \times 10^7$/ml. An equivalent growth rate and yield in HL5 medium without additives was reported by Cocucci and Sussman (1970) and by Schwalb and Roth (1970). Two other isolates, Ax-2 and Ax-3, were selected (Watts and Ashworth, 1970; W. F. Loomis, unpublished data) by the same procedure in modifications of HL5, and they are known to carry two recessive mutations, *axe*A and *axe*B (Williams *et al.*, 1974). Franke and Kessin (1976) have devised a defined medium that supports ample growth and indeed can be used for the selection of auxotrophic mutants.

Maeda (1983) has reported the successful axenic cultivation of wild-type *D. discoideum* with a doubling time of 12 hours and a yield of 4×10^6 cells/ml. We have confirmed both values. The major difference between his medium, MA, and HL5 lies in the source of the peptone (Oxoid Bacteriological peptone vs. Difco proteose peptone or BBL thiotone) and its concentration (30 g/liter vs. 12.5). Substitution of the Oxoid product with other peptones drastically reduced or eliminated the growth of the wild type (but not of Ax-3). Yeast extract is an essential nutrient of MA medium, but other additives (vitamins, trace salts) are only slightly stimulatory. The relatively low cell yield (4×10^6/ml) limits somewhat the utility of the system, but it does obviate the need for introducing the two mutant alleles into a strain to be cultivated.

Polysphondylium pallidum wild-type strain PP-1 (a clonal isolate derived from the Raper strain WS-320) can also be grown axenically. Medium A (Sussman, 1963), containing proteose peptone, lipid-free milk powder, and lecithin, provides a T_d of 3.5 hours and a yield of $1.5–2 \times 10^7$ cells/ml. A more elaborate medium (Hohl and Raper, 1963), defined except for the presence of bovine serum albumin, can be used but with lower rate and yield.

To start a culture from a bacterially grown stock plate, pick spores from several fruiting bodies with a needle or loop. Choose well-isolated fruits and make sure that the loop or needle touches only the sori and not the substratum. Inoculate a 5-ml volume of filter-sterilized HL5 containing 500 μg/ml of strep-tomycin sulfate. Incubate 2–4 days until culture becomes turbid. Check micro-scopically for contamination. Transfer at an initial density of $\sim 10^5$/ml to 10-ml volumes of HL5 in 125-ml Erlenmeyer flasks and incubate on a rotary shaker at 100 rpm. After 4 days at 22°C these stock cultures should have reached the stationary phase (2×10^7 cells/ml). Serial stock transfers should be made at least twice per week. Cultivation of cells for experiments demanding developmental synchrony should be conducted in 100- to 200-ml volumes in 1-liter flasks shaken at 100 rpm.

III. Strain Maintenance

Mass cultivation of *D. discoideum* in association with *K. aerogenes* involves 10–12 doublings and a stationary-phase population of $\sim 10^9$ amoebae per plate. In axenic medium, a typical passage involves ~ 6 doublings and a yield of $1–2 \times 10^{10}$ cells/liter. Given mutation rates of $10^{-6}–10^{-9}$/cell/generation and of ploidal variation ($10^{-4}–10^{-5}$), even a few serial passages must inevitably lead to the presence of a variety of mutants (or revertants) and of ploidal variants in the population. Those derivatives able to emerge from the lag phase more quickly than the parental type, grow faster in the exponential phase, or survive better in stationary phase, will increase differentially within the population. To combat this problem, which in practice is a serious one, requires two remedies: (1) eliminate mutants and ploidal variants by interspersing mass cultivations with frequent clonal passages, and (2) carry out the clonal reisolation under conditions permitting optimal expression of morphogenetic competence so that prototypic clones can be selected.

A. Maintenance of Bacterially Grown Stocks

In the regime employed in our laboratory for many years, the two requirements are met by sequential passages on three kinds of stock plates. See Fig. 1 for a sample protocol. Prototypic clones taken from *clonal plates* are used to inoculate a number of *holy mass plates* (holy because they are employed only to maintain stock purity, each being used once and then sacrificed). At weekly

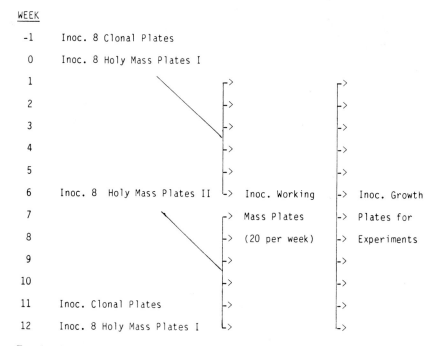

FIG. 1. A 12-week cycle for strain maintenance and supply of cells for experiments. The suggested numbers of plates have been found adequate to supply growth plates for four or five active investigators doing daily experiments, each involving 15–25 filters.

intervals, each serves to inoculate a series of *working mass plates,* used the following week to inoculate the NZCase growth plates that ultimately provide cells for experiments. After 10 weeks, new clonal plates are prepared, examined, and used to start a new set of holy mass plates. Note that within the 12-week cycle shown in Fig. 1 there are only two mass passages from clone to clone. For nonsporulating stocks, stab plates can be substituted. The clonal platings are carried out on two sets of four SM/5 agar plates at dilutions that will yield 1–10 clones on the first set and 20–30 on the second. The latter plates are examined under a dissecting scope after 3–4 days to assess the genetic homogeneity of the population and are then discarded. The former plates supply prototypic clones and are sealed and stored in the cold as insurance against contamination of the holy mass plates.

B.   Maintenance of Axenically Grown Stocks

A monthly clone-to-clone cycle is employed. Initially the stock is grown clonally with bacteria on SM/5 plates prepared and stored as described above. A prototypic clone is placed in axenic cultivation (see Section II,G) and is carried in 10-ml stock cultures by twice-weekly serial passages. To avoid contamination, these should be used only once and discarded. Serial passages should be continued for not more than 4 weeks. After 2 weeks, the stock is recloned (from an axenic culture). This will yield new axenic stock cultures 2 weeks later. In our experience, less fussy procedures open the way to unexpected invasions of the population by aggregateless, fruitless, and slugger mutants and to deliciously worded retractions of previously reported findings.

We have had occasion in the past to compare clonal isolates of Ax-2 and Ax-3 with our own Ax-1 in respect to growth rates and yields in HL5 medium and morphogenetic competence on filters and found no significant differences among them.

C.   Maintenance of Sporadically Used Stocks

SM/5 mass plates sealed with parafilm, placed in plastic bags, and stored in the cold can supply viable cells for at least 3–4 months.

IV.   Strain Preservation

Three methods are generally employed: lyophilization, silica gel desiccation, and frozen storage. Each has virtues and disadvantages.

Lyophilized preparations, stored in a refrigerator or even at room temperature, are, for practical purposes, immortal. Spores of *D. discoideum* lyophilized in 1941 are reported to have remained viable for >40 years (Raper, 1984). A multiport lyophilizer and vacuum pump are required, but the procedure is simple and reliable. However, only spores can survive the treatment. Fortunately, almost all morphogenetically deficient mutants can be made to fruit synergistically in combination with other mutants or the wild type (Sussman, 1954), and can be recovered after lyophilization by clonal reisolation.

Silica gel desiccation is also a simple, reliable procedure and requires no special apparatus (Reinhardt, 1966). Here too, only spores survive the treatment, but in the case of morphogenetically deficient mutants, synergistically fruited mixed populations can be employed. If stored properly (at 5°C in sealed containers with periodically renewed drierite), the desiccated preparations remain viable indefinitely.

Vegetative amoebae frozen with glycerol (Sussman, 1966) or dimethyl sulfoxide (DMSO) (Laine *et al.*, 1975) retain a satisfactory level of viability if stored at −73°C in a freezer or under liquid nitrogen. The procedure is simple and applicable to all strains. The only disadvantage is the possibility over the years of a power failure or that someone will forget to replenish the supply of liquid nitrogen.

My personal bias has been to lyophilize everything that can be induced to form spores.

A. Lyophilization

This procedure, a modification of the original technique of Raper and Alexander (1945) comes from our methods book. For tubes we use 1-ml Gold seal vacules from Wheaton Sci. Co., Millville, NJ 08332. Available surface area is increased by inserting a few 3-mm polished glass beads (Corning). To identify the contents, each tube receives a rectangle of Whatman filter paper bearing the strain number and date inscribed with soft lead pencil. The tubes are capped with aluminum foil and autoclaved. The spores should be taken from mass plates incubated 4–7 days (by tapping the inverted plates smartly on the bench top and collecting the spores from the lid). Viability is significantly reduced thereafter. The denser the spore suspension, the better. Each tube receives 3–4 drops from a Pasteur pipet and is then placed in an acetone–dry ice mixture to freeze the contents. Lyophilization is accomplished on a 24-port lyophil assembly fitted with Virtis Quick Seal taps. If all is well, the tube contents should become white and the exterior should be frost-covered. If the contents melt before drying, discard the tubes and find the leak. To ensure complete desiccation, continue the pumping until the tubes are no longer even cool to the touch. In practice, 1 hour is more than enough. The tubes are then sealed off with a twin-headed oxygen torch and are stored at room temperature or in a refrigerator.

To subculture, a tube is broken at the Gold seal and filled sequentially with two 0.5-ml aliquots of bacterial broth suspension. The contents are poured and spread on two SM/5 agar plates. It is prudent practice to test one tube of each freshly lyophilized batch for viability.

B.   Silica Gel Desiccation

The following procedure is taken from a work sheet supplied by P. C. Newell. It was adapted from the procedure of Perkins (1962) originally applied to *Neurospora conidia*. Screw-cap scintillation vials (Fisher No. 3-337-5) are half-filled with ~6 g of silica gel, 6–12 mesh grade 40 (Fisher S-684), and autoclaved 90 minutes. Vials and caps are sterilized separately and can be stored in closed containers over drierite. A 5% solution of Carnation instant nonfat dried milk is prepared and autoclaved. Spores are obtained as described above and suspended in 0.8 ml milk solution per plate. Suspension and vials are cooled for 30 minutes in an ice bath. (Silica gel generates considerable heat when wetted.) A 0.8-ml aliquot of suspension is rapidly pipetted onto the gel. The vial is shaken vigorously, capped, and returned to the ice bath for 5 minutes. The vials are stored over drierite in plastic boxes at 0–5°C. The drierite is redesiccated once or twice a year.

Subcultures are made by shaking a few granules of gel into 1 ml cold broth and spreading 0.2-ml aliquots with bacteria on SM agar plates.

C.   Frozen Storage

Vegetative cells are harvested at late log–early stationary phase and are centrifugally washed and suspended at high density ($\sim 10^8$/ml) in cold 5% (v/v) glycerol solution (Sussman, 1966). Aliquots of 0.5–1 ml are stored in foil-wrapped ampules under liquid nitrogen. For recovery, the cells are thawed rapidly in warm water, centrifuged and resuspended in water without glycerol, and plated on SM/5 agar with bacteria. In the procedure of Laine *et al.* (1975), the cells are washed centrifugally in Bonner's salt solution and suspended at 10^7/ml in HL5 medium containing 10% DMSO (unsterilized and added just before freezing). Aliquots of 0.5 ml are dispensed in 8-mm capsules and stored in racks at -73°C. For recovery the cells are rapidly thawed by shaking in a water bath at 45°C and plated on agar with bacteria. Peter Newell (private communication) suspends amoebae in serum or HL5 medium with 5–10% DMSO and freezes them 2 hours at -20°C before storage under liquid nitrogen. As containers he uses "Bull sperm straws," unspecified objects which I presume are available in the local Oxford ice cream parlors.

V.   Conditions for Synchronous Development

Under optimal conditions, vegetative cell populations harvested from nutrient medium and dispensed homogeneously on a solid substratum can be made to exhibit remarkably consistent morphogenetic behavior in respect to the timing and patterns of aggregation and of fruiting body construction. The requirements for this and the protocols that have been found useful to satisfy them are outlined below.

A.   Preparation of Cells

It is essential that at the time they are harvested from nutrient medium the myxamoebae be physiologically homogeneous. It has been shown (Newell *et al.*, 1971) that when cells, allowed to proceed partway through the morphogenetic sequence, are dissociated and redispersed on solid substratum, they can recapitulate previously accomplished development within a very brief period. This applies even to the earliest events in the acquisition of aggregation competence (Soll and Waddell, 1975; Soll, 1979). Hence, if at the time of harvest from nutrient medium the population includes even a relatively small contingent of cells that had already entered the stationary phase and begun to acquire aggregation competence, developmental synchrony suffers markedly. In the case of cells grown with bacteria on plates, this can be avoided by two stratagems. Initially, the inoculum must be spread carefully in order to yield a bacterial lawn in which the myxamoebae are homogeneously distributed. The end of the exponential phase will then be marked by an even as opposed to a patchy clearing of the bacterial lawn. Finally, the cells should be harvested before the lawn is completely cleared, ensuring that all the myxamoebae will still be surrounded by some bacteria. In axenic cultures the cessation of exponential growth is a gradual one occupying the last one or two doublings. Drastic physiological changes occur therein, including a progressive decrease in polysomal content (Cocucci and Sussman, 1970) and a concomitant drop in protein-synthetic capacity. When cells harvested at this stage are dispensed on solid substratum, many fail to aggregate and morphogenetic synchrony is very poor. To assure optimal results, the cells should be harvested when they reach a density of 5–6 $\times$ 10^6/ml.

Harvest plate-grown cells in cold dH_2O with a glass spreader. Rinse the plates, pool the suspensions, disperse by sucking and blowing through a 10-ml pipet, pour into 50-ml centrifuge tubes, and spin 3–4 minutes at ~200 g (1250 rpm in a Sorvall SS-34 rotor). This yields a firm amoebal pellet and leaves most of the bacteria in the supernatant. Decant the latter and resuspend the pellet in cold

H_2O. Wash twice and resuspend in ~2 ml/plate of buffered salt solution (LPS in Section VI). Count a 1:100 dilution and adjust to the desired density. Using the 24-hour NZCase agar regime described previously, the yield of washed amoebae should be $3–6 \times 10^8$ myxamoebae per plate depending on the growth phase at which they were harvested. Axenic cultures are also washed with cold H_2O but are spun at ~300 g to maximize recovery.

Occasionally it has been important to completely rid plate-grown myxamoebae of bacteria before beginning an experiment. This can be done by suspending the washed cells at a density of 2×10^7/ml in SM broth containing 500 µg/ml streptomycin sulfate. A 250-ml volume in a 1-liter Fernbach flask is shaken at 100 rpm on a rotary shaker for 1 hour. The cells are harvested, washed once, and dispensed on filters in the usual manner (Inselburg and Sussman, 1967).

B.   Synchronous Morphogenesis on Filters

This method, first described in 1965 by Sussman and Lovgren, has remained unchanged except that the Millipore filters have been replaced (except in special circumstances) by their paper counterparts. Briefly, washed cells are evenly dispensed on 42.5-mm paper or 47-mm Millipore filters over LPS (Sussman, 1966)-saturated pads in Petri dishes. In this condition, 10^8 D. discoideum cells enter (completely) into ~1000 equal-sized aggregates per filter, and these proceed through the morphogenetic sequence with a temporal deviation that at its lowest is 10–15 minutes between the earliest and latest structures. The population can be exposed to morphogens, drugs, metabolites, isotopically labeled compounds, or other entities at any stage and for any period simply by switching filters to new pads. Cells can be recovered almost quantitatively from the filters by gentle trituration. A detailed protocol follows.

Whatman No. 50 filters are generally used. Black Millipore HA filters were employed when it was necessary to monitor morphogenesis microscopically, particularly at early stages, but Whatman No. 29 filters (D. R. Soll, personal communication) are cheaper and more convenient. Millipore filters are still useful when one wishes to reduce the diffusive efficiency between cells and substratum as in slug migration. Boil the filters briefly in LPS, store in the cold under LPS, and rinse before use. Place each filter on one or two absorbent pads (Whatman No. 17 or Millipore AP 1004700 in a 60-mm plastic Petri dish. Saturate with 1.6 ml LPS per pad. Air bubbles between filter and pad should be removed by repositioning the filter with forceps. (Millipore blunt-end forceps are best). Dispense 0.5-ml aliquots of the cell suspension on the filters (which should be resting on an absolutely horizontal surface). Use a 1-ml blowout pipet and dispense slowly and evenly while tracing a spiral starting from the center of the filter. Allow 3–4 minutes for the excess fluid to seep out of the filter. Its surface should display an even matte finish. Remove the excess fluid from the dish with a Pasteur pipet under gentle suction. When the dish is held vertically, liquid should just be visible at the bottom of the pad as a small meniscus. Replace dish covers, place dishes in a Pyrex cake pan containing ~20 ml water (to maintain high humidity), seal with Saran wrap, and incubate. The entire procedure including harvesting the cells should take no more than 1 hour.

C. Control of Slug Migration on Filters

In *D. discoideum,* the newly formed aggregate can elect alternative morphogenetic pathways. It can immediately initiate fruiting body construction at the site of aggregation, or it can transform into a migrating slug. Subsequently, the latter can be induced to reenter the fruiting mode and complete it within 8 hours regardless of the time spent as a slug (Newell *et al.,* 1969). These morphogenetic alternatives are attended by profoundly different programs of gene expression that affect, for example, the accumulation and disappearance of a discrete set of proteins (Ellingson *et al.,* 1971). Hence it is of prime importance that the choices open to the developing aggregate be under strict experimental control. In this, a definitive event has been shown (Schindler and Sussman, 1977) to be the accumulation of NH_3 in and around the developing aggregate as determined by (1) the rate of endogenous generation, which, during the preaggregative period, approaches 1.5 μmol $NH_3 + NH_4^+/hr/10^8$ cells; (2) the rate of dissipation by diffusion into the substratum and evaporation; and (3) the the ambient pH (NH_3/NH_4^+ interconversion has a pK of 9.2). NH_3 accumulation over threshold induces the aggregate to transform into a slug. Its subsequent decrease below threshold induces the slug to reenter the fruiting mode.

The experimental manipulation of these three parameters can reproducibly alter the choice of pathways. The rate of NH_3 production can be controlled by varying the number of cells per filter between 5×10^7 and 2×10^8. Diffusion can be influenced by the number of absorbent pads employed and hence the volume of LPS supporting the filters. Evaporation is enhanced by taping two support pads to the underside of each Petri dish lid and saturating them with 1 *M* phosphate at pH 6.0 (UPS in Section VI). Ambient pH can be affected by varying the buffer concentration in the LPS (10–40 m*M*) or the initial pH (6.0–7.5). Figure 2 shows the timing and patterns of morphogenesis under three different experimental regimes.

Unless one is interested in studying the choice of morphogenetic alternatives or the slug mode per se, the ''well-buffered'' condition is best. The pads containing UPS can be reused for or five times before replacing them. They should be resaturated with fresh UPS before each use. Please note that the conditions stipulated refer only to *D. discoideum,* wild type. Slugger mutants (Sussman *et al.,* 1978; Newell, 1982; Newell and Ross, 1982) and other variants respond differently. Hence the conditions must be redefined for each new strain employed.

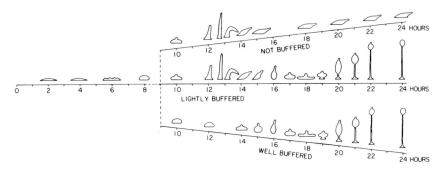

FIG. 2.   Schematic representation of the developmental programs observed without buffer, with a
lightly buffered substratum (0.01 M phosphate, pH 6.5), and with a well-buffered substratum (0.04
M phosphate, pH 6.5, with, in addition, two absorbent pads saturated with 1.0 M phosphate buffer,
pH 6.0, held a few millimeters above the developing aggregates).

D.   Giant, Mini, and Microfilters—Multifilter Assemblies

The filter system can be scaled up by using 11-cm-diameter Whatman No. 50
filters resting on 11.5-cm Whatman No. 17 pads inside 140-cm plastic Petri
dishes. The filters are loaded with 3.5 ml of LPS containing $7–10 \times 10^8$ cells.
Two pads affixed to the dish covers are saturated with UPS. The plates are
incubated in Saran wrap-sealed baking dishes. The system can be scaled down by
using quarter segments of 47-mm filters loaded with 0.1-ml aliquots of cell
suspension or miniaturized even further by using filter circles made with a
standard hole punch and loaded with 5- to 20-μl volumes of suspension. Multi-
filter assemblies are used when one needs to prepare very large numbers of
optimally synchronous aggregates. Each consists of a 185×310 cm stainless-
steel rectangle resting on 5-mm sidewalls (made by bending the long sides)
within a Pyrex baking pan. The rectangle has 28 punched-out holes, 39 mm in
diameter. The baking pan is filled with LPS to the level of the holes. The latter
are then covered with standard filters preloaded with cells while resting on a level
glass plate. The pan is sealed as usual with Saran wrap. We also occasionally use
a 9-hole version.

E.   Use of Filters in Labeling Experiments

As noted above, this can be done at any stage and for any period simply by
switching filters to fresh pads. However, the dilution of the label by the underlying LPS makes it difficult to attain high specific radioactivity. This can be
avoided in short-term labeling experiments. The filter is removed from its pad,

briefly rested on a paper towel to absorb excess moisture, and placed over 0.1 ml of labeling solution in a Petri dish. (A filter absorbed in this manner contains 0.1–0.15ml fluid.) After 30 minutes, the excess fluid can be removed with a pipet. Populations incubated in this manner can continue to develop normally for periods ≤3 hours.

F. The Valentich Chamber: An Alternative to Filters (Schindler and Sussman, 1977)

This is very useful for experiments in which one wants to sample or collect metabolites secreted by the cells in the course of development or in which the population must be continuously perfused. The apparatus, illustrated in Fig. 3, consists of a Lucite lid and reservoir with overall dimensions of 6 × 6 × 3 cm. A filter bearing the developing population rests on a square piece of dialysis membrane whose dimensions are slightly greater than those of the reservoir. The membrane is placed on a hollow square rubber gasket over the reservoir. The lid is added and the assembly is sealed by tightening nuts and bolts in extensions of lid and reservoir. This also serves to keep the membrane taut. The level of fluid in the reservoir is sufficient to maintain contact with the filter. Fluid can be added, removed, or perfused through the capped exit and entry ports. Black filters should be employed, enabling convenient monitoring of morphogenetic progress under a dissecting microscope. The chambers were designed and constructed by Frank Valentich, our departmental machinist.

G. Harvesting Cells from Filters and Redispersing the Cells

Remove the filter from its pad with forceps and stick on the inner wall of a centrifuge tube or beaker. Squirt 1 ml of cold harvesting fluid at it through a 5-ml pipet. Repeat twice using additional fluid. Wash twice more with the accumulated contents, making sure that all of the filter is covered. It is also helpful to score the filter surface gently with the pipet tip as the fluid is delivered. Multiple filters

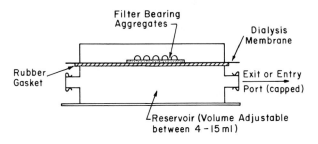

FIG. 3. The Valentich chamber. See text for details.

are harvested sequentially in one container. Each filter after the first is squirted with the previously accumulated contents and finally with fresh fluid. The general rule is to use 3 ml per filter for maximal recovery.

If one wants to redeposit the cells on fresh filters or to incubate them in shaken suspension or to perform cohesivity assays or other operations, it is necessary first to disperse the harvested aggregates. This is accomplished by repeatedly sucking up the suspension in a 10-ml pipet and blowing it out forcibly for ~5 minutes, by which time at least 50% of the cells are singlets. Centrifuge the cells for 5 minutes at 270 g and resuspend in cold dH_2O or LPS. The population must be mechanically agitated by pipet or by fast shaking to prevent very rapid reaggregation. The yield should be at least 70% of the cells originally deposited, all appearing healthy when checked at 400× magnification.

H.   Large-Scale, Synchronous Production and Manipulation of Migrating Slugs

I am at a loss to understand why molecular biologists interested in developmental regulation of gene expression have not applied their probes to migrating slugs. As noted above, newly formed (15–16 hours) aggregates that initiate fruit construction or alternatively transform into slugs exhibit profoundly different temporal and spatial patterns of protein accumulation and disappearance. Migrating slugs, induced to reenter the fruiting mode, immediately proceed through an invariant sequence of morphogenetic movements that results in the formation of a Mexican hat 156 ± 5 minutes after the start of the induction and a fruiting body within 8 hours (Newell et al., 1969; Deml-Rand and Sussman, 1983). Long slugs develop subsidiary signaling centers and form multiple fruits (Raper, 1941; Deml-Rand and Sussman, 1983). During reentry, the differential accumulation and disappearance of specific proteins occurs in step with the morphogenetic events (Ellingson et al., 1971; Sussman et al., 1978). Experiments involving actinomycin D have indicated that all of the changes that accompany slug formation and reentry are under transcriptional control. Slugger mutants whose aggregates are temporarily or indefinitely trapped in the slug mode under conditions that permit the wild type to eschew this pathway, provide remarkable material for comparative studies (Sussman et al., 1978; Newell and Ross, 1982).

Slug boxes provide a convenient system for production and manipulation of migrating slugs. The plastic boxes, 120 mm square, in which batches of 100 Millipore filters or pads are shipped can be used for this purpose. About 200 ml of 2% agar in water is poured into a box and allowed to set. A narrow agar shelf is created at one end by placing a Lucite bar on the firm agar and pouring molten agar between the bar and the wall of the box. After setting, the bar is removed. The boxes are used immediately. With a narrow-tipped micropipet (we use an Eppendorf), 100 μl of a washed cell suspension at 4×10^8/ml is deposited in a thin line at the bottom of the shelf. It is best to tilt the box by placing the end away from the shelf on a 10-ml pipet before adding the cells. This prevents the fluid from spreading as it is dispensed. After the excess fluid is absorbed, the box is covered and placed in a sleeve made of aluminum foil oriented so that the end away from the shelf is open. The stack of boxes is incubated with open ends exposed to a fluorescent lamp. The cells form a thin line of aggregates with normal timing. The high density and unbuffered substratum induce the aggregates to transform into slugs by 16 hours, and they begin to migrate orthogonally in a tight "battalion front" toward the light at the other end of the box, moving at a steady rate of 2 mm/hour. In doing so, they are made to tranverse black Millipore filters that had been placed in their path before the boxes were covered. To induce reentry the filters are removed and placed on pads saturated with LPS at pH 6.5, and the slugs are exposed to overhead light. The combination of low pH and omnidirectional light is particularly effective.

I.   Synergistic Morphogenesis on Filters

Morphogenetically deficient mutants, aggregateless, fruitless sluggers, and others can be made to complete the sequence in synergistic combinations with the wild type or other mutants. This is best carried out on filters. The two cell populations are harvested and washed separately and are mixed in varying proportions before deposition. Optimal development depends on two critical conditions: the proportions of the two genetic varieties in the population and their developmental synchrony. The first is determined by systematically varying the proportions between 0 and 100% (usually in steps of 20). The second is determined combining populations harvested from the growth plates in different states of developmental accomplishment.

VI.   Appendix: Some Useful Media and Solutions

Where not otherwise designated, the amounts of solid constituents are given in grams, liquid in milliliters. See text for pertinent literature references.

SM Medium

Glucose	10.0
Difco Bacto peptone	10.0
Yeast extract	10.0
$MgSo_4 \cdot 7H_2O$	1.0
KH_2PO_4	1.9
K_2HPO_4	1.0

H_2O 1000.0
Difco Bacto agar   20.0
Final pH = 6.4

Pour plates slightly more than one-half full. Store at room temperature for not more than 1 week. Desiccation of the surface produces physical changes in the agar that prevent even spreading of the inoculum.

SM/5 Medium

Reduce the amounts of glucose, Bacto peptone, and yeast extract each to 2, and $MgSO_4 \cdot 7H_2O$ to 0.2.

NZCase Medium

Glucose	10.0
NZCase peptone*	10.0
$MgSO_4 \cdot 7H_2O$	1.0
KH_2PO_4	2.2
K_2HPO_4	1.0
H_2O	1000.0
Bacto agar	20.0
Final pH = 6.4	

Pour plates slightly more than two-thirds full. Use after 1 day. Store at room temperature for not more than 1 week.

HL5 Medium for Axenic Growth of D. discoideum Ax Strains

Glucose	14.0
Yeast extract	7.0
Peptone	14.0
$Na_2HPO_4 \cdot 7H_2O$	0.95
KH_2PO_4	0.5
H_2O	1000.0
pH = 6.5	

Autoclave (a) the glucose, (b) yeast extract and peptone, and (c) Na and K phosphates in three separate solutions. The choice of peptone is critical. Thiotone (BBL) or B acteriological peptone (Oxoid) have given routinely satisfactory results. Some batches of Difco proteose peptone are excellent, while others are unusable, increasing the doubling time to ~20 hours and reducing the yield. David Soll (personal communication) uses the Difco product but pretests batches and then purchases large stocks thereof. HL5 medium can be stored at room temperature (Cocucci and Sussman, 1970; Soll et al., 1976).

MA Medium for Axenic Growth of D. discoideum Wild Type

Glucose	10.0
Peptone	30.0

---

*Manufactured by Sheffield Products, PO Box 398, Memphis, TN 38101.

Yeast extract	4.0
$Na_2HPO_4 \cdot 12H_2O$	1.28
KH_2PO_4	0.49
B_{12}	5 mg
Folic acid	4 mg
Biotin	1 mg
Riboflavin	1 mg
Thiamine·HCl	1 mg
$FeCl_3$	25 mM (final)
$ZnCl_2$	25 mM
$MgCl_2$	0.5 mM
H_2O	1000.0
pH = 6.5	

As noted in the text, the brand of peptone is crucial to the efficacy of this medium. Oxoid Bacteriological peptone gave the best results but considerable batch-to-batch variation was encountered. Daigo proteose peptone was less effective and Difco Bacto peptone or tryptone and Daigo Poly peptone supported little or no growth (Maeda, 1983).

A-Medium for Axenic Growth of Polysphondylium pallidum (Sussman, 1963)

Lipid-free milk powder (Starlac)	5.0
Proteose peptone	10.0
Animal lecthin (Glidden Products)	0.4
Na_2HPO_4	0.45
KH_2PO_4	0.92
$H_2O_{‰}$	1000.0

Note: The milk powder can be omitted, but rate and yield suffer.

Davis Minimal Medium

Add in the order shown below:

H_2O	1000.0
K_2HPO_4	7.0
KH_2PO_4	3.0
Na_3-citrate·$2H_2O$	0.5
$MgSO_4·7H_2O$	0.1
$(NH_4)_2SO_4$	1.0
Final pH = 6.9–7.1	

Autoclave the above and then add 10 ml/liter of a filter-sterilized 20% (w/v) glucose solution.

Lower Pad Solution (LPS)

KCl	12
Streptomycin sulfate	4
$MgCl_2·6H_2O$	4

$$NaH_2PO_4 \qquad 30$$
$$Na_2HPO_4 \qquad 27$$
$$H_2O \qquad 8000$$
$$pH = 6.4$$

Store at 0–5°C in a dispensing jug.

Upper Pad Solution (UPS)
$KH_2PO_4 \qquad 1 \; M$

Bonner's Standard Solution (Bonner, 1947)

NaCl	0.6
KCl	0.75
CaCl$_2$	0.3
H$_2$O	1000.0

REFERENCES

Balamuth, W. (1964). *J. Protozool.* **11** (Suppl.), 19–20.
Bonner, J. T. (1947). *J. Exp. Zool.* **106**, 1–26.
Ceccarini, C., and Cohen, A. (1967). *Nature (London)* **214**, 1345–1346.
Cocucci, S., and Sussman, M. (1970). *J. Cell Biol.* **45**, 399–407.
Deml-Rand, K., and Sussman, M. (1983). *Differentiation* **24**, 88–96.
Ellingson, J. S., Telser, A., and Sussman, M. (1971). *Biochim. Biophys. Acta* **224**, 388–395.
Franke, J., and Kessin, R. (1977). *Proc. Natl. Acad. Sci. U.S.A.* **74**, 2157–2161.
Franke, J., and Sussman, M. (1973). *J. Mol. Biol.* **81**, 173–185.
Gerisch, G. (1959). *Naturwissenschaften* **46**, 654–656.
Hohl, H. R., and Raper, K. B. (1963). *J. Bacteriol.* **86**, 1314–1320.
Inselburg, J., and Sussman, M. (1967). *J. Gen. Microbiol.* **46**, 59–64.
Laine, J., Roxby, N., and Coukell, M. B. (1975). *Can. J. Microbiol.* **21**, 959–962.
Maeda, Y. (1983). *J. Gen. Microbiol.* **129**, 2467–2473.
Newell, P. C. (1982). *In* The Development of *Dictyostelium discoideum* (W. F. Loomis, ed.), pp. 35–65. Academic Press, NewYork.
Newell, P. C., and Ross, F. M. (1982). *J. Gen. Microbiol.* **128**, 1639–1652.
Newell, P. C., Telser, A., and Sussman, M. (1969). *J. Bacteriol.* **100**, 763–768.
Newell, P. C., Longlands, M., and Sussman, M. (1971). *J. Mol. Biol.* **58**, 541–554.
Pardee, J. D., and Spudich, J. A. (1982). *Methods Cell Biol.* **24**, 271–289.
Perkins, D. G. (1962). *Can. J. Bot.* **8**, 591–594.
Raper, K. B. (1941). *Growth* **5**, 41–76.
Raper, K. B. (1984). "The Dictyostelids." Princeton Univ. Press, Princeton, NJ.
Raper, K. B., and Alexander, D. F. (1945). *Mycologia* **4**, 499–525.
Reinhardt, D. J. (1966). *J. Protozol.* **13**, 225–226.
Schindler, J., and Sussman, M. (1977). *J. Mol. Biol.* **116**, 161–170.
Schwalb, R., and Roth, R. (1970). *J. Gen. Microbiol.* **60**, 283–286.
Soll, D. R. (1979). *Science* **203**, 841–849.
Soll, D. R., and Waddell, D. R. (1975). *Dev. Biol.* **47**, 292–302.
Soll, D. R., Yarger, J., and Mirick, M. (1976). *J. Cell Sci.* **20**, 513–523.
Sussman, M. (1951). *J. Exp. Zool.* **118**, 407–418.
Sussman, M. (1954). *J. Gen. Microbiol.* **10**, 110–120.

Sussman, M. (1961). *J. Gen. Microbiol.* **25**, 375–378.
Sussman, M. (1963). *Science* **139**, 338.
Sussman, M. (1966). *Methods Cell Physiol.* **2**, 397–410.
Sussman, M., and Lovgren, N. (1965). *Exp. Cell Res.* **38**, 97–105.
Sussman, M., Schindler, J., and Kim, H. (1978). *Exp. Cell Res.* **116**, 217–228.
Sussman, R. R., and Sussman, M. (1963). *J. Gen. Microbiol.* **30**, 349–355.
Sussman, R. R., and Sussman, M. (1967). *Biochem. Biophys. Res. Commun.* **29**, 53–55.
Watts, D. J., and Ashworth, J. M. (1970). *Biochem. J.* **119**, 171–174.
Williams, K. L., Kessin, R., and Newell, P. (1974). *Nature (London)* **247**, 142–143.

Chapter 3

Genetic Tools for Dictyostelium discoideum

WILLIAM F. LOOMIS

Department of Biology
Center for Molecular Genetics
University of California, San Diego
La Jolla, California 92093

I.   Introduction

What sets *Dictyostelium* apart from other multicellular developing organisms is the opportunity it presents to isolate and analyze pertinent mutations with relative ease. This is a consequence of its microbial style of growth coupled to its growth-independent development (Loomis, 1975). The ability to grow from a single amoeba or spore permits analysis of genetically homogeneous clones of cells. These need not complete development to be propagated or stored as vegetative amoebae, and so mutations affecting development but not growth can all be considered conditional; i.e., they result in observable mutant phenotypes only under developmental conditions. Conversely, the absence of required cell division during development means that many genes vital for growth are dispensable for development. This was shown many years ago in the first analysis of temperature-sensitive mutations. While some of the dozen or so *ts* mutations affected both growth and development at nonpermissive temperatures (27°C), other *ts* mutations affected growth but not development, and yet another class of *ts* mutations exclusively affected development (Loomis, 1969).

Dictyostelium discoideum develops into fruiting bodies consisting of at least two distinct cell types, spores and stalk cells, all in the absence of required cell division, by virtue of the ability of cells to aggregate into groups of 10^4–10^5 cells after the nutrients have been exhausted. During the 24-hour period it takes to integrate and differentiate the cell types from vegetative amoebae, there are at least a dozen developmental stages that can be unequivocally recognized by morphological, physiological, biochemical, and molecular biological criteria.

Thus, there is much to look for when developmental mutations are to be analyzed. Using gross morphological aberrations as observed phenotypes, it has been estimated that ~300 genes are uniquely involved in producing the highly structured fruiting body (Loomis, 1980). This fairly small number of genes is amenable to intensive genetic analysis. So far, >50 such genes have been mutationally marked and mapped to unique loci on the seven chromosomes. The products of most of these genes have yet to be defined, but the consequences to their loss or aberrations have been carefully described in most cases (see Loomis, 1975, 1980, for reviews).

Dictyostelium discoideum grows and develops as a haploid as well as a diploid. The isolates from nature have been haploid, but diploids can be selected in the laboratory, as will be described below, and turn out to develop with identical timing and morphogenetic stages as the haploid parents (Loomis, 1969). This property makes *D. discoideum* even more attractive for genetic studies. Mutations can be induced in haploid cells and recognized in the clonal progeny with no need for backcrosses or specially constructed chromosomes as is required in obligatory diploid organisms such as *Drosophila*. Moreover, once a mutation of interest has been isolated it can be genetically analyzed for dominance and recessiveness as well as complementation by constructing diploids with other strains wild-type or mutant for the affected phenotype. Diploids can also be induced to segregate chromosomes randomly while generating recombinant haploid progeny. Cosegregation of genetic markers maps the mutation to a given linkage group. Further mapping can be done by analyzing the relative frequency of progeny that are the products of mitotic crossing-over events. In this way select genes have been ordered relative to the acrocentric centromeres on several linkage groups (Newell, 1982; Welker and Williams, 1982). Further genetic resolution is just appearing on the horizon as DNA-mediated transformation has finally become feasible and can be used to define the function of genes at the nucleotide level (Nellen *et al.,* 1984, this volume, Chapter 4; De Lozanne, this volume, Chapter 27; Knecht *et al.,* 1986).

In this chapter I will spell out in detail the steps in the generation, isolation, genetic analysis, and interpretation of mutations affecting development of *D. discoideum*. Reviews on genetic analysis have been provided by Newell (1982, 1984) and Godfrey and Sussman (1982).

II.   Mutagenesis

The function of a gene can only be shown unequivocally by analysis of the consequences of mutations in the gene. Since there is no convincing evidence in *D. discoideum* for directed mutagenesis as occurs in yeast, we must rely on random mutagenesis followed by isolation of the mutations of interest to give us marked genes.

Spontaneous mutations occur quite infrequently, at the rate of $\sim 10^{-6}$ per gene. The frequency can be increased 100-fold by treatment with UV irradiation or 1000-fold by chemical mutagens (Sussman, 1955; Yanagisawa *et al.*, 1967). It is essential to increase the rate of mutagenesis if only a few thousand clones can be realistically screened for the desired mutations. This is the case if one is visually screening for morphological mutations affecting aggregation or culmination and even more so if alterations in specific proteins such as enzymes or surface antigens are sought. There are drawbacks to heavy mutagenesis that we will go into later, but in many cases they can be evaded later, once you have the mutants in hand. A flowchart for mutagenesis is presented in Table I.

Mutagenesis with N'-methyl N'-nitro N-nitrosoguandine (NTG) or ethyl methyl sulfonate (EMS) are about equally effective in generating stable mutants without appreciable metabolic killing (Yanagisawa *et al.*, 1967). UV treatment, on the other hand, gives far more killing in relation to the generation of stable mutants. For historical reasons, NTG rather than EMS has been used in most

TABLE I

MUTAGENESIS OF *Dictyostelium*

Step 1: Wash cells. A culture of strain Ax-3 growing exponentially in HL5 medium is centrifuged at 300 g for 3 minutes and the pelleted cells resuspended in 20 mM phosphate buffer, pH 6.0, at 10^8 cells/ml. A 50-ml culture at 2×10^6 cells/ml in HL5 medium will give 10^8 cells total that can be resuspended in 1 ml buffer. Alternatively, cells growing on SM agar plates in association with *Klebsiella aerogenes* can be washed free of bacteria by centrifuging the amoebae at 300 g for 3 minutes and resuspending them in 15 ml cold distilled water before repelleting them (see Chap. 1, this volume). After three sequential centrifugations, the amoebae are free of most bacteria and can be suspended at 10^8 cells/ml in 20 mM phosphate buffer, pH 6.0. One or two 9-cm diameter SM plates inoculated with $\sim 10^5$ amoebae and 10^8 bacteria will have 10^8 *Dictyostelium* cells after 2 days incubation at 22°C.

Step 2: Mutagen. A solution of N'-methyl N'-nitro N-nitrosoguanidine (NTG) is prepared in distilled water at 10 mg/ml. Add 50 μl to 1 ml of cells suspended at 10^8/ml to give a final concentration of 500 μg/ml NTG. Incubate at 22°C for 20 minutes with shaking.

Step 3: Stop mutagenesis. The cells are centrifuged at 300 g for 3 minutes and the pellet resuspended in 10 ml HL5 medium. The guanine in the medium inactivates any remaining NTG. The suspension contains 10^7 visible cells/ml. However, only $\sim 0.1\%$ of the cells are viable after mutagenesis due to the presence of lethal mutations in most of the cells. Viability varies between 10^{-4} and 10^{-3} under these conditions, so the 10-ml suspension contains 10^3-10^4 viable cells/ml. Plate out 100 μl as well as 10 μl to each of a series of SM plates along with a drop of a dense bacterial culture. At one or the other inoculum, plates with ~ 100 viable cells will give 100 plaques after incubation at 22°C for 3 days. Each plaque will be a population derived from a single mutagenized cells.

Alternatively, the population can be inoculated into HL5 medium to which 100 units/ml penicillin and 500 μg/ml streptomycin are added to block growth of any contaminating bacteria. Of course, this approach is only used when mutagenizing an axenic strain such as Ax-3. The population will show visible increase in titer within a week, at which point cells can be cloned or subjected to selective conditions.

---

procedures with *D. discoideum,* so there is a larger body of evidence on its effects (Liwerant and Pereira da Silva, 1975; Dimond *et al.,* 1973).

Vegetatively growing cells are usually used for mutagenesis, although cells at various stages of development can be used equally well. At one time it was hoped that a higher proportion of mutations affecting genes being transcribed during aggregation might be recovered if aggregation-stage cells were mutagenized, but the proportion of aggregation mutations from such cells was found to be the same as in mutagenized populations of growing cells (Loomis, unpublished). Therefore, cells growing axenically in HL5 medium or in association with *Klebsiella aerogenes* are used for convenience and reproducibility. The cells are washed free of medium or associated bacteria and suspended at $\sim 10^8$ cells/ml in 20 mM phosphate buffer, pH 6.0, or buffered salts (PDF). It is important to use enough cells such that the number of viable cells following mutagenesis is greater than the number to be screened for mutations. It is also important that the number of cells lost due to attachment to the walls of the test tube (often $\sim 10^6$ cells) be a trivial proportion of the total number of cells.

NTG is an extremely dangerous compound because of its ability to cause mutations in cells of any organism including those of humans. NTG is supplied by Sigma in sealed ampules containing 10 mg each. The reddish powder is dissolved directly in the ampule by injecting 1 ml of sterile water with a syringe and allowing about an hour for the NTG to dissolve and form a yellow solution of 10 mg/ml. The solution is withdrawn with a syringe and diluted 20-fold in the cell suspension. Mutagenesis is allowed to proceed for 20–30 minutes with shaking at room temperature ($\sim 22°C$) before the cells are pelleted by brief centrifugation and the supernatant poured off into several milliliters of 1 M sodium thioguanine. The thioguanine instantaneously inactivates NTG and should be kept available to neutralize any accidental spills. This procedure keeps the NTG well contained throughout and should pose no danger to the investigator.

The cells are resuspended in fresh, sterile HL5 medium, which neutralizes any remaining NTG. They can then be appropriately diluted for the selection or screen in mind.

Under these conditions, mutations in individual genes that result in an observable phenotype occur at a frequency of 10^{-3} among the surviving cells. However, the population as a whole sustains about 7 mutations per cell, each of which is lethal. Only one in a thousand cells survives and does not carry any lethal mutations. Since it is easy to start with 10^8 cells, this level of killing leaves 10^5 viable cells that can be screened for the desired mutations. If the progeny of each of the 10^5 cells were fully tested, 100 independent mutants would be recovered from the population with alterations in a particular gene. However, in practical terms only a few thousand clones are usually analyzed and from these several independent mutations are recovered in the gene of interest.

The degree of mutagenesis by NTG is variable even among parallel cultures treated identically, probably as the result of nonlinear variables that have yet to be recognized. Thus, the proportion of survivors has been found to range from 10^{-3} to 10^{-5} under standardized conditions. Increasing the time of NTG treatment or the concentration of NTG increases the rate of mutagenesis and decreases the number of surviving cells. The variability in NTG-induced mutation rate poses problems for several screening procedures, since it takes 3–4 days to titer the surviving cells by plating a fraction on plates spread with a bacterial lawn and waiting for visible plaques to appear. One way around this problem is to incubate the complete mutagenized population in axenic medium (HL5) or in association with bacteria food source and wait 4–5 days for the surviving cells to multiply until they make up the majority of the population. Such grown-up cultures can be counted in a hemocytometer or a Coulter counter and the titer of viable cells accurately determined. However, clones with slight growth disadvantages will be proportionally reduced in the grown-up culture, and mutants recovered from such a culture need not necessarily carry independent mutations: they may be siblings of a single clone that received the mutation. The probability of recovering siblings is reduced if the original number of surviving cells is far greater than the number of cells screened. Nevertheless, when it is important to be certain that each mutant isolate represents an independent random mutational event, it is best to plate out the mutagenized population immediately following the removal of NTG in such a way that separate clones from each survivor can be analyzed. The uncertainty as to viable titer following mutagenesis can be compensated for by plating at several different dilutions and choosing the set that gives the greatest number of individual clones.

Those who have had experience in selecting mutants from bacterial populations often consider allowing a period following mutagenesis for phenotypic expression. This is seldom necessary for the type of mutations sought in *D. discoideum*. Developmental defects are not expressed until a cell has gone through many generations of growth to generate a clone of at least 10^6 cells, and the growth selections commonly used, such as drug or heavy-metal resistance, allow a generation or two of growth before establishing stasis. Mutations are usually expressed in haploid cells within one generation.

While it is important to mutagenize heavily so as to recover mutants at a realistic frequency, it can be predicted (and has been confirmed in dozens of cases) that a selected clone has about a 50% chance of also carrying a second mutation that is unrelated but affects development. This comes about because there are several hundred genes, mutations in any one of which will affect development. The number of such developmental genes is ~5% the number of genes vital for growth. Therefore, a population that receives 7 lethal mutations per cell also receives 0.35 morphological mutations per cell among the survivors as well as the dead. These unwanted second mutations can complicate analysis of the selected mutation but can be removed by backcrossing or by being covered by a wild-type allele in diploid strains as will be described below.

III. Selections

Microbial genetics of processes in bacteria and phage is rightfully respected for the ingenious selection procedures that permit only cells of a desired mutant type to grow under specific conditions. Many selections allow a mutant clone to be picked as the sole survivor from 10^9 or 10^{10} bacteria. Such selections are rare in *D. discoideum* studies due to the natural resistance of the organism to many common drugs and the fact that most mutations of interest affect developmental processes that are dispensable for growth. However, a few selections have been found to be convenient and useful.

A.   Methanol Resistance

While carrying out studies on the sensitivity to acriflavine, it was found that mutations that occurred in the *acr*A locus on linkage group II (LGII) conferred resistance not only to 100 μg/ml acriflavine but also to 2% methanol (Williams *et al.*, 1974). Mutations in this locus turn out to be the easiest to select of any known so far. It is often useful to mark a strain genetically by a mutation in *acr*A before embarking on reversion studies at another locus. The selection is so tight that mutagenesis is not needed. Spontaneous mutants arise at a frequency of 10^{-5}–10^{-6}.

Standard SM agar plates (40 ml/plate) can be made 2% methanol by pipetting 0.8 ml of methanol over the surface and allowing it to diffuse into the agar overnight. If the plate is wet the next day, the excess liquid can be allowed to evaporate by removing the lid for an hour or so. Each plate is then spread with a drop of an overnight culture of *K. aerogenes* and ~10^6 amoebae or spores. The bacteria grow well in the presence of 2% methanol, but wild-type (*acr*A$^+$) amoebae do not grow. Within 5 days, 1–10 plaques of *acr*A mutant clones grow up and can be picked for further analysis. These mutations are stable and always result in growth on methanol plates.

B.   Poison Resistance

Positive selection for growth in the presence of drugs or poisons at concentrations that stop growth of wild-type cells are listed in Table II. The procedures for selection are almost identical to those used with methanol, but usually muta-

TABLE II

DIRECT SELECTION PROCEDURES[a]

Gene	Linkage group	Selective agent	Comments
acrB	I	Acriflavine (100 μg/ml)	
cycA	I	Cycloheximide (500 μg/ml)	Mutants are rare ($\sim 10^{-7}$)
acrA	II	Methanol (2%) or acriflavine (100 μg/ml)	Keep drug plates in the dark
acrC	III	Acriflavine (100 μg/ml)	
acrD	IV	Acriflavine (100 μg/ml)	
ebrA	IV	Ethidium bromide (35 μg/ml)	
nysB	V	Nystatin (100 μg/ml)	
cobA	VII	Cobaltous chloride (300 μg/ml)	Range: 250–350 μg/ml

[a] References to these procedures can be found in Loomis (1980), Newell (1982).

genized populations are used because the spontaneous frequency of mutation to resistance is not much greater than the frequency at which cells escape growth repression in a nonhereditary manner (breakthrough growth).

A rather unique selection has been used to isolate mutants altered in their ability to ingest particles (Vogel *et al.*, 1980). Wild-type cells phagocytically engulf not only bacteria but also tungsten balls (1 μm diameter). Cells that have eaten tungsten balls are heavy and settle out rapidly. Mutants were selected from cells still in suspension from cultures allowed in ingest tungsten balls. Some of these were found to carry mutations in the *phg*A locus, which results in inability of cells to trap bacteria on their surfaces in the presence of glucose.

IV. Screening Techniques

The great majority of mutants of *D. discoideum* have been isolated by screening the survivors of a mutagenized population rather than by direct selection. Techniques have been developed that can recognize an altered phenotype among several thousand clones, so that mutants of interest can be readily recovered. It is only a matter of finding the proverbial needle in the haystack. A variety of different approaches have proved successful for different genes.

A. Genes Affecting Growth

The first selectable mutations were those that result in reduced growth in association with *K. aerogenes* (Loomis and Ashworth, 1968). Mutagenized populations were diluted to 2×10^6 visible cells/ml and 0.1 ml spread on each of 10 agar plates in association with *K. aerogenes*. Since the population was mutated with NTG to 10^{-3} survival, each plate received 200 viable cells. After 4 days incubation at 22°C, plaques covered almost the whole surface of each plate. Most plaques were about the same size, but a few of the plaques were much smaller than the others and were picked with sterlized toothpicks and touched to fresh plates in a regular gridwork pattern. A grid is conveniently made by placing the plate over a piece of paper on which a rectilinear pattern of 12 boxes has been drawn. A mark at the bottom of the pattern is aligned with a mark put on the side of the agar plate to set a unique orientation. Each box is then inoculated. The grid plates were incubated for a week at 22°C and plaque sizes observed. From 96 putative mutants (8 plates with 12 picks each), 10 small plaque-forming strains were recovered. These bred true in that clones of each gave rise to minute plaques on further subculture. Several of these *min* strains were used to select for diploid strains, since they all turned out to be recessive and noncomplementary (Loomis and Ashworth, 1968). The frequency of recovery of *min* mutations (10/2000) is higher than that for mutations in a single gene (10^{-3} of the survivors under these conditions) and appears to result from the fact that mutations in any of a dozen or so genes can give the small-plaque phenotype. Although these early experiments were not very quantitative, they suggest that there are about five *min* genes that can be mutated.

A variation on this technique has generated over a hundred mutants that are temperature-sensitive for growth (*tsg*) (Loomis, 1969). To screen for *tsg* mutations, plaques from the survivors of a mutagenesis are picked with toothpicks to two sets of 20 plates each. Pairs of plates are set over the grid pattern papers, which can be seen easily through the plates just spread with an inoculum of bacteria. A toothpick is lightly touched to a plaque on the original plate of survivors and then touched to identical grid positions on each of the two plates. It takes <1 hour to pick 240 clones to a set of replica plates. Matched plates in each set are numbered and then incubated at either 22° or 27°C. Five days later the plates are once again aligned over the grid patterns and the growth patterns compared. Clones that grew well at 22°C but not at all at 27°C are picked from the 22°C plate to a fresh set of replicas. Screening 240 clones usually generates 4–8 putative mutants, of which 1 or 2 turn out to carry stable mutations resulting in growth temperature sensitivity. There are probably several hundred potential *tsg* genes, but the frequency of mutating to a conditional state is at least 10-fold lower than the frequency of inactivating the gene product. Strains carrying *tsg* mutations can be crossed with other strains carrying conditional selectable markers as will be described below.

This technique of replica plating can be used in many different screening procedures. For instance, in a massive mutant hunt Newell *et al.* (1977) isolated a useful mutant from several thousand clones that were replicated to sets of plates spread either with *K. aerogenes* or *Bacillus subtilis*. One of the replicated clones grew on the gram-negative bacteria *K. aerogenes,* but not on the gram-positive bacteria *B. subtilis*. This strain carries a mutation in a gene referred to as *bsg*A (Ratner and Newell 1978). Since then two other genes, *bsg*B and *bsg*C, have been found that give the same phenotype when mutated (Morrissey *et al.,* 1980; Welker and Williams, 1980). Strains carrying a *bsg* mutation can be convenient-ly used in genetic crosses with strains carrying *tsg* mutations or a *bsg* mutation in one of the other loci.

Mutants of *D. discoideum* have been isolated that do not require a bacterial food source (Sussman and Sussman, 1967; Watts and Ashworth, 1970; Loomis, 1971). A liquid medium used for growth of the dinoflagellate *Nigleria* was provided by Prof. Chandler Fulton and found to keep wild-type *D. discoideum* cells apparently healthy for a week or two. However, the population did not grow during this period. About a month later the number of cells was found to be increasing and a strain, Ax-1, was cloned from the population which would grow in the rich medium in the absence of bacteria (Sussman and Sussman, 1967). Two other strains, Ax-2 and Ax-3, were selected in other laboratories in a slightly simpler medium, HL5 (Watts and Ashworth, 1970; Loomis, 1971). Both of these strains carry mutations in two unlinked loci, *axe*A and *axe*B, required for growth in HL5 medium (William *et al.,* 1975). These strains will also grow in a defined medium consisting of certain amino acids, vitamins, and salts, but only if inoculated at titers $>10^5$/ml (Franke and Kessin, 1977). If this minimal medium is supplemented with 5% dialyzed calf serum, growth from single cells can be achieved (Knecht and Loomis, unpublished results). This observation indicates that at titers $>10^5$/ml cells produce a growth factor that reaches a required level to permit growth. The growth factor can be supplied by calf serum. It is >5000 Da, heat-stable (autoclavable), and soluble in 65% saturated ammonium sulfate (Loomis, unpublished). Although the identity of the growth factor in calf serum is unknown, it permits a range of selection techniques.

Since growing *D. discoideum* cells do not all attach well to plastic Petri dishes, it is not possible to keep clones localized in a dish, since even gentle movement of the plates dislodges a few cells from each clone which float about and settle elsewhere. This problem can be overcome by growing clones in the wells of multititer plates (Dimond *et al.,* 1973). Each well is inoculated with 0.03–0.04 ml of HL5 medium in which cells are suspended at 12 cells/ml. In this way about half of the wells receive a cell. Since the presence of cells in a drop is random, Poisson distributions can be expected. At this inoculation frequency, about a third of the wells get a single cell, a third get more than one cell, and a third remain cell-free. The cells grow $\sim 10^6$ cells/well in 7–10 days. To facilitate the inoculation of thousands of wells, a convenient set of tools has been developed (Brenner *et al.,* 1976). The cell suspension is placed in a sterile 400-ml container and a dripper with 96 tubes arranged to match the wells of a multitest plate lowered into it (Fig. 1). The dripper is used like a pipet to dispense a single drop to each well. The procedure can be used to inoculate several thousand wells (30 multitest plates) in under an hour. These are incubated for a week or so, until visible populations can be seen on the bottoms of half of the wells. Half of these will be clones derived from a single initial cell. They can be picked from the wells or allowed to develop or lysed. However, it is best to replicate them first.

It is easy to replicate *D. discoideum* from multitest plates by using a "bed of nails" replicator (Fig. 1). A metal plate can be fitted with 96 screws or prongs in a pattern that fits the wells of a multitest plate. It can be sterilized by dipping it in ethanol, shaking off the liquid, and lighting it on a flame. If the bowl holding the ethanol is ignited by a burning drop from the replicator, as occasionally happens, it can be easily extinguished by covering it with a plate. The replicator gets quite hot as the ethanol burns off and will kill cells in mutitest plates unless cooled. It can be cooled in cold, sterile water before being used. When replicating ≥30 plates, two replicators are used. While one is cooling, the other can be used to replicate. The replicator is fitted into the wells of a plate to be replicated and then placed into the wells of a plate freshly dripped with medium. Several hundred cells from each well are passaged in this way. A plate can be sequentially replicated to several fresh plates if more than one screening is planned. It takes about an hour to number 30 master plates and each set of replicas, arrange them in proper orientation, and passage the cells. Clones grow up in the replicas in less than a week.

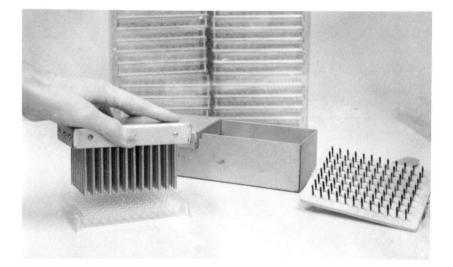

Fig. 1.   Sterile container and dripper with 96 tubes arranged to match the wells of a multitest plate.

The "bed of nails" can also be used to replicate clones onto the surface of large Petri dishes spread with bacteria. Mutants that grow in axenic medium but not on bacteria have been isolated in this way (Loomis, unpublished). Most have been found to ingest bacteria but to fail to digest them in phagocytic vesicles. These mutations most likely affect genes necessary for the initial steps in digesting bacteria.

The same approach could be used to isolate auxotropic mutants. By replicating clones from rich HL5 medium to minimal medium supplemented with dialyzed calf serum, it should be possible to screen for strains that require any one of eight amino acids, or purines, pyrimidines, vitamins, and other nutritional factors.

B. Genes Affecting Enzymes

About two dozen enzymes have been found that increase in specific activity at one stage or another during development of *D. discoideum* (Loomis, 1975). Many of these have been shown to be dispensable for growth under laboratory conditions by isolating mutants lacking one or another developmentally regulated enzyme. These mutations were screened for among the survivors of mutagenized populations grown in axenic medium in multitest wells.

Exponentially growing cells are treated with NTG and then plated out in 40 multitest plates on the assumption of 10^{-3} survivors. Often another 40 multitest plates are dripped at a threefold higher concentration in case the mutagenesis resulted in only 3×10^{-4} survivors. Usually one or the other set will have 30% empty wells after the clones have grown up, indicating that ~1400 wells contain cells derived from a single cell each. This is sufficient to recover one or two mutants lacking any given enzyme dispensable for growth under these conditions.

The appropriate set of multitest plates is replicated to two or more fresh sets that are then incubated for a week to allow the clones to grow. When temperature-sensitive mutations are sought, the plates are shifted to 27°C after being replicated. The first enzymes screened in this way were lysosomal acid hydrolases that are secreted from axenically growing cells. Therefore, the spent medium as well as the cells were screened for enzymatic activity.

Chromogenic substrates such as *p*-nitrophenyl phosphate are used to indicate the presence or absence of enzyme activity. To assay acid phosphatase the pH of the medium must be dropped to pH 5. This is done by dissolving the substrate in 0.1 *M* citrate, pH 4.8, containing 0.1% NP-40, a nonionic detergent that lyses the cells. Some mutations lower the affinity of the enzyme for the substrate without significantly affecting the catalytic activity at high substrate concentrations. Since K_m mutations are likely to be in the structural gene for the enzyme rather than in a regulator gene, they are useful for subsequent analyses. They can be recognized by screening the clones with subtrate at a concentration that gives half-maximal activity for the wild-type enzyme (at the K_m value). Mutations that raise the K_m only five-fold can be recognized as resulting in only 20% wild-type activity.

The desired substrate is dissolved in the appropriate buffer at the K_m value and added to one of the replica sets with the dripper. The reaction is allowed to proceed at the optimal temperature for the enzyme for several hours. Each well contains $\sim 10^6$ cells (~ 50 µg protein), and reaction times ≤ 20 hours are necessary for some enzymes such as β-glucosidase. For the acid hydrolases the reaction must be stopped by dripping in 1 M sodium carbonate to develop the yellow color of the product, p-nitrophenol. Alkaline phosphatase is assayed at pH 9 and so the course of the reaction can be observed as it proceeds. To get an idea of when sufficient product has been produced by an acid hydrolase, wells can be inoculated with wild-type cells at 10^6/well, assayed in the same manner as the mutagenized clones, and periodically stopped to see if enough product has been generated to be recognized.

When wild-type clones have produced sufficient product, the reactions are stopped and wells with little or no product noted. These are compared to an unreacted replica set, well by well, and those wells with visible cells that produced less product in the assayed set are picked to fresh wells as putative mutants. Depending on the enzyme being screened, from 10 to 100 wells are usually picked from 40 multitest plates.

The putative mutants are grown up, replicated, and rescreened. Since only a single multitest plate can hold all the putatives, this is not a laborious procedure and takes only a few days. Clones that score as mutants in the second screening are recloned and analyzed in depth. Mass cultures can be grown up in 6-liter flasks for enzyme purification, and the cells can be clonally spread on agar plates in association with bacteria to assess growth rates and morphogenetic capability.

TABLE III

MUTATIONS RECOVERED BY SCREENING ENZYME ACTIVITY[a]

Gene	Linkage group	Substrate	Comments
alpA	III	p-Nitrophenyl phosphate (pH 8)	Mutation affects K_m.
nagA	IV	p-Nitrophenyl N-acetylglucosamine	Mutation affects migration.
gluA	VI	p-Nitrophenyl β-D-glucoside	
manA	VI	p-Nitrophenyl α-D-mannoside	Mutation does not affect morphogenesis.
acpA	II	p-Nitrophenyl phosphate (pH 5)	Mutation affects two isozymes.
uppA	Unmapped	UDPG and linked enzymes	Mutation blocks terminal differentiation.

[a] References to these mutant hunts can be found in Loomis (1980) and Loomis and Kuspa (1984).

If more mutants are needed, the procedure can be repeated. In some cases only *ts* or K_m mutants have been recovered for a given enzyme, indicating it is essential for axenic growth. Alkaline phosphatase is an example of this sort (MacLeod and Loomis, 1979).

Many different enzymes can be screened in this way (Brenner *et al.*, 1976). Either chromogenic or fluorogenic substrates can be used. It should be possible to adapt the procedure to radioactive assays by using multitest plates with removable bottoms. If the reaction product can be fixed to the bottom, autoradiography will indicate which wells contained mutant cells. These clones can then be recovered from replicas.

A list of the enzyme screens that have been successful so far is given in Table III. With ingenuity and diligence the table could easily grow severalfold to give interesting new mutants.

C.   Genes Affecting Antigens

Many developmentally regulated components of *Dictyostelium* have been recognized immunologically rather than by enzymatic activity. Mutations affecting these components can be screened for in a manner similar to that used for enzymes. All one needs is monospecific antibodies. The increased use of procedures generating monoclonal antibodies has expanded the availability of antibodies that bind to only a single gene product.

The first steps in these screens are identical to enzyme screens: populations of cells are mutagenized with NTG, dripped into multitest wells, grown up, and replicated. The medium is shaken out and the cells dried to the bottoms. The wells are blocked with buffer containing 10 mg/ml bovine serum albumin (BSA) to saturate nonspecific protein-binding sites as in the Western procedure. After an hour the blocking buffer is replaced with antibody at 10 μg/ml in phosphate-buffered saline (PBS) containing 10 mg/ml BSA. After incubation for 4–20 hours at 22°C, the antibody is removed by shaking out the plates and the wells are washed twice with PBS. Then enzyme-linked second antibody, such as alkaline phosphatase goat anti-mouse IgG, is added to tag the initial antibodies bound to the cells. If rabbit antiserum is used, enzyme-linked goat anti-rabbit IgG is used. After an hour the second antibody is washed out, the wells rinsed, and either chromogenic or fluorogenic substrate added to show the presence of antibody. The wells are then visually screened, and those that did not bind antibody in the screened plates but have cells in the replicas are picked and retested before being genetically and biochemically analyzed.

Up to now this procedure has been used successfully only with epitopes that are added posttranslationally to specific proteins (Murray *et al.*, 1984; Loomis *et al.*, 1985; Knecht *et al.*, 1984; Loomis, unpublished). Mutations affecting the glycosylation of specific molecules, such as gp80, lysosomal hydrolases, and a spore coat protein, SP96, have been isolated in this way. The mutations appear to affect the enzymes involved in posttranslational modification rather than the structural genes for these glycoproteins. They were recovered at the same frequency as mutations affecting enzymes, namely, 10^{-3} of the clones screened. In two cases, the carbohydrate epitope on gp80 and the sulfated oligosaccharide on lysosomal enzymes, monoclonal antibodies were used in the screen. In the other case a polyclonal rabbit antiserum specific to SP96 was used (Loomis, unpublished). The mutant (HL250) isolated with the polyclonal anti-SP96, fails to add fucose to SP96 but produces the normal amount of the spore coat protein. The spores appear to be normal.

The SP96 gene is not expressed until after 14 hours of development (Orlowski and Loomis, 1979; Devine et al., 1982; Dowds and Loomis, 1984). To screen for mutants affecting SP96 it was necessary to let the cells develop. By shaking out growth medium from the wells and replacing it with phosphate buffer, development was initiated in the cells. Fortunately, most of the cells stick tightly to the flat bottoms of multitest plates. Incubation at 22°C overnight in buffer results in expression of genes that are normally expressed only after aggregation (late genes). At this stage the spore coat proteins are packaged in prespore vesicles inside cells (Devine et al., 1983). The cells can be lysed and their contents fixed to the plates by air-drying the wells and washing with 50% methanol before blocking buffer is added. The ability to develop cells directly in the wells opens up approaches to many late genes. However, one has to be aware of the potential pleiotropic effects of mutations in steps early in the developmental sequence affecting expression of later genes.

Antibodies specific to a single gene product can be prepared by immunizing rabbits or mice with purified proteins and then analyzing the sera or monoclonal antibodies with both purified protein and whole-cell extracts. Another approach that has proved useful is to sequence the N terminus of the protein or an internal fragment and then synthesize the peptide. When a 10- to 14-amino acid sequence is attached to BSA it acts as an excellent immunogen. The serum recovered is often monospecific for the peptide, but if the serum reacts with several proteins the desired antibodies can be easily purified on a column prepared from Sepharose to which the peptide is covalently attached.

Yet another approach now coming into wide use is to insert a cloned DNA sequence into the β-galactosidase gene of λgt11. When this phage grows in Escherichia coli, a fusion protein is synthesized which consists of β-galactoside tailed with the protein coded for by the cloned sequence. The ras gene product of Dictyostelium was recognized with antibodies prepared against such a fusion protein (Reymond et al., 1984). It is easy to purify the fusion protein, since it is one of the largest proteins in E. coli extracts and carries β-galactosidase activity. Mutants in the ras gene have not yet been recovered because it appears to be a vital gene. However, it may be possible to screen for conditional mutants.

D.   Genes Affecting Morphogenesis

The easiest of all genetic screens with *Dictyostelium* are for mutations that cause morphological aberrations. A mutagenized population need only be plated clonally and the fruiting bodies in plaques visually inspected. Plates should be spread such that each receives ~100 viable cells. These will grow up to give 100 discrete plaques within which each clone will form characteristic fruiting bodies.

TABLE IV

MORPHOLOGICAL GENES[a]

Gene	Linkage group	Defect
*agg*B	I	Agg⁻
*dev*A	I	Fingers
*mod*A	I	Squatty fruits
*spr*A	I	Round spores
*agg*A	II	Agg⁻
*agg*F	II	Agg⁻
*agg*I	II	Agg⁻
*dev*B	II	Fingers
*slg*E	II	Slug
*stk*A	II	Stalky
*stm*B	II	Streamer
*stm*D	II	Streamer
*spr*B	II	Long spores
*whi*A	II	White sorus
*slg*B	III	Slug
*slg*D	III	Slug
*slg*G	III	Slug
*stm*C	III	Streamer
*stm*E	III	Streamer
*whi*B	III	White sorus
*agg*J	IV	Agg⁻
*agg*L	IV	Agg⁻
*bwn*A	IV	Brown pigment
*frt*A	IV	Slippery stalk
*slg*C	IV	Slug
*slg*I	IV	Slug
*spr*H	IV	Round spore
*spr*J	IV	Immature spore
*whi*C	IV	White sorus
*stl*A	VI	Stalk defect
*slg*J	VII	Slug
*stm*A	VII	Streamer

[a] These mutations are referenced in Loomis (1980) and Newell (1982).

The majority (~70%) will be of wild-type morphology. The remainder will be split equally into those that fail to aggregate and those that make clearly aberrant fruiting bodies. There are several hundred genes, mutations in any one of which will result in mutant morphology.

Table IV gives some of the classes of morphological mutants that have been studied. Some are of use in genetic linkage studies such as the white and brown mutations, while others such as the stalky mutation tell a lot about cytodifferentiation of cell types. This approach requires the ability to predict accurately the phenotype expected from a given type of mutation. Unless imaginative insight into morphogenesis is used, one recovers only bizarre curiosities that do not explain the mechanisms of differentiation. But if one can predict the form that will result from a given mutation, say one in adenyl cyclase, then this approach is very powerful.

V.   Mutant Analysis

Once a mutant strain is isolated, there are several standard procedures that are followed. First, the isolate should be recloned and several clones tested to ensure that the line is pure. The strain should then be stored so as to have a stock for immediate as well as future analyses. If the strain forms spores, they can be stored for a year or so by drying them down on silica chips. However, permanent storage is far more reliable if a suspension of ~10^8 spores/ml in 5% nonfat dry milk is lyophilized in an ampule that is sealed off while still under vacuum. In this dry, airless state spores are viable for >20 years when kept at 4°C. If a strain carries a developmental mutation that blocks spore formation, amoebae can be frozen in 5% DMSO solution and stored at −70°C or under liquid nitrogen. Reasonable recovery (≥1% survival) has been found after 5 years.

Synchrononous development on Millipore filters often shows if the mutation has developmental consequences. Likewise, aggregation in low-density populations can show up other defects. As was mentioned above, heavy mutagenesis often generates unwanted second mutations in the isolated strain. The consequences of mutations in a given enzyme or antigen can be assessed by comparing the phenotype of a half-dozen or so independently isolated mutants selected for this trait. The least severe phenotype can be assigned to the selected mutations. While some mutants may have more problems, those that are not common to all of the isolates may not result from the gene under study. This analysis of the minimal syndrome can be complemented by determining the temperature-sensitive phenotype in strains carrying *ts* mutations in the selected enzyme or antigen.

When the K_m value of an enzyme is apparently modified by mutation or the activity is temperature-sensitive, it is worthwhile to purify the residual activity as much as possible and compare the biochemical properties of purified mutant gene product to purified wild-type gene product. This procedure can eliminate the possibility of ancillary gene products being responsible for the molecular phenotype.

Each gene presents its own challenges for analysis of mutants. However, this is the exciting stage in mutant hunting, when it is already clear the mutants are in hand. Some analyses are obvious while others follow unexpected surprises.

VI.   Genetic Exchange

A.   Diploid Selection

The first step in genetic analysis of specific mutations is to select for diploids. Since there is no obligatory sexual stage in the life cycle of *Dictyostelium,* rare fusions of normal haploid amoebae have to be selected. Diploids normally occur at a frequency of $\sim 10^{-5}$ during early aggregation (Loomis, 1969). Several procedures have been worked out where neither parent strain will grow while diploid progeny grow well. There are two general approaches:

1. Each haploid strain carries a different recessive mutation rendering growth conditional. Under nonpermissive conditions neither of the parent strains grows, but diploids proliferate due to the wild-type allele provided by one of the parents.
2. One strain need not carry a selectable marker, but the other carries two mutations, one a recessive block to growth under selective conditions and the other a dominant resistance to a poison. When such crosses are plated onto medium with the poison and incubated under nonpermissive conditions, only the diploids grow out.

The most extensively used technique involves crossing two strains each carrying different *tsg* mutations that preclude growth at 27°C without impairing growth at 22°C. Since there are hundreds of genes that can be mutated to become *tsg*, seldom if ever have two independently selected growth temperature-sensitive strains been found to carry *tsg* mutations in the same locus. All of the hundred or so *tsg* mutations studied have been found to be recessive. Therefore, the diploids are genetically *tsg* 1/+; +/*tsg* 2 and grow well at 27°C.

To carry out a *tsg*/*tsg* cross, a loopful of cells is taken from the growing edge of a plaque of each strain. With no special precautions, the cells are mixed in 10 mM phosphate buffer (pH 7) with 100 μg/ml streptomycin sulfate. When this is done in 1 ml held in 6-cm test tubes, there are $\sim 4 \times 10^6$ cells/ml. Crosses can also be carried out in 500 μl in the wells of multitest plates. The amoebae rapidly ingest residual bacteria and start to develop. After incubating overnight at 22°C, 0.1 ml can be spread on plates inoculated with *K. aerogenes* and placed immediately at 27°C. In about a week plaques appear that can be picked and restreaked on fresh plates. Depending on the particular *tsg* alleles used, there may be some breakthrough growth after a week. For this reason, it is best to pick plaques as soon as they appear and retest them for growth at 27°C. Most crosses generate ~10 diploids per plate. All you need to continue the analysis is a single diploid isolate. To increase the probability of success, about a dozen putative diploids are streaked on a plate to be incubated at 27°C and again on a plate to be incubated at 22°C. Strains that grow well at 27°C and fruit well at 22°C are inspected for spore size. Diploid spores are larger than haploid spores (Loomis and Ashworth, 1968; Sinha and Ashworth, 1969). A diploid strain is chosen for analysis, picked from the 27°C plate, and spread on a growth plate at 22°C.

The same general technique is used when the recessive selection marker results in lack of growth on *B. subtilis*. Three *bsg* loci have been found that give this phenotype. A diploid strain carrying (*bsg*A/+; +/*bsg*B) will grow on plates spread with *B. subtilis* (Morrissey *et al.,* 1980). Within a week plaques appear in the thin lawn of *B. subtilis*. For this selection it is essential to have streptomycin in the buffer to kill off the *K. aerogenes* on which the parental strains were grown. *Bacillus subtilis* is naturally resistant to streptomycin. An asporogenous mutant of *B. subtilis* is usually employed as the food source so as to keep down contamination of the lab, but the problem is not acute. A dozen putative diploids are picked in replica to plates spread with *B. subtilis* or *K. aerogenes*. Cells within plaques on *B. subtilis* plates are too sparse to fruit well.

It is often convenient to cross a haploid strain carrying a *tsg* marker with a strain carrying a *bsg* marker. The overnight mating is plated on *B. subtilis* and incubated at 27°C. Putative diploids are streaked on *B. subtilis* and retested for growth at 27°C and also streaked on *K. aerogenes* at 22°C for fruiting body inspection.

The only known dominant selectable marker is to resistance to cobalt (Williams, 1978). When a strain carries *cob*A^D and a *tsg* marker, diploids can be selected from crosses with any other strain on plates made with 100 μg/ml cobaltous chloride when incubated at 27°C. Only the diploids (*cob*A^D/+; *tsg*/+) grow. *Klebsiella aerogenes* does not grow well in the presence of cobalt, but resistant bacteria spontaneously arise at ~10^{-8} and can easily be selected by spreading a few cobalt plates each with 0.1 ml of an overnight culture of *K. aerogenes*. Resistant bacteria grow up in a day and can be subcultured in liquid for use in cobalt selections.

Cobalt plates are tricky to use because the metal tends to form complexes that come out of solution. While *cob*^{D}, *tsg* selections sound simple and general, in practice they are difficult to get to work. Often the *cob*A$^+$ cells grow slowly on cobalt plates while other times diploids fail to grow. The variables are not well understood.

Another technique with promise for convenience uses the phenotype conferred by *nys* mutations. These can be easily selected by plating on the drug, nystatin, that blocks growth of wild-type cells. Resistant mutants grow out and those that carry recessive mutations in *nys* loci A, B, or C are more sensitive than wild-type to a different drug, coumarin (Kasbekar *et al.*, 1983). This coumarin sensitivity can be used to block growth of the strain when it is a parent in a genetic cross with a *tsg* strain or a strain carrying a mutation in a different *nys* locus. Only diploids grow on coumarin (1.25 μg/ml) plates at 27°C. Since nystatin can be used as a positive selection for *nys*, this marker can be selected in any strain one has in mind. However, there is a drawback to this scheme: the concentration of coumarin that blocks growth of *nys* strains but is not toxic to *nys*⁺ strains is dependent on "genetic background," i.e., it must be determined for each strain. Nevertheless, this approach has been successfully used several times (Chadwick *et al.*, 1985).

B. Dominance and Recessiveness

When a strain carrying a new mutation is crossed with another strain wild type for that gene, the phenotype of the diploid indicates the dominance of the mutation. In almost every case analyzed to date in *Dictyostelium*, the mutations have been found to be recessive. That is, the phenotype of the diploid for the marker under study is wild type. To a certain extent this analysis depends on how the phenotype is measured. For instance, a mutation that leads to white fruiting bodies (*whi*A, B, or C) is recessive, since heterozygous diploids (*whi*/+) are yellow. But if the fruiting bodies produced only half the normal amount of the yellow pigment (a zeta carotenoid) it would be hard to tell. On the other hand, a mutation leading to thermolability of an enzyme can be thought of either as recessive or codominant, since heterozygous diploids produce both wild-type thermostabile enzyme and mutant thermolabile enzyme.

C. Complementation

One of the most useful genetic analyses is to determine whether two independently isolated mutants with similar phenotypes for a given characteristic carry mutations in the same gene. This question is answered in a straightforward manner by determining the phenotype of diploids selected between the two mutant strains. If the mutations are recessive and the diploid expresses the mutant phenotype, the mutations are said to be noncomplementing and to occur, by definition, in the same genetic locus. The mutations are given the same three-letter italicized symbol with an capital roman letter, such as *whi*A, followed by a specific allele number, such as *whi*A501. In this way each locus and allele is unequivocally labeled.

If the diploid from a cross of two mutant strains has a wild-type phenotype, the mutations are said to be recessive, complementing, and to occur in distinct loci.

When a temperature-sensitive mutant and an unconditional mutant are crossed, the phenotype must be checked at both permissive and nonpermissive temperatures. This was recently done when a *ts* stalky strain, HH31, was crossed to another stalky strain, HL16, that carries mutation *stKA*501 (C. West, personal communication). The diploid formed wild-type fruiting bodies at 22°C and stalky fruiting bodies at 27°C. The phenotype of HH31 at 27°C was not completely identical to that of HL16, perhaps because the *ts* stalky mutation resulted in partial loss of activity at 27°C. The complementation analysis settled the question and showed that HH31 carries a *ts* mutation in the *stk*A locus on linkage group II. Without genetics, this point would have remained unsettled for a long time.

Construction of diploids between two independently isolated strains that carry mutations in the same locus has the added advantage of extinguishing the effects of secondary recessive mutations that might have been inadvertently introduced into one or the other strain. The only locus homozygous for a mutation in such a diploid is the one for which the mutants were selected. The probability of the secondary mutations being in the same locus is vanishingly small, since they were not directly selected. The consequences of loss of function of a specific gene can be clearly analyzed in such mutants which have a wild-type genetic background except at the specific locus. This homozygous diploid technique has clarified the consequences to mutations in such loci as *mod*B that affect EDTA-resistant adhesion (Loomis *et al.*, 1985).

D. Haploid Selection

Mapping of loci to specific linkage groups, backcrossing to remove unlinked markers, and the construction of double mutants all require the segregation of haploid progeny from diploid strains. Haploidization occurs naturally at a frequency of 10^{-4}–10^{-6} per generation depending on the specific diploid. There is some evidence that when 1 chromosome of the 14 in a diploid is lost at a mitotic division, the resulting aneuploid (13 chromosomes) rapidly loses other chromosomes on a random basis. Stable aneuploids never seem to occur. A cell that has lost both homologs of a chromosome is not viable. Likewise, a cell that contains only 6 chromosomes is dead. In this way, haploids with random assortments of each of the 7 chromosomes are generated (Loomis and Ashworth, 1968; Sinha and Ashworth, 1969). With the exception of the domant *cob*A mutation, all mapped mutations are recessive to the wild-type genes.

Williams and Barrand (1978) found that the agricultural pesticide, ben late, dramatically increases the frequency of haploidization. This drug appears to destabilize the microtubule arrays necessary for mitotic separation of chromosomes. To isolate haploid segregants, 10^3 diploid cells are spread on nutrient agar plates containing 20 μg/ml ben late. The plates are spread with *K. aerogenes* as a food source and incubated at 22°C for 4 or 5 days. Several hundred plaques arise and grow outward. By picking from a point on the growing edge of each plaque and inoculating a grid, it can be be shown that 10–50% of the plaques gave rise to haploid strains. If the diploid is heterozygous for *acr*A (methanol and acriflavine resistance), most of the haploids turn out to carry the chromosome bearing the recessive mutation *acr*A. This is because the *acr*A mutations give rise to cross-resistance to ben late. Diploids that do not carry *acr*A appear to segregate chromosomes randomly after growing in the presence of ben late.

If a chromosome carrying a selectable recessive marker is sought such as

TABLE V

SOME MAPPING STRAINS AND SELECTION[a]

Linkage group	Strain					
	HL106	XP99	XP95	X9	HL51	HL204
I	*cyc*A	*cyc*A	*cyc*A	*cyc*A	*cyc*A	—
II	*whi*A *tsg*D	*acr*A *whi*A	*acr*A *whi*A	*acr*A *tsg*D	*axe*A *acr*A	*axe*A *acr*A
III	*bsg*A	*bsg*A *tsg*A	—	*axe*B *rad*C	*tsg*A *bsg*A	—
IV	*bwn*A	*bwn*A	—	—	*tsg*A *ebr*A	—
V	—	—	—	—	—	—
VI	*man*A	*man*A	*man*A	—	*man*A	—
VII	—	*cob*A	*cob*A	—	—	—

[a] Strain HL106 can be crossed with another strain carrying a *tsg* mutation or a *bsg*B or C mutation. Haploids can be selected from the resulting diploids by plating 10^5 cells on 2% methanol plates or growing in the presence of ben late and then plating a few hundred cells on plates containing 500 μg/ml cycloheximide. The methanol-resistant haploids will all be white, which helps in avoiding mitotic crossover diploids. They should all be temperature-sensitive for growth. Haploids can be analyzed for linkage groups I, II, III, IV, and VI by growth on cycloheximide plates, methanol plates, *Bacillus subtilis,* and the presence or absence of *N*-acetylglucosaminidase or α-mannosidase, respectively. Strain XP99 can be crossed with selectable strains other than those carrying *bsg*A. Segregants are analyzed as described above except that LGIV is recognized by production of a brown pigment due to the lack of homogentisic oxidase (Morrissey, 1983), and LGVII is scored on growth in the presence of cobaltous chloride. Strain XP95 has LGII marked and can be crossed with *bsg*A strains. Otherwise, it resembles strain XP99. Strain X9 resembles strain HL106 on LGI and II and is a healthy strain. Strain HL51 carries *axe*A and *axe*B and grows axenically. Diploids made from this strain and other axenic strains grow in HL5 medium. *acr*A segregants grow on methanol plates. LGII can be recognized by temperature-sensitive growth and radiation sensitivity. Strain HL204 has the same markers on LGI and II as strain HL51. Haploids resistant to ethidium bromide (*ebr*A cells grow in the presence of 250 μg/ml ethidium bromide) can be directly selected and will be temperature-sensitive for growth, since they carry *tsg*B on the same linkage group (IV).

resistance to cycloheximide (*cyc*A), the diploid can be grown on ben late plates to confluency, the plate harvested, and dilutions plated on growth plates containing the selective agent. When this is done for *cyc*A (LGI) from a diploid heterozygous for cycloheximide resistance (*cyc*A/+), ~10% of the cells plated on 500 μg/ml cycloheximide plates grow. These are haploids that all carry the chromosome marked with *cyc*A but have random assortments of the other six chromosomes.

Some recessive resistance genes are so easy to select for that growth in the presence of ben late is unnecessary. *acr*A is perhaps the easiest marker to select from a heterozygous (*acr*A/+) diploid. One merely grows the diploid at 22°C for a cycle (about 20 cell generations followed by fruiting), picks sori, and spreads ~10^5 spores on a nutrient plate made 2% in methanol. Within a week 10–20 *acr*A strains will form plaques that can be further analyzed.

Backcrossing is carried out by mating a newly isolated mutant strain with a standard strain, such as HL100, that only carries a mating mutation (*bsg*B500 in this case). Segregants of the diploids are screened for the mutation under study and then analyzed for possible consequences. Alternatively, the new mutant is crossed with a well-marked strain whose chromosomes can be directly selected for (Table V). This approach can rule out the requirement for a genetic background carried on specific chromosomes in the original mutant.

Double mutants can be constructed by crossing two strains that carry mutations of interest on different linkage groups. Haploid segregants of the resulting diploids can be screened for the two mutations or the genotype can be inferred by the presence or absence of markers known to be linked to the mutations under study (Table V).

E. Mapping

Mapping of loci to specific linkage groups greatly facilitates genetic manipulations and indicates whether a given phenotype depends on linked or unlinked mutations. The first such study was carried out by Sussman and Sussman (1963). They found that a white mutation (*whi*A) and a brown mutation (*bwn*A) segregated independently. For historical reasons these mutations have since been assigned to linkage group II and IV, respectively. Markers for each of the seven linkage groups have been slowly accumulated until now almost a hundred loci are mapped (Table VI).

With only rare exceptions, markers on the same linkage group are found together among haploid segregants of the heterozygous diploid while markers on separate linkage groups segregate independently in up to half the haploid derivatives.

To map a newly isolated mutation the mutant strain is crossed with a well-marked mapping strain (Table V). Resulting diploids are induced to segregate haploids and ~100 such segregants are individually scored for every marker present in the heterozygous diploid. Each segregant must be individually scru-

TABLE VI

GENETIC LINKAGE GROUPS OF *Dictyostelium* AND TENTATIVE MAPS[a]

	I	II	III	IV	V	VI	VII
Centromeres	●	●	●	●	●	●	●
Ordered by mitotic recombination	aggB	aggA	whiB	acrD			couA
	cycA	whiA	radB	bwnA			tsgK
	tsgE	acrA	bsgA	bsgC			bsgB
	sprA	stmG	acrC	ebrB			fgdB
	acrB	tsgD	radG				cobA
		tsgD					
Mapped but unordered	benA	aggF	acrD	aggJ	nysB	gluA	frtB
	cadA	aggI	alpA	aggL	nysC	manA	slgJ
	couB	arsA	axeB	couH	stmF	modB	stmA
	couI	arsB	couD	ebrA		stlA	tsgG
	devA	axeA	radC	frtA		tsgV	tsgK
	modA	cbpA	radE	minA			tsgM
	radA	clyA	rdeC	nagA			phgA
	tsgI	couC	slgB	phgB			
	tsgL	couE	slgD	pdsA			
	tsgQ	couF	slgG	radD			
		devB	stmC	radF			
		oaaA	stmE	rdeA			
		radH	tsgA	slgC			
		slgE	tsgC	slgI			
		sprB	tsgJ	sprH			
		stkA	tsgN	sprJ			
		stmB	tsgR	tsgB			
		stmD		whiC			
		tsgF					
		tsgH					
		tsgP					

[a] References can be found in Newell (1984), Welker and Williams (1982, 1985).

tinized and marginal phenotypes retested. It is sometimes necessary to reclone individual segregants because the putative haploid plaques may have arisen from two cells that by bad luck lay side by side. Usually the phenotype of 90% of the segregants is clear-cut, and it is the last 10% that require most of the effort to be sorted out. Depending on the markers used for each chromosome mapping can take anywhere from several weeks to several months to complete. A fictitious mapping experiment is given at the end of this chapter (Section XI) as an example of these techniques.

A mutation is mapped when it is found to be present in all haploid segregants carrying a known marker of a linkage group and absent in all haploid segregants lacking the marker. The marker might be either resistance to a poison such as *acr*A, or the absence of such resistance (*acr*A$^+$). The new mutation should also segregate randomly with markers on other linkage groups such that ideally half of the mutant segregants carry a specific unlinked mapping marker and half do not. In reality, the frequencies are often far off from 50:50 due to the nature of the techniques used for *Dictyostelium*.

Haploidization occurs as an infrequent event among a population, and so one sees only the progeny of the initial haploid cell after many generations of growth, not the cell itself. This is in contrast to yeast—*Neurospora* or *Aspergillus*— where the products of meiotic haploidization are directly analyzed. Parasexual segregation in *Dictyostelium* generates haploids, but they are picked only after 10–20 cell generations, and considerable skewing of the relative numbers can result from minor differences in growth rates of the individual segregants. Unintentional mutations with slight effects on vegetative growth are often carried by strains that affect the proportions of different genotypes recovered only because one chromosomal complement has a slight growth advantage over another. Thus, an unequal distribution of unlinked markers is more to be expected than not. If two markers separate in even one of a hundred haploid segregants, that is good evidence that they are unlinked.

Mapping experiments often give unequivocal linkage of a gene to markers on an established linkage group. However, translocation and other chromosomal rearrangements have been observed in some strains (Welker and Williams, 1985). These take longer to sort out but with genetic experience plausible solutions can be found.

To give a feeling for mapping, a flowchart for a mapping experiment may help (Table VII). Let us say a mutation was isolated in strain HL100, a bsg derivative of the isolate from the wild, strain NC-4. Let us say the phenotype of the mutant, strain HX6, is that it fails to aggregate (agg−) unless cAMP is added at 4 hours of development. To map the mutation, strain HX6 is crossed with strain HL106 (Table V) and diploids selected at 27°C on plates spread with *B. subtilis*. Since strain HL106 carries *bsg*A, it cannot grow on *B. subtilis* either. Diploids form plaques in a week and are streaked to get clones on another plate spread with *B. subtilis*. Single plaques are picked to plates spread with *K. aerogenes* and incubated 5 days at 22°C. The diploids form normal fruiting bodies at 22°C that have yellow sori and large (diploid) spores. The mutation *cam* is recessive. We are ready to map.

About 20 sori (10^6 spores) are picked from the 22°C *K. aerogenes* plates of the diploid and spread on 10 plates made 2% in methanol. In 4 days, 120 plaques come up on the methanol plates and the cells develop 2 days later. Almost all the methanol-resistant segregants that fruit have white sori (*whi*A is linked to *acr*A on LGII). However, 30% (40 of 120 clones) are unable to aggregate (agg−).

TABLE VII

FLOWCHART FOR MAPPING A GENE

Step 1: Select diploid from	HX6	HL106
LGI		*cyc*A1
LGII		*acr*A1 *whi*A1 *tsg*D12
LGIII		*bsg*A5
LGIV		*nag*A211
LGV		*man*A2
LGVII	*bsg*B500	
Unassigned	*cam*A500	

Step 2: Select haploid segregants on 2% methanol plates

Step 3: Pick methanol-resistant segregants to (1) *Bacillus subtilis,* (2) cycloheximide, (3) *Klebsiella aerogenes* 22°C, (4) *K. aerogenes* 27°C.

Step 4: Score segregants for tsg, bsg, cyc, nag, man.

Step 5: Look for linkage group missing segregants in opposition.

Potential results:

Phenotype	Number of segregants								
	Met^R	Bsg⁺	Bsg⁻	Cyc⁺	Cyc⁻	Nag⁺	Nag⁻	Man⁺	Man⁻
Cam⁺	80	18	62	50	30	80	0	35	45
Cam⁻	36	4	32	16	20	0	36	28	8

This confirms the diploidy of the strain recovered from the cross and shows that the *cam* mutation is not on LGII (*cam* from HX6; *acr*A from HL106).

Each of the methanol-resistant clones (agg⁻ and agg⁺) are picked separately to grid patterns on each of four plates: two normal plates spread with *K. aerogenes,* one incubated at 22°C, the other at 27°C; one plate spread with *B. subtilis;* one plate containing 500 μg/ml cycloheximide. They are incubated for a week.

None of the fruiting methanol-resistant segregants with white sori grows at 27°C, since they carry *tsg*D on LGII. Almost none of the agg⁻ methanol-resistant segregants grow at 27°C, but 4 of them do. These are discarded as probably being diploids that are homozygous for *acr*A due to mitotic crossing over. That leaves 36 agg⁻ methanol-resistant *tsg*D segregants. Four of them grow on *B. subtilis,* while 24 do not. That shows that the *cam* mutation is not on LGVII, since some (4) are bsg⁺. It still could be on LGI, III, IV, V, or VI. Of the 80 fruiting methanol-resistant segregants, 50 grow on cycloheximide plates. Of the 28 agg⁻ haploids, 16 grow on cycloheximide. That eliminates the possibility that *cam* is an LGI.

The haploid segregants are scraped individually from the grid plaques that grew up at 22°C on *K. aerogenes* plates and suspended in 116 test tubes containing 10 m*M* acetate, pH 5, containing 0.1% NP-40 to lyse the cells. Half of each tube is incubated with *p*-nitrophenyl *N*-acetylglucosamine and the other half with *p*-nitrophenyl α-mannoside. The activities of *N*-acetylglucosaminide (*nag*A LGIV) and α-mannosidase (*man*A LGVI) are scored 1 and 5 hours later, respectively. Of the 80 fruiting segregants, all have *N*-acetylglucosamine activity, 35 have α-mannosidase activity, and 45 do not. Of the 28 agg⁻ segregants, none have *N*-acetylglucosaminidase activity, 28 have α-mannosidase activity, and 8 do not. These results show that *cam* is on LGIV and segregates independently of LGVI. This completes the mapping of *cam* to a linkage group (IV) and is backed up by independent segregation from LGI, II, VI, and VII. It is not necessary to prove independent segregation from LGIII, or V, although this is to be expected.

The flowchart (Table VII) presents just one of various strategies that can be followed. The most appropriate mapping strain should be used depending on the genotype of the strain carrying the new mutation. Sometimes several mapping strains are used.

VII. Mitotic Crossing Over

Recombination occurs in all diploid somatic cells, albeit at far lower frequency than in cells that have evolved meiotic division. The frequency at which two linked markers separate is an indication of how much chromosomal material separates them. This is the basis for ordering loci on all genetic maps including those of *Drosophila* and humans. In *Dictyostelium*, recombination is usually observed as the reappearance of a recessive marker from a heterozygous diploid in segregants that still retain the full diploid complement.

Attempts in several labs to observe high-frequency recombination following meiotic division have not succeeded. Cells of different mating types, such as NC4 and V12, will form the sexual structures known as macrocysts when they coaggregate in submerged cultures. It is likely that the cells fuse to form a diploid and divide meiotically before germinating. However, no one has found conditions for reproducible germination of the macrocysts. Therefore, meiotic recombination has never been clearly shown for *D. discoideum,* and high-frequency recombination is not available for genetic analyses at this time. We have to rely on mitotic recombination to order genes on the linkage groups.

Mitosis of a diploid cell separates the newly replicated copies of a chromosome by splitting the shared centromere. If mitotic crossing-over occurs near the centromere, all distal markers become homozygous in the daughter cells. Since mitotic recombination is rare, the possibility of two recombinational events can be ignored. Therefore, if one is screening for diploids homozygous for a centromere proximal marker—for example, *whi* on LGII—one finds that all of them are homozygous for distal markers such as *acr*A on LGI. The converse is not true: diploids homozygous for *acr*A are sometimes but not always homozygous for *whi*. A large number of markers have been tentatively ordered in this way (Table VI) (Welker and Williams, 1982).

The level of resolution of mitotic ordering is not sufficient to be of practical value in recognizing pairs of independent mutations that might lie in the same locus, and it is far too crude to be able to position a mutation at the 5' or 3' end of a gene. Molecular dissections are needed for that degree of resolution. But mitotic maps are useful when one wants to construct double mutants from parental strains each carrying a desired mutation on the same linkage group. Luckily, this has not been necessary in past constructions, since it requires an inordinate amount of effort.

VIII. Mapping Cloned Sequences

Since the mid-1970s a considerable number of *Dictyostelium* genes have been cloned and sequenced (Kimmel and Firtel, 1982). The nucleotide sequences have indicated some of the molecular signals used in this organism and have given the amino acid sequences of the protein products. It would be nice to know where these genes map on the various linkage groups.

Restriction-site polymorphisms near the sequences have been observed in separate isolates from nature. For instance, the restriction fragments carrying actin genes differ in length when prepared from DNA of strain NC4 than when prepared from strain V12. Diploids can be selected between genetically marked derivatives of these strains and then induced to segregate haploids with various combinations of chromosomes from each strain. The genetic complement is determined by the presence or absence of markers on the chromosomes of the different strains. DNA prepared from the set of haploid segregants can then be cut with restriction enzymes and analyzed by the Southern blotting technique. Characteristic fragments should be found to cosegregate with unique genetic complements. For instance, actin 8 cosegregates with LGI, while actin 12 and actin M6 segregate with LGII (Welker *et al.*, 1986). It should be possible to carry out this analysis with almost any cloned fragment but there may be unforeseen pitfalls.

IX. DNA Transformation

After several aborted starts, conditions necessary for transformation of *Dictyostelium* with cloned DNA have been found (Nellen *et al.*, 1984; Knecht *et al.*,

1986; Nellen *et al.*, this volume, Chap. 4). A vector, B10, was constructed from the bacterial plasmid pBR322 into which was inserted a sequence of ~600 bases that precedes the actin 6 gene of *Dictyostelium*. At the point following the eighth codon of actin 6, the *neo*R gene of bacterial transposon Tn5 was inserted. Transcription and translation from the vector uses the actin 6 signals, proceeds through about 12 bases of sequence upstream of the *neo*R gene, and then reads the whole *neo*R gene. *Dictyostelium* cells that contain B10 can be selected in medium containing 20 μg/ml G418. By using the calcium phosphate precipitate technique, B10 DNA will transform *Dictyostelium* cells to G418 resistance at a frequency of ~10^{-5}.

Transformation has demonstrated that the actin 6 gene signals are equally well recognized and obeyed when carried on a plasmid as when in the normal genomic environment. At least 200 bases upstream of the transcriptional start site are required for expression of the gene. In the near future, transformation studies should yield a wealth of detailed results as many labs use these techniques. It is possible to attach *Dictyostelium* genes to the 3' end of the *neo*R gene and get expression. It is also possible to put other promoters ahead of the *neo*R gene and determine whether they signal transcription or not (Knecht *et al.*, 1986). An endogenous plasmid of *Dictyostelium,* Ddp1, can be cotransformed with B10 and will be retained extrachromosomally in the recipient (Firtel *et al.*, 1986). It is an attractive candidate as a shuttle vector.

Transformation technology holds promise for quite sophisticated genetics. It should be possible to isolate cloned sequences that when carried on a transformation vector complement recessive mutations previously isolated in *Dictyostelium*. This may be the only way we will ever know what the products are of a large number of developmental genes. Moreover, by inserting the sequence of a known gene backwards into a vector, transcription of the antistrand can be driven by the promoter on the vector. Antistrand RNA has been shown to interfere with translation of its complementary mRNA in several systems as well as in *Dictyostelium* (Crowley *et al.*, 1985; Knecht and Loomis, submitted). This opens a whole new approach to observing the consequences to loss of function of a specific gene. These and other techniques are discussed in detail (Nellen *et al.*, this volume, Chapter 4; De Lozanne, this volume, Chapter 27).

X.   Discussion

The application of genetics to developmental problems has been highly fruitful in a variety of organisms, including *Drosophila melanogaster, Caenorhabditis elegans, Dictyostelium discoideum,* and *Saccharomyces cerevisiae.* Each has its own strengths and weaknesses. The strengths of *Dictyostelium* lie in its relative simplicity and rapid growth rate. Its weaknesses lie in its lack of metazoan tissue systems. Mutations have been recovered in the structural genes of a fairly large number of developmentally regulated proteins, and interesting morphological mutants have been characterized.

There are some genes that are not amenable to classical mutant hunts. Genes that exist in multiple copies will give only a minor mutant phenotype when one member is mutated. For instance, over a dozen actin genes are expressed in *Dictyostelium* (Romans and Firtel, 1985; Knecht *et al.*, 1986). Loss of one would not be expected to result in an easily observable phenotype. Genes vital for growth are difficult to mutate, since only conditional mutations can be recovered. Considerable ingenuity and brute force is necessary to mark such genes.

Mapping loci to linkage groups has been of use in constructing complex genotypes and has a formal elegance. However, unto itself neither mapping nor arranging loci on a chromosome tells us anything about the mechanisms of development.

On the other hand, complementation studies have shed considerable light on developmental genes. Analysis of multiple alleles that have been shown to fall in the same locus is far more convincing than the analysis of a single mutant strain. The homozygous mutant diploid technique overcomes many problems raised by unwanted mutations present in any given strain.

The greatest potential of all genetic techniques lies in transformation. If even one of the procedures being tested, complementation by cloned sequences, antistrand interference, or insertional mutagenesis (see De Lozanne, this volume, Chapter 27) become standard routines, a wealth of fascinating problems will be open for inspection. Moreover, it should be possible to modify genes, base by base, and reinsert them to determine the crucial as well as dispensable characteristics of developmental genes.

For a long time, development of complex organisms was thought of as a somewhat mystical process outside the range of biochemical analysis. Now we know that it may be complex but that it is driven by the same processes that control cellular functions. It is a rich storehouse for exploration with genetic tools.

XI. Sample Genetic Study

The following fictitious report is presented as a summary of present genetic tools. It describes the search for mutations in the structural gene coding for myosin heavy chain, their isolation and genetic characterization, and the molecular basis for the mutations. While this is all wishful thinking, it is based on experience in genetic techniques built up over the years, and hopefully serves to review the points brought up in this chapter. To avoid tedium, the report is written in a form appropriate for this chapter. More complete descriptions of real results can be found in papers describing similar studies by Knecht *et al.* (1984), Murray *et al.* (1984), and Loomis and Kuspa (1984).

Isolation of Mutations in the Carboxy-Terminal Portion of the Myosin Heavy-Chain Gene of *Dictyostelium:* A Fictitious Paper

Dictyostelium discoideum contains a single gene coding for myosin heavy chain (DeLozanne *et al.*, 1985). While it might be expected to be an essential gene for growth, mutations affecting certain specific portions of myosin might not disrupt its functions so drastically as to preclude growth. The availability of monoclonal antibodies to seven different portions of the myosin heavy chain (Peltz *et al.*, 1985; Flicker *et al.*, 1985) prompted us to screen for mutations affecting one or another of these portions. Two mutations were recovered that result in loss of the epitope recognized by monoclonal antibody My4 without affecting reactivity to the other four myosin-specific monoclonal antibodies. The mutations are recessive, noncomplementing, and map to LGII. Molecular analyses show that the mutations cause alterations in the carboxy-terminal region of the heavy chain of myosin. Strains carrying either of these mutations aggregate but fail to complete development.

Methods

Organism

Strain M4-3 derived from strain Ax-3 of *D. discoideum* was chosen for the screen because it grows axenically, and it carries the *man*A4 mutation which marks LGIV as well as an unmapped *tsg* mutation convenient for constructing diploids in subsequent genetic analyses (Free *et al.*, 1976). Strain M3-4 was grown in HL5 medium.

Mutagenesis

10^8 exponentially growing cells were collected from 100 ml of HL5 medium by centrifugation, washed in 20 mM phosphate buffer, pH 6.5, and suspended in 100 µg/ml N'-methyl N-nitronitrosoguanidine for 30 minutes at 22°C. Cells were collected by centrifugation and diluted to 10^4 cells/ml in 10 liters of HL5 medium. They were immediately deposited into wells of 150 multitest plates using a dripper (Brenner *et al.*, 1976). The plates were incubated at 22°C for 10 days before being replicated to six sets of 150 plates each.

Monoclonal Antibodies

My1, 2, 3, 4 have been previously shown to react specifically with *D. discoideum* myosin (Peltz *et al.*, 1985; Flicker *et al.*, 1985). They were used at 1

$\mu g/ml$ in the mutant screenings and at 10 $\mu g/ml$ for Western analysis. The immunological screens were carried out as described by Knecht *et al.* (1984).

Results

Mutagensis of a population of strain M4-3 resulted in 10^{-3} survivors. 1.5×10^4 wells were inoculated with the mutated population. Cells grew in 10^4 wells, indicating that half of these received a single viable cell while the rest received more than one viable cell. Thus, 5000 clones of the mutagenized population could be effectively screened. The plates were replicated and the clones screened after 4 days of growth at 22°C.

Four different monoclonal antibodies (My1–4) that recognize different epitopes on myosin heavy chain were separately used to screen fixed, lysed cells directly in the wells. After developing the tagging enzyme-linked rabbit anti-mouse IgG antibody, 40 putative wells were found that reacted with three of the monoclonal antibodies but failed to react with one. Sixteen failed to react with My2, 14 failed to react with My3, and 10 failed to react with My4. Cells from these wells were repicked from the remaining unscreened replica set and inoculated into six replica multitest plates.

CHARACTERIZATION OF HX-320 AND HX-321

The putative mutants were rescreened with each of the monoclonal antibodies after the cells had grown for 4 days. Two strains were found that failed to react with My4 while reacting strongly with the other monoclonal antibodies. The remainder of the putative mutants reacted with each of the monoclonal antibodies and had been incorrectly scored in the first screen due to failure of the dripper to deliver a reagent, loss of cells from the wells, failure of the cells to grow in the wells due to contaminating microorganisms, misreading of the plates, or unrecognized problems. It is also possible that some had been correctly screened but reverted to wild type during the subsequent growth period. In any case, the two strains that passed the rescreen were subcultured from the unassayed replica and named HX-320 and HX-321.

Both strains grow well in HL5 medium (MDT = 12 hours) and grow in association with *K. aerogenes* at 18°C but not at 27°C, as is to be expected since the progenitor strain carries a *tsg* mutation. Neither strain forms fruiting bodies, although strain HX-320 aggregates to form tipped aggregates before morphogenesis stops. Strain HX-321 fails to aggregate. Immunofluorescence using monoclonal antibodies My1 or My2 showed that myosin is far less ordered in the cytoskeleton of strains HX-320 and HX-321 than in wild-type cells (unpublished results). Monoclonal antibody My4 gave no specific immunofluorescence with cells of strains HX-320 and HX-321, although it reacts with the cytoskeleton of wild-type cells. Extracts of strains HX-320 and HX-321 separated on 6% polyacrylamide SDS gels were transferred to nitrocellulose and immunostained with monoclonal antibodies My1, My2, and My4. No specific reactivity was observed with My4, while both the other monoclonal antibodies recognized a band at a position corresponding to 210 kDa in extracts from both mutant strains. Myosin (210 kDa) of wild-type cells is recognized on Western blots by each of these monoclonal antibodies (Peltz *et al.*, 1985).

GENETIC ANALYSIS OF STRAIN HX-320

Strain HX-320 was crossed with strain HL-50 (*acr*A1, *bsg*500) and diploids selected by growth on *B. subtilis* at 27°C. Strain HX-320 will not grow under these conditions, since it carries a *tsg* mutation precluding growth at 27°C. Strain HL-50 will not grow on *B. subtilis* due to the *bsg*B mutation. A diploid, DX-300, was selected for analysis. It grows well on *B. subtilis* at 27°C and develops in a manner indistinguishable from wild-type strains. Monoclonal antibody My4 reacts with material at ~200,000 Da on Western blots of extracts of strain DX-300 and can be used to stain the cytoskeleton of this diploid. Therefore, it appears that the mutation carried by strain HX-320 is recessive with respect to reactivity with My4 and morphogenesis past the tipped-aggregate stage.

Haploidization of strain DX-300 was induced by plating 10^3 spores on each of five plates containing 25 μg/ml ben late. A week later cells were picked from the growing edge of 96 plaques and spotted in replica grid patterns on plates spread with either *K. aerogenes, B. subtilis,* or *K. aerogenes* on 2% methanol. The segregants were scored for reactivity with My4, as well as for the genetic markers *acr*A, *man*A, and *bsg*B. Results on 67 segregants that could be unequivocably shown to be haploid are given in Table VIII.

The pattern of segregation of haploids from strain DX-300 showed that the mutation resulting in lack of reactivity to My4 (*myo*A320) cosegregated with

TABLE VIII

PHENOTYPES OF HAPLOID SEGREGANTS OF DX-300[a]

	Phenotypes					
	acrA	acrA⁺	manA	manA⁺	bsgB	bsgB⁺
myoA	0	6	3	3	1	5
myoA⁺	61	0	24	37	15	46

[a] *myo*A haploids failed to react with My4; *acr*A haploid grew on 2% methanol; *man*A haploids had <1 unit/mg protein of α-mannosidase activity; *bsg*B haploids failed to grow on *Bacillus subtilis*.

acrA $^+$ on LGII. None of the haploids received the *myo*A mutation from HX-320 as well as the *acr*A marker from HL-50; nor did any receive the *myo*A $^+$ locus of HL-50 and the *acr*A $^+$ locus of HX-320. The lack of these classes of segregants from >50 haploids tested is evidence for assigning *myo*A to LGII. The skewing toward *acr*A (61 of 67 haploids) is due to the advantage of such strains when grown on ben late. *myo*A segregated independently of *man*A on LGVI and *bsg*B on LGVII, showing that there is no reason to suspect a requirement of any specific genetic background for expression of *myo*A. The slight skewing toward *bsg*B haploids is not statistically significant.

The six *myo*A320 segregants were tested for growth on *K. aerogenes* at 27°C; three grew and three did not, indicating that the *tsg* mutation of strain HX-320 is not on LGII. One of the temperature-resistant strains also carried *man*A4 and *bsg*B500. This strain, HX-322, was crossed with strain HX-321 to test for complementation of the myosin mutations. Diploids were selected for growth at 27°C on *B. subtilis*. Strain HX-322 grows at 27°C on *K. aerogenes* but fails to grow on *B. subtilis* at any temperature due to the *bsg*B mutations. Strain HX-321 carries the *tsg* mutation of the original strain M4-3 and so does not grow at 27°C. A diploid, DL301, was recovered that grew well at 27°C on *B. subtilis*. It aggregates well but stops morphogenesis before forming a migrating slug. Cells of strain DL301 failed to react with monoclonal antibody My4. These results show that the mutation in strain HX-321 fails to complement *myo*A320 and thus by definition is in the same locus. The mutation in this strain, *myo*A321, also mapped to LGII in a cross with strain HL-50 (unpublished results).

MOLECULAR CHARACTERIZATION OF *myo*A320 AND *myo*A321

Northern blots of total RNA isolated from strains M4-3, HL-50, DX-300, DX-301, HX-320, and HX-321 were probed with a cloned sequence of the myosin heavy-chain gene, *pJS101* (DeLozanne *et al.*, 1985). A 7.1-kb mRNA was found in each of these strains, indicating that the *myo*A mutations did not affect transcription or stability of myosin mRNA, as could be predicted from the fact that each of these strains accumulated the same amount of the modified myosin as judged with monoclonal antibodies My1, My2, or My3 (data not shown).

Transformation of strains HX-320 and HX-321 with a neo^R plasmid carrying a complete myosin heavy-chain gene resulted in transformants that fruited normally and reacted with My4. These results directly indicate that the *myo*A mutations lie in the structural gene coding for myosin heavy chain. Sequence analysis of the cloned myosin gene from strains HX-320 and HX-321 shows that the former carries a mutation in the 1605th codon (Lys → Ser), while the latter carries a mutation in the 1624th codon (Gly → Arg) that account for the loss of the My4 epitope.

Discussion

The isolation of two independent mutations affecting a portion of the myosin heavy chain confirms molecular analyses that only a single gene for this protein occurs in *D. discoideum*. Myosin is a very large gene (>6 kb) and so can be expected to be more subject to mutagenesis than a small gene. However, our screen turned up mutants at a frequency of only ~5 × 10^{-4} among the survivors. This is undoubtedly due to using monoclonal antibodies for the screen. Only mutants affecting the My4 epitope were recovered, perhaps because mutations affecting the other epitopes on myosin are lethal. Mutations affecting the carboxyl-terminal portion of myosin (a.a. 1605 and 1624) result in a block to morphogenesis after aggregation, perhaps because of an increased demand on myosin integration into the cytoskeleton at subsequent stages. These results indicate how different regions of a protein play specific roles at different stages in the life cycle. The inability of strain HX-321 to aggregate appears to result from a secondary mutation unrelated to *myo*A321, as shown by the fact that strain DX-301 aggregates well yet still lacks the My4 epitope. Segregants from a diploid constructed between HX-321 and HL-50 separated an agg$^-$ mutation from the *myo*A321 mutation (data not shown).

REFERENCES

Brenner, M., Dimond, R., and Loomis, W. (1976). *Methods Cell Biol.* **14,** 187–194.
Chadwick, C., Collodi, P., Kasbekar, D., Katz, E., and Sussman, M. (1985). *Dev. Genet.* **6,** 59–74.
Crowley, T., Nellen, W., Gomer, R., and Firtel, R. (1985). *Cell* **43,** 633–641.
DeLozanne, A., Lewis, M., Spudich, J., and Leinwand, L. (1985). *Proc. Natl. Acad. Sci. U.S.A.* **82,** 6807–6810.
Devine, K., Morrissey, J., and Loomis, W. F. (1982). *Proc. Natl. Acad. Sci. U.S.A.* **79,** 7361–7365.
Devine, K., Bergmann, J., and Loomis, W. F. (1983). *Dev. Biol.* **99,** 437–446.
Dimond, R., Brenner, M., and Loomis, W. (1973). *Proc. Natl. Acad. Sci. U.S.A.* **70,** 3356–3360.
Dowds, B., and Loomis, W. F. (1984). *Mol. Cell Biol.* **4,** 2273–2278.
Firtel, R., Silan, C., Ward, T., Howard, P., Metz, B., Nellen, W., and Jacobson, A. (1986). *Mol. Cell Biol.* **5,** 3241–3250.
Flicker, P., Peltz, G., Sheetz, M., Parham, P., and Spudich, J. (1985). *J. Cell Biol.* **100,** 1024–1029.
Franke, J., and Kessin, R. (1977). *Proc. Natl. Acad. Sci. U.S.A.* **74,** 2157–2161.
Free, S., Schimke, R., and Loomis, W. F. (1976). *Genetics,* **84,** 159–174.
Godfrey, S., and Sussman, M. (1982). *Annu. Rev. Genet.* **16,** 385–404.
Kasbekar, D., Madigan, S., and Katz, E. (1983). *Genetics* **104,** 271–277.
Kimmel, A., and Firtel, R. (1982). *In* "The Development of Dictyostelium" (W. F. Loomis, ed.). Academic Press, San Diego.
Knecht, D., Dimond, R., Wheeler, S., and Loomis, W. F. (1984). *J. Biol. Chem.* **259,** 10633–10640.
Knecht, D., Cohen, S., Loomis, W., and Lodish, H. (1986). *Mol. Cell Biol.* **6,** 3973–3983.
Liwerant, I., and Pereira da Silva, L. (1975). *Mutat. Res.* **33,** 135–146.

Loomis, W. F. (1969). *J. Bacteriol.* **99**, 65–69.
Loomis, W. F. (1971). *Exp. Cell. Res.* **64**, 484–486.
Loomis, W. F. (1975). "Dictyostelium discoideum: A Developmental System." Academic Press, New York.
Loomis, W. F. (1980). *In* "The Molecular Genetics of Development" (T. Leighton and W. Loomis, eds.). Academic Press, San Diego.
Loomis, W. F., and Ashworth, J. (1968). *J. Gen. Microbiol.* **53**, 181–186.
Loomis, W. F., and Kuspa, A. (1984). *Dev. Biol.* **102**, 498–503.
Loomis, W. F., Wheeler, S., Springer, W., and Barondes, S. (1985). *Dev. Biol.* **109**, 111–117.
MacLeod, C., and Loomis, W. F. (1979). *Dev. Genet.* **1**, 109–121.
Morrissey, J. (1983). *J. Gen. Microbiol.* **39**, 1127–1130.
Morrissey, J., Wheeler, S., and Loomis, W. F.(1980). *Genetics* **96**, 115–123.
Murray, B., Wheeler, S., Jongens, T., and Loomis, W. F. (1984). *Mol. Cell. Biol.* **4**, 514–519.
Nellen, W., Silan, C., and Firtel, R. (1984). *Mol. Cell. Biol.* **4**, 2890–2898.
Newell, P. (1982). *In* "The Development of *Dictyostelium*" (W. Loomis, ed.). Academic Press, San Diego.
Newell, P. (1984). *In* "Genetic Maps, 1984" (S. J. O'Brien, ed.), Vol. 3, pp. 248–251. Cold Spring Harbor Laboratory, Cold Spring Harbor, New York.
Newell, P., Henderson, R., Mosses, D., and Ratner, D. (1977). *J. Gen. Microbiol.* **100**, 207–212.
Orlowski, M., and Loomis, W. F. (1979). *Dev. Biol.* **71**, 297–307.
Peltz, G., Spudich, J., and Parham, P. (1985). *J. Cell Biol.* **100**, 1016–1023.
Ratner, D., and Newell, P. (1978). *J. Gen. Microbiol.* **109**, 225–236.
Reymond, C., Gomer, R., Mehdy, M., and Firtel, R. (1984). *Cell* **39**, 141–148.
Romans, P., and Firtel, R. (1985). *J. Mol. Biol.* **183**, 311–326.
Sinha, V., and Ashworth, J. (1969). *Proc. R. Soc. Edinburgh, Sect. B* **173**, 531–540.
Sussman, M. (1955). *J. Gen. Microbiol.* **13**, 295–309.
Sussman, M., and Sussman, R. (1962). *J. Gen. Microbiol.* **28**, 417–429.
Sussman, R., and Sussman, M. (1967). *Biochem. Biophys. Res. Commun.* **29**, 53–55.
Vogel, G., Thilo, L., Schwartz, K., and Steinhart, R. (1980). *J. Cell Biol.* **86**, 456–465.
Watts, D., and Ashworth, J. (1970). *Biochem. J.* **119**, 171–174.
Welker, D., and Williams, K. (1980). *FEMS Microbiol. Lett.* **9**, 179–183.
Welker, D., and Williams, K. (1982). *Genetics* **102**, 691–710.
Welker, D., and Williams, K. (1985). *Genetics* **109**, 341–364.
Welker, D., Hirth, P., Romans, P. Noegel, A., Firtel, R., and Williams, K. (1986). *Genetics* **112**, 27–42.
Williams, K. (1978). *Genetics* **90**, 37–48.
Williams, K., and Barrand, P. (1978). *FEMS Microbiol. Lett.* **4**, 155–159.
Williams, K., Kessin, R., and Newell, P. (1974). *J. Gen. Microbiol.* **84**, 68–78.
Yanagisawa, K., Loomis, W., and Sussman, M. (1967). *Exp. Cell. Res.* **46**, 328–334.

Chapter 4

Molecular Biology in Dictyostelium: Tools and Applications

W. NELLEN,[1] S. DATTA, C. REYMOND,[2] A. SIVERTSEN, S. MANN, T. CROWLEY, AND R. A. FIRTEL

Department of Biology
Center for Molecular Genetics
University of California, San Diego
La Jolla, California 92093

I. Introduction

Since 1970 molecular biology has contributed substantially to the better understanding of development and cell differentiation in *Dictyostelium discoideum*. Approaches include the general examination of nuclear and cytoplasmic RNA, the analysis of changes in the transcriptional pattern, and the isolation and characterization of developmentally regulated genes. DNA-mediated transformation has subsequently provided the means to delineate the regulatory regions of differentially regulated genes, and to investigate the effects of "antisense mutagenesis," i.e., the specific inhibition of gene expression by transformation with an antisense construct.

Since molecular biological techniques have become indispensable for *Dictyostelium* workers, the goal of this chapter is to describe some of these methods and their application to *Dictyostelium*. It is not a replacement for books such as the Cold Spring Harbor Cloning Manual (Maniatis *et al.*, 1982), but rather a complement emphasizing modifications in protocols and giving new ones specific for *Dictyostelium*. Where appropriate, a method will be presented in detail and

[1]Present address: Department of Cell Biology, Max-Planck-Institut for Biochemistry, D-8033 Martinsried, Federal Republic of Germany.
[2]Present address: ISREC, 1066 Epalinges s./Lausanne, Switzerland.

67

can be used as a "bench protocol"; in other cases, we will mainly point out differences with respect to the commonly used procedures.

The major factors influencing molecular biological techniques in *Dictyostelium* are its genome organization and DNA composition. *Dictyostelium* grows normally as a haploid organism and has a haploid DNA size of 50,000 kb (Firtel and Bonner, 1972). Approximately one-sixth of the genome consists of rDNA composed of ~90 copies of an 88-kb extrachromosomal palindrome. Each half encodes one copy of the 36 S ribosomal RNA precursor and one copy of the 5 S rRNA. The organization of the ribosomal genes has been described in detail, including restriction and transcription maps and chromatin structure (Firtel *et al.*, 1976; Maizels, 1976; Frankel *et al.*, 1977; Taylor *et al.*, 1977; Cockburn *et al.*, 1978; Ness et al., 1983; Edwards and Firtel, 1984). Similar to the chromosomal portions of the *Dictyostelium* genome, the GC content of the ribosomal DNA is highly skewed. The coding regions of the 5, 17, and 26 S mature ribosomal RNAs are ~50% G/C, fairly low compared to most ribosomal RNAs from other organisms but high considering the composition of the remainder of the *Dictyostelium* genome. The nontranscribed spacer regions on the other hand contain a proportion of A/T >80%. The chromosomal DNA also shows an extremely skewed G/C composition: protein-coding regions on the average are ~38% G/C, which is close to the minimal 36% necessary to encode an average protein. As expected, the codon utilization is also biased toward the use of A and T. Codon usage for several *Dictyostelium* genes has been tabulated (Kimmel and Firtel, 1983; see also Warrick, this volume, Appendix). Certain codons have never been observed in *Dictyostelium* genes, but we presume that the appropriate tRNAs are present, since bacterial genes which contain these codons are expressed and properly translated after transfection into *Dictyostelium* (Nellen *et al.*, 1984a,b; Datta *et al.*, 1986). It is, however, possible that the translation efficiency of these genes is lower. Finally, the unusually strong bias in codon usage simplifies deriving synthetic oligonucleotides from a known protein sequence. Therefore good homology is achieved when these probes are used in hybridization experiments. Even when DNA is extracted from purified nuclei, some contamination with mitochondrial DNA cannot be avoided in most cases. The ~50-kb circular mitochondrial DNA is present at 200–250 copies per cell, thus making up approximately one-third of the cellular DNA (Firtel and Bonner, 1972).

II. Nucleic Acid Preparations

A. Large-Scale DNA Preparations

Because of the relatively high cellular content of carbohydrate and RNA, DNA extraction procedures which are used for other cell types are not readily applicable. To eliminate the majority of cellular proteins, large-scale purification of DNA (1–2 mg) is made from purified nuclei. Separation of DNA from carbohydrate, RNA, and residual protein is performed on CsCl gradients in the presence of ethidium bromide (Firtel and Bonner, 1972). Under these conditions, RNA will be pelleted, and protein will float on top of the gradient while carbohydrates (ρ = 1.65–1.7 g/cm^3) are clearly separated from DNA (ρ = 1.50–1.55 g/cm^3).

This method should be used for 2–20 liters of axenically grown cells (10^{10}–10^{11} cells) (Firtel *et al.*, 1976). Cells are harvested at a density of ~5 × 10^6 cells/ml (late log phase) by centrifugation at 2000 rpm for 5 minutes (Beckman J-6 centrifuge), are resuspended in 1:40 volume of ice-cold HMN; 0.5% NP-40 is added, and cells are lysed by repeatedly shaking or vortexing the suspension at 0°C for ~5 minutes. The solution clears during this process, and lysis can be monitored using a phase microscope. Nuclei are sedimented at 10,000 rpm for 10 minutes (Beckman J-2B). The pellet is resuspended in 1:80 volume of the HMN–NP-40 solution, shaken, repelleted, and resuspended in 1:100 volume of ice-cold HMN. The resuspended nuclei are dripped into 1:80 volume of 0.1 M EDTA, pH 8.5, 2% sarcosyl, while gently swirling. After incubation at 65°C for 15 minutes, the volume is measured and 0.4 mg ethidium bromide (stock solution of 10 mg/ml) and 0.95 g of CsCl are added per milliliter, yielding a final density of 1.50–1.55 g/ml. To avoid DNA breakage, CsCl is dissolved by carefully swirling the solution at room temperature or at 40°C. The solution is allowed to sit and if a particulate protein film is formed, this is removed after centrifugation at 10,000 rpm for 10 minutes. The mixture is then poured into ultracentrifuge tubes (Beckman Quick Seal tubes are convenient) and centrifuged for 24–36 hours at 48,000 rpm, 15°C. The top of the tubes are carefully cut off with a sharp razor blade, and the red protein layer and mineral oil are removed by suction. The DNA band is pulled from the top using a wide-bore pipet. If the DNA is in a large volume or contaminated with particulate protein material, further purification can be obtained by a second CsCl equilibrium centrifugation. DNA is then diluted with an equal volume of water and carefully extracted with an equal volume of water-saturated phenol–chloroform (1:1). In a beaker or Erlenmeyer flask, the aqueous phase is overlaid with two volumes of ethanol. DNA is wound out from the interphase with a glass rod, rinsed in 70% ethanol, and redissolved in "6,6,.2." This may take ~24 hours at room temperature or overnight at 40°C. DNA is stored at 4°C with a drop of chloroform to prevent bacterial contamination.

When DNA is prepared from nonaxenic strains, it is essential to remove the bacterial food source. This is done by four to five differential centrifugations and washes with PDF buffer. The amoebae are finally resuspended in PDF at a concentration of ~5 × 10^{10} cells/ml and shaken for 2–5 hours. Remaining bacteria will be taken up and digested during this period of time.

The size of DNA in the preparation is usually >50 kb. The quality is examined by a restriction digest and subsequent gel electrophoresis. The restriction pattern of the ribosomal DNA is known (Firtel et al., 1976) and serves as a marker for purity of the preparation and the degree of DNA degradation. Restriction fragments of the 50-kb mitochondrial DNA circles may sometimes be found in the digests, indicating some cytoplasmic contamination in the nuclei preparation. The yield of DNA in these preparations is variable but up to 40% of the theoretical amount may be obtained. Major factors are the yield of nuclei in the first step and the concentration of DNA in the CsCl gradient. In both cases lower concentrations of nuclei and DNA, respectively, appear to increase the recovery of DNA.

A different nuclei preparation method is used to isolate chromatin (Edwards and Firtel, 1984). Cells are washed and resuspended at a concentration of 2×10^7/ml in A buffer (Hewish and Burgoyne, 1973) with 0.5 mM phenylmethylsulfonyl fluoride (PMSF) and 0.5 mM N-ethylmaleimide. NP-40 (20%) is added to a final concentration of 0.2%, and cells are lysed by vigorously shaking for 2 seconds. Nuclei are collected by centrifugation for 8 minutes at 10,000 g. The pellet is then resuspended in 30% Percoll, 1 M sorbitol, 0.5 mM CaCl$_2$, 10 mM MES (pH 6.5), and nuclei are banded by density centrifugation in a Beckman Ti 60 rotor (42,000 rpm, 15 minutes). The band is collected, washed with eight volumes of 1 M sorbitol, 0.5 mM CaCl$_2$, 10 mM MES (pH 6.5), and nuclei are pelleted by centrifugation at 10,000 g for 8 minutes.

This method has mainly been used for chromatin analysis but might also prove useful for DNA preparations, nuclear run-on transcription, or other experiments when very clean nuclei are desired.

B. DNA Minipreps

For analytical methods, an easy and rapid DNA isolation procedure has been developed (Nellen et al., 1984b). Though rather crude, this DNA extract is sufficiently pure for restriction analysis and Escherichia coli transformation (see Section VI). The miniprep method is especially useful for the analysis of Dictyostelium transformants, when large numbers of small cultures have to be examined. The procedure may be used for DNA preparations from $<10^6$ cells up to 5 $\times$ 10^8 cells. Volumes and concentrations given below should be adjusted appropriately. Cells (5×10^6) are pelleted and resuspended in 2 ml of 10 mM Tris, 10 mM EDTA (pH 7.5). Then, 2 ml of 10 mM Tris, 0.7% SDS, containing proteinase K (100 μg/ml) is added. It is important to resuspend the cells completely before the addition of detergent, since cell aggregates or clumps will not be readily lysed, which results in DNA degradation and decreased yields.

After incubation for 1–2 hours at 65°C the cells are left at 37°C for several hours. (Both incubation times may be reduced to 30 minutes.) A 1:10 volume of 1 M Tris (pH 8.4) and 8 M LiCl each are added and the solution is gently extracted with Tris (pH 8.4)-saturated phenol. (Reextraction of the interphase with 50 mM Tris (pH 8.4) may increase the yield.) DNA is precipitated with 2.5 volumes of ethanol, the pellet is resuspended in "6,6,.2," 0.8 M LiCl at 65°C and again precipitated with ethanol. The DNA is finally resuspended in 0.5–1 ml of "6,6,.2." Since the solution contains large amounts of polysaccharides and RNA, a rapid determination of the DNA concentration is not possible, usually 20–40 μl of the solution are sufficient for a restriction digest (0.5–2 μg). The addition of spermidine (2.3 mM) and RNase A (0.1 μg/μl) is essential for proper digestion. Since it is sometimes difficult to resuspend ethanol-precipitated mini-prep DNA, the entire restriction mixture is loaded on a gel.

DNA isolated with this procedure is of large molecular size (>20 kb) and almost no degradation products are detected on ethidium bromide-stained gels.

C. RNA Extractions

RNA exractions in *Dictyostelium* from whole cells, nuclei, and cytoplasm are easy and straightforward, since the cells contain a relatively high RNA/DNA ratio (~40:1) (see Firtel and Lodish, 1973). There are, however, a few considerations that should be taken into account.

RNA (and DNA) should be isolated from fresh material, since thawing of a frozen cell pellet results in rapid degradation of nucleic acids. An exception are very small pellets (<10^7 cells) quick-frozen in dry ice–ethanol. Following the procedure of Mehdy *et al.* (1983), cells are thawed in the presence of DEP, SDS, and H_2O-saturated phenol–chloroform (1:1). Since the majority of the cells are in immediate contact with the detergent and phenol, lysis is rapid and degradation negligible.

Crude RNA preparations contain large amounts of carbohydrates and also some DNA. For many experiments, especially when working with *in vivo* ^{32}P-labeled material, it is essential to remove these contaminants, since they become highly labeled. High molecular weight RNA can be precipitated in high salt. Low molecular weight RNA (tRNA, 5 S RNA, etc.), double-stranded DNA, carbohydrate, and most other compounds are not precipitated under these conditions (Schimke *et al.*, 1974).

While isolated polysomes (see Section II,D) are fairly stable, their dissociation with EDTA results in substantial degradation of mRNA (Firtel, unpublished). We assume that this is due to the activation of ribosome-bound nucleases.

For total RNA, washed and pelleted cells are resuspended in 50 mM Tris (pH 8.4) at a density of 1–5 × 10^7 cells/ml. Keeping the suspension on ice, DEP is added to 1% and SDS to a final concentration of 2%. The suspension is briefly shaken and an equal volume of phenol–chloroform (1:1) is added. After vigorously vortexing (two times, 15 seconds each), phases are separated by centrifugation and the aqueous phase is reextracted twice with phenol–chloroform.

For the last extraction, sodium acetate (pH 4.7) is added to a final concentration of 0.4 M. This excludes a substantial amount of DNA from the aqueous phase; in addition, the combination of low pH–moderate salt appears to precipitate proteins, modified by DEP. When, however, sodium acetate is added in the first or second phenol extraction, substantial losses of RNA may occur (Firtel and Lodish, 1973).

The aqueous phase is finally precipitated with 2.5 volumes of ethanol for 2–3 hours or overnight at −20°C. For further purification, RNA may then be resuspended in water and precipitated by addition of 1 volume of 8 M LiCl for at least 5 hours at −20°C. Due to the high salt, the solution will not freeze. Since this precipitation is concentration-dependent, volumes should be kept small and the estimated RNA concentration should be between 0.2 and 2 mg/ml. RNA is pelleted at 10,000 rpm for 15 minutes; the LiCl precipitation may be repeated once. The RNA is then precipitated at −20°C with 2.5 volumes of ethanol in the presence of 0.8 M LiCl, resuspended in "6,6,.2," and quantitated by measuring the A_{260}.

Using the cell fractionation method described in the DNA section (Section II,A), RNA can be isolated from nuclei or cytoplasm by essentially the same procedure as outlined above. To obtain a purer cytoplasmic fraction, nuclei are pelleted at 11,000 rpm for 12 minutes. The nuclear fraction is further purified by one or two more treatments with NP-40 and subsequent centrifugation at 10,000 rpm for 10 minutes. It is evident at this point that a gain in purity is accompanied by a loss in yield. It should also be noted that since DNA is abundant and nuclear RNA concentrations are low, the high-salt precipitation is essential but should be carried out in a very small volume. When RNA is further purified by oligo(dT) cellulose or poly(U) Sepharose chromatography, high-salt precipitations are not necessary, since impurities in the preparation do not bind to the matrix. RNA preparations from later stages in development, especially from mature spores, are technically more difficult and should be done as described by McLeod *et al.* (1980).

When RNA is prepared from vegetative cells grown on bacterial plates, amoebae should be isolated in the early log phase of growth. Although this results in a much lower yield of cells, it is necessary to ensure that early expression is not initiated in a subpopulation of the cells. The bacterial food source is removed by a series of differential centrifugations. The material which has been washed off the plates is centrifuged at 1700 g for 7 minutes. The pellet is resuspended in ice-cold MES–PDF and washed four more times at decreased centrifugal force (750 g in the last centrifugation). The purity of the cell preparation is examined in the microscope and RNA is extracted as described above. The yield of RNA from vegetative cells is ~1000 A_{260} units/10^{10} cells; later in development approximately half of this can be obtained. This is due in part to the fact that *Dictyostelium* metabolizes nucleic acids as well as proteins during development, and also to the less efficient RNA extraction from later developmental stages. *Dictyostelium* RNA is analyzed either after glyoxylation (Thomas, 1980) on agarose gels or on formaldehyde–agarose gels (Lebrach *et al.*, 1977). Upon staining with ethidium bromide, the two major bands observed correspond to 17 S and 26 S ribosomal RNAs (1900 and 4100 nucleotides, respectively). In addition, two minor bands are seen, which correspond to the mitochondrial ribosomal bands (~1500 and ~3900 nucleotides). Higher percentage gels may resolve the 4.5 S and 5 S bands also found in other eukaryotes. For RNA blots, it is recommendable not to stain the gel since the sensitivity of hybridization may be reduced.

D.   Polysome Isolation

Analysis of polysomes is useful to determine the fraction of a particular mRNA which is actively being translated and thus provides a subfractionation of cytoplasmic RNA. Approximately 5×10^7 cells give sufficient material to examine the size distribution of polysomes and to examine RNA from polysome fractions by gel analysis. Cycloheximide may be added to the medium 5–10 minutes before cells are harvested.

To isolate polysomes, vegetative or developing cells are quickly washed and resuspended in lysis buffer at a concentration of 2×10^8/ml (Kimmel, pers. commun.). NP-40 is added to a final concentration of 0.5%, and cells are lysed on ice by repeatedly swirling the suspension. Lysis is completed when the suspension has cleared. If clearing takes longer than 1–2 minutes, the concentration of detergent should be increased (up to 2%). Centrifugation at 10,000 rpm for 3 minutes pellets nuclei and mitochondria. The supernatant is carefully removed and loaded immediately on a precooled 7–50% sucrose gradient (Beckman SW41) and centrifuged at 25,000 rpm, 4°C, for 3 hours. Under these conditions 15–20 mers are found approximately in the middle of the gradient. After removing 0.5–1 ml from the top of the tube (mainly lipids), the gradient is fractionated by carefully taking 0.5- to 1-ml fractions from the top. Alternatively, polysomes are fractionated with an Isco or equivalent fractionator using a UV monitor.

The quality of a polysome preparation—i.e., the yield of material and the ratio of monosomes to polysomes—depends mainly on how fast cells are worked up from the lysis to loading of the sucrose gradient. One should also keep in mind that the quantity and size of polysomes decreases during development.

1. *Polysome Stock Buffer* (HKM)
   50 mM HEPES, pH 7.5
   40 mM MgCl$_2$
   20 mM KCl `

2. *Lysis Buffer*
   Stock plus 5% sucrose
3. *20% NP-40*
   (Two phases after autoclaving, let cool, then shake)
4. *Gradient Solutions (solutions are w/w)*
   Stock containing 7% sucrose
   Stock containing 50% sucrose
   (*Note:* changes in salt concentration have no negative effect. Use RNase-free sucrose.)

The above solutions are treated with DEP and autoclaved before the addition of sucrose; glassware is sterilized and polyallomer tubes for centrifugation are DEP-treated. All solutions should be ice cold. The preparation should be done quickly and always be kept at $0°-4°C$. Gradient fractions may be either precipitated with EtOH or an aliquot (5–20 μl) may be diluted with three volumes of a solution of 75% formamide, 9% formaldehyde, $1.5\times$ RNA gel buffer heated for 5 minutes at $60°C$, cooled on ice, and loaded immediately on a formaldehyde–agarose gel.

III.   Recombinant DNA Libraries

A.   Genomic Libraries

Cloning and sequencing in *Dictyostelium* has proved to be a more difficult enterprise than originally expected, primarily due to the unusual base composition of *Dictyostelium* DNA. Overall, *Dictyostelium* DNA consists of ~78% A/T, but when the base composition of coding (62% A/T) and noncoding (90% A/T) regions are considered, the peculiarity is even more pronounced (Kimmel and Firtel, 1983). As a result of this nucleic acid structure, the 6-base restriction-enzyme recognition sites usually used for library construction tend to be distributed in a nonrandom fashion. Many of these sites are apt to be found in coding regions, resulting in the isolation of "half-genes" (Datta and Firtel, unpublished observations). Another phenomenon which may be linked to *Dictyostelium's* penchant for A's and T's is the increased rate of instability of large (>8 kb) genomic fragments, which is observed in both λ banks and plasmid banks. Cloned genomic fragments, therefore, must always be cross-checked for size against a genomic DNA blot hybridized with the appropriate probe in order to determine whether deletions have occurred.

One of the approaches which has been used to circumvent some of the problems inherent in genomic cloning has been the construction of banks in λ Charon 35, λL47.1 (Maniatis *et al.*, 1982) or EMBL3 and 4 (Frischauf *et al.*, 1982) using partial digests. When screening for a specific genomic DNA fragment (e.g., isolation of a genomic clone harboring a specific cDNA fragment or isolation of a genomic fragment adjacent and partially overlapping an already cloned piece of DNA), by far the most successful method has been the construction of plasmid banks which are tailor-made for each genomic clone desired. To determine the restriction sites to be used in the bank construction, genomic DNA is analyzed by a series of single and double digestions. From the hybridization pattern on a Southern blot, the enzyme combination resulting in a suitable fragment with heterologous ends is chosen.

Appropriately cleaved pUC vectors are usually chosen for the cloning. Background colony formation is largely reduced by separating the double-digested vector from the small internal fragment either on 15–30% sucrose gradients with ethidium bromide (30,000 rpm, 8 hours) or on low-melting agarose gels (Maniatis *et al.*, 1982).

Optimal ligation conditions are determined by ligating 0.1 μg of vector DNA to dilutions of digested *Dictyostelium* DNA ranging from 0.1 to 2 μg. One-fiftieth to 1/25 of each reaction mix is plated to estimate the number of recombinant colonies. For high-density screens, the optimal ligation is subdivided and spread to give ~5000 colonies per plate. For subsequent filter hybridization, positive and negative controls should be included. Purified DNA fragments free of vector sequences have to be used as hybridization probes. Purification is done either on low-melting agarose gels or by electroelution onto DEAE paper (Maniatis *et al.*, 1982).

Low-density screening, and subsequent subcloning and mapping of the inserts are done by standard procedures (Maniatis *et al.*, 1982). Since plasmid libraries are rather unstable and difficult to maintain, it is in most cases easier to construct a new bank than to rescreen a stored one.

B. cDNA Libraries

cDNA libraries have been used successfully for the isolation and characterization of developmentally regulated genes and of heat shock genes from *Dictyostelium* (Rowekamp and Firtel, 1980; Mehdy *et al.*, 1983; Rosen *et al.*, 1983). Synchronous development and the ease of inducing gene expression by various exogenous signals allow for the isolation and characterization of differentially expressed genes via cDNA libraries in *Dictyostelium*. For this method, RNA should be very pure; however, isolation of poly(A)$^+$ RNA is not required, since the priming specificity of oligo(dT) seems to be sufficient to exclude ribosomal RNA.

The reaction for the first-strand synthesis (Faust *et al.*, 1973; St. John and Davis, 1979) contains 10 μg of total RNA (10 μl), which is denatured by methyl mercury hydroxide (1 μl of 1 *N*) at room temperature for 10 minutes. Two microliters of β-mercaptoethanol (0.7 *M*) is added to inactivate the methyl mercury hydroxide and 20 units of RNasin (1 μl) are added to prevent degradation of the RNA. cDNA is synthesized by the addition of:

10 μl 5× RT buffer (0.5 M Tris pH 8.3, 0.7 M KCl, 50 mM MgCl$_2$)
1 μl oligo(dT) (10 mg/ml)
2.5 μl of each dNTP (19 mM)
4 μl reverse transcriptase (15 units/μl)
1 μl ^{32}P-dATP (3000 Ci/mmol)
10 μl distilled water

Actinomycin D may be used to increase full-length synthesis.

The reaction is performed at 42°C for 1–3 hours. Aliquots are taken at time intervals, TCA-precipitated, and counted to follow the extent of the reaction. Once no additional incorporation is observed, the reaction is stopped by the addition of 2.5 μl EDTA (0.2 M) and 12.5 μl NaOH (1 N). The RNA is hydrolyzed at 65°C for 30 minutes, and the solution is then neutralized by adding 12.5 μl of 1 N HCl (or see Gubler and Hoffmann, 1983). cDNA is separated from unincorporated nucleotides on a Sephadex G50 spin column, equilibrated with "6,6,.2," and lyophilized. (Spin columns are prepared by loading Sephadex gel in a 1-ml tuberculin syringe stoppered at the bottom with glass wool or a glass fiber disk. The column, which fits in a 15-ml corex tube, is centrifuged for 5 minutes at 2000 rpm. The sample is loaded onto the almost dry column and the exclusion volume is collected in an Eppendorf tube by centrifugation at 2000 rpm for 5 minutes.)

Second-strand synthesis is either done by the RNase H procedure, in which DNA synthesis is primed by short RNA fragments (Gubler and Hoffmann, 1983), or by self-priming of the 5′ ends as described here. The reaction (10 μl) is carried out in

0.1 M HEPES (pH 6.9)
10 mM MgCl$_2$
2.5 mM DTT
70 mM KCl
0.5 mM each dNTP
40 units Klenow fragment of DNA polymerase I per microgram of cDNA

A fraction of one nucleotide is ^{32}P-labeled to help monitor nucleotide incorporation. To ensure efficient priming the reaction is done at 14°C overnight. Further incubation at 42°C for 1 hour after addition of 20 more units of polymerase (per microgram of cDNA) increases the incorporation of nucleotides by 10–30%. The reaction is stopped by the addition of EDTA (final concentration, 15 mM). The samples are phenol–chloroform extracted and concentrated by lyophilization.

In order to clone the synthesized cDNA fragments, linkers are added onto the ends. To prevent internal cuts in the cDNAs during this procedure, the restriction sites corresponding to the linkers (usually *Eco*RI) are protected by methylation. For 1 μg of cDNA, methyl groups are specifically added by incubating the DNA in 100 mM Tris, pH 8.0, 10 mM EDTA, 6 μM S-adenosylmethionine, and 40 units of *Eco*RI methylase. The reaction is carried out at 37°C for 1 hour. The samples are again prepared for the next step by phenol–chloroform extraction and lyophilization.

The cDNAs are then blunt-ended by S_1 digestion. *Dictyostelium* DNA, because of its AT-rich untranslated regions, requires special precautions. We recommend a pilot reaction with 1:100 of the sample to optimize the S_1 concentration. In our hands, 0.1 units of S_1 per microgram of double-stranded cDNA is sufficient to obtain blunt ends with minor overall losses of incorporated counts. This indicates that only short regions at the ends of the DNA fragments are removed. (For detailed reaction conditions see Section V,A). The addition of linkers is performed by standard procedures (Maniatis *et al.*, 1982).

Cloning of the cDNA is achieved by ligating the cDNA into a properly restricted vector. Different vectors have been used: λgt11 phage libraries are required for expression (Young and Davis, 1983), but we have experienced difficulties when amplifying such libraries. *Dictyostelium* cDNAs seem to be more stable in plasmid vectors, and most cloned cDNAs were isolated from plasmid libraries, even though such libraries are technically more difficult to screen. To prevent vector religation without insert, terminal phosphates are removed by treatment with calf intestine alkaline phosphatase (10 units/μg DNA, in 10 mM Tris, pH 8, 30 minutes at 37°C, 30 minutes at 65°C followed by one phenol–chloroform extraction and ethanol precipitation). In ligation reactions, the volume seems critical for insertion into phages. possibly due to the requirement of concatamers for efficient packaging. The volumes we used were 3–5 μl containing 1 μg of phage DNA and 0.1–0.05 μg of cDNA. When using plasmid vectors, a larger volume is used to ensure a high proportion of single cDNA insertion events (2 mg/ml plasmid DNA, 0.2 mg/ml insert). The libraries are then propagated in the appropriate bacterial strains.

The success in constructing these libraries depends on many factors. The incorporation of label during both strand syntheses gives an estimate for the efficiency of the reactions. Linker additions and ligation can only be indirectly monitored. Special attention is required for the preparation of the vectors. Test ligations and transformations (plasmid vector) or packaging (phage vectors) should be carried out to ensure efficiency of the phosphatase treatment and fidelity of the cloning sites.

IV. Mapping and Sequencing of *Dictyostelium* Genes

Mapping and subcloning of cloned DNA fragments is basically done by standard techniques; however, several peculiarities should be mentioned here. Due to a lack in restriction sites, subcloning of fragments is not always straightforward.

In addition, some sequences, especially AT-rich noncoding regions, are very unstable in *E. coli*. Switching to a different vector system or a different host strain may help in some cases. Some fragments are difficult to clone in both orientations. Sometimes use of the pUC or M13 vectors with the opposite polylinker orientation eliminates the problem. Quite often, restriction enzymes recognizing four base pair sites have to be used to generate the desired fragments for subcloning; blunt-ending is done with the Klenow fragment of DNA polymerase I, S_1 nuclease, or T_4 DNA polymerase (Maniatis *et al.*, 1982), and the fragment is cloned into the flush ends of a *Hinc*II or *Sma*I site.

For sequence analysis, the dideoxy technique (Sanger *et al.*, 1977) is usually preferred over the Maxam–Gilbert (Maxam and Gilbert, 1980) method, because it can be carried out easier and faster. However some precautions have to be taken into consideration: the polymerization reaction by the Klenow fragment tends to stop or "stutter" at homopolymer runs, which are frequently found in noncoding regions. When sequencing reactions are carried out at higher temperatures (up to 55°C), DNA synthesis usually proceeds past such homopolymer regions. Since the polymerase loses activity at these temperatures, fresh enzyme is added for the "chase" reaction. Sequencing samples are fairly unstable and should be loaded on a gel immediately following the reactions. When 70 mM 2-mercaptoethanol is included in the reaction buffer, stability of the fragments is increased and breakdown products in later gels are minimized (Gomer *et al.*, 1985a). For convenience 10× *Hinc*II restriction buffer may be used for the polymerization reaction.

V. Analysis of Transcripts

Different techniques for the mapping of 5′ ends, 3′ ends, and introns, and for quantitation of mRNA levels and transcription levels will be described. It has to be decided from case to case which of these methods is the most suitable.

A. S_1 Protection Assays

In order to map 5′ and 3′ ends of a mRNA, 5′ or 3′ end-labeled restriction fragments are hybridized to RNA and S_1 digestion is performed as described elsewhere (Favaloro *et al.*, 1980). Alternatively, single-stranded, uniformly labeled DNA may be used as a hybridization probe in an S_1 protection assay. In contrast to an end-labeled fragment, introns will be detected with a uniformly labeled probe because two or more protected fragments will be generated. However, since the fragments can only be characterized by their size, the location of intron and 5′ (or 3′) end cannot be unambiguously determined in one experiment. Two or more assays using overlapping DNA probes are required to map the positions with certainty. The relative ease of synthesizing and isolating single-stranded probes with this method is a strong advantage. A similar technique making use of *in vitro* transcribed RNA will be described in Section V,C.

The labeled complementary strand is synthesized by the large fragment (Klenow) of DNA polymerase I using single-stranded (SS) M13 DNA containing the cloned fragment as a template and the 15-bp sequencing primer. The reaction is done as follows:

ss DNA template	1 μl(~0.1 μg)
15-bp sequencing primer (BRL)	2 μl (10 ng)
10× *Hinc*II buffer	2 μl
Distilled water	5 μl

The sample is denatured at 100°C for 2 minutes, *quickly* brought to 45°C, and allowed to anneal for 40 minutes. The following is then added:

Mixture of dTTP, dGTP, dCTP (0.125 mM each)	5 μl
[^{32}P]dATP (3000 Ci/mmol)	5 μl
dATP (25 μM)	1 μl
Klenow fragment of polymerase I (5 units/μl)	1 μl

The mixture is incubated at 45°C for 45 minutes. The reaction is stopped by adding EDTA to 15 mM and 100 μl of phenol-saturated "6,6,.2," the sample is passed over a Sephadex G50 spin column, and is ethanol precipitated. The sample is digested with an appropriate restriction enzyme which recognizes a unique site flanking the insert opposite to the sequencing primer. The linearized M13 (>6.4 kb) is easily separated from the labeled single-stranded fragment of interest by electrophoresis of the denatured DNA on a urea–acrylamide gel (6%).

The fragment is detected by a brief autoradiographic exposure and purified by electroelution and BND cellulose, eluting onto DE81 or NA45 paper, or other suitable procedures (Maniatis *et al.*, 1982).

DNA fragments shorter than the expected size are often seen on gels. These probably reflect stop signals for DNA polymerase within the fragment. Raising the incubation temperature decreases the proportion of such shortened fragments but also the overall yield.

The following hybridization reaction is then set up in capillary tubes:

ss DNA (~50,000 cpm)	1 μl
RNA (10 μg of total RNA)	2 μl
NaCl (4 M)	1 μl
HEPES (0.1 M)–EDTA (10 mM)	1 μl
Distilled water	5 μl

The capillary tubes are incubated at 100°C for 3 minutes, then at 65°C for 20 hours.

The content of each tube is diluted into 90 μl of S$_1$ buffer (ice cold), and 5–25 units of S$_1$ are added.

S_1 buffer   0.28 *M* NaCl
          0.05 *M* Na acetate, pH 4.6
          4.5 m*M* $ZnSO_4$
          20 µg/ml carrier ss DNA (salmon sperm)
The reaction is incubated at 37°C for 25 minutes and stopped by a phenol–chloroform extraction. Protected fragments are precipitated twice with ethanol and thoroughly rinsed with 80% ethanol (-20°C) to eliminate salts before loading on a sequencing-type gel.

Some nonspecific degradation of the labeled fragment occurs in all samples. It is therefore important to compare the protected fragments to a control assay containing no RNA or nonspecific RNA (e.g., yeast tRNA).

The optimal amount of S_1 has to be determined by using a range of enzyme concentrations. Different S_1 batches were found to react with *Dictyostelium* DNA to different extents. When a ladder of fragments below the protected fragment is observed, the S_1 concentration needs to be reduced.

B.   cDNA Extension

The lack of suitable restriction sites (due to A/T-rich intergenic regions) often does not allow the isolation of appropriate fragments for S_1 protection assays. In an alternative method a single-stranded cDNA fragment or a synthetic oligonucleotide is annealed to RNA and serves as a primer for reverse transcriptase (Faust *et al.*, 1973; St. John and Davis, 1979). When the primer is extended up to the 5' end of the RNA, a discrete-size fragment is detected on a sequencing-type gel and the position of the mRNA 5' end can be calculated. However, if introns are located in this part of the gene, the 5' end calculation will be misleading. Synthesis of the second strand, cloning, and sequencing will overcome this problem: comparison of cDNA and genomic sequences will provide information on the location of introns.

Several factors were found to influence the efficiency of the primer extensions: the purity of the RNA, the purity of the DNA primer, and the source of reverse transcriptase. Single-stranded primers isolated by electroelution were found to work erratically, probably because of contaminating gel matrix. The reverse transcriptase obtained from Life Sciences, Inc. (St. Petersburg, FL), gave the best results. However, "ultrapure" reverse transcriptase (Promega) and the cloned product (Boehringer Mannheim), which were recently introduced, have not been tested. RNasin (Promega) is included in the reactions, since reverse transcriptase batches may contain RNase activities.

Typically 400 ng of primer and 25 µg of total RNA are heated at 100°C for 2 minutes in 10 µl of Ext-buffer:
    100 m*M* KCl
    20 m*M* MgCl

200 m*M* Tris, pH 8.6

20 m*M* DTT

Hybridization is carried out between room temperature and 65°C, depending on the primer length (20 mer ~42°C) (see Craik, 1985) for at least 1 hour. Twenty microliters of each dNTP (5 m*M*) is added, together with 20 units of reverse transcriptase, 20 units of RNasin, and distilled water to a final volume of 20 μl. After incubation at 42°C for 2 hours (lower temperature for shorter primers), the reaction is stopped by the addition of EDTA to a final concentration of 15 m*M*.

Series of oligo(A)s or oligo(T)s in the untranslated regions of *Dictyostelium* mRNAs can sometimes cause extension stops. We found that raising the incubation temperatures up to 55°C increased the proportion of full-length extension products. The RNA is hydrolyzed by adding NaOH to 0.2 *M* and incubating at 65°C for 30 minutes. The solution is neutralized by addition of an equal molar amount of HCl. To remove unincorporated nucleotides, the reaction is passed over a Sephadex G50 spin column equilibrated in 10 m*M* Tris (pH 7.5), 1 m*M* EDTA. cDNA is then either ethanol-precipitated or passed over a second G50 column (equilibrated in 100 μ*M* Tris, pH 7.5, 10 μ*M* EDTA), and lyophilized.

The primer-extended cDNA can then be analyzed on a sequencing-type gel, or a second strand can be synthesized to clone and sequence the fragment (see Section V,B).

C.   SP6-Derived RNA Probes

For many hybridization-type experiments, radioactively labeled RNA probes may be used as an alternative to the corresponding DNA (see Melton *et al.*, 1984; Nellen *et al.*, 1984a,b; Zinn *et al.*, 1983). RNA probes have some advantage in that strand-specific probes are easily synthesized and RNA–RNA hybrids are more stable than DNA–RNA hybrids. The SP6 system, developed by Melton and co-workers (1984), is used to generate RNA transcripts from any DNA fragment. Vectors (SP64 and SP65, Promega Biotec) have been constructed in which a promoter from the *Salmonella typhimurium* phage SP6 is located adjacent to a multiple-cloning site. Insertion of a DNA fragment in either orientation allows for the synthesis of RNA transcribed from either strand. In a second vector construction (Gemini 1 and 2, Promega Biotec) the promoter region from phage T_7 has been added in opposite orientation adjacent to the multiple-cloning site such that, depending on the polymerase used in the experiments, either strand of the DNA insert can be transcribed from the same vector.

Such probes can be used for many purposes, including hybridization to Southern and Northern blots. Similar to S_1 protection experiments, they can be used for RNase protection mapping to determine 5′ or 3′ ends of transcripts and the size and position of introns. Quantitation of relative mRNA levels for a specific gene at various times in development or for individual genes in a multigene family has been done using these probes. In addition, RNA can be synthesized for subsequent *in vitro* translation or microinjection. Since SP6 polymerase is very stable and reinitiates transcription efficiently, microgram quantities of a specific RNA can be easily synthesized.

In the absence of a strong SP6 terminator, this system makes run-off transcripts. Therefore, the vector must be linearized at a restriction site close to the distal end of the sequence of interest. The choice of enzyme is important, as 5' overhangs give little background, blunt ends slightly more background, and 3' overhangs can give very high background levels. The background is in part caused by "nonspecific" transcription initiation at the free end. It is suggested that 3' overhangs be blunt-ended by treatment with T_4 polymerase or S_1 nuclease. It is essential to have very clean DNA to make templates: 20 μg of DNA is digested in 100 μl with the appropriate buffer and enzyme. After addition of 100 μl 0.1 M Tris, pH 8.4, the digest is extracted once with phenol saturated with 0.1 M Tris, pH 8.4, once with a 1:1 phenol–chloroform mixture, and finally once with ether followed by two ethanol precipitations. If ends must be repaired, this is done at this point and then the DNA is reextracted and precipitated. The pellet is air-dried, dissolved in 15 μl sterile "6,6,.2" buffer for a final concentration of 1.3 μg/μl, and stored frozen.

It is extremely important that all solutions, pipet tips, and tubes are sterile and RNase-free. Add to an Eppendorf tube at room temperature in the following order:

1. 5 μl [^{32}P]UTP (this must be fresh)
2. 1 μl cold nucleotide mix: 5 mM CTP, ATP, GTP; 75 μM UTP (can be frozen in small aliquots, degrades with a few freezings and thawings)
3. 2.2 μl fresh 5×B-2 buffer: 48 μl 5×B buffer (180 mM Tris, pH 7.5, 30 mM MgCl$_2$, 10 mM spermidine—can be kept frozen in aliquots) + 2.2 μl 1 M DTT + 0.5 μl 2 mg/40 μl BSA)
4. 1 μl RNasin (30 units/μl, Promega Biotec)
5. 0.7 μl (1 μg) linearized template DNA
6. 1.1 μl SP6 RNA polymerase (4 units)

Mix by tapping, then spin briefly in a microfuge. Incubate at 37°–42°C for 1 hour. At room temperature add 1 μl RNase-free DNase (use either ultrapure DNase RQ-1 from Promega or further purify other "RNase-free" DNase by passing it over a UMP–agarose column). Incubate at 37°C for 30 minutes. Stop the reaction by putting it on ice and adding 130 μl phenol-saturated "6,6,.2" buffer and 3 μl 5 mg/ml tRNA that has been phenol-extracted and ethanol-precipitated. Put over a sterile G50 spin column and determine [^{32}P]UTP incorporation.

For *Dictyostelium*, we use [^{32}P]αUTP, taking advantage of the high AT content of *Dictyostelium* DNA. We have used [^{32}P]αCTP for certain experiments to limit incorporation to specific regions of our probes (see Romans *et al.*, 1985). The reaction is prepared at room temperature to eliminate precipitation of the DNA by spermidine. For the same reason, buffer is added in the order described.

The quality of the probe is determined by running a small aliquot on a sequencing gel (usually as a control lane together with the experimental samples). *Dictyostelium* inserts (especially larger ones) often generate variable amounts of nonfull-length transcripts. This is presumably caused by termination within AT-rich sequences and can sometimes (but not always!) be overcome by changing the reaction temperature (37°–42°C).

Solution hybridization should be done under sterile conditions with freshly made probe. In an Eppendorf tube, 2×10^6 cpm labeled RNA probe and 5–50 μg RNA, the amount depending on the abundance of the message of interest, are mixed, dried down, and redissolved in 30 μl hybridization buffer (80% deionized formamide, 40 mM NaPIPES, 400 mM NaCl, 1 mM EDTA, pH 7.2). The sample is denatured at 85°C for 5 minutes and incubated overnight at an appropriate temperature up to 45°C. Because of the high AT content of *Dictyostelium* DNA, we usually incubate 4–6 hours at 37°C, then overnight at 32°C.

For RNase treatment of the resulting hybrids, both RNase A (C and U specific) and RNase T$_1$ (G specific) can be used, although RNase A is usually sufficient for *Dictyostelium* transcripts. To each hybridization tube, 350 μl of 350 mM NaCl, 10 mM Tris, pH 7.5, 5 mM EDTA, RNase A (30 μg/ml), and RNase T$_1$ (2 μg/ml) are added. The time and temperature of the RNase digestions can significantly influence the signal/noise ratio, so optimal conditions for a particular probe should be determined. For many cases, a 30-minute incubation at 30°C is satisfactory. The reaction is stopped by adding 10 μl of 20% SDS to each tube, followed by 50 μl of a fresh 1 mg/ml proteinase K solution in "6,6,.2" plus 0.5% SDS, incubated at 37°C for 30 minutes, and extracted once with phenol–chloroform. The aqueous phase is precipitated with ethanol, and 80 μl 5 M ammonium acetate and 15 μg of tRNA are added as carrier. Protected fragments are analyzed on a sequencing-type gel. As mentioned above, an aliquot of the probe should be run as a control. In addition, the probe should be hybridized to nonspecific RNA (e.g., yeast tRNA), RNase treated as described and run on the same gel; a further control is a hybridization sample, mock-treated with RNase.These controls are essential to avoid misinterpretation of nonspecific bands in the experimental samples.

One application of this technique is to determine the relative abundance of a message with respect to mRNAs transcribed from other members of the same gene family. The actin genes in *Dictyostelium* are, for example, almost identical in their coding regions while the 5′ untranslated regions are divergent. A probe covering both regions of a specific actin gene will hybridize to transcripts of the homologous gene and to transcripts of all other genes. Upon RNase digestion however, two different fragments will be recovered: the larger one was protected by the homologous message, the shorter one by the coding region of all other actin mRNAs. When the probe is in excess, the radioactivity in the two bands can be used to calculate the abundance of the specific gene transcripts relative to the amount of message transcribed from the entire gene family.

D. Nuclear Transcription Assays

Transcriptional regulation of specific genes in *Dictyostelium* has been studied by analysis of RNA steady-state levels *in vivo* (Mehdy *et al.*, 1983; Barklis and Lodish, 1983), by pulse labeling *in vivo* (Rowekamp and Firtel, 1980), and by *in vitro* transcription with isolated nuclei (run-on assays) (Landfear *et al.*, 1982; Williams *et al.*, 1980). *In vitro* transcription of purified gene templates by soluble *Dictyostelium* extracts has, however, not been successful (Takiya *et al.*, 1984).

The method described here has been successfully applied to give evidence for transcriptional regulation of gene expression in *Dictyostelium*. When nuclei are incubated in this assay, transcription primarily results from extension of nascent RNA chains; few new rounds of transcription are initiated by the polymerases. Incorporation of ribonucleotides in the assay is therefore a direct measure of total transcriptional rate, and the relative transcriptional activity of single genes can subsequently be quantitated by filter hybridization with specific DNA in excess.

Using this assay and looking specifically at genes that are expressed temporally during the *Dictyostelium* life cycle, we and others have obtained evidence for regulation at the level of transcriptional initiation (Landfear *et al.*, 1982; Williams *et al.*, 1980).

The method for isolation of nuclei is essentially a modification of that described by Jacobson *et al.* (1974). Nuclei are isolated from *Dictyostelium* amoebae grown axenically in rich medium (HL5), from cells developed on filters, or from cells developed in shaking cultures. A typical nuclei preparation is done with $\sim 10^8$ cells, corresponding to one filter's cells plated for development. Cells grown axenically in HL5 are harvested while still growing exponentially, and cells developed in shaking cultures are likewise maintained at low density until harvest (5×10^6 cells/ml). Following a wash in phosphate buffer, cells are resuspended at a density of $\sim 5 \times 10^7$ cells/ml and lysed with NP-40. The exposure of nuclei to the detergent is limited by an immediate change of buffer. Following sedimentation, the nuclei are finally suspended in a glycerol-containing buffer, quick-frozen on dry ice, and may be stored at $-70°C$ for several months without significant loss of transcriptional activity.

Hoescht staining of nuclei prepared by this relatively quick procedure shows some contamination of the nuclei with cell debris and cytoplasm.

The standard transcription assay is done with nuclei obtained from 2×10^7 cells. Using an input of 100 μCi of [^{32}P]UTP average incorporation per reaction is 10^7 cpm. The transcripts synthesized range in size from 500 to 2000 bases by comparison to a DNA molecular weight marker [the mean size of *Dictyostelium* mRNA labeled *in vivo* is ~1200 bases (Jacobson *et al.*, 1974)].

E.   Analysis of Transcription Products

The abundance of specific transcripts produced in the assay is analyzed by hybridization to filter strips of a plasmid Southern. Included on each filter is a DNA insert homologous to the transcript being studied as well as an internal standard corresponding to a transcript (e.g., actin) whose total rate of synthesis is assumed to be the same in the nuclei preparation being compared. The inclusion of this internal standard is essential for interpretation of the data, since the overall efficiency of hybridization has been observed to vary significantly even within supposedly identical strips of a single Southern blot.

Reduction of background hybridization caused by nonspecific sticking to filters has been achieved by purification of the probe on affinity columns (NEN-sorb). A final treatment of filters with RNase (1 μg/ml) in the presence of high salt (3× SSC) has also proved useful in this respect.

Nuclear run-on transcription is a very sensitive technique to analyze transcriptional regulation in *Dictyostelium*. It has been successfully applied to study the transcription of genes that are expressed at relatively low levels (Sivertsen, Reymond, Nellen, and Firtel, unpublished). Analysis of gene expression in *Dictyostelium* transformants is easy, since transcripts from genes present in high copy number are very abundant. Sometimes, steady-state RNA of a certain gene fusion cannot be detected in transformants either because it is heterologous in size or unstable. In the case of a discoidin antisense construct, for example, transcription could only be proven by nuclear run-on transcripts (Crowley *et al.*, 1985).

Protocol

Isolation of Nuclei

1. Harvest cells at 600 *g*, and wash with cold phosphate buffer.
2. Resuspend cells in 1:10 original volume of lysis buffer II.
3. Add NP-40 to final concentration of 1% and mix well.
4. Pellet nuclei at 3000 *g*, 5 minutes.
5. Resuspend pellet in 1:10 original volume of lysis buffer II and spin down unlysed cells at 150 *g*, 5 minutes.

6. Pellet nuclei from supernatant at 3000 g, 5 minutes, and wash once in 1:20 original volume of lysis buffer I (3000 g, 5 minutes).
7. Quick-freeze nuclei suspension in a dry ice–ethanol bath and store at $-70°C$.

Buffers

Lysis Buffer I
50 mM HEPES, pH 7.5
40 mM MgCl$_2$
20 mM KCl
0.15 mM Spermidine
5% Sucrose
14 mM Mercaptoethanol
0.2 mM PMSF

Lysis Buffer II
Lysis buffer I + 10% Percoll

Storage Buffer
40 mM Tris, pH 8.0
10 mm MgCl$_2$
1 ml EDTA
50% Glycerol
14 mM Mercaptoethanol

Transcription Assay

1. Set up reaction:
   34 μl H$_2$O
   20 μl 5× transcription buffer
   5 μl each 4 mM ATP, GTP, CTP
   1 μl RNasin
   10μl [^{32}P]UTP (100 μCi)
   20 μl nuclei (from 2×10^7 cells)
   Incubate at 23°C for 30 minutes.
2. Stop reaction by addition of:
   10 μl 1 M Tris, pH 8.4
   10 μl 0.2 M EDTA
   10 μl 20% SDS
   80 μl H$_2$O
   200 μl phenol–chloroform
3. Extract and remove unincorporated nucleotides by passing the aqueous phase over a Sephadex G-50 spin column. Probe may be stored at $-20°C$.

Buffer

5× Transcription Buffer
200 mM Tris, pH 7.9
1.25 M KCl
50 mM MgCl$_2$
25% Glycerol
0.5 mM DTT

Analysis of Run-On Transcription Products

1. Digest plasmid DNAs with restriction enzymes such that the following three classes of fragments can be distinguished by size:
   a. A fragment homologous to transcript of interest
   b. A fragment homologous to transcript present at constant level in the nuclei preparations under investigation (internal standard)
   c. A fragment nonhomologous to transcribed *Dictyostelium* sequences (nonspecific hybridization control, e.g., pBR322)
2. Separate DNA fragments by agarose gel electrophoresis; ~0.1 μg of each fragment should be transferred per filter strip. It is advantageous to load DNA in a wide slot on a short gel (4–5 cm) to ensure equal amounts and the same banding pattern on each strip.
3. Transfer DNA to filter, bake, prehybridize, and hybridize with ^{32}P-labeled run-on transcripts using standard procedures.
4. Wash filters at 37°C:
   a. 1 × 1 hour in fresh hybridization solution
   b. 2 × 1 hour in 0.1× SSC, 0.2% SDS
   c. 1 × 1 hour in 1 mM EDTA, 0.2% SDS
   d. (optional)
      1 × 1 hour in 3× SSC, 1 μg/ml RNase A
5. Filters are air-dried and exposed for autoradiography.

Solutions

Prehybridization Solution
  50% Formamide
  3× SSC
  10 mM EDTA, pH 7.2
  0.06 M phosphate buffer, pH
  0.2% SDS
  4× Denhart
Hybridization Solution
  Same as prehybridization solution, plus 200 μg/ml yeast RNA

VI. DNA-Mediated Transformation

Introducing cloned genes into an organism has become a powerful tool to study the regulation of gene expression and for isolating genes that complement mutant phenotypes. In *Dictyostelium*, a large number of differentially regulated genes have been isolated and characterized (see Kimmel and Firtel, 1982; Mehdy

et al., 1983; Barklis and Lodish, 1983; Reymond et al., 1984; Romans and
Firtel, 1985), and several of the physiological parameters regulating their expression have been identified (see Kimmel and Firtel, 1982; Mehdy and Firtel,
1985). Since a transformation system for Dictyostelium has been developed, the
molecular basis of differential gene expression can be investigated.

A.   Transformation Vectors

The Dictyostelium transformation vectors are based on the bacterial neomycin
resistance gene which encodes a phosphotransferase (APHII) (Nellen et al.,
1984a,b). The enzyme inactivates the related aminoglycoside drugs neomycin,
kanamycin, and G418. Vegetative Dictyostelium cells are sensitive to G418 at
doses as low as 1–5 µg/ml. In order to obtain sufficient expression, the neo^R
gene (from transposon Tn5) was set under the control of a Dictyostelium promoter. A DNA fragment containing ~600 bp of 5′ flanking sequences, the 5′
untranslated region, and 24 bp of the coding region of the Dictyostelium actin 6
gene was fused in frame to the coding region of the neo^R gene (Nellen et al.,
1984a,b). Actin 6 is known to be expressed at moderate to high levels during
vegetative growth. Later in development, actin 6 mRNA levels are substantially
reduced (McKeown and Firtel, 1981; Romans and Firtel, 1985). A pBR322
vector harboring the actin 6–neo^R fusion (pB10) can be used to transform Dictyostelium cells to G418 resistance. Transformed cell populations contain vector
DNA at an average of 2–5 copies/cell, and cells can be subsequently selected
that will grow at drug concentrations of at least 20 µg/ml (see below). In these
cells the average copy number is ~150/cell. pB10 and its derivatives (see below)
form large tandem arrays and are apparently integrated into the genome of
transformed cells (Nellen and Firtel, 1985). Up to now, there is no indication that
integration into the genome results from homologous recombination. The pB10
vector shares homology with genomic DNA only in the actin 6 5′ region, and
homologous recombination would probably disrupt the gene; this, however, has
not been observed. pB10 transformants express the actin 6 gene in a way indistinguishable from untransformed cells (Nellen et al., 1984b).

Since no functional 3′ end is supplied at the end of the coding region, the
fusion gene is transcribed as a large RNA with a predominant termination site in
the actin 6 5′ flanking region. Northern analysis of the transcripts reveals a smear
below the predominant band, indicating either minor termination sites or instability of the large transcript. The transformation frequency of this vector is
about 10^{-4} to 10^{-5} (see below and Table I). The transformation vector pB10S
contains the 3′ end of the Dictyostelium actin 8 gene fused to the end of the neo^R
coding region. A discrete-size mRNA of ~1.2 kb is transcribed from this fusion
gene. Transformation frequency with this vector is 5–10 times higher compared
to B10. The gene fusion has also been cloned in pXF3, pUC vectors, and M13

TABLE I

PROPERTIES OF TRANSFORMATION VECTORS

Vector (site)	Copy no.	State	Single cloning sites	Transformation frequency	Remarks	References
B10 (5.7 kb)	100–200 after selection on G418	Integration	SalI, BamHI, EcoRI	$\sim 10^{-5}$	pBR322-Derived, read-through transcript from actin 6 promoter of ~5 kb[a]	Nellen et al. (1984a); Nellen and Firtel (1985)
B10S (6.1 kb)	100–200 after selection	Integration	SalI, BamHI, EcoRI	$\sim 10^{-4}$	pBR-Derived, 1.2-kb actin 6-neo^R transcript, minor amounts of readthrough have been observed	Nellen and Firtel (1985)
B10SX (4.8 kb)	100–200 after selection	Integration	EcoRI, SalI, BamHI, HindIII	$\sim 10^{-4}$	pUC19-Derived, properties like B10S	Nellen and Firtel (1985)
Ddp1-20 (22 kb)	~100	Extrachromosomal	BglII[b], KpnI[c], BamHI	$\sim 10^{-3}$	Contains actin 6-neo^R gene fusion from B10S	Firtel et al. (1985)
pBMW1,2,3,4	~50	Extrachromosomal	BglII[b,c], BamHI[c], SphI[e], KpnI[b,c,e], SalI[c,e]	See note d	Does not contain neo^R gene, for cotransformation only	Firtel et al. (1985)

[a] If an additional DNA fragment is inserted into B10, this might be included in the readthrough transcript. Depending on the orientation, this may give rise to an antisense RNA of the fragment.

[b] From deletion studies using pBMW1, 2, 3, and 4, these should represent usable sites (Howard and Firtel, unpublished). Cloning into these sites will result in a small deletion which we expect not to affect replication or stability.

[c] Can be used in pBMW3 or 4.

[d] The frequency of cotransformation appears to be close to 100% with these vectors.

[e] Can be used in pBMW1 or 2.

without significantly affecting its transformation properties (Nellen and Firtel, 1985, Nellen et al., 1986).

Ddpl-20 (Firtel et al., 1985) is a transformation vector derived from the endogenous Dictyostelium plasmid Ddpl (Metz et al., 1983). It contains a Col E1 replicon, the bacterial amp^R gene, and the gene fusion from B10. In contrast to the B10-derived vectors, Ddpl-20 is maintained extrachromosomally in transformants. It therefore serves as a shuttle vector which can replicate in E. coli and in Dictyostelium and which has selectable markers for both organisms (Firtel et al., 1985).

B. Gene Expression in Transformants

The Dictyostelium genes examined so far in transformation experiments are regulated in a manner similar to their genomic counterparts: the actin 6–neo^R fusion gene is downregulated during development, as is expected for the actin 6 promoter (Nellen et al., 1984b); genes fused to a discoidin I promoter are induced early in development and are downregulated later and after treatment with cAMP (Crowley et al., 1985; Nellen and Firtel, 1985); the Dd-ras gene is induced by cAMP in shaking cultures under starvation conditions (Reymond et al., 1985), and a fusion of the prestalk-specific cathepsin (pst-cath) gene and the β-glucuronidase coding region is developmentally and spatially regulated in aggregates and by cAMP as expected for the pst-cath promoter (Datta et al., 1986). First experiments have been done to identify regulatory regions in Dictyostelium genes. Deletion analysis of the actin 6 promoter region shows that an upstream activator sequence (UAS) located ~250 bp upstream from the ATG initiation codon is necessary for proper expression of the actin 6–neo^R gene fusion. Since gene constructs containing the deleted upstream fragment are still transcribed in a regulated manner (downregulation during development), sequences controlling the differential expression are supposedly located further downstream. In addition to the UAS, a positive element which enhances gene expression ~10-fold has been located ~500 bp upstream from the ATG (Nellen et al., 1986). Prestalk-specific genes, including pst-cath and Dd-ras, appear to have a cis-acting regulatory element controlling a homologous sequence (Howard, Reymond, Spann, and Firtel, unpublished; Datta and Firtel, unpublished).

In several cases it has also been shown that the fusion gene mRNA is translated into a stable protein (Nellen et al., 1984b; Datta et al., 1986), which is only found in the appropriate cell type (Datta et al., 1986).

Termination of transcription has been studied in several transformation vector constructs. As mentioned before, the actin 6 5' flanking region provides an efficient termination signal in the B10 vector. The transcript, however, is either unstable or minor termination sites within the pBR322 part of the vector are being used (Nellen et al., 1984b). The same is true for the discoidin I–α gene 5' flanking region, although transcripts terminating in this sequence seem to be even more unstable (Nellen and Firtel, 1985). When the *Dictyostelium* gene *D11* (Barklis *et al.*, 1985) is cloned into B10 downstream from the actin 6–*neo*^R fusion, transcription of the *neo*^R gene is efficiently terminated within the 5' flanking region of *D11* (Knecht and Loomis, unpublished).

In the transformation vector B10S, the actin 8 3' end has been cloned downstream from the actin 6–*neo*^R fusion, resulting in efficient termination within this region. The 3' end fragment in this construct is, however, inverted with respect to the coding region, suggesting that the termination signal functions in both orientations (Nellen and Firtel, 1985). Similar results have been obtained with the *D11* 3' end (Knecht and Loomis, unpublished). When *D11* is cloned into B10 in the opposite orientation in respect to the actin 6–*neo*^R fusion, the *neo*^R gene transcript is terminated within the *D11* 3' end (Knecht and Loomis, unpublished).

In contrast to these observations, functional 3' ends from other eukaryotic genes do not provide termination sites in *Dictyostelium*: the 3' end of the *Acanthamoeba* actin gene (Nellen and Firtel, 1985) and the 3' end of the SV40 T-antigen are not recognized (Knecht and Loomis, unpublished). These fragments are relatively rich in G and C residues, while both 5' and 3' noncoding regions of *Dictyostelium* genes are >90% AT. Since no obvious sequence homologies have been found in the regions that function as termination sites in *Dictyostelium*, it might well be that termination does not require a specific sequence but that an AT-rich region alone is sufficient. It should, however, also be remembered that introns in *Dictyostelium* genes are >85% AT (Kimmel and Firtel, 1983).

C. Antisense Mutagenesis

Antisense mutagenesis can be used to study the phenotypic consequences of "gene repression": fragments of a cloned gene can be ligated in reverse orientation to their own promoter, thus generating the regulated expression of an antisense mRNA. In high copy number transformants, this antisense message will be present in ~150-fold excess over the sense message transcribed from the corresponding single-copy endogenous gene. *In vivo* hybridization of sense and antisense message probably leads to rapid degradation of the double-stranded RNA, since neither sense nor antisense RNA is accumulated in the transformed cells. This results in a transformant exhibiting the phenotype of a mutant in the respective gene (Alexander *et al.*, 1983; Springer *et al.*, 1984; Crowley *et al.*, 1985). *Dictyostelium* appears to be different from another eukaryotic system transformed with an antisense construct: L-cells transformed with an antisense TK gene fragment show no reduction of the endogenous TK mRNA levels. The message is, however, accumulated in the nucleus as a sense–antisense hybrid, thus indicating an inhibition of transport or processing rather than degradation of the double-stranded RNA (Kim and Wald, 1985).

D.   Transformation Procedure

The strain KAx-3 (see Poole and Firtel, 1984; Datta et al., 1986), kept in our laboratory for ~4 years, or strain Ax-2 is used for transformation. Transformation experiments using strain LAx-3 (see Poole and Firtel, 1984) give significantly different results (higher drug sensitivity, lower transformation frequency). The choice of the strain might therefore be important. Cells are grown in HL5 medium to a density not exceeding 2×10^6 cells/ml. Fresh cultures are started from spores approximately every 6 weeks.

Cells ($1-2 \times 10^7$) are plated in HL5. When the cells have settled and attached to the plastic, the medium is removed and replaced by MES–HL5. After 3–8 hours a calcium phosphate precipitate of DNA is applied and fresh medium is added.

After 4 hours the medium is removed and cells are treated with 18% glycerol. The cells round up during this treatment, and attachment to the plastic is weakened; therefore, the glycerol solution should be removed very carefully. Fresh HL5 is applied and cells are allowed to express for 6–8 hours; then medium is changed and the cells are kept in HL5 with 20 μg G418/ml for 5 days with one change of medium after 2–3 days. (The change of medium seems not to be crucial.) During the selection the cell number on the plate decreases, the remaining cells are small and round, and attachment to the plate is weak.

Cells are harvested by pelleting in conical tubes, tubes are drained, and the pellet is resuspended in a slurry of autoclaved *Klebsiella aerogenes* containing 40 μg G418/ml, and carefully spread on a Millipore filter supported by three layers of Whatman 3 MM in drugged HL5 (40 μg G418/ml). Plates are sealed with parafilm and left at room temperature until plaques form in the bacterial lawn (3–6 days).

The number of plaques gives an estimate on the transformation frequency. However, since it is not known whether some transformants are capable of multiplying during the 5 days of selection on the Petri dish, one plaque cannot be strictly correlated to one initial transformation event.

E.   Cotransformation

In the transformation procedure, cells apparently take up multiple DNA molecules which are subsequently maintained by the cells. The option to introduce different vectors simultaneously by cotransformation increases the application of the *Dictyostelium* transformation system (Nellen and Firtel, 1985; Crowley et al., 1985; Datta and Firtel, unpublished).

Due to a limited number of cloning sites and incompatibility of some *Dictyostelium* DNA fragments with B10 vectors, the possibility of engineering and

modifying gene constructs in the transformation vector is limited. This is facilitated by using a different vector, for example, pUC vectors. We have also cloned
Ba131 deletions in M13 phage containing the neo^R gene. In this case the phage
and the replicative form can be isolated from the same culture. The single-
stranded DNA can then be used for sequencing reactions, thus mapping the
deletion, while the double-stranded DNA is available for *Dictyostelium* transformation (Nellen *et al.*, 1986).

In addition, it could be advantageous to transform with DNA fragments contained in different vectors in order to study trans-acting factors. This is made
possible by cotransformation with B10 derivatives and pBR322 (or pUC)-based
vectors. Using a 1:1 mass ratio of both plasmids for transformation, cells contain
100–150 copies of both vectors after selection in G418. In some cases a lower
copy number of the second vector is observed. Since this appears to depend on
the gene contained on this plasmid, it is probably due to a growth disadvantage
conferred by high copy number of this gene or its products. Both vectors are
apparently integrated into the genome and form separate tandem arrays. In DNA
from single colonies, additional hybridizing DNA fragments are observed which
might be due to recombination of the two plasmids. However, the vast majority
of the vector DNA maintains its original restriction pattern (Nellen and Firtel,
1985).

Cotransformations have also been done using B10 and pBMW vectors, which
contain the *Dictyostelium* plasmid Ddpl (Metz *et al.*, 1983) cloned in pBR322
(Firtel *et al.*, 1985).

After a period of selection, recombination is observed between pBMW and
B10S in cotransformation. Vectors Ddpl-20 and Ddpl-11 are results from such
recombinations (Firtel *et al.*, 1985). These vectors probably took over the population of transformants because they had a selective advantage, while this seems
not to be the case in recombinants between pBR derivatives and B10 derivatives.
Vector Ddpl-20 carries the actin 6–neo^R gene fusion recombined into pBMW3.
In cotransformation experiments between pBMW3 and a B10S vector carrying
another *Dictyostelium* promoter-driven gene fusion, extrachromosomal vectors
can be isolated containing Ddpl plasmid sequences, actin 6–neo^R, and the other
Dd-gene fusion (Firtel and Silan, unpublished observations).

F.   Growing Transformants

The entire population of transformants, including part of the bacterial lawn, is
sucked off the black filter and transferred to 10–20 ml of HL5 without G418.
The cell titer is monitored periodically and cells are diluted so that the titer is
10^5–2×10^6 cells/ml. When DNA is prepared from this population, the average
copy number is 2–5 per cell.

An aliquot or the entire population can be stepwise-adapted to growth under
selective conditions by adding G418 to a final concentration of 5 μg/ml. Cell

growth usually slows down and the titer may decrease. When the cells have resumed exponential growth, the drug concentration may be stepwise-increased to 20 µg G418/ml. In order to isolate lines derived from single transformants, aliquots of the population grown in G418 are plated on SM plates with *K. aerogenes*. Single colonies are picked into shaking cultures with HL5 medium containing G418, piperacillin, and streptomycin sulfate. Cells derived from these clonal isolates or from the whole population after growth under selective conditions contain vector DNA at 100–200 copies/cell (Nellen and Firtel, 1985). When single colonies are isolated from the black filter, vector DNA is detectable in 5–10% of the clones at high copy number; another 5–20% contain ≤2 copies. Since untransformed cells do not survive the initial selection on Petri dishes and black filters, it is suggested that many transformants are not stable and are expressing the *neo*[R] gene transiently. During growth in shaking culture with G418, cells that are stably transformed and contain a high copy number of vector DNA are selected. It is, however, likely that in addition to the selection, transformants also amplify the vector DNA during growth in G418.

No significant change in copy number is detected when high copy number transformants are grown without selection. An exception are transformants carrying a mutated *ras* gene (Reymond et al., 1986). A large number of experiments have shown that even after several months of growth in the absence of selection, there is no change in the copy number of DNA in clonal isolates. In several transformants, however, reduction of copy number has been observed when selective pressure was removed. We assume that in these cases the high copy number transformants had a growth disadvantage compared to cells with reduced copy number. Therefore, transformants are usually grown in the presence of G418.

Stocks of transformants are kept using standard procedures: vegetative cells are frozen at −70°C in HL5 containing 10% DMSO. Spores can be kept in two different ways: for long-term storage, spores are resuspended in 5% skim milk and absorbed to silica gel. These stocks are viable for many years. To start a new culture, a grain of silica gel is shaken in 1 ml of an exponentially growing *K. aerogenes* culture and plated on SM plates.

Spores may also be resuspended in spore storage buffer (SSB: 10 mM KCl, 10 mM NaCl, 2.7 mM CaCl$_2$) and kept at 4°C. These spores are viable for at least 1 year and will germinate when placed in HL5 medium. Transformants selected for growth in G418 will germinate in the presence of the drug.

G.   Recovery of Vector DNA from Transformants

In *Dictyostelium* cells transformed with B10 derivatives, the vector DNA is apparently integrated into the genome in tandem array. In undigested DNA, vector sequences hybridize at the position of the bulk chromosomal DNA. When cut with an enzyme that would linearize the vector, a single band the size of linearized vector DNA is observed. In partial digests, a ladder of hybridizing bands the size of vector multimers is detected. When undigested cellular DNA from transformants is used to transform *E. coli,* no colonies are recovered. (Using DNA from high copy number transformants, occasionally some transformed *E. coli* colonies can be isolated. It is suggested that the bacteria take up large tandem repeats of the vector DNA and circularize these molecules by recombination.) When DNA is digested with an enzyme that cuts once within the vector, ligated, and then used for *E. coli* transformation, recombinant bacteria are recovered and most of these contain the vector in the original configuration. It should be noted, however, that the efficiency of transformation in *E. coli* is significantly reduced by the presence of genomic *Dictyostelium* DNA. DNA prepared by the miniprep method can be readily used for *E. coli* transformation. DNA from Ddpl-20 transformants or similar shuttle vectors can be used to directly transform *E. coli,* since Ddpl-20 is extrachromosomal.

H.   Perspective

The *Dictyostelium* transformation system provides the tool to study gene regulation in *Dictyostelium* at the molecular level. Gene fusions carrying the promoter regions of developmentally regulated *Dictyostelium* genes are expressed and regulated in a manner similar to their genomic counterparts. Up to now this has been shown for an actin gene (Nellen *et al.,* 1984b), two discoidin genes (Nellen and Firtel, 1985; Crowley *et al.,* 1985), the *Dictyostelium ras* gene (Reymond *et al.,* 1985), and *pst-cath,* a prestalk-specific gene (Datta *et al.,* 1986). The first major application of the transformation system will probably be the mapping of promoters and sequences essential for regulated gene expression. Trans-acting factors may be investigated using cotransformation.

It is possible that extrachromosomal vectors containing the promoter of a differentially regulated gene could be isolated at time points during development while maintaining their chromatin structure. The investigation of the protein organization in the promoter region on such a DNA would give insight into the DNA protein interactions during gene regulation.

In addition, antisense transformation provides means to study the phenotype of cells deficient in a known gene product, thus mimicking a loss of function mutant.

Until recently, it was not known whether gene substitution via homologous recombination (like in yeast) could be achieved. Homologous recombination has, however, now been reported with the myosin II heavy chain gene (see De Lozanne, this volume, Chapter 27). It is also not known yet whether mutant rescue by complementation is feasible. Since most *Dictyostelium* mutants are derived from nonaxenic strains, a modified transformation system has to be worked out. Preliminary results indicate, however, that wild-type strains can be transformed to G418 resistance though at much lower frequency.

I. Transformation Protocol

1. SOLUTIONS

HBS (HEPES-buffered saline)
　4 g NaCl
　0.18 g KCl
　0.05 g NaHPO$_4$ or 0.062 g Na$_2$HPO$_4$·H$_2$O
　2.5 g HEPES
　0.5 g Dextrose, pH 7.05, with NaOH (pH is important) in 250 ml for 2× HBS
　Filter-sterilize. Keep frozen in 50- to 100-ml aliquots.
　2 *M* CaCl$_2$—store at −20°C (filter-sterilize)
　60% Glycerol (autoclave)
　H$_2$O (distilled, not DEP-treated)—autoclave
　Piperacillin, 25 mg/ml in H$_2$O (filter-sterilize) (Lederle)
　G418, 4 mg/ml in 10 m*M* HEPES, pH 7.2 (filter-sterilize). (Schering, 39.7%
　　active drug, now available from Gibco)

2. METHOD

1. Grow cells in HL5 (pH 6.3–6.4) to ~10^6 cells/ml.
2. Plate 10–12 ml on Falcon Optilux Petri dishes the morning before the transformation is to be done. Medium should be 250 μg/ml piperacillin.
3. Remove medium after ≥8 hours and replace with 10–12 ml MES–HL5 (pH 7.0–7.1). Some cells will not attach; on the average 1–1.5 × 10^7 cells will remain on the plate. Allow to sit for 25–30 minutes (20 minutes).
4. 12 μg DNA in 0.6 ml 1× HBS are placed in a 5-ml sterile glass tube.
5. Add 2 *M* CaCl$_2$ to 125 m*M* (38 μl per 0.6 ml) while vortexing for 10 seconds. Let sit 25 minutes at room temperature.
6. Remove medium from plate as completely as possible without losing cells. Dropwise add DNA to cells. Rock plates very gently to spread. Let sit on cells for 25 minutes at room temperature, rocking gently periodically.
7. Without removing DNA, add 10 ml of MES–HL5 (pH 7.0, 250 μg/ml piperacillin). Let sit for 4–5 hours (3 hours).
8. Remove medium. Add 2 ml of 18% glycerol in 1× HBS carefully (18% = 3 ml 60% glycerol, 2 ml sterile H$_2$O, 5 ml 2× HBS).
9. Let sit 5–9 minutes at room temperature.
10. Remove glycerol. Add 10 ml HL5, pH 6.3, containing 250 μg/ml piperacillin. Let sit 4–5 hours (3 hours).
11. Remove medium. Add 10 ml HL5 with 250 μg/ml piperacillin and 20 μg/ml G418.

12. Wrap plates with parafilm. Let sit at room temperature for 4–5 days total. Change medium (HL5 with G418 and piperacillin on day 2 or 3).

Note: One must be extremely careful not to lose cells from plates. A point should be marked on the plate and all removals and additions of medium should be made from that point. Exception: DNA is added in center of dish, gently! Times in parentheses may be used for convenience but might result in lower transformation frequency.

3.   PREPARATION FOR HARVEST

Cut Whatman 3-MM paper $2\frac{1}{2}$ in. $\times$ $2\frac{1}{2}$ in., trim corners. Autoclave:

1. 3-MM paper squares, Millipore pads No. AP10047SO, and black filters, Millipore No. HABP 047 00, 20 minutes, slow exhaust.
2. Rack with Kimwipes for draining.
3. Freeze-dried Ka* (*Klebsiella aerogenes*) resuspended in HL5 (1 g/10 ml), autoclave in small beaker covered with aluminum foil for 15 minutes, slow exhaust, compensate volume with HL5 after autoclaving.

4.   HARVEST

1. Set up plates.
   a. Add 8 ml drugged HL5 to 100 × 15 Petri plate (piperacillin 250 μg/ml, G418 40 μg/ml).
   b. Place two Whatman squares, 1 round disk, and finally, black Millipore filter in Petri dish.
2. Harvest plates by swirling and resuspending carefully with plastic pipet, transfer into 15-ml Falcon conical tubes No. 2095. Then rinse plate with 17 mM phosphate buffer.† Keep tubes on ice.
3. Spin 1200 rpm, 5 minutes.
4. Pour off and drain in sterile rack.
5. Resuspend in 0.8 ml of Ka slurry (drugged same as HL5), and transfer to black filter. Seal plates with parafilm. It usually takes 4–5 days before colonies come up.

VII.   Appendix

A.   Buffers

1. *HMN Buffer*
   0.01 M Magnesium acetate
   0.01 M Sodium chloride

---

*Hirth *et al.* (1982).

†10× solution: 123.8 g KH$_2$PO$_4$, 36.9 g K$_2$HPO$_4$, H$_2$O to 5 liters.

   0.03 *M* HEPES
   10% sucrose
2. *6,6,.2*
   6 m*M* sodium chloride
   6 m*M* Tris (pH 7.5)
   0.2 m*M* EDTA
3. *PDF*
   20 m*M* Potassium chloride
   1.2 m*M* Magnesium sulfate
   7.5 m*M* MES
   ph. 6.5
4. *HL5*
   5 g/liter Yeast extract
   5 g/liter Tripticase peptone
   5 g/liter Proteose peptone
   10 g/liter Glucose (anhydrous)
   1.2 g/liter KH_2PO_4
   0.35 g/liter Na_2HPO_4
   pH 6.5
5. *RNA-Gel Buffer*
   20 m*M* MOPS
   5 m*M* Sodium acetate
   1 m*M* EDTA
   pH 8
6. *10× HincII Buffer*
   500 m*M* Sodium chloride
   100 m*M* Tris, pH 7.5
   100 m*M* Magnesium chloride
   1 m*M* DTT

B. Abbreviations

Ax-3, The two *D. discoideum* strains used in the experiments are both derived from Ax-3 and should, theoretically, be identical. However, both have been kept isolated for many years and obviously have diverged in several respects. To avoid a controversy on which strain is closer to the original Ax-3, we suggest denoting the one derived from a strain obtained from Richard Kessin's lab as KAx-3 and the one from William Loomis' lab LAx-3.
DEP, Diethylpyrocarbonate
dNTP, All four deoxyribonucleotides
DTT, Dithiothreitol
HEPES, *N*-2-hydroxyethylpiperazine-*N'*-ethanesulfonic acid

Ka, *Klebsiella aerogenes*
kb, kilobase pairs (1000 base pairs)
MES, 2-(*N*-morpholino)ethanesulfonic acid
MOPS, Morpholinopropane sulfonic acid
NP-40, Nonidet P40

ACKNOWLEDGMENTS

We would like to thank M. Mehdy and A. Kimmel for contributions and J. Roth for typing the manuscript. Work on this chapter was supported by USPHS grants GM24279 and GM30693 to Richard A. Firtel.

REFERENCES

Alexander, S., Shinnick, T. M., and Lerner, R. A. (1983). *Cell* **34**, 467–475.
Barklis, A., and Lodish, H. F. (1983). *Cell* **32**, 1139.
Barklis, E., Pontius, B., and Lodish, H. F. (1985). *Mol. Cell Biol.* **5**, 1473–1479.
Cockburn, A. F., Newkirk, M. J., and Firtel, R. A. (1976). *Cell* **9**, 605–613.
Cockburn, A. F., Taylor, W., and Firtel, R. A. (1978). *Chromosoma* **70**, 19–29.
Craik, C. S. (1985). *BioTechniques* **Jan/Feb**, 12–19.
Crowley, T. E., Nellen, W., Gomer, R. H., and Firtel, R. A. (1985). *Cell* **43**, 633–641.
Datta, S., Gomer, R., and Firtel, R. A. (1986). *Mol. Cell. Biol.* **6**, 811–820.
Edwards, C. A., and Firtel, R. A. (1984). *J. Mol. Biol.* **180**, 73–90.
Faust, C. H., Diffelman, H., and Mach, B. (1973). *Biochemistry* **12**, 925–931.
Favaloro, J., Treisman, R., and Kamen, R. (1980). *In* "Methods in Enzymology" (L. Grossman and K. Moldave, eds.), Vol. 65, pp. 718–749. Academic Press, New York.
Firtel, R. A., and Bonner, J. T. (1972). *J. Mol. Biol.* **66**, 339–361.
Firtel, R. A., and Lodish, H. F. (1973). *J. Mol. Biol.* **79**, 295–314.
Firtel, R. A., Cockburn, A. F., Frankel, G., and Hershfield, V. (1976). *J. Mol. Biol.* **102**, 831–852.
Firtel, R. A., Silan, C., Ward, T. E., Howard, P., Metz, B., Nellen, W., and Jacobson, A. (1985). *Mol. Cell. Biol.* **5**, 3241–3250.
Frischauf, A.-M., Lehrach, H., Poustka, A., and Murry, N. (1983). *J. Mol. Biol.* **170**, 827–842.
Frankel, G., Cockburn, A. F., Kindle, K. L., and Firtel, R. A. (1977). *J. Mol. Biol.* **109**, 539–558.
Gomer, R. H., Datta, S., and Firtel, R. A. (1985a). *Focus* **7**, 6–7.
Gomer, R. H., Datta, S., Mehdy, M. C., Crowley, T., Sivertsen, A., Nellen, W., Reymond, C., Mann, S., and Firtel, R. A. (1985b). *Cold Spring Harbor Symp. Quant. Biol.* **L**, 801–812.
Gubler, V., and Hoffman, B. J. (1983). *Gene* **25**, 263–269.
Hewish, D. R., and Burgoyne, L. A. (1973). *Biochem. Biophys. Res. Commun.* **52**, 504–510.
Hirth, K. P., Edwards, C. A., and Firtel, R. A. (1982). *Proc. Natl. Acad. Sci. U.S.A.* **79**, 7356–7360.
Jacobson, A., Firtel, R. A., and Lodish, H. (1974). *J. Mol. Biol.* **82**, 213–230.
Kim, S. K., and Wold, B. J. (1985). *Cell* **42**, 129–138.
Kimmel, A., and Firtel, R. A. (1982). *In* "The Development of *Dictyostelium discoideum*" (W. F. Loomis, Jr., ed.), pp. 233–324. Academic Press, New York.
Kimmel, A. R., and Firtel, R. A. (1983). *Nucleic Acids Res.* **11**, 541–552.
Landfear, S., Lefebvre, P., Chung, S., and Lodish, H. (1982). *Mol. Cell. Biol.* **2**, 1417–1426.
Lebrach, H., Dimond, D., Wozney, J. M., and Boedker, H. (1977). *Biochemistry* **16**, 4743–4751.
McKeown, M., and Firtel, R. A. (1981). *Cell* **24**, 799–807.

McLeod, C. L., Firtel, R. A., and Papkoff, J. (1980). *Dev. Biol.* **76,** 263–274.

Maizels, N. (1976). *Cell* **9,** 431–438.

Maniatis, T., Fritsche, E., and Sambrook, J. (1982). "Molecular Cloning, a Laboratory Manual." Cold Spring Harbor Laboratory, Cold Spring Harbor, New York.

Maxam, A. M., and Gilbert, W. (1980). *In* "Methods in Enzymology" (L. Grossman and K. Moldave, eds.), Vol. 65, pp. 499–560. Academic Press, New York.

Mehdy, M. C., and Firtel, R. A. (1985). *J. Mol. Cell. Biol.* **5,** 705.

Mehdy, M. C., Ratner, D., and Firtel, R. A. (1983). *Cell* **32,** 763–771.

Melton, D. A., Krieg, P. A., Rebagliti, M. R., Maniatis, T., Zinn, K., and Green, M. R. (1984). *Nucleic Acids Res.* **12,** 7035–7056.

Metz, B. A., Ward, T. E., Welker, D. L., and Williams, K. L. (1983). *EMBO J.* **2,** 515–519.

Nellen, W., and Firtel, R. A. (1985). *Gene* **39,** 155–163.

Nellen, W., Silan, C., and Firtel, R. A. (1984a). *In* "Molecular Biology of Development" (E. H. Davidson and R. A. Firtel, eds.), Vol. 19, pp. 633–645. Liss, New York.

Nellen, W., Silan, C., and Firtel, R. A. (1984b). *Mol. Cell. Biol.* **4,** 2890–2898.

Nellen, W., Silan, C., Saur, U., and Firtel, R. A. (1986). *EMBO J.,* **5,** 3367–3372.

Ness, P. J., Labhart, P., Banz, E., Koller, T., and Parish, R. W. (1983). *J. Mol. Biol.* **166,** 361–381.

Poole, S. J., and Firtel, R. A. (1984). *Mol. Cell. Biol.* **4,** 671–680.

Reymond, C. D., Gomer, R. H., Mehdy, M. C., and Firtel, R. A. (1984). *Cell* **39,** 141–148.

Reymond, C. D., Nellen, W., and Firtel, R. A. (1985). *Proc. Natl. Acad. Sci. U.S.A.* **82,** 7005–7009.

Reymond, C. D., Gomer, R. H., Nellen, W., Theibert, A., Devreotes, P., and Firtel, R. A. (1986). *Nature (London)* **323,** 340–343.

Romans, P., and Firtel, R. A. (1985). *J. Mol. Biol.* **183,** 311–326.

Romans, P., Firtel, R. A., and Saxe, C. L., III (1985). *J. Mol. Biol.* **186,** 337–355.

Rosen, E., Sivertsen, A., and Firtel, R. A. (1983). *Cell* **35,** 243–251.

Rowekamp, W., and Firtel, R. A. (1980). *Dev. Biol.* **79,** 409–418.

St. John, T. P., and Davis, R. W. (1979). *Cell* **16,** 443–452.

Sanger, F., Nicklen, S., and Coulson, A. (1977). *Proc. Natl. Acad. Sci. U.S.A.* **74,** 5463–5467.

Schimke, R. T., Palacios, R., Sullivan, D., Kiely, M. C., Gonzales, C., and Taylor, J. M. (1974). *In* "Methods in Enzymology" (L. Grossman and K. Moldave, eds.), Vol. 30, pp. 631–648. Academic Press, New York.

Springer, W. R., Cooper, D. N. W., and Barondes, S. H. (1984). *Cell* **39,** 557–564.

Takiya, S., Tabata, T., Twabuchi, M., Hirose, S., and Suzuki, Y. (1984). *J. Biochem.* **95,** 1367–1377.

Taylor, W. C., Cockburn, A. F., Frankel, G. A., Newkirk, M. J., and Firtel, R. A. (1977). *ICN-UCLA Symp. Mol. Cell Biol.* **VII,** 309–313.

Thomas, P. (1980). *Proc. Natl. Acad. Sci. U.S.A.* **77,** 5201–5205.

Young, R. A., and Davis, R. W. (1983). *Proc. Natl. Acad. Sci. U.S.A.* **80,** 1194–1198.

Williams, J. G., Tsang, A. S., and Mahbubani, H. (1980). *Proc. Natl. Acad. Sci. U.S.A.* **77,** 7171–7175.

Yanisch-Perrou, C., Vieira, J., and Messing, J. (1985). *Gene* **33,** 103–119.

Zinn, K., DiMaio, D., and Maniatis, T. (1983). *Cell* **34,** 865–879.

Part II.   Vesicular Traffic in Cells

This section is concerned with approaches to membrane traffic in cells and begins with a chapter by Goodloe-Holland and Luna which describes several different methods for membrane isolation from *Dictyostelium*, the pros and cons of these methods, and their usefulness for particular considerations. Besides comparing plasma membranes obtained by different methods, this chapter describes methods for vesiculation and sealing of membranes and determination of sidedness of sealed vesicles.

Chapter 6 by Vogel deals with endocytosis and recognition mechanisms. Assays for phagocytosis, pinocytosis, adhesion, aggregation, and methods for isolation of related mutants are presented. Cardelli *et al.* in Chapter 7 deal with methods for defining intracellular membrane shuttling pathways. They provide methods for carrying out pulse-chase labeling, subcellular fractionation, *in vitro* protein synthesis, and isolation and characterization of relevant mutants. The final chapter of Part II, by Clarke and Kayman, combines many of the considerations discussed to this point in the book and relates axenic mutations to endocytosis. General considerations in choosing strains and growing cells and methods for monitoring the behavior of cells affected by axenic mutations are presented.

Chapter 5

Purification and Characterization of *Dictyostelium discoideum* Plasma Membranes

CATHERINE M. GOODLOE-HOLLAND[1]
AND ELIZABETH J. LUNA

Department of Biology
Princeton University
Princeton, New Jersey 08544

I. Introduction

Dictyostelium discoideum is a eukaryote that can be grown axenically in large quantities with the ease usually associated with prokaryotes (Loomis, 1971; Sussman, Chap. 2, this volume). This organism also exists either as a free-living amoeba or as part of a multicellular aggregate that undergoes differentiation and morphological change (reviewed in Bonner, 1967; Loomis, 1975, 1982). Because of these characteristics, *D. discoideum* is a popular system for the study of many important problems in membrane biology, such as endocytosis, chemotaxis, and cell–cell adhesion (reviewed in Loomis, 1982). Cell surfaces of vegetative amoebae contain at least two phagocytic receptors (Vogel *et al.*, 1980), chemotactic receptors for folic acid and pterin (Pan *et al.*, 1972; Van Haastert *et al.*, 1982), and an EDTA-sensitive cell–cell adhesion system (Beug *et al.*, 1973; Chadwick and Garrod, 1983) which also may be involved in phagocytosis (Chadwick *et al.*, 1984). During early development, new membrane proteins appear on the surfaces of aggregating amoebae. These proteins mediate cell–substrate attachment (Springer *et al.*, 1984; Gabius *et al.*, 1985), form an EDTA-stable cell–cell adhesion system (Beug *et al.*, 1973; Müller *et al.*, 1979),

---

[1]Present address: Department of Biochemistry, University of Missouri, Columbia, Missouri 65211.

and are involved in the initiation, reception, and amplification of cAMP pulses (reviewed by Devreotes, 1982). During late development, other adhesion molecules appear on the surfaces of the prespore and prestalk cells in the motile aggregate (Geltosky *et al.*, 1979; Steinemann and Parish, 1980; Lam *et al.*, 1981; Saxe and Sussman, 1982).

Much information also is available regarding the cytoplasmic face of the *D. discoideum* plasma membrane. For example, actin and myosin usually coisolate with purified plasma membranes (Spudich, 1974; Clarke *et al.*, 1975; Condeelis, 1979; Jacobson, 1980a; Spudich and Spudich, 1982; Das and Henderson, 1983; Luna *et al.*, 1981, 1984; Spudich, this volume, Chapter 11). If the endogenous actin is removed, these plasma membranes bind exogenous F-actin (Jacobson, 1980a; Luna *et al.*, 1981). Integral, perhaps transmembrane, proteins have been implicated in this binding (Luna *et al.*, 1981, 1984), which appears to involve primarily the sides, rather than the ends, of the actin filaments (Goodloe-Holland and Luna, 1984; Bennett and Condeelis, 1984). In addition, a number of cytoskeletal proteins that bind to actin and that may act, directly or indirectly, at the cytoskeleton–membrane interface have been identified (reviewed in Taylor and Fechheimer, 1982). Also, techniques for the genetic modification of this organism (reviewed by Loomis, Chapter 3, this volume) and mutants defective in motile functions (Vogel *et al.*, 1980; Clarke, 1978; Kayman *et al.*, 1982) are available. See other chapters in this volume for further discussion of the points raised above.

To facilitate the study of membrane functions, many researchers have devised methods for purifying *D. discoideum* plasma membranes (see, for example, Cardelli *et al.*, this volume, Chapter 7). Others have reviewed these purification schemes and the properties of the membranes obtained with them (Murray, 1982; Parish, 1983). Very few comparative studies exist, however, in which different membrane preparations have been evaluated in the same laboratory. Thus, we have compared three plasma membrane preparations with respect to structure, composition, and integrity. First, to ensure the isolation of all plasma membrane subdomains, we have prepared crude membranes by the method of Spudich (1974). Second, as a representative grind-and-find method, we have followed the procedure of Das and Henderson (1983). Third, we have isolated membranes using an extension of the concanavalin A (Con A)-stabilization, Triton-extraction method described by Parish and Müller (1976) and modified by Condeelis (1979). Knowledge of the properties of the different plasma membrane preparations permits the selection of a membrane isolation procedure appropriate for any experiment.

The comparison of *D. discoideum* plasma membrane preparations presented here is not exhaustive, since we have not examined membranes isolated by all of the many published methods. For instance, we have not lysed cells using digitonin (Riedel and Gerisch, 1968) or amphotericin B (Rossomando and Cutler, 1975). Also, since membranes isolated by two-phase polymer partitioning (Brunette and Till, 1971) exhibit SDS gel profiles that are similar to the gel profiles of our crude membranes (Siu *et al.*, 1977; Müller *et al.*, 1979), we assume that our crude membranes are approximately representative of these preparations. Similarly, we assume that the procedure of Das and Henderson (1983) is representative of other straightforward methods in which membranes are purified on sucrose gradients (Green and Newell, 1974; Rossomando and Cutler, 1975; Gilkes and Weeks, 1977; Sievers *et al.*, 1978). Although these membranes may be purified further on gradients formed from metrizamide (Serrano *et al.*, 1985) or Renografin (McMahon *et al.*, 1977), such extensions of the basic Das–Henderson technique have not been explored (but see Luna *et al.*, 1981, for a comparison of membranes purified by the techniques of Spudich, 1974, and McMahon *et al.*, 1977). Also, conspicuously absent from our analysis are membranes prepared from cells bound to polycation-coated beads (Jacobson, 1980b) or to cationic colloidal silica (Chaney and Jacobson, 1983). Although membranes isolated by these techniques may be ideal for studies involving only the cytoplasmic face of the *D. discoideum* plasma membrane, harsh reagents such as SDS are required to remove membrane components from cationic supports. This characteristic limits the flexibility of these membrane preparations and precludes their use as a source of undenatured membrane proteins.

Highly purified plasma membranes can be prepared either as large sheets with both membrane surfaces exposed to exogenous reagents or as vesicles sealed to small molecules. Sealed vesicles, produced by sonication, do not retain the native orientation of cell surface and cytoplasmic components but contain mixed orientations of these components within each vesicle. Therefore, in addition to our comparison of the three representative plasma membrane preparations, we present a characterization of the purest of the preparations with respect to membrane form, permeability, and sidedness.

II. Membrane Purification Procedures

A. Cell Culture and Harvest

Dictyostelium discoideum amoebae, strain Ax-3, are grown in HL5 medium (Cocucci and Sussman, 1970). Cells are harvested at a density of 10^7 cells/ml by centrifugation in 250-ml bottles at 650 g_{max} for 2 minutes and are washed twice in an appropriate buffer. Although repeated centrifugations are performed in each bottle during cell harvest, pellets are always resuspended between centrifugations.

We have found that successful recovery of unproteolyzed plasma membranes depends on keeping the membranes ice-cold at all times. Therefore, all membrane isolation procedures are carried out at $0°–2°C$ unless otherwise specified.

B. Crude Membranes

The easiest method for obtaining membranes from *D. discoideum* is the method of Spudich (1974). In this procedure, cells are lysed by mechanical disruption and a crude membrane fraction is collected from discontinuous sucrose gradients. Cells may be lysed by homogenization, by freezing and thawing, by nitrogen decompression, or by forced passage through 5-μm Nucleopore filters. Although vigorous Dounce homogenization is a classical technique for lysing cells and results in 80–90% lysis, this method is time-consuming and physically demanding. Cell lysis by freezing in liquid nitrogen and thawing at room temperature is easy and complete, but allows the release of proteases and other lytic enzymes from subcellular organelles. Another method that results in 90–95% cell lysis and that is convenient for lysing large volumes of cells involves the equilibration of cells with nitrogen in a Parr Cell Disruption Bomb (Parr Instrument Co., Moline, IL) for 25 minutes at 4°C and the controlled release of the lysate into a beaker at atmospheric pressure. The optimum nitrogen pressure is the lowest pressure that produces nearly complete cell lysis. This pressure varies with the osmotic strength of the lysis buffer and is ~1100 psi for the sucrose lysis buffer described below. However, the controlled release of the lysate must be discontinued before the last of the lysis solution is expelled from the Parr bomb to prevent the pressurized nitrogen from blowing the lysate out of the beaker. Perhaps the easiest way to obtain nearly 100% cell lysis is to push the cells through the cylindrical 5-μm pores of a Nucleopore filter (Das and Henderson, 1983).

To increase the efficiency of cell lysis and decrease proteolysis of membrane proteins, we have modified slightly the Spudich (1974) crude membrane isolation procedure (Fig. 1). About 3×10^{10} cells, washed in 14.6 mM KH$_2$PO$_4$, 2.0 mM Na$_2$HPO$_4$, pH 6.1 (Sorensen's phosphate buffer), are resuspended with an approximately equal volume of ice-cold sucrose lysis buffer containing protease inhibitors [30% sucrose, 40 mM sodium pyrophosphate, 2 mM EDTA, 1 mM EGTA, 5.1 mM 1,10-phenanthroline, 0.5% ethanol, 0.57 mM phenylmethylsulfonylfluoride (PMSF), 2 mM N-carbobenzoxyphenylalanine, 0.4 mM dithiothreitol, 0.02% sodium azide, 10 mM Tris-HCl, pH 7.6]. PMSF and N-carbobenzoxyphenylalanine are first dissolved in ethanol and added to the lysis buffer just before use. Resuspended cells are lysed immediately, since prolonged soaking in this buffer induces cell shrinkage, making the mechanics of lysis more difficult.

After the cells are broken, crude membranes are collected by centrifugation at 38,700 g_{max} for 15 minutes and washed once in sucrose lysis buffer containing protease inhibitors. The membranes then are resuspended, with gentle Dounce homogenization, in 30–50 ml of 20 mM sodium phosphate, pH 6.8, and layered onto six step-gradients, each of which contains 15–20 ml of 35% sucrose over 5

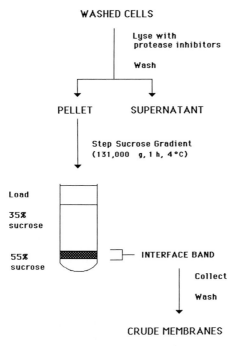

WASHED CELLS

Lyse with
protease inhibitors

Wash

PELLET   SUPERNATANT

Step Sucrose Gradient
(131,000 g, 1 h, 4°C)

Load

35%
sucrose

55%   —⊏— INTERFACE BAND
sucrose

Collect

Wash

CRUDE MEMBRANES

FIG. 1.   Flow diagram for the preparation of crude membranes by the method of Spudich (1974). For a complete explanation of the protocol, refer to Section II,B of the text.

ml of 55% sucrose in the phosphate buffer. After centrifugation at 131,000 g_{max} for 1 hour, the thick membrane band at the 35/55% sucrose interface is collected by hand. These membranes are washed twice at 26,900 g_{max} for 10 minutes in 20 mM sodium phosphate, pH 6.8, and are stored on ice in the same buffer with 0.02% sodium azide added to retard microbial growth. About 50 mg of membrane protein are recovered for each 10^{10} cells lysed.

C.   Das–Henderson Method

Das and Henderson (1983) have devised a membrane isolation procedure that leads to the preparation of relatively pure *D. discoideum* plasma membranes. In this method, cells are lysed through 5-μm Nucleopore filters and the cell lysate is fractionated by equilibrium density gradient centrifugation. Though technically similar to the method described above, the glycine lysis buffer used in this method (0.5 mM CaCl$_2$, 0.5 mM MgCl$_2$, 5 mM glycine, pH 8.5 at room temperature and pH 8.9–9.0 at 0°C) is quite different from the protease inhibitor-enriched lysis buffer used to prepare crude membranes. The divalent cations in the glycine lysis buffer stabilize the plasma membranes. The high pH of the buffer decreases nonspecific binding of contaminants and inhibits the activity of proteases with low pH optima.

In our preparatively scaled version of the Das–Henderson protocol (Fig. 2), 1.5×10^{10} cells are harvested and washed in glycine lysis buffer at 650 g_{max}. Washed cells are resuspended to $3–5 \times 10^{7}$ cells/ml in the same buffer and are allowed to warm slightly prior to lysis. The cell suspension is poured into the barrel of a 50-ml syringe fitted to the luer slip of a 47-mm Nucleopore Swin-Lok holder–5 μm Nucleopore filter assembly. The cells (10–15 μm in diameter) are forced through the filter with the plunger of the syringe into clean, chilled 50-ml centrifuge tubes. The filter must be changed as soon as an increase in resistance to the passage of the cells is felt to prevent tearing or clogging of the filter. In general, one filter is needed for each 200–300 ml of cell suspension. The cell lysate is centrifuged immediately at 5900 g_{max} for 20 minutes. These conditions are important because, although greater centrifugal force may increase the plasma membrane yield, higher speeds at this point also pellet contaminants which later tend to blur the banding pattern in sucrose gradients. The membrane pellet

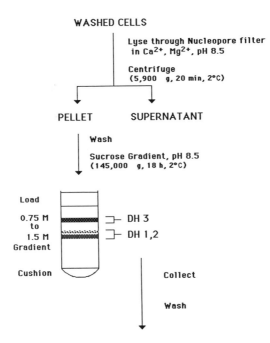

FIG. 2.   Flow diagram for the preparation of plasma membranes by the method of Das and Henderson (1983). For a complete explanation of the protocol, refer to Section II,C of the text.

is washed several times, with gentle Dounce homogenization, in cold glycine lysis buffer at 38,700 g_{max} for 10 minutes and resuspended in this buffer.

Resuspended membranes are layered onto chilled gradients containing equal volumes of 0.75 M and 1.5 M sucrose in 50 mM glycine, pH 8.5 at 22°C, formed over a cushion of 1.8 or 2.5 M sucrose in the same buffer. One preparative gradient (2-ml load on 13 ml of each sucrose solution over a 2-ml cushion) or two smaller gradients (1-ml load on 4.5 ml of each sucrose solution over a 1-ml cushion) are recommended for every 3 × 10⁹ lysed cells. Gradients, overlayered with glycine lysis buffer, are centrifuged at 145,000 g_{max} for 18 hours. Three plasma membrane bands sediment just above the middle of the gradient. DH 1, the most dense of the three bands, and DH 2, the middle band, are usually very close together and we do not try to collect them separately. DH 3, the least dense band, is found just below a light, diffuse band of contaminants. Usually, DH 3 is collected separately. Occasionally, perhaps due to irregularities in the gradient, the DH 1, DH 2, and DH 3 bands almost overlap and the three bands must be collected together. This band overlap problem is sometimes observed in preparative gradients and is always observed when too much cell lysate is loaded. In our laboratory, Das–Henderson membranes are washed and pelleted at 38,700 g_{max} for 30 minutes in 20 mM sodium phosphate, pH 6.8, and are stored on ice in the phosphate buffer plus 0.02% sodium azide. Between 20 and 25 mg of membrane protein are recovered for each 10¹⁰ cells lysed.

D.   Con A-Stabilization, Triton-Extraction Method

The plasma membrane isolation procedure that we use most often is an extension of the Con A-stabilization, Triton-extraction procedure first described by Parish and Müller (1976), modified by Condeelis (1979), and extended by Luna and colleagues (1981, 1984). In this procedure (Fig. 3), *D. discoideum* cells are coated on their extracellular surfaces with Con A. During the patching and capping of the bound Con A (Condeelis, 1979), actin and myosin become associated with the cytoplasmic surface of the plasma membrane domains that are stabilized by Con A (Condeelis, 1979; Luna *et al.*, 1984). The cells are lysed with Triton X-100, and the Con A-stabilized plasma membrane domains are separated from less dense intracellular membranes and other cellular components on a sucrose gradient. The density of the plasma membrane domains then is decreased by the removal of Con A and by the dissociation of coisolating actin and myosin. The decrease in density of the plasma membranes allows their separation from the dense, contaminating, cellular components on a second sucrose gradient. This density shift method is convenient for large numbers of cells and produces highly purified plasma membranes. However, this procedure takes at least 4 days to complete, and the final membrane preparation may not contain all subdomains of the native plasma membrane.

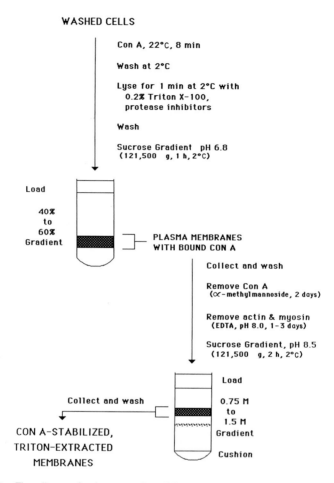

Fig. 3.   Flow diagram for the preparation of Con A-stabilized, Triton-extracted membranes by the density shift technique of Luna *et al.* (1984). For a complete explanation of the protocol, refer to Section II,D of the text.

1.   Con A-Induced Patching

About 10^{11} cells are washed twice at room temperature in Sorensen's phosphate buffer, pH 6.1. Washed cells are resuspended to $1–8 \times 10^7$ cells/ml in bottles containing 200 ml Sorensen's phosphate buffer at room temperature. Con A stock solution (10 mg/ml Con A, 1 M NaCl, 0.25 mM CaCl$_2$, 0.25 mM MnCl$_2$, 40 mM sodium phosphate, pH 7.0) is added such that 25 μg/ml Con A is available for every 10^7 cells/ml. For example, if cells are resuspended to 4×10^7

cells/ml in each of several bottles, Con A stock solution is added to a final concentration of 100 μg/ml in each bottle. The cells are swirled at room temperature for 5–8 minutes with Con A. The length of this incubation depends on the actual room temperature, since the Con A-induced patching and capping is temperature-dependent. At 22°C, a 5- to 6-minute Con A incubation results in cell surfaces coated with Con A patches. Longer incubations increase the amount of capped and internalized Con A, reducing the surface area covered by the lectin, and, thus, drastically reducing the yield of plasma membranes.

2. Cell Lysis with Triton X-100

Immediately after the Con A incubation, cells are chilled to 2°C, harvested at 650 g_{max} for 2 minutes, washed once in ice-cold Sorensen's phosphate buffer, and resuspended to $1–8 \times 10^7$ cells/ml in ice-cold extraction buffer (40 mM sodium pyrophosphate, 0.57 mM PMSF, 2 mM EDTA, 1 mM EGTA, 5 mM 1,10-phenanthroline, 2 mM N-carbobenzoxyphenylalanine, 0.5% ethanol, 0.4 mM dithiothreitol, 0.02% sodium azide, 10 mM Tris-HCl, pH 7.6). The suspension is brought to 0.2% Triton X-100 by adding 10% Triton X-100 stock solution, and the cells are shaken vigorously with the detergent for exactly 1 minute. After centrifugation at 4000 g_{max} for 10 minutes, the pellet contains a large, flocculent, yellow-white mass of membrane. Occasionally, however, a small button of unlysed cells lies under the membrane pellet. We usually collect only the flocculent pellet, although the unlysed cells may be collected separately and reextracted with 0.2% Triton X-100 in extraction buffer to increase yield, if desired. After a wash in extraction buffer at 14,500 g_{max} for 5 minutes and a wash in 1 mM EGTA, 5 mM Tris-HCl, pH 7.6, at 14,500 g_{max} for 10 minutes (both resuspensions facilitated by gentle Dounce homogenization), the plasma membranes with bound Con A are centrifuged on sucrose gradients. Resuspended membranes are layered onto chilled, 28-ml linear gradients of 40–60% sucrose in 0.02% sodium azide, 20 mM sodium phosphate, pH 6.8 (5-ml load on 14.7 ml of 40% sucrose and 13.3 ml of 60% sucrose). The gradients are centrifuged at 121,500 g_{max} for 1 hour, and the thick, white, sticky membrane bands near the bottom of the gradients are collected by hand and pooled. The membranes are washed several times at 26,900 g_{max} for 10 minutes in 20 mM sodium phosphate, pH 6.8, with resuspensions effected using gentle Dounce homogenization.

3. Con A Removal

To remove Con A, plasma membranes are resuspended in equal volumes of 20 mM sodium phosphate, pH 6.8, and 0.5 M α-methyl-D-mannoside, 2 mM

$MgCl_2$, 2 mM ATP, 0.02% sodium azide, 100 mM Tris-HCl, pH 8.5. After an overnight incubation on ice, the membranes are diluted with the phosphate buffer, pelleted at 38,700 g_{max} for 30 minutes, resuspended in 0.5 M α-methyl-D-mannoside, 1 mM $MgCl_2$, 1 mM EDTA, 1 mM ATP, 0.02% sodium azide, 100 mM Tris-HCl, pH 8.5, and again incubated overnight on ice. Con A removal is completed by washing the membranes twice at 38,700 g_{max} for 30 minutes, with homogenization, in 20 mM sodium phosphate, pH 6.8.

4. REMOVAL OF ACTIN AND MYOSIN

To remove cytoskeletal components from the membranes, we depolymerize and denature the membrane-bound actin by dialysis against dilute EDTA. Myosin also is removed from the membranes by this procedure, since it apparently is associated with the membranes only through its interaction with actin (Spudich, 1974). First, the membranes are diluted, pelleted (38,700 g_{max}, 30 minutes), and resuspended with homogenization in ice-cold distilled water. Then, the membranes are dialyzed overnight against 0.1 mM EDTA, 0.2 mM sodium phosphate, pH 7.9–8.3. After a wash in excess dialysis buffer, most of the actin and myosin is extracted from the membranes. Complete removal of actin and myosin from log phase cells usually is obtained after an additional 1–2 days of dialysis, with daily washes and resuspensions in dialysis buffer. We also find that a short (5–10 seconds) sonication of membranes in dialysis buffer potentiates the removal of the cytoskeletal proteins. However, if the initial cell density is much greater than 1 × 10[7] cells/ml (late log phase or early stationary phase), residual actin may remain even after 3 days of dialysis. Membranes from the final dialysis are pelleted and homogenized in 0.5 mM $CaCl_2$, 0.5 mM $MgCl_2$, 5 mM glycine, pH 8.9–9.0 at 0°C. A 5-ml portion of membrane suspension is layered onto each of six, chilled sucrose gradients (13 ml each of 0.75 M and 1.5 M sucrose in 50 mM glycine, pH 8.5 at 22°C, formed on a cushion of 2 ml 1.8 M sucrose in the same buffer). The gradients are centrifuged at 121,500 g_{max} for 2 hours. Usually, we observe two membrane bands on these gradients—one thick, white band near the middle of the gradient above a thin, yellowish band. The occurrence of the lower band is variable. Its appearance correlates with different amounts of contaminants and with incomplete removal of actin and myosin. We collect the top, purified plasma membrane band by hand and wash it several times at 38,700 g_{max} for 45 minutes with 20 mM sodium phosphate, pH 6.8. If the lower band is large, it is collected separately, redialyzed, and recentrifuged. Purified plasma membranes are stored on ice in 20 mM sodium phosphate, 0.02% sodium azide, pH 6.8. From 1 to 2 mg of membrane protein is recovered for each 10[10] cells lysed.

III. Comparison of Plasma Membranes Obtained by Different Methods

A. Electron Microscopy

The purity of the *D. discoideum* plasma membranes produced by the three procedures described above may be assessed by transmission electron microscopy. Briefly, as described by Luna *et al.* (1984), freshly prepared samples are fixed for 30 minutes at room temperature in 1% glutaraldehyde, 2% tannic acid, 50 mM sodium cacodylate, pH 7.0. Fixed samples are washed, incubated in 0.8% potassium ferrocyanide, 50 mM sodium cacodylate, pH 7.0, for 30 minutes at 0°C, and postfixed in the buffered ferrocyanide plus 0.5% OsO_4 for 30 minutes at 0°C. Inclusion of the ferrocyanide in this step greatly enhances membrane contrast (Karnovsky, 1971; McDonald, 1984; Neiss, 1984). After a buffer rinse and two or three washes with distilled water, the samples are stained for 3–16 hours in 2% aqueous uranyl acetate and further processed for electron microscopy.

Electron micrographs of crude membranes, Das–Henderson membranes, and membranes prepared from Con A-stabilized, Triton-extracted cells are shown in Fig. 4. It is clear that crude membranes prepared from filter-lysed cells by the modified Spudich method (C) contain remnants of intracellular organelles but it is unclear from which organelles these contaminants arise. Das–Henderson membranes (DH) contain no visible contaminating organelles, but they do contain at least two morphologically distinct vesicle populations. Membranes prepared from Con A-stabilized, Triton-extracted cells (T) contain few, if any, small vesicles and consist almost entirely of large vesicles and long membrane sheets.

B. Polyacrylamide Gels

Membrane purity also may be evaluated by polyacrylamide gel electrophoresis under denaturing conditions (SDS–PAGE). We use 1.5-mm-thick, 6–16% gradient slab gels prepared and run according to the method of Laemmli (1970). These gels resolve polypeptides with molecular weights ranging from >300,000 to ~15,000 (if electrophoresis is stopped immediately after the dye front runs off the gel).

SDS–PAGE profiles of the three membrane preparations described above are presented in Fig. 5A. Crude membranes (lane C), Das–Henderson DH 1 plus 2 (lane DH 1,2), and Das–Henderson DH 3 (lane DH 3) exhibit very similar Coomassie blue-stained profiles; relatively minor differences in band intensities

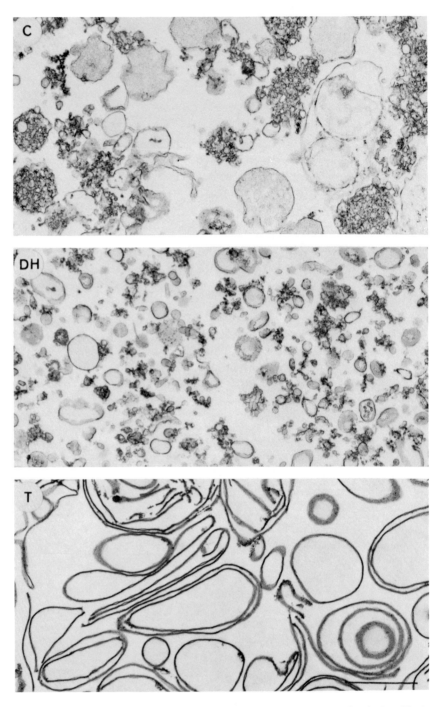

FIG. 4. Transmission electron micrographs comparing crude membranes (C) with Das–Henderson membranes DH 1, 2 + DH 3(DH), and Con A-stabilized, Triton-extracted membranes after removal of Con A, actin, and myosin (T). Bar ≃ 1 μm.

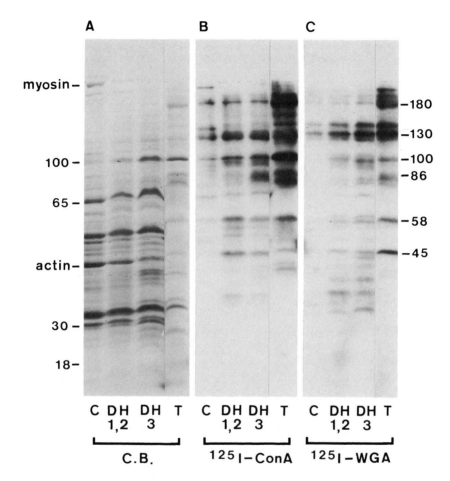

Fig. 5. A comparison of plasma membrane preparations by Coomassie blue-stained SDS–PAGE (A) and by autoradiography of blots from the same gel labeled with ^{125}I-Con A (B) or ^{125}I-WGA (C). Each panel shows crude membranes prepared by the method of Spudich (lane C), Das–Henderson DH 1 plus DH 2 (lane DH 1,2), Das–Henderson DH 3 (lane DH 3), and Con A-stabilized, Triton-extracted membranes after removal of Con A, actin, and myosin (lane T). Each lane was loaded with 60 μg of membrane protein. Glycoconjugates present in *D. discoideum* plasma membranes probably cause the arcing of proteins observed in polyacrylamide gels (Wilhelms *et al.*, 1974; West and McMahon, 1979). Numbers to the left of the figure represent the molecular weights (in thousands) of molecular weight standards. Numbers to the right of the figure indicate the molecular weights of major glycoprotein bands visible in the figure. Autoradiographs were exposed for 17 days (^{125}I-Con A) or for 28 days (^{125}I-WGA).

are observed. Membranes prepared by the Con A-stabilization, Triton-extraction method (lane T) lack many of the bands evident in the other preparations; other bands are intensified. All four membrane preparations contain Coomassie blue-stained bands at about 130, 100, 58, and 30 kDa. Bands present in the first three lanes and absent in the fourth lane in Fig. 5A may represent contaminants in the Spudich and Das–Henderson membrane preparations. Alternatively, some of these bands may represent plasma membrane polypeptides removed during Triton extraction of Con A-stabilized cells. Actin and myosin, obviously, are missing from lane T in Fig. 5A, since these proteins are removed during the preparation of these membranes.

C.   Lectin Blots

The glycoprotein compositions of the different plasma membrane preparations are compared by probing nitrocellulose blots of electrophoretically separated membrane proteins with radiolabeled lectins. The transfer of proteins from SDS–PAGE slab gels to nitrocellulose paper follows standard procedures (Towbin *et al.*, 1979). We preincubate blots with 2% BSA (w/v) in 0.15 M NaCl, 10 mM sodium phosphate, pH 7.4, for 2 hours at room temperature or for 8–16 hours at 4°C and label them, with agitation, in 5–10 ml of 4% BSA (w/v), 150 mM NaCl, 0.02% Triton X-100, 50 mM Tris-HCl, pH 7.2, containing 200,000 cpm/ml of ^{125}I-Con A (prepared according to Chang and Cuatrecasas, 1976) or wheat germ agglutinin (WGA) radiolabeled with ^{125}I-Bolton–Hunter reagent (Bolton and Hunter, 1973). To determine the specificity of lectin binding, a sugar known to bind the lectin is included in the labeling solution with a duplicate, control blot. We use 0.2 M α-methyl-D-mannoside as a control for ^{125}I-Con A binding and 0.2 M N-acetyl-D-glucosamine as a control for ^{125}I-WGA binding. Blots are washed for a total of 2 hours with four changes each of 150 mM NaCl, 0.02% Triton X-100, 50 mM Tris-HCl, pH 7.2, and 0.1% Triton X-100, 50 mM Tris-HCl, pH 7.2. The increased Triton concentration in the second series of washes reduces background binding. No binding is observed on control blots.

The ^{125}I-Con A blots (Fig. 5B) and the ^{125}I-WGA blots (Fig. 5C) show that most lectin-binding glycoproteins are found in all three plasma membrane preparations. The major differences among the lanes are the intensities with which ^{125}I-Con A and ^{125}I-WGA bind to glycoproteins with apparent molecular weights of 80,000–200,000. These glycoproteins predominate in lane T, perhaps because the use of Con A in the isolation of these membranes selects for glycoprotein-rich plasma membrane domains. Alternatively, since the gel lanes are loaded with equal amounts of total protein, the relative intensities of lectin binding may reflect the relative enrichments for plasma membranes in these preparations. Glycoproteins of low molecular weight found only in the Das–Henderson membrane preparations are either polypeptides peculiar to this method of membrane purification or represent fragments of larger glycoproteins generated by proteolysis during purification. The hypothesis that these bands are proteolytic fragments is supported by the observation, using two-dimensional peptide mapping, that cleavage of the myosin heavy chain occurs in the Das–Henderson preparation (E. J. Luna, unpublished).

D. Surface Labeling

Perhaps the best way to assess plasma membrane purity is to examine membranes from cells labeled with a side-specific, cell surface probe. We find that the external face of the *D. discoideum* plasma membrane is modified easily with a variety of probes. Although we occasionally radioiodinate cell surface tyrosine residues with Na^{125}I and lactoperoxidase–glucose–glucose oxidase (Hubbard and Cohn, 1975; Luna *et al.*, 1984), our preferred method of surface labeling cells is to use sulfo-*N*-hydroxy-succinimidobiotin (sulfo-NHS-biotin) to label cell surface amino groups (Staros, 1982). Biotinylated membrane polypeptides are visualized on autoradiographs of nitrocellulose blots incubated with ^{125}I-labeled avidin.

Sulfo-NHS-biotin has several practical advantages over ^{125}I as a cell surface label. First, since some of the lysis techniques used to generate these membranes can be messy (especially if large numbers of cells are involved), radioactive contamination of laboratory equipment and the investigator is minimized if biotin, rather than ^{125}I, is bound to the cells prior to lysis. Second, since the radiolabel is used only in the final step of the sulfo-NHS-biotin procedure, this method generates far less radioactive waste and, therefore, is less expensive than cell surface iodination. Finally, it is likely that the lysine side chains and amino sugars labeled by sulfo-NHS-biotin are more numerous and more representatively distributed on the cell surface than are the tyrosine residues labeled by ^{125}I.

To label cell surfaces with sulfo-NHS-biotin, $1–5 \times 10^{10}$ cells are washed twice with Sorensen's phosphate buffer, pH 6.1, and are resuspended with 20 mg of sulfo-NHS-biotin (Pierce) in 100 ml of Sorensen's phosphate buffer, pH 8.0. (The pH of Sorensen's phosphate buffer is raised to 8.0 using a few drops of 6 *N* NaOH.) To prevent hydrolysis, the probe is dissolved in buffer immediately before use. Cells are incubated with the probe at 22°C on a gyrotory shaker at 150 rpm for no more than 30 minutes (longer incubations increase label incorporation but also increase internalization of the probe). After incubation, cells are pelleted at 650 g_{max} for 2 minutes at 2°C and are washed four times in Sorensen's phosphate buffer, pH 6.1. To check for mechanical damage, surface-labeled cells are viewed in a phase contrast microscope. Then, the cells are lysed and plasma membranes are isolated as described above. As a control for the specificity of the label, some membranes are labeled after cell lysis. Unlabeled membranes isolated from $1-5 \times 10^{10}$ cells are labeled at 22°C for 30 minutes with 2 mg sulfo-NHS-biotin in 10 ml of Sorensen's phosphate buffer, pH 8.0. Surface-labeled membranes and membranes labeled on both sides are washed thoroughly and centrifuged separately on 40–60% sucrose gradients. Membranes are collected, washed, and stored on ice in 0.02% sodium azide, 20 mM sodium phosphate, pH 6.8.

Biotinylated polypeptides in membranes labeled with sulfo-NHS-biotin are best compared by loading equal amounts of biotin in each lane of an SDS gel. To quantify the biotin concentrations in our membrane preparations, we perform an avidin-binding assay on each sample. In this assay, 1- to 5-μg aliquots of membranes are incubated with 150,000 cpm ^{125}I-avidin for 2 hours at 22°C with gentle mixing in 50 mM sodium borate buffer, pH 10.0, in a total volume of 110 μl. Ninety microliters of the reaction mixture are pelleted at 27,600 g_{max} for 30 minutes in a 400-μl Eppendorf tube containing 250 μl of 5% sucrose, 50 mM sodium borate, pH 10.0. After centrifugation, the tubes are frozen quickly and the membrane pellets, plus the lower 100 μl of the sucrose solution, are cut off with a wire cutter. The pellets and the supernatants (the remainder of the solution in each tube) are counted in a γ counter. By plotting the amount of bound ^{125}I-avidin versus the amount of membrane protein in the assay, we generate linear binding curves from which the avidin-binding activity (the counts per minute of ^{125}I-avidin per microgram of membrane protein) of each sample is ascertained. SDS gels are loaded with a maximum of 75 μg of the sample with the lowest avidin-binding activity. Other samples are loaded onto the gels in proportionally lower amounts, based on their avidin-binding activities.

To visualize biotinylated cell surface polypeptides, nitrocellulose blots of SDS gels are incubated with ^{125}I-avidin at pH 10.0–10.5. The pH is critical to the success of this protocol. At neutral pH, avidin binds nonspecifically to many membrane proteins, probably via electrostatic interactions, since avidin's isoelectric point is ~10.0 (Wooley and Longsworth, 1942). Nitrocellulose blots are incubated with 2% (w/v) BSA, 150 mM NaCl, 50 mM sodium borate, pH 10.0–10.5, for 2 hours at 22°C or overnight at 4°C. Biotinylated polypeptides are labeled by incubating the blots with 0.5 μCi of ^{125}I-Bolton–Hunter-labeled avidin in 5 ml of 150 mM NaCl, 4% BSA (w–v), 0.02% Triton X-100, 50 mM sodium borate, pH 10.0–10.5, while shaking at 300 rpm for 2 hours at 22°C. Labeled blots are washed for a total of 2 hours with four changes of 150 mM NaCl, 0.01% Triton X-100, 50 mM sodium borate, pH 10.0–10.5, and four changes of 0.1% Triton X-100, 50 mM sodium borate, pH 10.0–10.5. Labeled polypeptides are visualized by exposing Kodak XAR-2 film to air-dried blots.

Figure 6 shows Coomassie blue-stained gel profiles (Fig. 6A) and ^{125}I-avidin blots (Fig. 6B) of Con A-stabilized, Triton-extracted plasma membranes isolated either from unlabeled cells (lanes U) or from cells labeled only on their extracellular surfaces (lanes S). Similarly prepared membranes labeled on both mem-

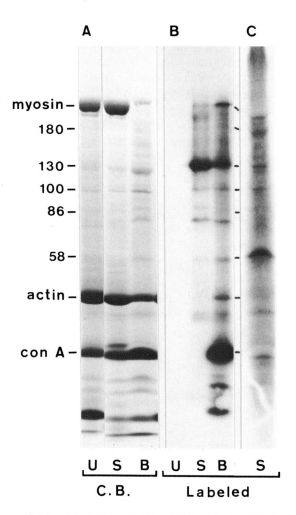

FIG. 6. Coomassie blue-stained SDS gels (A) and ^{125}I-avidin blot (B) of unlabeled Con A-stabilized, Triton-extracted plasma membranes (lanes U) or similarly isolated membranes labeled with sulfo-NHS-biotin only on their extracellular surfaces (lanes S) or on both membrane faces (lanes B). The Coomassie blue-stained lanes were loaded with equal amounts of membrane protein (60 μg per lane). The lanes on the ^{125}I-avidin blot were loaded such that 67,000 cpm of ^{125}I-avidin would bind to each lane (60 μg of protein were loaded in lanes U and S and 7.5 μg of protein were loaded in lane B). The lane in C is an autoradiograph of a dried gel containing 60 μg of membranes from cells surface-labeled with [125]I. Numbers to the left of the figure indicate the apparent molecular weights (in thousands) of some of the major labeled polypeptides. Autoradiographs were exposed for 5 days (panel B) or for 7 days (panel C).

brane faces are shown in lanes B. Since the cytoplasmic proteins, actin and myosin, are labeled only if the cells are lysed prior to the addition of sulfo-NHS-biotin, we conclude that this probe does not penetrate the plasma membrane and, therefore, that sulfo-NHS-biotin is a good cell surface label. Because Con A is used to stabilize the membranes after surface labeling but before cell lysis, it is tagged only when membranes are labeled after cell lysis. For comparison, a gel profile of membranes from cells surface-labeled with ^{125}I, using lactoperoxidase, glucose, and glucose oxidase (Hubbard and Cohn, 1975; Luna *et al.*, 1984), is shown in Fig. 6C. Although the two surface-labeling methods appear to label the same proteins, differences are evident in the relative intensities of label

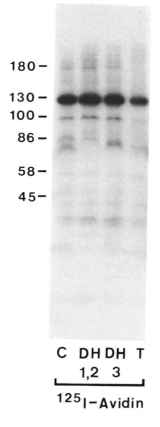

FIG. 7.   Biotinylated cell surface polypeptides present in crude membranes prepared by the method of Spudich (lane C), Das–Henderson DH 1 plus DH 2 (lane DH 1,2), Das–Henderson DH 3 (lane DH 3), and Con A-stabilized, Triton-extracted membranes after removal of Con A, actin, and myosin (lane T). The lanes were loaded such that 8500 cpm of ^{125}I-avidin would bind to each lane. The autoradiograph was exposed for 29 days.

incorporated into individual polypeptides. Such differences are expected, since the labels react with different amino acid residues.

Figure 7 shows the biotinylated cell surface polypeptides present in the three different plasma membrane preparations described above. Although a few polypeptides present in the crude membranes and in the Das–Henderson membranes are absent from the Con A-stabilized, Triton-extracted membranes, the overall impression is that the surface-labeled polypeptides, like the glycoproteins visualized with ^{125}I-labeled lectins (Fig. 5), are the same in all three membrane preparations. However, Con A-stabilized, Triton-extracted membranes (with bound Con A, actin, and myosin) have three to four times more biotin surface label, expressed as the ratio of biotin to membrane protein, than do the crude membranes. Das–Henderson membranes contain intermediate amounts of cell surface biotin. The Con A-stabilized, Triton-extracted membranes (depleted of Con A, actin, and myosin) also are the only membranes for which we observe a close correspondence between the Coomassie blue-stained gel profile (Fig. 5A) and the gel profiles obtained either after staining with lectins (Fig. 5B and C) or after cell surface labeling (Fig. 7). Therefore, we conclude that essentially all of the Con A-stabilized, Triton-extracted membranes are derived from the plasma membrane. However, since not all of the surface-labeled polypeptides found in the crude and Das–Henderson preparations are found in the Con A-stabilized, Triton-extracted membranes (Fig. 7), this last membrane preparation may not contain all portions of the intact plasma membrane.

IV. Vesiculation and Sealing of Plasma Membranes

For many experiments the large, open plasma membrane sheets (Fig. 4, C) produced by the Con A-stabilization, Triton-extraction method are an ideal preparation. Other experiments (e.g., transport studies), require purified plasma membranes in the form of sealed vesicles. We find that sonication fragments Con A-stabilized, Triton-extracted membranes (devoid of Con A, actin, and myosin) into small vesicles (average diameter of 110 nm) and disk-like fragments (average diameter of 96 nm; Luna *et al.*, 1984). Recently, we have characterized these sonicated membranes and find that many of the vesicles are sealed to small molecules and that unsealed vesicles and disk-like fragments can be separated from the sealed vesicle population.

A. Sonication

We sonicate 500 μl of plasma membrane suspension in a 5-ml tube for a total of 2 minutes (20-second bursts at 10-second intervals) with the microtip of a

Branson Model 200 Sonifier set at power level 2. To prevent heat denaturation, the tube is immersed in an ice-water bath during sonication. When membranes are sonicated in 100 mM 5(and 6)-carboxyfluorescein (Molecular Probes), 10 mM sodium phosphate, pH 7.1, many of the resulting vesicles seal around the carboxyfluorescein, an easily observed, membrane-impermeant dye (Weinstein *et al.*, 1977). To investigate whether the dynamics of membrane sealing in this system are similar to those described for the erythrocyte membrane (Steck and Kant, 1974; Lieber and Steck, 1982b), we add 1 mM MgCl$_2$, 70 mM NaCl (salt solution), to the membranes and incubate them for 1–2 hours at 37°C. We find that incubation of *D. discoideum* membranes with salt solution produces little or no sealing in addition to that produced by sonication. Furthermore, incubation of unsonicated, Con A-stabilized, Triton-extracted membranes with carboxyfluorescein and salt solution at 37°C does not induce membrane sealing to carboxyfluorescein.

B.   Vesicle Permeability

Free carboxyfluorescein is separated from membranes by passing the sample through a 1 × 20 cm column containing 15 ml of packed Sephacryl S-1000 (Pharmacia) equilibrated with 1 mM MgCl$_2$, 70 mM NaCl, 10 mM sodium phosphate, pH 7.4. If the column is run no faster than 0.5 ml/min, the first orange peak, which contains membrane vesicles, can be separated by eye from the second, intensely red-orange peak of free dye, since the dilute free dye at the beginning of the second peak appears green rather than orange. However, if the column is run too quickly, the membrane peak may be masked completely by excess dye. The orange color of the first peak indicates that the carboxyfluorescein dye remains at a high local concentration and thus, that the dye must be sealed inside at least some of the membrane vesicles.

Sealed vesicles can be separated from unsealed membranes by equilibrium density centrifugation through gradients formed from a large, dense solute, such as Dextran T70 (Steck and Kant, 1974; Lieber and Steck, 1982a,b). Vesicles with holes larger than the Stokes radius of the solute sediment to their inherent density, which is about 1.16–1.17 g/cm^3 for *D. discoideum* plasma membranes (Smart and Hynes, 1974; Luna *et al.*, 1984). Vesicles impermeable to the solute equilibrate at a density that is much less than their inherent density, since they retain an internal aqueous compartment with a density of only ~1.00 g/cm^3.

To prepare Dextran T70 gradients, we use 11.5-ml cellulose tubes (Sarstedt) in which we prepare gradients from 4.5 ml each of 5% and 25% (w/w) Dextran T70 (Pharmacia; MW ~70,000; Stokes radius of 5.8 nm) in 1 mM MgCl$_2$, 70 mM NaCl, 20 mM sodium phosphate, pH 7.4, over a 0.3-ml cushion of 30% (w/w) Dextran T70 in the same buffer. To ensure that the highly viscous dextran solutions are mixed thoroughly and that the gradients are as similar as possible, we use an Isco Model 570 Gradient Former set at the lowest speed. Two milliliters of membrane vesicles from Sephacryl S-1000 columns are loaded onto each gradient and the gradients are centrifuged at 210,000 g_{max} for 18 hours at 2°C.

Reproducible fractionation of the gradients is ensured by using an Isco Model 640 Density Gradient Fractionator with an attached Model UA-5 Absorbance Detector. This machine pumps a very dense solution (Fluorinert FC-40) into the bottom of a punctured centrifuge tube, pushing a virtually undisturbed gradient up through the flow cell of the absorbance monitor into a fraction collector. Equipped with 280-nm filters and an event marker, the monitor continuously reads and records the optical density at 280 nm. Since concentrated Dextran T70 has substantial optical density, optical densities of fractions from a blank gradient are subtracted from the optical densities of corresponding fractions from

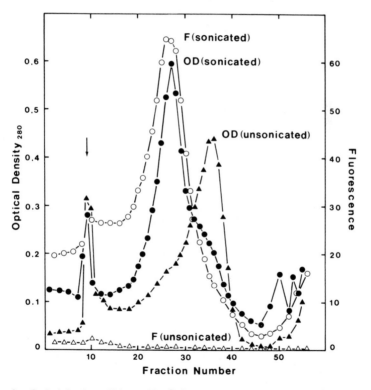

Fig. 8.   Optical density at 280 nm (OD; ●,▲) and relative fluorescence (F; ○,△), in arbitrary units, of 200-μl fractions collected from Dextran T70 gradients. The gradients were loaded with plasma membranes prepared by the Con A-stabilization, Triton-extraction method. These membranes were sonicated (○,●) or incubated without sonication (△,▲) in the presence of 5 (and 6)-carboxyfluorescein. The top of the centrifuge tube is to the left of the figure. The arrow at about fraction 10 indicates the interface between the loaded volume and the upper part of the gradient.

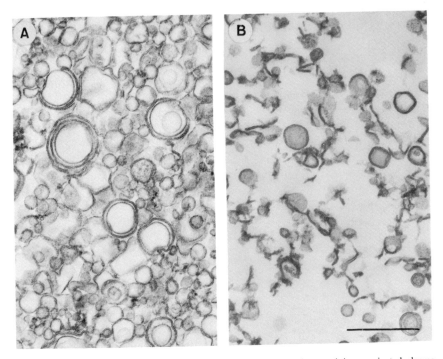

FIG. 9.   Transmission electron micrograph of carboxyfluorescein-containing, sonicated plasma membranes from a gradient similar to those shown in Fig. 8. The sealed membranes obtained from the upper band (A) and the unsealed membranes from the lower band (B) are shown. Bar ≃ 0.5 μm.

gradients containing membranes. The relative amount of carboxyfluorescein in each fraction is determined by adding 0.7 ml of 1% SDS, 50 mM sodium borate, pH 9.0, to each fraction and measuring the fluorescence of the sample in a Perkin–Elmer MPF-66 spectrofluorimeter.

Figure 8 shows the optical density at 280 nm and the relative fluorescence of fractions from Dextran T70 gradients loaded with either sonicated, carboxyfluorescein-containing plasma membranes or unsonicated plasma membranes incubated at 37°C with 5(and 6)-carboxyfluorescein and salt solution. Unsonicated membranes band at about fraction 36 and clearly do not contain trapped carboxyfluorescein. On the other hand, sonicated membranes, which band at about fraction 27 (upper band), are orange to the naked eye and, when solubilized with SDS, release a large amount of trapped carboxyfluorescein which is strongly fluorescent. Therefore, these membranes are sealed both to Dextran T70 and to carboxyfluorescein. The appearance in this sample of a minor, nonfluorescent peak at fraction 50 (lower band) is variable.

Differences in sealing between the upper band and the lower band from Dextran T70 gradients also are observed by electron microscopy. Figure 9A shows a transmission electron micrograph of pooled, sonicated, carboxyfluorescein-containing membranes from the upper band of a gradient similar to the one shown in Fig. 8. Most, if not all, of the membranes in this fraction appear to be sealed, since no open vesicles or disk-like fragments are apparent in the micrograph. Figure 9B is a transmission electron micrograph of the pooled, nonfluorescent fractions containing the lower band of the same gradient. This micrograph clearly shows unsealed membrane sheets and disk-like membrane fragments.

C. Sidedness of Sealed Vesicles

For many experiments, it is advantageous to have plasma membrane vesicles with the same orientation as the intact cell (right-side-out vesicles, or ROVs) separated from plasma membrane vesicles with an inverted orientation in which the cytoplasmic surface is exposed (inside-out vesicles, or IOVs). Con A-affinity chromatography is a classical technique for the separation of ROVs and IOVs (Zachowski and Paraf, 1974; Walsh *et al.*, 1976; Lindsay *et al.*, 1981; Virtanen *et al.*, 1981). While sonicated *D. discoideum* plasma membranes can be separated into fractions that do and do not bind to Con A-affinity columns, closer scrutiny of the two vesicle populations shows that at least part of this fractionation is due to the size exclusion properties of the gel filtration matrix.

In cosedimentation experiments on Dextran T70 gradients, we find that all sealed vesicles bind Con A, WGA, and actin. Since WGA and, to a lesser extent, Con A bind specifically to the *D. discoideum* cell surface (Ryter and Hellio, 1980), and since actin binds only to the cytoplasmic face of isolated membranes (Condeelis, 1979; Luna *et al.*, 1981, 1984), we conclude that all the sealed, sonicated vesicles must have a "scrambled" sidedness. This conclusion is consistent with theoretical models which propose that vesicles are formed in a two-step process during sonication. Membranes are believed to fragment initially into open planar disks which then fuse spontaneously into vesicles (Finer *et al.*, 1972; Fromherz, 1983; Fromherz and Rüppel, 1985). The fusion of membrane fragments of opposite polarity would generate vesicles with the properties observed.

V. Choosing a Membrane Isolation Procedure

An ideal plasma membrane preparation (1) contains only plasma membranes as judged by electron microscopy and by comparison of surface-labeled and Coomassie blue-stained gel profiles, (2) contains no proteolyzed polypeptides detectable on gels or by comparative peptide mapping, and (3) contains all subsets of the intact plasma membrane as monitored by recovery of cell surface label. None of the *D. discoideum* membrane preparations described herein meets all three of these criteria. Crude membranes prepared by the method of Spudich (1974) appear to be unproteolyzed and to contain all cell surface polypeptides, but they clearly contain other cell components as well. Membranes prepared by the method of Das and Henderson (1983) are purer than the crude membranes and are representative of the plasma membrane as a whole, especially if all three plasma membrane bands from the sucrose gradient are combined. However, some proteolysis of membrane polypeptides occurs during the isolation of membranes by this procedure, perhaps because the divalent cations used to stabilize the plasma membranes stimulate endogenous calcium-activated proteases. Membranes prepared by the Con A-stabilization, Triton-extraction procedure are highly purified and apparently unproteolyzed, but they lack some of the surface-labeled polypeptides found in the other two preparations.

In the absence of an ideal method for the purification of plasma membranes, researchers must choose their membrane isolation protocol based on the requirements of each experiment. We find that the highly purified and unproteolyzed membranes prepared by the Con A-stabilization, Triton-extraction procedure are an excellent choice for the study of actin–membrane interactions (Luna *et al.*, 1981, 1984; Goodloe-Holland and Luna, 1984). However, Triton X-100 solubilizes or inactivates some plasma membrane enzymes, such as adenylate cyclase, alkaline phosphatase, and glycosyl transferase (Sievers *et al.*, 1978; Devreotes, 1982; Hagmann, 1985). Obviously, if the biochemical activity of interest is impaired by Triton X-100, another isolation method should be employed. However, the various advantages and disadvantages of the many *D. discoideum* plasma membrane preparations ensure that there is an appropriate membrane isolation procedure for any experimental application.

ACKNOWLEDGMENTS

We thank Hilary Ingalls and Gail Barcelo for assistance with surface-labeling procedures, membrane isolations, and manuscript preparation. We also thank George Q. Daley for initiating the experiments on sidedness of *D. discoideum* membrane vesicles. This work was supported by National Institutes of Health grant GM33048, March of Dimes grant 1-981, and a Faculty Research Award (FRA-289) from the American Cancer Society to E.J.L.

REFERENCES

Bennett, H., and Condeelis, J. (1984). *J. Cell Biol.* **99**, 1434–1440.
Beug, H., Katz, F. E., and Gerisch, G. (1973). *J. Cell Biol.* **56**, 647–658.
Bolton, A. E., and Hunter, W. M. (1973). *Biochem. J.* **133**, 529–539.
Bonner, J. T. (1967). "The Cellular Slime Molds." Princeton Univ. Press, Princeton, NJ.
Brunette, D. M., and Till, J. E. (1971). *J. Membr. Biol.* **5**, 215–224.
Chadwick, C. M., and Garrod, D. R. (1983). *J. Cell Sci.* **60**, 251–266.
Chadwick, C. M., Ellison, J. E., and Garrod, D. R. (1984). *Nature (London)* **307**, 646–647.
Chaney, L. K., and Jacobson, B. S. (1983). *J. Biol. Chem.* **258**, 10062–10072.

Chang, K.-J., and Cuatrecasas, P. (1976). *In* "Con A as a Tool" (H. Bittiger and H. P. Schnebli, eds.), pp. 187–189. Wiley, London.

Clarke, M. (1978). *In* "Cell Reproduction" (E. R. Dirksen, D. M. Prescott, and C. F. Fox, eds.), pp. 621–629. Academic Press, New York.

Clarke, M., Schatten, G., Mazia, D., and Spudich, J. A. (1975). *Proc. Natl. Acad. Sci. U.S.A.* **72**, 1758–1762.

Cocucci, S. M., and Sussman, M. (1970). *J. Cell Biol.* **45**, 399–407.

Condeelis, J. (1979). *J. Cell Biol.* **80**, 751–758.

Das, O. P., and Henderson, E. J. (1983). *Biochim. Biophys. Acta* **736**, 45–56.

Devreotes, P. N. (1982). *In* "The Development of *Dictyostelium discoideum*" (W. F. Loomis, ed.), pp. 117–168. Academic Press, New York.

Finer, E. G., Flook, A. G., and Hauser, H. (1972). *Biochim. Biophys. Acta* **260**, 49–58.

Fromherz, P. (1983). *Chem. Phys. Lett.* **94**, 259–266.

Fromherz, P., and Rüppel, D. (1985). *FEBS Lett.* **179**, 155–159.

Gabius, H.-J., Springer, W. R., and Barondes, S. H. (1985). *Cell,* in press.

Geltosky, J. E., Weseman, J., Bakke, A., and Lerner, R. A. (1979). *Cell* **18**, 391–398.

Gilkes, N. R., and Weeks, G. (1977). *Biochim. Biophys. Acta* **464**, 142–156.

Goodloe-Holland, C. M., and Luna, E. J. (1984). *J. Cell Biol.* **99**, 71–78.

Green, A. A., and Newell, P. C. (1974). *Biochem. J.* **140**, 313–322.

Hagmann, J. (1985). *Cell Biol. Int. Rep.* **9**, 491–494.

Hubbard, A. L., and Cohn, A. Z. (1975). *J. Cell Biol.* **64**, 438–460.

Jacobson, B. S. (1980a). *Biochem. Biophys. Res. Commun.* **97**, 1493–1498.

Jacobson, B. S. (1980b). *Biochim. Biophys. Acta* **600**, 769–780.

Karnovsky, M. J. (1971). *Proc. Annu. Meet. Am. Soc. Cell Biol., 11th* **51**, 146.

Kayman, S. C., Reichel, M., and Clarke, M. (1982). *J. Cell Biol.* **93**, 705–711.

Laemmli, U. K. (1970). *Nature (London)* **227**, 680–685.

Lam, T. Y., Pickering, G., Geltosky, J., and Siu, C.-H. (1981). *Differentiation* **20**, 22–28.

Lieber, M. R., and Steck, T. L. (1982a). *J. Biol. Chem.* **257**, 11651–11659.

Lieber, M. R., and Steck, T. L. (1982b). *J. Biol. Chem.* **257**, 11660–11666.

Lindsay, J. G., Reid, G. P., and D'Souza, P. (1981). *Biochim. Biophys. Acta* **640**, 791–801.

Loomis, W. F. (1971). *Exp. Cell Res.* **64**, 484–486.

Loomis, W. F. (1975). "*Dictyostelium discoideum.* A Developmental System." Academic Press, New York.

Loomis, W. F., ed. (1982). "The Development of *Dictyostelium discoideum.*" Academic Press, New York.

Luna, E. J., Fowler, V. M., Swanson, J., Branton, D., and Taylor, D. L. (1981). *J. Cell Biol.* **88**, 396–409.

Luna, E. J., Goodloe-Holland, C. M., and Ingalls, H. M. (1984). *J. Cell Biol.* **99**, 58–70.

McDonald, K. (1984). *J. Ultrastruct. Res.* **86**, 107–118.

McMahon, D., Miller, M., and Long, S. (1977). *Biochim. Biophys. Acta* **465**, 224–241.

Müller, K., Gerisch, G., Fromme, I., Mayer, H., and Tsugita, A. (1979). *Eur. J. Biochem.* **99**, 419–426.

Murray, B. A. (1982). *In* "The Development of *Dictyostelium discoideum*" (W. F. Loomis, ed.), pp. 71–116. Academic Press, New York.

Neiss, W. F. (1984). *Histochemistry* **80**, 231–242.

Pan, P., Hall, E. M., and Bonner, J. T. (1972). *Nature (London) New Biol.* **237**, 181–182.

Parish, R. W. (1983). *Mol. Cell. Biochem.* **50**, 75–95.

Parish, R. W., and Müller, U. (1976). *FEBS Lett.* **63**, 40–44.

Riedel, V., and Gerisch, G. (1968). *Naturwissenschaften* **55**, 656.

Rossomando, E. F., and Cutler, L. S. (1975). *Exp. Cell Res.* **95**, 67–78.

Ryter, A., and Hellio, R. (1980). *J. Cell Sci.* **41**, 75–88.

Saxe, C. L., III, and Sussman, M. (1982). *Cell* **29**, 755–759.

Serrano, R., Cano, A., and Pestaña, A. (1985). *Biochim. Biophys. Acta* **812**, 553–560.

Sievers, R., Risse, H.-J., and Sekeri-Pataryas, K. H. (1978). *Mol. Cell. Biochem.* **20**, 103–110.

Siu, C.-H., Lerner, R. A., and Loomis, W. F., Jr. (1977). *J. Mol. Biol.* **116**, 469–488.

Smart, J. E., and Hynes, R. O. (1974). *Nature (London)* **251**, 319–321.

Springer, W. R., Cooper, D. N. W., and Barondes, S. H. (1984). *Cell* **39**, 557–564.

Spudich, J. A. (1974). *J. Biol. Chem.* **249**, 6013–6020.

Spudich, J. A., and Spudich, A. (1982). *In* "The Development of *Dictyostelium discoideum*" (W. F. Loomis, ed.), pp. 169–194. Academic Press, New York.

Steck, T. L., and Kant, J. A. (1974). *In* "Methods in Enzymology" (S. Fleischer and L. Packer, eds.), Vol. 31, part A, pp. 172–180. Academic Press, New York.

Steinemann, C., and Parish, R. W. (1980). *Nature (London)* **286**, 621–623.

Taylor, D. L., and Fechheimer, M. (1982). *In* "Calmodulin and Intracellular Ca+ + Receptors" (S. Kakuichi, H. Hidaka, and A. R. Means, eds.), pp. 349–373. Plenum, New York.

Towbin, H., Stahelin, T., and Gordon, J. (1979). *Proc. Natl. Acad. Sci. U.S.A.* **76**, 4350–4354.

Van Haastert, P. J. M., De Wit, R. J. W., and Konijn, T. M. (1982). *Exp. Cell Res.* **140**, 453–456.

Virtanen, I., Lehto, V.-P., and Aula, P. (1981). *In* "Lectins: Biology Biochemistry, Clinical Biochemistry" (T. C. Bøg-Hansen, ed.), Vol. 1, pp. 215–220. De Gruyter, New York.

Vogel, G., Thilo, L., Schwarz, H., and Steinhart, R. (1980). *J. Cell Biol.* **86**, 456–465.

Walsh, F. S., Barber, B. H., and Crumpton, M. J. (1976). *Biochemistry* **15**, 3557–3563.

Weinstein, J. N., Yoshikami, S., Henkart, P., Blumenthal, R., and Hagins, W. A. (1977). *Science* **195**, 489–492.

West, C. M., and McMahon, D. (1979). *Exp. Cell Res.* **124**, 393–401.

Wilhelms, O.-H., Lüderitz, O., Westphal, O., and Gerisch, G. (1974). *Eur. J. Biochem.* **48**, 89–101.

Woolley, D. W., and Longsworth, L. G. (1942). *J. Biol. Chem.* **142**, 285–290.

Zachowski, A., and Paraf, A. (1974). *Biochem. Biophys. Res. Commun.* **57**, 787–792.

Chapter 6

Endocytosis and Recognition Mechanisms in Dictyostelium discoideum

GÜNTER VOGEL

Bergische Universität, Gesamthochschule Wuppertal
Fachbereich 9-Naturwissenschaften II, Biochemie
D-5600 Wuppertal 1, Federal Republic of Germany

I. Introduction

Endocytosis is the uptake of fluid (pinocytosis) or particles (phagocytosis) by a eukaryotic cell from the extracellular environment into the cytoplasm via plasma membrane-derived vesicles. The first step in phagocytosis involves recognition and binding of a particle to the cell surface. Binding seems to trigger a transmembrane signal which leads to extension of pseudopodia and subsequent internalization by membrane fusion (for reviews, see Silverstein *et al.*, 1977; Besterman and Low, 1983). *Dictyostelium discoideum* is particularly well suited as an experimental system to study mechanisms of phagocytosis and cell recognition. These amoebae are professional phagocytes and can be grown in large quantities for biochemical studies. Classical genetic methods are established and make this organism particularly attractive (see Godfrey and Sussman, 1982; Newell, 1982; Loomis, Chap. 3, this volume, for recent reviews).

The major experimental problem in measuring initial rates of endocytosis is to separate cells containing ingested material from the bulk of noningested material. This problem is overcome by centrifuging the cell suspension through a column of highly viscous polyethylene glycol. Extracellular fluid and small particles with a high surface/volume ratio remain on top of the column, whereas the large phagocytes are found on the bottom. Fluorescently labeled dextrans are found to be suitable fluid-phase markers to measure pinocytosis (Thilo and Vogel, 1980; Berlin and Oliver, 1980). Various types of synthetic latex beads or fluorescently

METHODS IN CELL BIOLOGY, VOL. 28

labeled bacteria can serve as test particles for measuring phagocytosis. All of these endocytotic substrates are commercially available or easy to prepare, and all of them can be quantified by optical methods.

Isolation of mutants with altered phagocytotic properties can help to dissect the complex process of phagocytosis into individual steps. Mutants of *D. discoideum* selected for changes in motility which are consequently impaired in phagocytosis have been reported (Kayman *et al.*, 1982; Clarke and Kayman, Chap. 8, this volume). Another selection procedure to isolate mutants with altered phagocytotic properties is based on the fact that wild-type cells engulf heavy tungsten beads and nonphagocytosing mutants can be separated on the basis of their lower density (Vogel *et al.*, 1980). Further methods to isolate mutants with altered recognition processes involved in phagocytosis and other cohesive properties of the amoebae are described here.

II.   Quantitation of Endocytotic Uptake

A.   Phagocytosis Assays

1.   TEST PARTICLES

Hydrophobicity, charge, and chemical composition of the surface of a particle determine its interaction with the phagocytotic cell. Therefore, various particles are used to study the influence of different surface properties on phagocytotic recognition. Fluorescein-labeled bacteria (FITC-bacteria) and different types of polystyrene latex beads are convenient test particles which can be sensitively quantified by optical methods.

FITC-bacteria are prepared by suspending exponentially grown bacteria in 50 mM sodium phosphate, pH 9.2, (OD_{420} = 20) and adding 0.1 mg/ml of fluorescein isothiocyanate (Isomer 1, Sigma) from a stock solution (10 mg/ml). After incubation for 3 hours at 37°C on a rotary shaker (100 rpm), bacteria are collected and washed by centrifuging at 5000 g for 10 minutes in 17 mM potassium phosphate, pH 6.0 (standard buffer), until no fluorescence is detectable in the supernatant. Under these conditions, the dye/bacteria ratio is similar in different batches of cells. Routinely, we use a B/r strain of *Escherichia coli* but other strains of *E. coli* work equally well.

Hydrophobic polystyrene latex beads and more hydrophilic carboxylated and amino-containing latex beads are commercially available in different sizes (Polysciences). Amino-containing latex beads can be readily coupled with various ligands. For example, glucosylated particles can be prepared by coupling 1-thio-D-glucose using the bifunctional reagent *N*-succinimidyl-3-(2-pyridyldithio)propionate (SPDP, Pharmacia) (Vogel, 1983).

2. PHAGOCYTOSIS IN SHAKEN SUSPENSIONS

By incubation of samples on a rotary shaker, high shear forces are generated by shaking, and therefore strong adhesion of test particles to the cells is a prerequisite for successful engulfment. Amoebae are suspended at a density of $2–5 \times 10^6$ cells/ml in the chosen medium and incubated on a rotary shaker (routinely at 100 rpm). Test particles are added at a ratio of particles to amoebae of $\sim$200:1. Phagocytosis is stopped by diluting 1-ml aliquots at various times into 2 ml of ice-cold standard buffer. To separate amoebae from noningested particles, the cell suspension is layered over 10 ml of an aqueous solution of polyethylene glycol ($M_r = 6000$, Serva) 20% (w/w) in a centrifuge tube and centrifuged for 10 minutes at 2000 g in a swing-out rotor. Noningested particles are mainly contained in the top fluid layer. The supernatant is carefully suctioned off using a Pasteur pipet, the tip of which has been bent and to which a small piece of silicon tubing has been attached. This ensures the thorough removal of material sticking to the glass surface, a problem which occurs sometimes when using hydrophobic latex particles. Pelleted amoebae are washed once by resuspension and centrifuging in 3 ml of 50 mM sodium phosphate, pH 9.2, and finally resuspended in 3 ml of the same buffer. The cell number is determined with an electronic particle counter (e.g., Model Z_{BI}, Coulter Electronics) or using a hemocytometer. After counting, cells are lysed by addition of Triton X-100 (0.2% final concentration).

For FITC-bacteria as substrate particles, the fluorescence intensity of the solution is determined by a fluorescence spectrophotometer using excitation and emission wavelengths of 470 and 520 nm, respectively. The number of bacteria phagocytosed is determined by comparison with a standard curve. The standard curve is constructed by lysing a defined number of bacteria in a 1% sodium dodecyl sulfate solution with heating for 2 minutes at 90°C. The fluorescence intensity is determined in aliquots of this solution diluted in 50 mM sodium phosphate, pH 9.2. The additional treatment of bacteria for standardization is necessary, since noningested bacteria, in contrast to ingested ones, cannot be lysed with 0.2% Triton X-100. The treatment does not cause a change in quantum yield. Fluorescein fluorescence is strongly pH-sensitive, and it is therefore important to ensure that the pH is >9, a condition which gives constant and maximal dye fluorescence.

When using latex beads as test particles, the light scattering at 560 nm is measured and the number of ingested particles is determined by comparison with a standard curve.

Uptake of bacteria and latex beads can be determined simultaneously in the same batch of cells. After incubation as described above and lysis of amoebae with Triton X-100, the number of ingested latex beads is determined by measuring the light scattering. Subsequently, the beads are removed by centrifuging 5 minutes at 2000 g and the fluorescence of the supernatant is determined. This is possible because FITC-bacteria are lysed by treatment with Triton X-100 and the fluorescently labeled components remain in solution in the supernatant after centrifugation.

3. PHAGOCYTOSIS ON FILTERS

Uptake of particles can be measured in the absence of shear forces by incubation of cells on filters. Between 2 and 5×10^6 amoebae and 2×10^9 test particles are rapidly mixed in 2 ml of the medium chosen. The suspension is uniformly deposited on filters (AABP04700, $0.8\text{-}\mu\text{m}$ pore size, 47-mm diameter; Millipore), which are then rested on absorbent support pads presoaked with the same medium. Samples are incubated in Petri dishes at the desired temperature in a moist atmosphere.

After various times, cells are harvested by placing the filter in a centrifuge tube containing 4 ml of ice-cold medium and cells are resuspended by vigorous shaking. Subsequently, the procedures described above are followed.

Phagocytic uptake of the various test particles reaches a saturation level. Maximal rates of initial uptake are obtained at a ratio of particles to amoebae of ~200:1 in shaken cultures. Control incubations at 20°C yield negligible background levels for bacteria and hydrophilic latex beads. In contrast, some batches of hydrophobic latex beads yield comparatively high background levels. This is probably due to nonspecific adsorption rather than true ingestion. Therefore, bacteria and carboxylated latex beads are most suitable for routine phagocytosis assays.

B.   Pinocytosis Assay

Fluorescein-labeled dextrans (FITC-dextrans) of high molecular weight are convenient fluid-phase markers for measuring pinocytosis. They are commercially available (e.g., FD-70 S, Sigma).

Cells are suspended at a density of $2-5 \times 10^6$ cells/ml in the chosen medium and incubated on a rotary shaker at the appropriate temperature. FITC-dextran is added from a stock solution (20 mg/ml) to a final concentration of 2 mg/ml. To stop pinocytosis, samples of 1 or 2 ml are diluted 1:5 in ice-cold standard buffer.

Cells are collected by centrifuging at 200 g for 5 minutes and resuspended in 1 ml of the above buffer. For complete removal of external FITC-dextran, the cell suspension is centrifuged through a polyethylene glycol solution as described for the phagocytosis assay. Cells are washed and resuspended in 2 ml of 50 mM sodium phosphate, pH 9.2. Instead of the centrifugation through polyethylene glycol, amoebae can be washed twice by suspension and centrifugation in 8 ml of standard buffer, and finally suspended in 2 ml of 50 mM sodium phosphate, pH 9.2. The cell number is counted and subsequently, cells are lysed by addition of

Triton X-100 (0.2% final concentration). The fluorescence intensity is measured as described above and the pinocytosed volume is determined by comparison with a standard curve.

FITC-dextran is suitable as a fluid-phase marker to measure pinocytosis according to several criteria. It is easily available, nontoxic to the cells, and can be analyzed fluorimetrically in small amounts. Uptake is proportional to its concentration in the range between 0.5 and 20 mg/ml and proceeds linearly with time for at least 90 minutes. No uptake is observed at 2°C. Nonspecific binding is low and the marker is only seen in pinocytotic vesicles.

III.   Adhesion Assay

A total of 4×10^6 cells suspended in 4 ml of the desired medium are placed into tissue culture flasks (plane area 25 cm², volume 50 ml), shaken 10 minutes to allow the cells to recover after pipeting, and incubated for 40 minutes without shaking. Subsequently, the flasks are agitated for 3 minutes on a rotary shaker at 60 rpm, and the percentage of nonadherent cells is estimated by counting cells in the supernatant.

IV.   Aggregation Assay

Amoebae are harvested, washed once, and suspended in standard buffer to a final density of $3-5 \times 10^6$ cells/ml. Cells are incubated on a rotary shaker at 100 rpm, and at various times 0.1-ml samples are taken by use of a wide-bore plastic pipet tip to avoid disaggregation of cell agglutinates by shear forces. They are diluted 1:100 into the same buffer contained into Accuvette sample beakers, and aggregation is measured by determining the decline in the number of single cells with an electronic particle counter.

V.   Isolation of Mutants with Altered Cohesive Properties

A selection procedure for isolating mutants defective in phagocytosis has been devised using tungsten beads as particulate prey. Nonphagocytosing cells can be isolated on the basis of their lower density. Using this technique, a number of mutants have been found which are defective in their recognition of certain types of test particles. These mutants are also defective in adhesion to surfaces and in cell–cell cohesion of vegetative cells (Vogel et al., 1980). In addition, methods are described below to obtain nonadhesive and nonaggregating mutants, and revertants of these mutants.

A. Mutagenesis

Amoebae (strain Ax-2) are grown axenically to a density of 2–4 × 10⁶ cells/ ml, harvested by centrifugation, and washed twice with standard buffer. They are suspended to a concentration of ~3 × 10⁶ cells/ml in the above buffer, and the mutagen N-methyl-N'-nitro-N-nitrosoguanidine from a freshly prepared stock solution (50 mg/ml) in dimethyl sulfoxide is added to a final concentration of 1 mg/ml. Cells are incubated on a rotary shaker at 120 rpm at 20°C for 30 minutes. To stop the reaction, the cell suspension is chilled and cells are washed twice in standard buffer. Viability tests reveal variable survival rates usually in the range of 1–5%. Cells are resuspended in axenic medium at a concentration of ~10⁶ cells/ml and separated into different batches.

Amoebae require ~4 days for recovery and are subsequently grown to a cell density of about 2–4 × 10⁶ cells/ml. The selection procedures are carried out separately for each batch of cells to ensure that mutants from different batches are of independent origin. Mutagenesis is reviewed by Loomis in this volume (Chap. 3).

B. Isolation of Nonadhesive Mutants

Mutagenized cells are suspended at a density of ~2 × 10⁶ cells/ml in 10 ml of the medium desired and placed into Petri dishes (9 cm diameter) or into tissue culture flasks (plane area 75 cm²; volume 250 ml). After incubation at 20°C for 2 hours without agitation, the cell suspension is gently shaken and the supernatant containing nonadherent cells is removed. Cells in the supernatant are harvested by centrifugation in axenic medium and grown again to a density of 2 × 10⁶ cells/ml in 100-ml Erlenmeyer flasks. The selection procedure is repeated until most cells are found to be nonadhesive as examined by microscopic observation (10–15 cycles). Finally, cells are plated clonally on nutrient agar plates seeded with bacteria, and nonadhesive strains are stored for further characterization. This technique can be used to isolate variant strains with altered adhesive properties by choosing an appropriate selective surface, e.g., more hydrophobic Petri dishes or more hydrophilic tissue culture dishes. Furthermore, polyacrylamide gels derivatized with various mono- and disaccharides are described. Dictyostelium discoideum adhesion to these gels seems to be mediated by at least three different receptors specific for glucose, N-acetylglucosamine, and mannose (Bozarro and Roseman, 1983). These types of gels can be employed to isolate mutants defective in these receptors.

C. Isolation of Nonaggregating Mutants

After mutagenesis, cells are harvested and suspended in 20 ml of standard buffer in a 100-ml Erlenmeyer flask to a final density of $2-3 \times 10^6$ cells/ml. During incubation on a rotary shaker at 100 rpm for 40 minutes they form large agglutinates. The cell suspension is carefully transferred into a test tube using wide-bore glass pipets to avoid shear forces and incubated for 20 minutes without shaking. After this time most of the agglutinates are sedimented and mostly single cells remain in the supernatant. A 15-ml portion of the supernatant cell suspension is then removed, and cells are harvested by centrifugation and grown axenically to a density of $2-3 \times 10^6$ cells/ml. The procedure is repeated until only weak or no agglutination is observed (about 10–15 times). Remaining cells are cloned and analyzed as described above.

D. Isolation of Adhesive Revertants

Revertants are isolated in a similar way to that described for nonadhesive mutants. Mutant cells are suspended in 10 ml of axenic medium at a density of $2-3 \times 10^6$ cells/ml and placed into a tissue culture flask (plane area 75 cm^2, volume 250 ml). After incubation for 90 minutes without agitation, the cell suspension is shaken on a reciprocal shaker (140 strokes/min; stroke 2.5 cm) for 15 minutes, and the supernatant is discarded. The bottom of the flask is washed once by flushing with 10 ml of axenic medium. After addition of 10 ml of fresh medium, the remaining adherent cells are grown to a density of about $1-2 \times 10^6$ cells/ml by incubating the tissue culture flasks on a rotary shaker. The selection procedure is repeated until microscopic observation reveals that most of the cells adhere and spread like wild-type cells (about 15–20 cycles). Subsequently, cells are cloned and stored.

VI. Characterization of Mutants

Using the selection procedure with tungsten beads, a number of mutants have been found which are altered in cell–particle binding in phagocytosis and which are additionally impaired in cell adhesion and cell cohesion. The phenotype will be referred to as ''Phag.'' The same type of mutants is obtained using procedures to select for nonadhesive or nonaggregating cells. Detailed analysis of the mutant phenotype reveals the following differences in comparison to wild-type amoebae:

1. Hydrophilic particles such as protein-coated latex beads, amino-containing latex beads or bacteria without terminal glucose residues cannot be pha-

gocytosed by mutant cells, whereas ingestion by wild-type amoebae remains normal. However, hydrophobic latex beads are phagocytosed by mutant and wild-type amoebae equally well.

2. Glucosylated latex beads and bacteria containing terminal glucose residues (e.g., *E. coli* B/r) are phagocytosed by mutant cells normally. However, in contrast to wild-type cells, uptake is specifically and competitively inhibited by glucose and carbohydrates containing glucose on their nonreducing termini.

3. Vegetative mutant amoebae do not form EDTA-sensitive aggregates. Furthermore, wild-type cells adhere tightly and spread on the surface of tissue culture plates, whereas mutant cells adhere only loosely or remain in suspension.

Revertants selected for the ability to adhere on plastic surfaces again ingest the various test particles like wild-type amoebae and form EDTA-sensitive aggregates. The code *phg* is given to the loci determining the phenotype described. Genetic analysis shows that there are at least two recessive loci, *phgA* assigned to linkage group VII and *phgB* assigned to linkage group IV. Both lead to the expression of the mutant phenotype. As shown by complementation analysis, a further 12 independently derived phagocytosis mutants fell into one of these two complementation groups (Duffy and Vogel, 1984). Genetic analysis of the adhesive revertants reveals that these carry extragenic suppressor mutations which are dominant and map into linkage group IV but are discrete from the *phgB* locus, which is also on linkage group IV (K. T. I. Duffy, unpublished).

VII. Conclusions and Discussion

Analysis of the Phag mutant phenotype reveals the presence of at least two membrane receptors which recognize different features of a foreign surface. A "lectin-type" receptor recognizes glucose residues and a second nonspecific receptor recognizes the relative hydrophobicity of a surface, which may be a phagocytosable particle, a solid surface, or the surface of another amoeba (see van Oss, 1978, for review). Mutants defective in the nonspecific receptor can be isolated using various selection procedures, and all of the mutants independently derived display a similar phenotype. They are impaired in cohesive properties such as particle binding, cell adhesion, and cell agglutination. It seems likely that a common structural basis exists for these properties. This is confirmed by the selection of revertants adhesive to plastic surfaces again. All of them redisplay the wild-type phenotype for phagocytosis and EDTA-sensitive cell agglutination (Vogel, 1983). However, it has been reported that a mutant HL 220, which lacks a modification of several membrane proteins (Murray *et al.*, 1984) and does not form vegetative cell agglutinates, shows strengthened adhesion to glass and Teflon surfaces via uropods (Gerisch *et al.*, 1985). Different experimental conditions could be the reason for this discrepancy, or possibly there exist subtle differences between cell agglutination and adhesion to solid surfaces not detected in our system.

Since genetically distinguishable loci are involved in the expression of the Phag phenotype, it is likely that the nonspecific receptor is not a single molecule but is composed of several components which could form "patches" in the membrane to mediate hydrophobic interactions. The strength of hydrophobic binding in this case would be dependent on the size of these patches, which in turn is determined by lateral weak interactions between these components. A model of this kind explains the fact that mutants with different adhesive strengths can be isolated.

REFERENCES

Berlin, R. D., and Oliver, J. M. (1980). *J. Cell Biol.* **85,** 660–671.

Besterman, J. M., and Low, R. B. (1983). *Biochem. J.* **210,** 1–13.

Bozarro, S., and Roseman, S. (1983). *J. Biol. Chem.* **258,** 13882–13889.

Duffy, K. T. I., and Vogel, G. (1984). *J. Gen. Microbiol.* **130,** 2071–2077.

Gerisch, G., Weinhardt, U. Z., Berthold, G., Claviez, M., and Stadler, J. (1985). *J. Cell Sci.* **73,** 49–68.

Godfrey, S., and Sussman, M. (1982). *Annu. Rev. Genet.* **16,** 385–404.

Kayman, S. C., Reichel, M., and Clarke, M. (1982). *J. Cell Biol.* **92,** 705–711.

Murray, B. A., Wheeler, S., Jongens, T., and Loomis, W. F. (1984). *Mol. Cell. Biol.* **4,** 514–519.

Newell, P. (1982). *In* "The Development of *Dictyostelium discoideum*" (W. F. Loomis, ed.), pp. 35–70. Academic Press, New York.

Silverstein, S. C., Steinmanm, R. M., and Cohn, Z. A. (1977). *Annu. Rev. Biochem.* **46,** 669–722.

Thilo, L., and Vogel, G. (1980). *Proc. Natl. Acad. Sci. U.S.A.* **77,** 1015–1019.

Van Oss, C. J. (1978). *Annu. Rev. Microbiol.* **32,** 19–39.

Vogel, G. (1983). *In* "Methods in Enzymology" (S. Fleischer and B. Fleischer, eds.), Vol. 98, pp. 421–430. Academic Press, New York.

Vogel, G., Thilo, L., Schwarz, H., and Steinhart, R. (1980). *J. Cell Biol.* **86,** 456–465.

Chapter 7

Defining the Intracellular Localization Pathways followed by Lysosomal Enzymes in Dictyostelium discoideum

JAMES A. CARDELLI

Department of Microbiology and Immunology
Louisiana State University Medical Center
Shreveport, Louisiana 71130

GEORGE S. GOLUMBESKI,[1] NANCY A. WOYCHIK, DAVID L. EBERT, ROBERT C. MIERENDORF,[2] AND RANDALL L. DIMOND[2]

Department of Bacteriology
University of Wisconsin
Madison, Wisconsin 53706

I. Introduction

We are using genetic and biochemical approaches to define the role of proteolytic processing and posttranslational modification in the proper localization of lysosomal enzymes in *Dictyostelium discoideum*. *Dictyostelium* lacks phosphomannosyl receptors (J. A. Cardelli, unpublished results) which in many cells bind mannose 6-phosphate residues attached to lysosomal enzymes and mediate the transport of these proteins to lysosomes (reviewed in Sly and Fischer, 1982). Similarly, certain cells in patients with I-cell disease contain normal amounts of lysosomal enzymes, even though the cells are deficient in *N*-acetylglucosaminyl phosphotransferase, the enzyme which covalently attaches phosphate to proteins

[1]Present address: Department of Molecular, Cellular, and Developmental Biology, University of Colorado, Boulder, Colorado 80309.
[2]Present address: Promega-Biotec, Madison, Wisconsin 53711.

METHODS IN CELL BIOLOGY, VOL. 28

destined to arrive in lysosomes. Thus, our studies using *Dictyostelium* may provide general information concerning the molecular mechanisms involved in the localization of lysosomal enzymes by a phosphomannosyl receptor-independent pathway.

Dictyostelium represents an excellent system in which to investigate the biosynthesis and function of lysosomal enzymes (reviewed in Cardelli and Dimond, 1987). The majority of vesicles in growing cells are thought to be part of an extensive lysosomal system (Dimond *et al.*, 1981), and many of the lysosomal enzymes are 10- to 1000-fold higher in concentration than in animal cells. This has facilitated the purification of these enzymes (Mierendorf *et al.*, 1983; Knecht and Dimond, 1981) and the production of enzyme-specific monoclonal antibodies (Mierendorf *et al.*, 1983). Finally, and most importantly, the haploid nature of the genome simplifies the isolation and characterization of mutants defective in lysosomal function. Specifically, we have isolated and are currently studying mutants affected in the modification (Knecht *et al.*, 1984; Freeze and Miller, 1980), localization (Woychik *et al.*, 1986), processing (Livi *et al.*, 1985a), and secretion (Dimond *et al.*, 1983) of lysosomal enzymes. In this chapter, we describe the biochemical and genetic methods currently being used to define both the intracellular pathways followed by lysosomal enzymes and the molecular nature of the "sorting" signals found on these proteins.

II.   Pulse–Chase Labeling Analyses of Lysosomal Enzymes

The lysosomal enzymes α-mannosidase (α-man) and β-glucosidase (β-glu) have been purified to homogeneity, and monoclonal antibodies specific for each enzyme have been generated (Mierendorf *et al.*, 1983; Golumbeski and Dimond, 1986). Purified α-man is a heterotetramer consisting of subunits of 60K and 58K, while β-glu is a dimer consisting of subunits of 100K. Both enzymes are synthesized as precursor polypeptides (140K for α-man and 105K for β-glu) which are proteolytically processed to the mature forms (Mierendorf *et al.*, 1983; Pannell *et al.*, 1982; Cardelli *et al.*, 1986; Golumbeski, unpublished results). The following pulse–chase labeling protocol has been used to determine the rate at which these precursors are processed (Mierendorf *et al.*, 1983). The wild-type parental strain Ax-3 is grown axenically in TM media (doubling time of 8 hours) at 21°C on a rotary shaker at 200 rpm. TM medium consists of 10 g trypticase, 5 g of yeast extract, 1 g glucose, 0.35 g Na_2HPO_4, and 1.2 g KH_2PO_4 per liter of distilled H_2O. Logarithmically growing cells are collected by centrifugation (1000 g for 3 minutes) and resuspended at 10^7 cells/ml in fresh TM medium containing 0.5–1.0 mCi/ml ^{35}S-Met (Amersham, Arlington Heights, IL, 1200 Ci/mmol). After incubation at 21°C for 10–15 minutes, the cells are quickly collected by centrifugation in a Beckman microfuge and resuspended at 10^7 cells/ml in fresh TM (chase). The ^{35}S-Met-containing TM medium is stored at $-70°C$ and can be reused up to five times with little decrease in the efficiency of incorporation of ^{35}S-Met into cellular proteins. At various times during the chase period, 10^7 cells (1 ml) are collected by centrifugation and resuspended in 200 μl of 0.5% Triton X-100. The extracellular medium is also adjusted to 0.5% Triton X-100 by the addition of 50 μl of 10% Triton X-100. At this time samples can be further processed or frozen at $-20°C$.

Labeled α-man and β-glu polypeptides are immunoprecipitated with specific monoclonal antibodies as follows (Mierendorf *et al.*, 1983), with all procedures being carried out at 4°C. Samples are incubated for 15 minutes with 50 μl washed Pansorbin (Calbiochem-Behringer Corp., San Diego, CA), following the addition of 1:5 volume of 5 × C (250 mM Tris, pH 7.6, 25 mM Na$_2$EDTA, 750 mM NaCl, 10 mM methionine, 5 mM Na azide, and 2.5% NP-40). Pansorbin is washed by centrifugation (10,000 g × 3 minutes) and resuspension in an equal volume of 1 × C. Following incubation, samples are centrifuged at 10,000 g for 3 minutes in a Beckman microfuge, supernatants are transferred to new microfuge tubes, and excess monoclonal antibodies specific for α-man and β-glu are added. After a 60- to 90-minute incubation, 100 μl of Pansorbin is added for another 60 minutes. The Pansorbin–antibody complex is collected by centrifugation and washed three times in 1 × C. The final pellet is resuspended in 50 μl of gel sample buffer (125 mM Tris-HCl, pH 6.8, 2% SDS, 10% β-mercaptoethanol plus 15% sucrose) and heated to 80°C for 3 minutes. Samples are centrifuged (10,000 g for 3 minutes), and the eluted proteins are subjected to SDS–PAGE (7.5% acrylamide) according to the method of Laemmli (1970). Following electrophoresis, gels are fixed in 10% TCA, treated with Enhance following manufacturer's instructions (New England Nuclear, Boston, MA), dried under a vacuum, and exposed to Kodak XAR-5 film at $-70°C$. Using the procedures we described in this section, we have determined that the majority of the 140K α-man precursor begins to be proteolytically processed 20–25 minutes after synthesis and is completely processed to mature forms by 60 minutes (Mierendorf *et al.*, 1983). In contrast, the 105K β-glu precursor begins to be processed within 10 minutes and by 30 minutes all of the precursor has been converted to the mature polypeptides (Golumbeski and Dimond, 1986; Cardelli *et al.*, 1986). A small percentage of both precursors are secreted from cells.

III. Subcellular Fractionation

The following procedure was developed to aid in determining the subcellular organelles involved in the localization of lysosomal enzymes (Mierendorf *et al.*,

1985). Growing cells are harvested by centrifugation (1000 *g* for 3 minutes), washed once in a solution containing 50 m*M* Tris-HCl, pH 7.6, 25 m*M* KCl, 5 m*M* MgCl$_2$ plus 0.25 *M* sucrose (0.25 *M* STKM), and resuspended to 2–4 × 10^8 cells/ml in the same solution. The following steps are all performed at 0°–4°C. When radioactively labeled Ax-3 or mutant cells are to be fractionated (usually a total of 1–5 × 10^7 cells), a 10- to 100-fold excess of the α-man and β-glu structural gene mutant GM1 is added to facilitate handling. This mutant lacks any detectable α-man and β-glu protein. Resuspended cells are disrupted by 15–20 strokes in a tight-fitting Dounce and centrifuged at 1000 *g* for 5 minutes to remove nuclei and unbroken cells. The supernatant is carefully removed and transferred to a new tube while the pellet is resuspended in one-third of the original volume of 0.25 *M* STKM, followed by homogenization and centrifugation as described above. This procedure consistently results in the breakage of 70–80% of the cells. Pooled postnuclear supernatants are carefully layered on 7.0 ml of 0.5 *M* STKM containing a 0.25-ml cushion of 2.0 *M* STKM and centrifuged on an SW40 rotor at 38,000 rpm (182,000 *g*) for 45 minutes at 4°C. Following centrifugation, the contents of the tubes are carefully withdrawn from the top down to the crude membranes which have pelleted on top of the 2.0 *M* STKM cushion. The membranes and cushion are resuspended by gentle vortexing in 1.8 ml TKM, which adjusts the sucrose concentration to ~0.25 *M*. The membrane sample (~2 ml) is then layered on a discontinuous sucrose gradient consisting of 2 ml each of 1.8 *M* STKM, 1.5 *M* STKM, 1.3 *M* STKM, 1.1 *M* STKM, 0.8 *M* STKM, and 1 ml of 0.5 *M* STKM. Following centrifugation at 38,000 rpm (182,000 *g*) for 2.5 hours on an SW40 rotor, fractions are collected from top and either assayed for membrane marker enzymes (see below) or diluted with an equal volume of TKM. Diluted samples are centrifuged in a 50 Ti rotor at 40,000 rpm (105,000 *g*) for 45 minutes and the resultant pellets are resuspended in 0.5 ml 1.0% Triton X-100. Samples are either immunoprecipitated as described in Section II or they are frozen at −20°C. The lysosomal marker enzymes that are assayed are *N*-acetylglucosaminidase and α-mannosidase. The endoplasmic reticulum (ER) marker enzymes assayed are α-glucosidase-2 (α-glu-2) and NADPH-cytochrome *c* reductase. See Mierendorf *et al.* (1985) for details and references concerning the enzyme assays. A schematic flowchart describing the subcellular fractionation procedure is shown in Fig. 1 (for additional information regarding fractionation of *Dictyostelium* membranes, see Goodloe-Holland and Luna, this volume, Chapter 5).

A photograph of an SW40 centrifuge tube containing membranes fractionated as described above is shown in Fig. 2. Four distinct membrane bands are observed at the interfaces between the various sucrose cushions. Shown in Fig. 3A are the distribution profiles on these gradients of the enzymes α-man, α-glu-2, and NADPH-cyt *c* reductase. Lysosomal vesicles (α-man activity) are distributed throughout the gradient but are primarily concentrated at the 1.3 *M*/1.5 *M* sucrose cushion interphase. In contrast, the peak of the ER marker enzyme

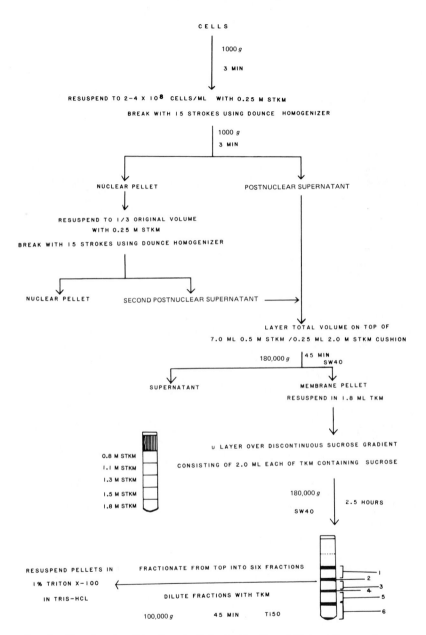

FIG. 1. Flowchart describing the subcellular fractionation of *Dictyostelium* membranes.

– EDTA   **+EDTA**

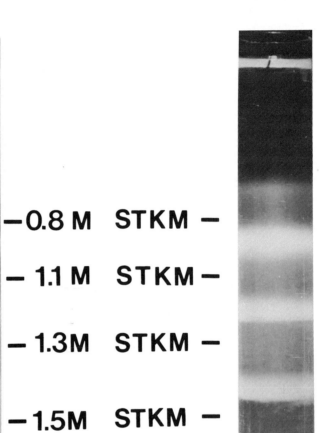

–0.8 M STKM –

– 1.1 M STKM –

– 1.3M STKM –

–1.5M STKM –

–1.8M STKM –

FIG. 2. Photograph of a centrifuge tube containing fractionated membranes. Postnuclear supernatant membranes were fractionated on discontinuous sucrose gradients as described in Section III. One tube contains membranes treated with 10 mM EDTA prior to centrifugation.

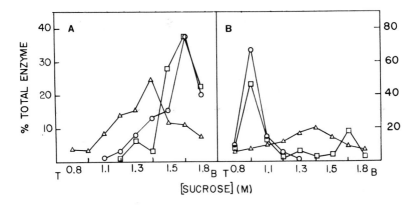

FIG. 3.   Fractionation of cytoplasmic membranes on discontinuous sucrose gradients. Growing cells were broken using a Dounce homogenizer followed by fractionation of the postnuclear membrane vesicles as described in Section III. After centrifugation, the gradients were fractionated from the top and samples assayed for the enzymes NADPH-cytochrome c reductase (□), α-glucosidase-2 (○), and α-mannosidase (△). Panel B represents membranes treated with 10 mM EDTA prior to centrifugation. T and B indicate the top and bottom of the gradient, and the interfaces between different sucrose concentrations are denoted by the dividing marks.

activity (α-glu-2 and NADPH-cyt c reductase) is found in the 1.5 M and 1.8 M sucrose cushions. In fact, the vast majority of the ER membranes in *Dictyostelium* sediment into or through the 1.5 M sucrose and are therefore most likely associated with dense ribosomes (Fig. 3A). Consistent with this are the results presented in Fig. 3B. Prior to centrifugation, membranes were incubated for 15 minutes with 10 mM EDTA, a Mg^{2+}-chelating agent which affects the removal of ribosomes from rough ER membranes. Following EDTA treatment, the majority of the ER membranes now sediment as a band at the 1.1 M sucrose interphase consistent with the removal of dense ribosomes. Lysosomes, however, are relatively unaffected by EDTA treatment as evidenced by little change in their distribution on sucrose gradients.

   None of the standard Golgi membrane marker enzymes, such as galactosyl transferase, are detectable in extracts prepared from axenically growing cells. Consistent with this is the finding that *Dictyostelium* glycoproteins lack sugars such as galactose and sialic acid which comprise the complex N-linked oligosaccharide structure attached to animal cell glycoproteins. However, lysosomal enzymes in *Dictyostelium* are phosphorylated and sulfated on their N-linked oligosaccharide side chains (Freeze *et al.*, 1983a; Mierendorf *et al.*, 1985; Cladaras and Kaplan, 1984) by enzymes which presumably reside in the Golgi (Sly and Fischer, 1982; Green *et al.*, 1983). Furthermore, >90% of the sulfated α-man precursor is located in the low-density region of the gradient following subcellular fractionation of cells pulse-labeled with $^{35}SO_4$ (Mierendorf *et al.*, 1985). This suggests that membranes which accumulate at the 1.1 M sucrose

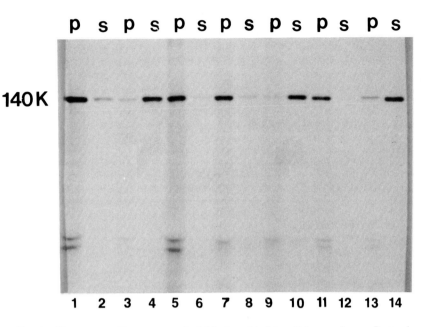

FIG. 4.   The α-mannosidase precursor is tightly bound to intracellular membranes. Postnuclear supernatants were prepared as described in Section III from cells pulse-labeled with ^{35}S-Met for 30 minutes. The supernatants were adjusted to various salt and H$^+$ ion concentrations by the addition of stock solutions. Following a 30-minute incubation on ice, membranes (p) were recovered by centrifugation (105,000 *g* for 30 minutes) and resuspended in 0.5% Triton X-100. Membranes and supernatants (s) were immunoprecipitated and subjected to SDS–PAGE followed by fluorography. Lanes 1–4 contain immunoprecipitated α-mannosidase from untreated (control) extracts. The other lanes contain α-mannosidase immunoprecipitated from postnuclear supernatants adjusted to the following concentrations: lanes 5–6, 0.5 *M* KCl; lanes 7–10, 0.5 *M* KI; lanes 11–14, 10 m*M* acetic acid, pH 3.0. Samples shown in lanes 3, 4, 9, 10, 13, and 14 were adjusted to 0.5% Triton X-100 prior to centrifugation to determine if the α-mannosidase precursor polypeptides precipitated out of solution following incubation under the conditions described above.

interface (Fig. 2) and contain the labeled precursor are enriched in sulfotransferase activity, indicating they are most likely part of the Golgi complex.

Recoveries of marker enzyme activities from disrupted cells fractionated on sucrose gradients are usually >75%. Most of the lost enzyme activity is recovered in the nuclear pellet. Increasing the centrifugation time during subcellular fractionation to 5 hours has little effect on the distribution of marker enzymes across the gradient, suggesting that organelles are near equilibrium densities after 3 hours of centrifugation.

Greater than 90% of the mature lysosomal enzymes remain in the supernatant following centrifugation of extracts prepared from cells homogenized under hypotonic conditions (Mierendorf *et al.*, 1985). For these experiments, cells are disrupted and fractionated in H_2O as opposed to 0.25 M STKM as described above. This result suggests that these enzymes are soluble in the lumen of lysosomal vesicles which burst under hypotonic conditions. In contrast, >90% of the ^{35}S-Met pulse-labeled α-man and β-glu precursor polypeptides sediment with membranes prepared from hypotonically disrupted cells (Cardelli *et al.*, 1986; Mierendorf *et al.*, 1985). Furthermore, the α-man precursor remains membrane-associated following treatment for 30 minutes at 0°C with 0.5 M KCl, 0.5 M KI, and 10 mM acetic acid, pH 3.0 (Fig. 4). These preliminary results suggest that the α-man and β-glu precursor polypeptides may interact with the ER and Golgi lipid bilayer as integral membrane proteins until they are proteolytically processed and released into the lumen of lysosomes.

IV. Synthesis *in Vitro* of Lysosomal Enzymes

The cellular precursor and mature forms of α-man and β-glu are extensively modified glycoproteins (Freeze *et al.*, 1983a; Mierendorf *et al.*, 1985; Cardelli *et al.*, 1986). To determine the initial modification events involved in their synthesis as well as the identity of the primary translation products, mRNA is translated *in vitro* in the presence and absence of dog pancreas microsomes. Translation mixtures contain a final volume of 28–30 μl, 50% (v/v) micrococcal nuclease (P.L. Biochemicals, Milwaukee, WI)-treated rabbit reticulocyte lysate, 1.25 mM ATP, 0.25 mM GTP, 85 mM KCl, 1.3 mM Mg acetate, 0.12 mM of 19 amino acids minus Met, 14 mM creatine phosphate, 1.5 μg creatine phosphokinase, 1.0 mCi/ml ^{35}S-Met, 1 unit RNasin (Promega-Biotec, Madison, WI), and 1–5 μg RNA. The untreated rabbit reticulocyte lysate is prepared as described in Housman *et al.* (1970) using serum from phenylhydrazine-treated rabbits. The lysate is divided into 250-μl aliquots and frozen at −70°C. Lysate is also available commercially from Amersham. Details for the purification of cytoplasmic RNA, membrane-bound polysomal RNA, and free polysomal RNA have been published (Cardelli *et al.*, 1981; Cardelli and Dimond, 1981). In brief, the procedure involves multiple phenol–chloroform extractions of cells or polysomes followed by ethanol precipitation. The precipitated RNA is dried and resuspended to 1 mg/ml in double-distilled H_2O.

Lysates are prepared for translation as follows. Hemin (7.5 μl at 4 mg/ml) and creatine phosphokinase (5 μl at 5 mg/ml) are added to one tube (250 μl) of frozen lysate, which is placed on ice to thaw. When the lysate has thawed, 3.75 μl 0.1 M $CaCl_2$ are added, followed by 3 μl of micrococcal nuclease (12,000 units/ml in 50 mM glycine, pH 9.2, 5 mM $CaCl_2$, 50% glycerol), and the tube is incubated at 21°C for 5 minutes. Following incubation, 3.75 μl of 0.2 M EDTA is added. This chelates the Ca^{2+} and inactivates the Ca^{2+}-dependent micrococcal nuclease. Finally, 25 µl of energy mix is added to the contents of the tube, which is then gently vortexed. Energy mix consists of 24.9 mM ATP, 4.9 mM GTP, 0.28 mM creatine phosphate, 1.7 M KCl, 34.2 mM Mg acetate, and 2.3 mM of 19 amino acids minus methionine. To initiate a typical translation reaction, the following are added to a 250-µl microfuge tube: 8.0 µl H$_2$O, 5.0 µl RNA (1 mg/ml), 1.0 µl RNasin (Promega-Biotech, Madison, WI), 14 µl prepared lysate, and 1.5 µl ^{35}S-Met (20 µCi). Tubes are gently mixed by flicking with a finger and incubated at 30°C for 60 minutes. Reactions can be supplemented with 1–3 µl dog pancreas microsomes (2 A_{260}/ml) prepared as described (Shields and Blobel, 1978). Microsomes are also available from Amersham.

Labeled α-man and β-glu polypeptides are immunoprecipitated from translation mixture as described in Section II. In axenically growing cells, α-man and β-glu mRNAs represent <0.1% of the total amount of mRNA. Consequently, a number of nonspecific labeled proteins often contaminate immunoprecipitates of the labeled α-man and β-glu polypeptides. To reduce the level of contaminating proteins, 30 µl of translation mixture are adjusted to 0.5% SDS by the addition of 1.5 µl of 10% SDS. Tubes are heated at 55°C for 3 minutes, diluted by the addition of 500 µl of 1 × C buffer, and incubated with antibodies as described in Section II.

By *in vitro* translation, we have demonstrated that the primary translation product of the α-man and β-glu mRNAs are 120K and 94K polypeptides, respectively (Fig. 5). When these mRNAs, which are predominantly associated with membrane-bound polysomes in growing cells (Cardelli *et al.*, 1986), are translated in the presence of dog pancreas microsomes, new polypeptides of 140K (α-man) and 105K (β-glu) are generated. These proteins are not synthesized if microsomes are added posttranslationally. Furthermore, these *in vitro*-modified polypeptides are identical in molecular weight to the authentic cellular precursor polypeptides. When the integrity of the microsomal vesicles are maintained, only the *in vitro*-modified forms of the precursor are protected from trypsin digestion (Fig. 5). This suggests that the modified proteins reside on the inside of the microsomal vesicle and that they are synthesized in a manner similar to the vectorial discharge of mammalian secretory proteins.

To determine whether the change in apparent weight of the precursor polypeptides dependent on dog pancreas microsomes is due to N-glycosylation, the 140K and 105K species generated *in vitro* are treated with endoglycosidase H (endo H, Miles, Elkhart, IN) as follows. The washed immunoprecipitates, prepared as described in Section II, are resuspended in 20 µl of 2% SDS, 10% 2-mercaptoethanol and heated for 3 minutes at 80°C. The samples are centrifuged at 10,000 g for 3 minutes and the supernatants mixed with an equal volume of 100 mM Na citrate, pH 5.5. Following the addition of 2 µl of endo H (1 unit/ml in 50 mM phosphate buffer, pH 5.5), samples are incubated at 37°C for 16 hours. After incubation, samples (40 µl) are mixed with 40 µl of 2× gel sample buffer

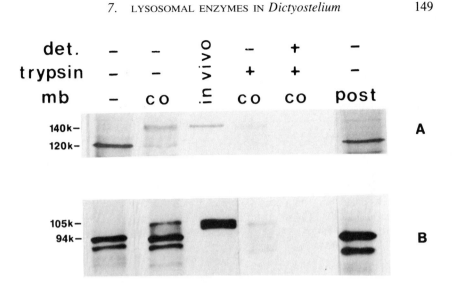

Fig. 5. Cotranslational translocation and modification of α-mannosidase and β-glucosidase by dog pancreas microsomal membranes. RNA prepared from growing cells was translated *in vitro* in the absence or presence (co) of 2 A_{260} units of microsomal membranes (mb). Following translation, the indicated samples (30 μl) received 7.5 μg trypsin, 7.5 μg trypsin plus 1.5 μl 10% Triton X-100 (det), or 2 A_{260} units of membranes (post). Following a 1-hour incubation on ice, samples containing trypsin received 75 μg of $α_1$-antitrypsin. α-Mannosidase (A) and β-glucosidase (B) polypeptides were immunoprecipitated as described in Section II and subjected to SDS–PAGE followed by fluorography. One lane contains immunoprecipitates prepared from cells pulse-labeled for 30 minutes with ^{35}S-Met.

and subjected to gel electrophoresis as described in Section II. Treatment in this manner of the immunopurified *in vitro*-modified and cellular precursor polypeptides generates polypeptides similar in molecular weight to the primary translation products of the α-man and β-glu mRNAs. Thus, these results suggest that lysosomal enzymes in *Dictyostelium* are synthesized as precursor polypeptides on membrane-bound polysomes, cotranslationally translocated into the lumen of the ER, and modified by the addition of N-linked oligosaccharide side chains (Cardelli *et al.*, 1986).

V.   Isolation and Characterization of Mutants Altered in the Modification, Processing, and Secretion of Lysosomal Enzymes

A genetic approach is potentially the most powerful method one can use in defining the molecular mechanisms involved in the intracellular lysosomal transport pathway. By direct screening, we have isolated mutants altered in the secretion and/or accumulation of lysosomal enzymes (reviewed in Cardelli and Dimond, 1987). Details concerning the methods used to isolate these mutants can be found in published references (Livi et al., 1985a; Dimond et al., 1983). Briefly, cells are grown in TM medium to titer of $1-2 \times 10^6$ cells/ml and mutagenized with N-methyl-N'-nitro-N-nitrosoguanidine as described (Dimond et al., 1983). Freshly mutagenized cells are transferred to 96-well tissue culture trays which are incubated at 21°C for 3–4 weeks. Following growth, master trays are replica-plated to new trays containing growth medium. After growth, replicated clones are assayed for the presence of a particular lysosomal enzyme in the extracellular medium (Livi et al., 1985a). Following enzyme assays, half of the contents of each well containing potential regulatory or secretory mutants are diluted into 2 ml of TM medium in a test tube and incubated on a reciprocating shaker at 21°C. The other half of the well is plated in association with Klebsiella pneumoniae on nutrient agar plates. These primary screens recognize both regulatory and structural gene mutants as well as secretory mutants. The 2-ml TM broth cultures are used to identify secretory mutants in a secondary screen. After growth, cells are separated from the medium by centrifugation and both fractions assayed for particular lysosomal enzymes. Strains differing by >5 in the ratio of extracellular to cellular activity when compared to wild type are selected for further study.

Using these methods, nine mutants have been isolated that do not accumulate α-man enzyme during axenic growth and development. One of these mutants, HMW 404, is blocked very early in development and does not accumulate α-man mRNA (Livi et al., 1985b) or a number of other developmentally regulated mRNAs (Cardelli et al., 1985). The other eight mutants synthesize the 140K α-man precursor polypeptide at rates ranging from 7 to 70% that of wild-type cells but do not accumulate >1% the normal levels of α-man activity (Livi et al., 1985a). Four of these mutants do not proteolytically process the α-man precursor, which may explain why they fail to accumulate functional enzyme. One of these strains, HMW 437 (Fig. 6), has been more fully characterized using the methods described in Section II. This strain, which carries a mutation in the α-man structural gene (C. Singleton, unpublished data), synthesizes an α-man precursor polypeptide that is not processed to the mature subunits and is altered in conformation as determined by sedimentation analysis and protease sensitivity (Woychik et al., 1986).

To determine the sedimentation rate of the mutant and wild-type α-man precursor polypeptides, ^{35}S-Met pulse-labeled cells are treated with Triton X-100 (final concentration 0.5%) and the postnuclear extracts layered on 5–20% linear sucrose gradients containing 0.1% Triton X-100. Following centrifugation (SW 60 rotor) at 100,000 g for 10 hours at 4°C, twenty 0.15-ml fractions are collected from the bottom using a polystaltic pump. Each fraction is diluted with 0.75 ml of 1.2 × C before immunoprecipitation as described in Section II.

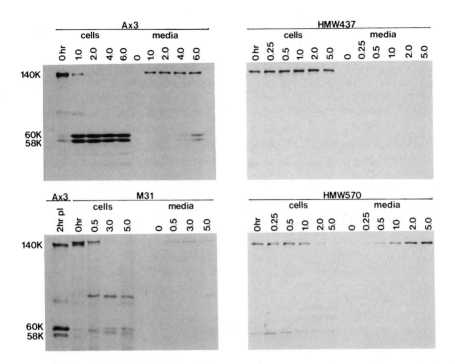

FIG. 6. Processing and secretion of α-mannosidase in mutant strains. Axenically growing cells were pulse-labeled with ^{35}S-Met and chased for the indicated times in growth medium lacking the isotope. Cells and medium were separated by centrifugation, and the labeled α-mannosidase immunoprecipitated from both fractions as described in Section II. Immunoprecipitates were subjected to SDS–PAGE followed by fluorography. Strains were pulse-labeled for the following times: Ax-3, 20 minutes; M31, 30 minutes; HMW 437, 60 minutes; HMW 570, 60 minutes.

To determine the sensitivity of the α-man mutant and wild-type precursors to protease treatment, labeled immunopurified polypeptides are resuspended in 30 μl of a solution containing 10 mM Tris-HCl, pH 7.6, and 10 mM calcium chloride. Trypsin is added to various concentrations (1–200 μg/ml) and the samples allowed to incubate on ice for 1.5 hours. Following centrifugation to remove the Pansorbin, the supernatants are subjected to electrophoresis as described in Section II. Using this procedure we have shown that *in vitro* trypsin treatment of the wild-type α-man precursor generates polypeptides similar in molecular weight to polypeptide subunits making up the mature form of the enzyme, while treatment of the mutant precursor generates a large number of very small fragments presumably as a result of the conformational change exposing cryptic protease sites. Most interestingly, subcellular fractionation studies (Section III) have revealed that the unprocessed precursor polypeptide remains localized in the RER. Perhaps the conformation assumed in the RER by the mutant precursor polypeptide renders it unrecognizable by an RER-to-Golgi transport-mediating receptor.

The largest group of mutants isolated, numbering >80, are altered in their ability to secrete lysosomal enzymes (Dimond *et al.*, 1981). Secretion is an energy-dependent process and is thought to involve the fusion of lysosomes with the plasma membrane followed by release of the enzymes. Two classes of secreted enzymes are observed in wild-type cells; one class consists of the efficiently secreted enzymes α-man, β-glu, and a few others, while the second consists of acid phosphatase, which is secreted linearly and to a lesser extent.

Three major classes of secretory mutants have been identified: one class undersecretes all the lysosomal enzymes examined; a second group undersecretes all the enzymes except acid phosphatase; and the third group oversecretes acid phosphatase and shows a variety of alterations in the secretion of the other glycosidases. Furthermore, many of the secretion mutants are altered in the modification of lysosomal enzymes, as revealed by isoelectric-focusing gel electrophoresis and thermolability studies (Ebert *et al.*, submitted).

Many of the strains which are altered in secretion and modification may actually be defective in localizing (targeting) lysosomal enzymes to lysosomes. We have therefore begun to characterize these mutants using the methods outlined in Section II in order to identify the modifications that are missing or altered which may be important in targeting. One of these mutants, HMW 570, does not proteolytically process the pulse-labeled α-man precursor and instead secretes this precursor polypeptide in an enzymatically active form (Fig. 6). The phenotype of this mutant is reminiscent of the properties of I-cell fibroblasts isolated from humans with mucolipidosis II disease. I-Cell fibroblasts are lacking an enzyme which attaches phosphate to the N-linked mannose residues. Thus, the lysosomal enzymes of I-cell fibroblasts do not contain mannose 6-phosphate, which acts as a lysosomal sorting signal in fibroblasts. In most cells in patients with I-cell disease, enzyme precursors without this moiety are secreted instead of being packaged into lysosomes. Perhaps lysosomal enzymes in HMW 570 also lack a critical modification necessary for localization to lysosomes.

The mutant strain M31, originally isolated as a modification mutant (Free *et al.*, 1978), secretes α-man and β-glu more slowly than wild-type cells. Furthermore, the mutant is deficient in ER-associated α-glu-2 activity (Freeze *et al.*, 1983b), and N-glycosylated proteins retain two of the three glucose residues normally removed by this enzyme. Pulse-labeling experiments with this mutant indicate that the α-man precursor is proteolytically processed to mature subunits only slightly slower in this mutant than in wild-type cells (Fig. 6). The precursor in M31 also has an apparent molecular weight of ~142K, consistent with the presence of glucose residues on the N-linked oligosaccharide side chains. These results suggest that in *Dictyostelium* glucosylated N-linked carbohydrate side chains do not significantly delay the transport and/or processing kinetics of lysosomal enzymes, a result at odds with recent reports concerning the transport of secretory proteins and lysosomal enzymes in animal cells (Lodish and Kong, 1984; Lemansky *et al.*, 1984).

VI. Summary and Future Perspectives

The mechanism by which cells sort and direct proteins to different locations remains unknown. The genetic and biochemical approaches described here are being successfully employed in studying the localization of lysosomal enzymes in *Dictyostelium* (summarized in Fig. 7). Using biochemical methods, we have determined that the lysosomal enzymes α-man and β-glu in *Dictyostelium* are synthesized as precursor polypeptides that are cotranslationally translocated into the lumen of the ER and N-glycosylated. The precursors move at different rates from the RER to the Golgi complex, where they are sulfated and sorted into two classes. One class, containing the majority of the precursor polypeptides, is

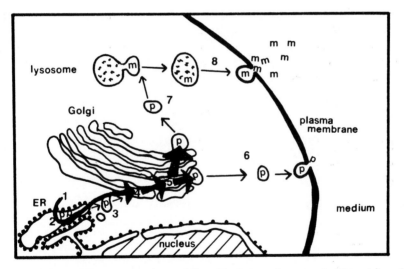

Fig. 7.   Intracellular transport pathways followed by lysosomal enzymes in *Dictyostelium*. (1) Synthesis of precursor (p) polypeptides on polysomes attached to the ER. (2) Cotranslational addition of N-linked oligosaccharides to membrane-associated precursor molecules residing in the lumen of the ER. (3) Transport of precursor polypeptides to the Golgi complex. (4) Sulfation and phosphorylation of precursor polypeptides on N-linked oligosaccharides. (5) Branch point in the transport pathways. (6) Secretion of the intact precursor into the extracellular medium. (7) Transport of intact precursor (p) to lysosomes, proteolytic cleavage of precursor to mature (m) subunits within pre-lysosomal or lysosomal vesicles. (8) Secretion of the mature (m) lysosomal enzymes 3–4 hours after synthesis.

directed to lysosomes where they are rapidly cleaved to mature forms of the enzymes. The other class of precursors is rapidly secreted from cells. The different intracellular transport rates for the α-man and β-glu precursors support the existence of a transport-mediating receptor (Lodish *et al.*, 1983; Fitting and Kabat, 1982; Cardelli *et al.*, 1986).

The precursor forms of the enzymes are membrane-associated until they are secreted or proteolytically processed and released into the lumen of the lysosomes. Present studies are being directed at determining if the precursors associate with the lipid bilayer as integral or peripheral membrane proteins. Future studies will involve immunocytochemical and electron microscopic methods to define further the transport pathway for lysosomal enzymes.

Our genetic approach has involved the isolation and characterization of mutants altered in the secretion, modification, processing, and/or localization of lysosomal enzymes. Through biochemical analysis of these mutants, we hope to identify and characterize the molecular components involved in the processing and transport of this group of enzymes. As an additional genetic approach in collaboration with C. Singleton (Vanderbilt University), the α-man and β-glu genes will be inserted into one of the vectors developed by Firtel *et al.* (1985) and reintroduced into α-man and β-glu structural gene mutants by DNA-mediated transformation. *In vitro* mutagenesis of the cloned genes followed by transformation and analysis of the processing and transport of α-man and β-glu polypeptides will help to define the regions of the precursor proteins critical in the localization process.

REFERENCES

Cardelli, J. A., and Dimond, R. L. (1981). *Biochemistry* **20**, 7391–7398.
Cardelli, J. A., and Dimond, R. L. (1986). *J. Cell Biol.* **102**, 1264–1270.
Cardelli, J. A., Knecht, D. A., and Dimond, R. L. (1981). *Dev. Biol.* **82**, 180–185.
Cardelli, J. A., Knecht, D. A., Wunderlich, R., and Dimond, R. L. (1985). *Dev. Biol.* **110**, 147–156.
Cardelli, J. A., Mierendorf, R. C., and Dimond, R. L. (1986). *Arch. Biochem. Biophys.* **244**, 338–345.
Cardelli, J. A., and Dimond, R. L. (1987). "Protein Transfer and Organelle Biogenesis." Academic Press, Orlando (in press).
Cladaras, M. H., and Kaplan, A. (1984). *J. Biol. Chem.* **259**, 14165–14169.
Dimond, R. L., Burns, R. A., and Jordan, K. B. (1981). *J. Biol. Chem.* **256**, 6565–6572.
Dimond, R. L., Knecht, D. A., Jordan, K. B., Burns, R. A., and Livi, G. P. (1983). *In* "Methods in Enzymology" (S. Fleischer and B. Fleischer, eds.), Vol. 96. pp. 815–828. Academic Press, New York.
Fitting, T., and Kabat, D. (1982). *J. Biol. Chem.* **257**, 14011–14017.
Free, S. J., Schimke, R. T., Freeze, H. H., and Loomis, W. F. (1978). *J. Biol. Chem.* **253**, 4102–4106.
Firtel, R. A., Silan, C., Ward, T. E., Howard, P., Metz, B. A., Nellen, W., and Jacobson, A. (1985). *Mol. Cell. Biol.* **5**, 3241–3250.

Freeze, H., and Miller, A. L. (1980). *Mol. Cell. Biochem.* **35,** 17–27.
Freeze, H., Yeh, R., Miller, A.L., and Kornfeld, S. (1983a). *J. Biol. Chem.* **258,** 14880–14884.
Freeze, H., Yeh, R. Miller, A. L., and Kornfeld, S. (1983b). *J. Biol. Chem.* **258,** 14874–14879.
Golumbeski, G. S., and Dimond, R. L. (1986). *Anal. Biochem.* **154,** 373–381.
Green, E. D., Gruenbaum, J., Bielinksa, M., Baenziger, J. U., and Boime, I. (1983). *Proc. Natl. Acad. Sci. U.S.A.* **81,** 5320–5324.
Housman, D., Jacobs-Lorena, M., Rajbhandary, U. L., and Lodish, H. F. (1970). *Nature (London)* **227,** 913.
Knecht, D. A. and Dimond, R. L. (1981). *J. Biol. Chem.* **256,** 3564–3575.
Knecht, D. A., Dimond, R. L., Wheeler, S., and Loomis, W. F. (1984). *J. Biol. Chem.* **259,** 10633–10640.
Laemmli, U. K. (1970). *Nature (London)* **227,** 680–685.
Lemansky, P., Gieselmann, V., Hasilik, A., and Von Figura, K. (1984). *J. Biol. Chem.* **259,** 10129–10135.
Livi, G. P., Cardelli, J. A., and Dimond, R. L. (1985a). *Differentiation* **29,** 207–215.
Livi, G. P., Cardelli, J. A., Mierendorf, R. C., and Dimond, R. L. (1985b). *Dev. Biol.* **110,** 514–520.
Lodish, H. F., and Kong, N. (1984). *J. Cell Biol.* **98,** 1720–1729.
Lodish, H. F., Kong, N., Snider, M., and Strous, G. (1983). *Nature (London)* **304,** 80–83.
Mierendorf, R. C., Cardelli, J. A., Livi, G. P., and Dimond, R. L. (1983). *J. Biol. Chem.* **258,** 5878–5884.
Mierendorf, R. C., Cardelli, J. A., and Dimond, R. L. (1985). *J. Cell Biol.* **100,** 1777–1787.
Nellen, W., Silar, C., and Firtel, R. A. (1984). *Mol. Cell. Biol.* **4,** 2890–2898.
Pannell, R., Wood, L., and Kaplan, A. (1982). *J. Biol. Chem.* **257,** 9861–9865.
Shields, D., and Blobel, G. (1978). *J. Biol. Chem.* **253,** 3753–3756.
Sly, W. S., and Fischer, H. D. (1982). *J. Cell Biol.* **18,** 67–85.
Woychik, N. A., Cardelli, J. A., and Dimond, R. L. (1986). *J. Biol. Chem.* **261,** 9595–9602.

Chapter 8

The Axenic Mutations and Endocytosis in Dictyostelium

MARGARET CLARKE AND SAMUEL C. KAYMAN

Department of Molecular Biology
Albert Einstein College of Medicine
Bronx, New York 10461

I. Introduction

A powerful adjunct to studies of endocytosis is the isolation and analysis of mutants with altered endocytotic capabilities. In *Dictyostelium*, such mutants can be readily obtained and analyzed by genetic, biochemical, and ultrastructural techniques. This chapter will focus on mutations in *Dictyostelium* that confer upon these normally phagocytotic cells the ability to feed by pinocytosis. Other mutations that alter endocytosis in *Dictyostelium* have been described elsewhere (Clarke, 1978; Kayman *et al.*, 1982; Vogel *et al.*, 1980; Vogel, Chap. 6 this volume; Duffy and Vogel, 1984) and will not be considered here. This chapter is based on information available through 1985.

Eukaryotic cells are separated from their environment by a selective barrier, the plasma membrane. Although some ions and small molecules pass through this membrane, most macromolecules and larger particles are brought into the cell by invaginations of the plasma membrane, leading to the formation and internalization of endocytotic vesicles. This process involves not only the accumulation of particles and macromolecules within the cell, but also a massive flow of membrane, and the sorting of internalized material destined for lysosomes from that to be cycled back to the cell surface (reviewed by Steinman *et al.*, 1983; Besterman and Low, 1983; Brown *et al.*, 1983; Farquhar, 1983). Two types of endocytosis are recognized: phagocytosis and pinocytosis. Phagocytosis

METHODS IN CELL BIOLOGY, VOL. 28

is the substrate-induced internalization of particulate matter. Pinocytosis is the ingestion of fluid and solutes.

Endocytosis is a remarkably dynamic process. *Dictyostelium* cells internalize through pinocytosis an amount of membrane equal to their entire surface area every 40–45 minutes (Thilo and Vogel, 1980; de Chastellier *et al.*, 1983). Macrophages internalize membrane equivalent to their surface area every half hour, while fibroblasts do so every 2 hours (Steinman *et al.*, 1978). Although endocytosis appears to take place at sites scattered all over the cell surface, there are data for fibroblasts indicating that new membrane is inserted primarily at the periphery of spreading cells (Bretscher, 1983; Hopkins, 1985) and at the leading edge of migrating cells (Bergmann *et al.*, 1983). Directional flow of surface membrane has been postulated to be instrumental in cell locomotion (Abercrombie *et al.*, 1970; Harris and Dunn, 1972; Bergman *et al.*, 1983; Bretscher, 1984) and capping (Bretscher, 1984).

Wild-type *Dictyostelium* cells grow only by phagocytosis of bacteria, yet mutant derivatives, called *axenic,* are capable of growth on liquid nutrients. The axenic mutations are remarkable in that they confer a *gain* of function. These mutations offer the opportunity of manipulating endocytotic behavior at both the genetic and environmental levels. In particular, they provide a means of exploring the regulatory mechanisms governing pinocytosis and phagocytosis, and the relationship between endocytosis and other aspects of cell motility.

Although a great deal of data has been collected about the behavior of axenic cells, there is relatively little information comparing wild-type and axenic strains or separating the constitutive effects of the axenic genotype from the effects induced by axenic growth. Earlier studies of endocytosis in *Dictyostelium* have been almost entirely restricted to established axenic cultures of strains Ax-2 and Ax-3. For such cells, membrane and receptors have been followed by ultrastructural and biochemical methods during phagocytosis (Ryter and de Chastellier, 1977; Ryter and Hellio, 1980; Hellio and Ryter, 1980; Favard-Sereno *et al.*, 1981; de Chastellier *et al.*, 1983) and pinocytosis (Thilo and Vogel, 1980; de Chastellier *et al.*, 1983; Schwarz and Thilo, 1983). The stimulation and inhibition of fluid uptake have also been studied (Lee, 1972; Turner *et al.*, 1979; Rossomando *et al.*, 1981; North, 1983). Measurements of the uptake of a variety of small molecules and macromolecules have provided no evidence for any solute uptake mechanism other than pinocytosis (North, 1983).

There is relatively little information in the *Dictyostelium* literature concerning endocytosis by wild-type cells. Previous studies include measurements of the uptake of ε-aminocaproic acid by wild-type and axenic strains (North and Williams, 1978), a description of changes in phagocytosis and exocytosis during differentiation of wild-type cells (Yamamoto *et al.*, 1981; Takeuchi *et al.*, 1983), and a report (discussed below) of the growth of wild-type cells on liquid medium (Maeda, 1983).

Our laboratory is employing a combination of genetic and biochemical techniques to characterize the axenic mutations and their effects. This chapter describes several methods that we have found useful in this analysis. These methods have distinguished constitutive and inducible properties dependent on the axenic genotype, and have suggested that a close relationship exists between the axenic mutations and cell motility.

II.   Methods

A.   General Considerations in Choosing Strains and Growing Cells

1.   AXENIC AND NONAXENIC STRAINS

The mutations that confer the ability to grow axenically are recessive; that is, a diploid formed between an axenic strain and a wild-type strain is unable to grow axenically. Because diploids between the commonly used axenic strains Ax-2 and Ax-3 can grow axenically, these strains are presumed to carry identical complements of the recessive axenic mutations (Williams *et al.*, 1974a). An early analysis of Ax-3 led to the conclusion that linkage groups II and III carried mutations essential for axenic growth, whereas linkage groups I and IV were not involved (Williams *et al.*, 1974b).

Chapter 3 by Loomis in this volume contains the genetic techniques that are employed in examining the contributions of various linkage groups to axenic growth. Markers (mutations) are used to monitor the segregation of individual linkage groups from each parent in a cross, and the axenic growth properties of segregants carrying different combinations of linkage groups are determined. To ascertain which linkage groups are needed for axenic growth, it is essential that the derivation (i.e., axenic vs. nonaxenic) of all linkage groups in the parental strains be known. The nonaxenic strains commonly used for mapping studies have some axenic background, and, in general, all seven linkage groups were not monitored in the crosses used to construct these strains. Consequently, most of these strains are not suitable for use in a genetic analysis of the axenic mutations. Therefore, rather than choosing a conveniently marked strain that is unable to grow axenically as the nonaxenic parent in a cross, one must begin with a strain that is close to wild type, and cross in or select appropriate markers. Suitable starting strains are TS12 and M28; both of these strains carry a *tsg* (temperature-sensitive growth on bacteria) mutation that can be used as a selector, as well as other markers. We have used HU49, which carries markers on linkage groups I and IV from its parent M28, and a spontaneous *cob*A358 (cobalt-resistance) mutation on VII.

An additional factor must be considered in designing crosses to analyze axenic growth. The *bsg* mutations, which render cells unable to grow on *Bacillus subtilis* and are often used as selectors and markers in mapping studies, appear to be inhibitors of axenic growth. The *bsgB*500 and *bsgC*350 mutations, found on linkage groups VII and IV, respectively, were originally isolated in nonaxenic backgrounds (Morrissey *et al.*, 1980; Welker and Williams, 1980). Several groups (Morrissey, Ross and Newell, and Welker, personal communications) attempted without success to move these mutations into a strain that could grow axenically. We subsequently found that strains carrying axenic linkage groups II and III and either *bsgB*500 or *bsgC*350 were incapable of axenic growth, even when diploids between these strains and an axenic strain could grow axenically. The removal of *bsgC*350 by backcrossing restored the ability to grow axenically. These data indicate that the *bsg* mutations are recessive suppressors of axenic growth. Consequently, other markers should be chosen for crosses designed to examine the effects of the axenic mutations. Although the *bsgA* mutation has not been tested in this way (because it lies on linkage group III), the data for *bsgB* and *bsgC* suggest that *bsgA* may also suppress axenic growth and thus should be avoided.

2. GROWTH ON AXENIC MEDIUM

a. Choice of Medium.   The methods used to measure axenic growth can affect whether a strain is scored as capable or incapable of growth. The first important variable is the growth medium. The axenic medium on which *Dictyostelium* cells are commonly grown is HL5, a mixture of proteose peptone, yeast extract, and salts (see Loomis, 1975; Sussman, this volume, Chapter 2). Wild-type strains do not grow on HL5. This medium is convenient for examining the axenic growth capabilities of a strain.

A liquid medium, MA, has been described that permits the growth of wild-type cells (Maeda, 1983). We have examined the mechanism of cell growth on this medium, which contains a very high concentration of Oxoid proteose peptone, and have found that the growth of wild-type cells is a function of particulate components present in the medium. In one such experiment, cells or spores from two different wild-type strains (DdB and HU49) were inoculated at a density of 1 $\times$ 10^5/ml into filtered and unfiltered MA medium, and into HL5. During 8 days of observation, the wild-type strains grew well on unfiltered MA but exhibited little or no growth on HL5 or filtered MA. An axenic strain used as a control grew at similar rates on all three media.

It seems likely that the growth differences between wild-type and axenic cells on filtered MA medium reflect differences in their endocytotic accumulation of medium components, although this has not been measured directly. If this is the case, then endocytosis in wild-type cells may occur only in the presence of particulate matter. In contrast, cells carrying the axenic mutations take up nutrients by pinocytosis in the absence of particulate matter. This difference is an interesting area for further analysis. However, the particulate components present in MA medium make it unsuitable for assaying the axenic growth capabilities of a strain.

b. Measurement of Growth. A second important variable is the method used to measure growth. The most accurate method is to count the cells in a cell-counting chamber or hemacytometer, which permits observation of the condition of the cells as well as determination of their number. For this purpose, a phase counting chamber, which is thinner than those commonly used for bright-field work, gives superior resolution. It is also important to maintain a culture for 2–3 weeks, especially when dealing with slow-growing strains. This permits an accurate determination of growth rate and an assessment of the range of cell densities over which exponential growth occurs; the upper end of that range may vary from 1×10^6/ml to 1×10^7/ml for different strains in the same medium. Of course, cells growing exponentially are needed for measurements of doubling time.

We have applied these methods to strains carrying different complements of the axenic mutations. Our studies have confirmed the lack of involvement of linkage groups I and IV in axenic growth, and have also provided preliminary evidence that linkage group VII is not involved (unpublished observations). However, in contrast to an earlier study (Williams *et al.*, 1974b), we find that an axenic linkage group II is not essential for axenic growth. Figure 1 shows

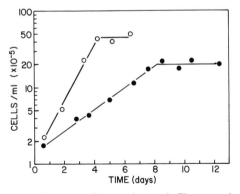

FIG. 1.   Effect of axenic linkage group II on axenic growth. The two strains whose growth curves are shown here are segregants from a cross between an axenic strain and a nonaxenic strain. In 21 segregants examined, the ability to grow axenically segregated independently of all linkage groups except III. (Linkage group VI was not monitored.) The two segregants shown carry an axenic linkage group III and identical linkage groups I, IV, and VII. They differ in linkage group II. Strain MC342 (○) (doubling time ~18 hours) carries an axenic linkage group II, and strain MC341 (●) (doubling time ~50 hours) carries a nonaxenic linkage group II. Both strains had been maintained in axenic culture for >2 weeks and had been diluted to low density at least twice before these growth data were collected. A slower growth rate and lower density at entry into stationary phase are characteristic of segregants lacking an axenic linkage group II.

representative data for a strain that carries an axenic linkage group III and a nonaxenic linkage group II, and for a strain that carries axenic linkage groups II and III. Strains with the former genotype typically have doubling times of 30–50 hours; on the average, the presence of an axenic linkage group II decreases the doubling time to 12–20 hours. The slow growth rate of the former group makes growth difficult to detect by less sensitive methods.

In the original study that reported a requirement for axenic linkage group II (Williams *et al.*, 1974b), amoebae were inoculated from bacterial plates into axenic medium in the wells of multititer plates; growth was monitored over a period of a week by inspection with an inverted microscope. By this assay, it is likely that the slow-growing strains lacking axenic linkage group II would be scored as negative. Thus the behavior of the cells in the two studies may have been similar, and different conclusions were reached because of different criteria used for axenic growth.

c. Inoculating Cultures. Axenic cultures may be started without danger of bacterial contamination by collecting spores from sori raised above the surface of a bacterial plate. However, cultures can be established much more rapidly by using amoebae collected from the growing rim of a plaque. In this case, some bacteria will be transferred as well. If one uses a streptomycin-sensitive bacterial strain and adds dihydrostreptomycin sulfate (50–100 μg/ml) to the HL5 medium, then only an occasional culture will become overgrown with bacteria (see Williams *et al.*, 1974b). This method is convenient for screening large numbers of strains and is particularly useful for studies of developmental mutants that fail to produce viable spores.

3. Growth on Bacteria

For many studies, it is necessary to examine cells grown on bacteria. The plates that are normally used to maintain cultures are a good source for modest numbers of cells (Sussman, this volume, Chapter 2). For this purpose, we plate 5 $\times$ 10^4 to 2 $\times$ 10^5 cells plus 50–100 μl of bacterial suspension on 35 ml of SM agar (Loomis, 1975) in 100-mm Petri plates, and incubate at 21°C for 36–48 hours. The plates are inoculated with either spores or cells, which have been suspended in buffer and counted. As bacteria, we use a *Klebsiella aerogenes* strain that carries a cobalt-resistance mutation; the mutation is convenient for our mapping studies but presumably irrelevant here. Amoebae are harvested from a plate at the earliest time at which thinning of the bacterial lawn is detectable, before actual clearing, so that developmental changes are unlikely to have been induced. Each plate yields ~10^8 cells. Cells are rinsed from the plate using phosphate buffer (17 mM potassium phosphate, pH 6.5) and are mixed vigorously before being centrifuged (110 g, 4 minutes). The pellet is resuspended and washed three times in phosphate buffer; 1 mM EDTA is present in the second wash. The final cell pellet is resuspended in the desired buffer or growth medium and titered by counting. The washes may be performed at 4°C to prevent the cells from responding to the removal of nutrients and to avoid possible adverse effects of high cell concentration. Cells prepared in this manner are essentially free of bacteria.

A drawback to obtaining cells from bacterial plates is that it is difficult to control the time when a particular plate will be ready for harvest. This problem can be circumvented by growing cells on a suspension of washed bacteria. With bacterial suspension cultures, cell growth can be monitored and the cells can be diluted as needed. The bacterial suspension is prepared by rinsing bacteria as described above from SM plates that have been incubated at 21°C for 30–72 hours. The bacteria are not pelleted, but are vigorously mixed and then diluted to a final volume of 40 ml of phosphate buffer per plate. This density of bacteria will support exponential cell growth to $>10^7$ cells/ml. We usually inoculate suspension cultures at 10^5/ml and expand them as needed. In suspension culture, most strains have doubling times of 3–4 hours. Bacteria prepared in this manner do not form clumps, and free bacteria can be readily separated from the *Dictyostelium* cells by differential centrifugation.

If large quantities of bacterially grown cells are desired, it is possible to use bacteria that have been grown on broth, then washed using sterile technique and resuspended in phosphate buffer. However, such bacteria will form clumps that interfere with an accurate determination of cell number and also prevent effective separation of cells from bacteria until the *Dictyostelium* cells have consumed most of the bacteria. Consequently, this method is less suitable for biochemical experiments that require a clean separation of cells from bacteria.

Because bacteria adhere to the surface of *Dictyostelium* cells, the absence of free bacteria does not necessarily mean that all have been removed. Glynn (1981) reported that addition of 2 m*M* sodium azide to the wash buffer is very effective at removing adherent bacteria from axenically grown cells; we have not tested whether this method also works for bacterially grown cells. The usual means of circumventing this problem is to wash cells as free as possible of bacteria by differential centrifugation and then to incubate them in buffer for 30–60 minutes to permit the cells to internalize any adsorbed bacteria.

4.   CONSTITUTIVE AND INDUCIBLE DIFFERENCES BETWEEN WILD-TYPE AND AXENIC CELLS

It is important to distinguish properties of axenic cells that are constitutive (expressed during growth on either bacteria or liquid medium) from those that are induced by growth on axenic medium. Commonly observed differences between wild-type and axenic cells in mass, composition, and rate of growth are dependent on growth conditions, not genotype. (See Loomis, 1975, for details of the effects of growth conditions on cell size and macromolecular composition.) When growing on bacteria, wild-type and axenic cells closely resemble each other in all of those respects. On the other hand, the axenic mutations do have significant constitutive effects on certain cell properties, particularly those related to cell–substratum interactions, as described in the next section.

A consequence of the differences induced by growth conditions is that methods developed for use with axenically grown cells are not necessarily suitable for bacterially grown cells. For example, Prem Das and Henderson (1983) have described a gentle lysis technique for isolating plasma membranes from *Dictyostelium* cells; these workers noted that the method worked well for axenically grown cells, but poorly with vegetative bacterially grown cells.

As summarized in the introductory section, almost all of the data concerning endocytosis in *Dictyostelium* have been obtained with Ax-2 and Ax-3 cells grown axenically. These data cannot safely be extrapolated to other cell types or growth conditions. In future studies of the endocytotic behavior of *Dictyostelium* cells, it will be important to examine at least three cell populations: wild-type and axenic cells growing on bacteria, and axenic cells growing in axenic culture.

B.   Monitoring Behaviors Affected by the Axenic Mutations

1.   ENDOCYTOSIS

a.   Pinocytosis.   A convenient assay for measuring the uptake of fluid by *Dictyostelium* cells is the use of dextran labeled with fluorescein isothiocyanate (FITC-dextran), as described by Vogel and co-workers (1980; see also Vogel, Chap. 6, this volume). We have employed this assay to compare pinocytosis rates for wild-type and axenic cells that had been transferred from bacterial growth plates to nutrient medium (Kayman and Clarke, 1983). These studies showed that the two cell types initially had similar, very low rates of dextran uptake. Over a period of 24 hours, the Ax-3 cells increased their rate of pinocytosis more than 100-fold, to a level similar to that of established axenic cultures (Fig. 2). Thus the axenic mutations permit a cell to accumulate solutes very rapidly, but this ability is expressed only when the cell is placed in nutrient medium, and several hours are required for this induction.

It has been demonstrated for several cell systems (though not yet examined in *Dictyostelium*) that there are two pathways that endocytosed fluid can follow. One pathway leads to lysosomes and the other returns directly to the cell surface, where the endocytosed fluid is released without having been altered (Besterman *et al.*, 1981; Tietze *et al.*, 1982; Adams *et al.*, 1982; Tolleshaug *et al.*, 1981; Daukas *et al.*, 1983). The latter pathway or compartment, which turns over very rapidly, can account for more than half of the fluid that is taken up. The relative size of the two compartments and the flow of material between them are subject to modification by nutrient availability (Besterman *et al.*, 1983).

The FITC-dextran assay employed in the study described above does not

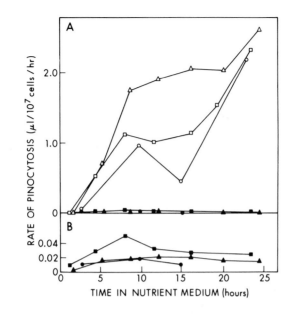

FIG. 2.   Pinocytosis in wild-type and axenic strains. Wild-type (DdB) and axenic (Ax-3) cells were collected from bacterial growth plates, washed, and suspended in HL5 medium at a density of ~5 × 10⁶/ml. At intervals after the shift to HL5, pinocytosis was measured using the FITC-dextran assay (Vogel *et al.*, 1980). The results of three separate experiments, each represented by a different symbol, are shown; in each experiment, the open symbols represent Ax-3 and the filled symbols DdB. (A) DdB data plotted on the same scale as Ax-3 data. (B) A 10-fold expanded scale. From Kayman and Clarke (1983), courtesy of Rockefeller University Press.

actually measure the rate of fluid uptake, but rather the amount of the fluid-phase marker retained by the cell, which is the difference between uptake and release. The magnitude of the increase in the rate of FITC-dextran accumulation as *Dictyostelium* cells carrying the axenic mutations adapt to axenic growth conditions argues that the uptake rate is affected, but this remains to be demonstrated. One means of calculating the true uptake rate is to measure the rate of release of endocytosed material and correct the measured uptake rate accordingly. An assay originally described for this purpose in macrophages appears also to be suitable for *Dictyostelium* cells. Macrophages were allowed to ingest [¹⁴C]sucrose for a short interval, then washed, and placed in isotope-free medium so that the release of labeled sucrose could be monitored (Besterman *et al.*, 1981). Similar experiments are needed to establish whether *Dictyostelium* cells utilize a two-compartment mechanism of fluid uptake as do mammalian cells, and whether transfer between the two compartments or their relative size changes during the onset of rapid pinocytosis.

    b. Phagocytosis.   Several methods suitable for measuring the uptake of particles by *Dictyostelium* cells have been described by Dr. Vogel (Vogel *et al.*, 1980; Vogel, 1983; Vogel, Chap. 6, this volume) and will not be repeated here. His laboratory and others have studied phagocytosis by axenically grown cells. For all types of particles examined, including bacteria, the rate of uptake by axenically grown Ax-2 cells began to decline within 10–30 minutes after the cells were mixed with the substrate particles (Vogel *et al.*, 1980; Vogel, 1983). A similar result was obtained by Glynn (1981), who examined the uptake of metabolically labeled *Escherichia coli* by axenically grown Ax-2 cells. He found that ingestion was linear only for ~20 minutes, and then continued at a steadily decreasing rate, reaching a plateau at ~60 minutes. The generation time of *Dictyostelium* cells is too long for this plateau to represent uniform labeling of *Dictyostelium* proteins; it may instead represent discontinuous feeding. Ultra-structural studies have also indicated that uptake of yeast particles by axenically grown cells occurs discontinuously (Ryter and de Chastellier, 1977).

We have observed a different pattern of uptake by bacterially grown Ax-3 cells (Kayman *et al.*, 1982). In our study, the decrease in optical density of a bacterial suspension was used to measure the rate of phagocytosis. We found that bacterially grown Ax-3 cells did not reach a plateau in their uptake, but exhibited continuous feeding (Fig. 3). Starving the cells for 1 hour before initiating the assay did result in a more rapid initial rate for the first 10–15 minutes, but this dropped smoothly by ~50% into an apparently steady-state rate that was maintained for at least 2 hours.

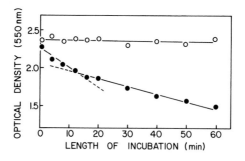

Fig. 3.   Phagocytosis by bacterially grown Ax-3 cells. Ax-3 cells were collected from bacterial growth plates, washed and resuspended in phosphate buffer, and incubated with shaking for 1 hour. They were then mixed with a washed suspension of *Klebsiella pneumoniae* in the same buffer. The Ax-3 cells were at a density of 6.6 × 10⁶ cells/ml, and the assay was conducted at 22°C. At intervals, aliquots were withdrawn and diluted into four volumes of cold phosphate buffer, then centrifuged for 2 minutes at 100 *g* to sediment the amoebae. The bacteria, which remained in suspension, were collected. The phagocytosis rate was calculated as the decrease in the optical density (550 nm) of the bacterial suspension with time. Uptake was continuous, although an initial more rapid rate was observed for cells that had been starved (●). Both rates appeared to represent uptake and not merely binding, because both were blocked at 4°C (not shown) and by dinitrophenol, an uncoupler of oxidative phosphorylation (○, 83 μ*M* dinitrophenol present in assay mixture). From Kayman *et al.* (1982), courtesy of Rockefeller University Press.

It therefore appears that growth conditions have a profound effect on the phagocytotic behavior of axenic cells. Further studies using a single strain and a single assay method are needed to characterize these differences and explore their physiological basis. In addition, it remains to be determined how the phagocytotic behavior of wild-type cells compares to these two patterns of uptake exhibited by axenic cells.

2. CELL MOTILITY

We used an enrichment for phagocytosis-defective cells to isolate mutants of Ax-3 with temperature-sensitive defects in motility (Clarke, 1978); surprisingly, the axenic growth capability of several of these mutants was altered even at permissive temperature (Kayman *et al.*, 1982). These results implied the existence of common pathways involved in motility and axenic growth. Indeed, as noted in the introductory section, data for other cell types have suggested that the membrane-cycling patterns that accompany endocytosis may contribute to cell spreading and locomotion. The axenic mutations, which greatly increase the rate of endocytosis by cells in axenic medium, offer a means of exploring the relationship between endocytosis and motility.

Two assays for cell motility are described in this section. The first is a direct assay for cell locomotion. This assay is relatively time-consuming; it provides detailed information concerning the motile behavior of individual cells. The second assay measures locomotion indirectly as track formation by cells on a surface coated with gold particles. The latter assay provides a convenient means of screening the behavior of multiple cell populations. In addition, track morphology is a sensitive indicator of differences in cell–substratum interactions between wild-type and axenic cells.

a. Measurement of the Rate of Cell Locomotion. Because *Dictyostelium* cells translocate relatively rapidly, 30 minutes of observation is sufficient for determining the rate of cell movement. However, cells placed on a microscope slide under a coverslip will deplete their oxygen supply and die within a few minutes, so special precautions must be taken to prepare the cells for observation during even this short time period. This problem can be circumvented by using a water immersion lens (e.g., Zeiss Achromat 40/0.75 Ph2 water) that is placed directly into the medium in which the cells are being incubated. Another approach is the use of a suitable observation chamber. A simple chamber that can be made in a few minutes from materials on hand is described below. This chamber was originally designed by Dr. John Sternfeld.

Two coverslips are used, one 22 mm square, and the other 22 × 30 mm. (A slide may be substituted for the larger coverslip if desired.) The smaller coverslip will serve as the lid of the chamber. It is coated on one side with a drop of a 10 mg/ml solution of bovine serum albumin (BSA) and set aside to air-dry. Meanwhile, the larger coverslip is prepared. A narrow line of stopcock grease is extruded from an 18-gauge syringe needle onto this coverslip to form the perimeter of a square 18–20 mm on a side. Microcaps (10 μl) are broken to appropriate lengths and pressed into the stopcock grease; they will serve as the sides of the chamber. A thin layer of grease is added to the top of the microcap fragments and extra grease is placed at each corner.

Approximately 100 μl of cell suspension is deposited within the square and spread evenly, taking care to avoid contact with the grease at the perimeter. We use cells from bacterial suspension culture or HL5 that have been washed and suspended at the desired density in phosphate buffer just before use. The cells are left undisturbed for 5 minutes, then excess buffer is drawn off. The small BSA-coated coverslip is placed coated-side-down onto the microcap supports and pressed gently to seal the chamber. The BSA coating is hygroscopic and collects moisture without permitting the formation of droplets that would interfere with observation of the cells. Oxygen sufficient for at least 1 hour is trapped within this chamber. The chamber is placed on the microscope stage, and the movement of individual cells is recorded during a 30-minute period by one of the methods described below. If desired, a temperature-controlled microscope stage may be used during this period. It is in any case essential to include a heat filter in the light path.

Cell movement can be recorded using a microscope equipped with phase-contrast optics and either a video camera or a 35-mm camera. If a video camera and recorder are used, a field containing several cells is videotaped for 30 minutes. During playback the videotape is stopped at 1-minute intervals, and the path of each cell is traced onto a piece of transparent plastic overlaid directly on the screen of the monitor. If a video system is not available, an alternative system is a motor-driven 35-mm camera connected to an inexpensive intervalometer (e.g., Spiratone Surefire Intervalometer). Pictures are taken at 1-minute intervals, and the negative images are projected onto a sheet of $8\frac{1}{2} \times 11$ in. paper using a slide projector. For this method, the coverslip that comprises the bottom of the incubation chamber should be scored lightly in a grid pattern with a diamond pen before the cells are added. These scratch marks provide reference points for the alignment of successive 35-mm frames. Finally, the cell paths are traced, and the path lengths are measured using a digitizing tablet.

We have examined the rate of movement of bacterially and axenically grown Ax-3 cells, using a 35-mm camera and the observation chamber described above. A temperature-sensitive motility mutant, MC2, was included in the same study. Several interesting findings emerged from this study (Table I). First, the rate of translocation of axenically grown Ax-3 cells was only about one-fifth that of bacterially grown cells. (This effect of growth conditions on motility was previously described in Kayman and Clarke, 1983.) Though not shown here, other experiments indicated that bacterially grown NC4 and Ax-3 cells moved at similar rates. Second, the slower movement of axenically grown cells was a

TABLE I

EFFECT OF GROWTH CONDITIONS AND INCUBATION TEMPERATURE ON THE RATE OF MOVEMENT OF CELLS OF THE AXENIC STRAINS AX-3 AND MC2[a]

	Time at 27°C (hours)	Movement (±SD)	No. of trials	No. of cells	No. of motile cells	Movement of motile cells
Cells grown axenically						
Ax-3	0	1.6 ± 0.6	2	27	27	1.6 ± 0.6
MC2	0	0.5 ± 0.8	3	49	17	1.4 ± 0.6
Cells grown on bacteria						
Ax-3	0	7.0 ± 2.1	7	32	31	7.1 ± 2.0
	1	8.1 ± 0.2	2	32	28	9.3 ± 0.7
	2	8.9 ± 0.7	2	35	35	8.9 ± 0.7
	24	4.2 ± 1.6	2	37	35	4.4 ± 1.3
MC2	0	5.6 ± 1.6	7	79	76	5.7 ± 1.5
	0.5	1.2 ± 1.5	1	13	5	2.2 ± 0.9
	1	0.6 ± 0.6	4	30	12	1.1 ± 0.9
	2	0.7 ± 0.3	4	49	17	2.3 ± 0.6
	4	0.3 ± 0.5	1	11	5	1.3 ± 0.3
	24	0.4 ± 0.9	2	23	8	1.2 ± 1.4

[a] The rate of translocation of two axenic strains, Ax-3 and MC2, was measured; MC2 is a temperature-sensitive motility mutant derived from Ax-3 (Kayman et al., 1982; Biswas et al., 1984). Cells grown axenically were examined only at 22°C, the normal temperature for Dictyostelium growth. Bacterially grown cells were also examined at 27°C, the restrictive temperature for the motility mutant. Growth plates were shifted to 27°C for the indicated period of time before the cells were transferred to the incubation chamber described in the text. The incubation chamber was placed on a temperature-controlled microscope stage, and photographs were taken at 1-minute intervals for a period of 30 minutes. Cells were considered "motile" if they moved >1 cell diameter during the 30-minute incubation period; translocation of <1 cell diameter was scored as 0. The units in the table are arbitrary ones from the digitizing tablet; they may be converted to millimeters per hour by multiplying by 0.07.

consequence of a reduction in the average rate of movement of all cells, not of a change in the fraction of motile cells. The data also indicated that Ax-3 is slightly temperature-sensitive for movement. After 24 hours at 27°C, the average rate of translocation of Ax-3 cells had diminished by ~40%; again, this change was manifested by all cells. The behavior of the motility mutant was different in that a shift either to axenic growth conditions or to restrictive temperature produced many nonmotile cells as well as decreasing the rate of movement of motile cells.

We have not yet examined the time course of the change in the rate of movement of Ax-3 cells shifted from one growth condition to another. It will be particularly interesting to monitor the rate of movement of Ax-3 cells at intervals after they have been shifted from bacterial culture to HL5. Cells tested, for example, 3 hours and 24 hours after the shift, would represent a 10-fold difference in pinocytosis rate (Fig. 2). If there is a direct relationship between the rates of endocytosis and translocation, it should be evident in these cell populations.

b. Track Formation Assay. Albrecht-Buehler and Goldman (1976) reported that fibroblasts would collect gold particles within the vicinity of the cell, and that this behavior could be used to provide a record of cell movement (Albrecht-Buehler, 1977). We have found that *Dictyostelium* cells exhibit a similar behavior, although, unlike fibroblasts, they do not internalize the particles (Kayman and Clarke, 1983). This provides a convenient assay for monitoring cell translocation and cell–particle interactions in *Dictyostelium.*

We have followed with minor modifications the protocol originally described by Albrecht-Buehler (1977) for preparation of gold-coated surfaces. We have specified here the sources of our reagents and materials, because at least for the coverslips and the serum albumin, other sources have proved less suitable. Glass-distilled water is used throughout. For most experiments, we have used coverslips (Corning No. 1) as the substratum, although plastic Petri dishes (Falcon 1008) and tissue culture chamber/slides (Lab-Tek 4802) are also suitable. The coverslips are first coated with a 1% solution of BSA (Schwarz-Mann crystallized) that has been filtered through a 0.45-μm filter. Excess BSA is removed by touching the coverslip to absorbent paper. The coverslip is rinsed in absolute ethanol until it drains cleanly and then dried in a stream of hot air. Each coverslip is placed in an individual 35-mm plastic Petri dish. (Other types of substrata are similarly treated.)

The gold particles are prepared by mixing 1.8 ml of 14.5 mM HAuCl$_4$·3H$_2$O (Fisher), 6 ml of 36.5 mM Na$_2$CO$_3$, and 11 ml of H$_2$O in a 250-ml flask. The mixture is heated gently over a Bunsen burner while swirling. As soon as the boiling point is reached, 1.8 ml of 0.1% formaldehyde is added (prepared from Fisher 37% stock). To facilitate rapid addition of the formaldehyde, we dump a premeasured aliquot from a test tube. The flask is swirled without further heating while the precipitate forms. A good preparation appears dense brownish orange in bulk and light blue in a thin film. A thinner, grayer bulk suspension will usually also work. A preparation that turns pink during precipitation, or in which the precipitate appears purple, will not settle or adhere properly. Our success rate at this point rarely exceeds 30–50%. We have best results using a fresh flask for each precipitation, although some workers are more successful using a single flask repeatedly.

While still hot (80–90°C), 2.5 ml of the gold particle suspension is pipeted into each Petri dish containing a coverslip. The gold particles are allowed to settle for about an hour. The coverslip is then dipped into phosphate buffer to rinse it, placed in a fresh Petri dish containing phosphate buffer plus 100 μg/ml dihydrostreptomycin sulfate, and stored at 4°C until use. We normally use a coverslip within 48 hours of its preparation. For the track formation assay, the buffer is removed by aspiration and replaced with a cell suspension in the desired buffer or medium. We have obtained best results using HL5, in which the cells do not initiate development and do translocate well. Although the optimal cell density will vary with the strain and the intended length of incubation, a good starting point for short incubations is 1000 cells/ml, 4 ml per Petri dish.

The process of particle collection and track formation can be visualized using a water immersion lens and the tissue culture chamber/slides listed above. Alternatively, using coverslips, a permanent record of the tracks may be obtained by inverting the wet coverslip on a drop of Aquamount and allowing it to air-dry. In this case, the tracks may be inspected using a dissecting microscope with side illumination. For photography, better results are obtained using a standard microscope with a low-power objective and dark-field illumination. We use a 2.5× objective and a phase 2 condenser fitted with a 15-mm phase ring to provide dark-field illumination.

This assay has revealed a striking constitutive difference between axenic and

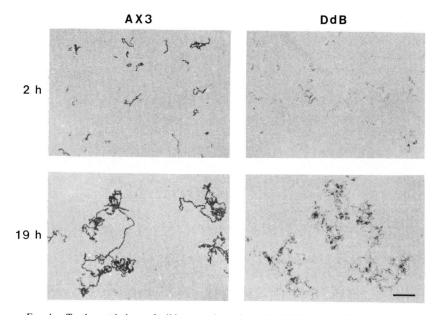

FIG. 4.   Track morphology of wild-type and axenic strains. Wild-type (DdB) and axenic (Ax-3) cells were collected from bacterial growth plates, washed, and incubated on gold-coated coverslips in nutrient medium as described in the text. At the end of the incubation period, the coverslips were treated with Aquamount and photographed using dark-field illumination. The figure shows tracks made by the two cell types during 2 hours and 19 hours of incubation at 22°C. Axenic cells produced clear, sharp tracks; wild-type cells (and other nonaxenic cells) produced fuzzy, indistinct tracks. Bar = 0.5 mm. From Kayman and Clarke (1983), courtesy of Rockefeller University Press.

nonaxenic cells. Wild-type (DdB) and axenic (Ax-3) cells were grown in association with bacteria, then washed, and placed on gold-coated coverslips as described above. The Ax-3 cells made clear, sharp tracks, whereas the wild-type cells made fuzzy, indistinct tracks (Fig. 4). This difference proved to be true of a wide variety of axenic and nonaxenic strains (Kayman and Clarke, 1983). Observation of living cells in the process of track formation revealed that axenic cells collected all gold particles within reach into large masses on their dorsal surface. At intervals, these masses were sloughed (Fig. 5A). Wild-type and other nonaxenic cells moved over most gold particles without disturbing them (Fig. 5B). These differences were manifested by cells within the first half hour after they had been collected from bacterial culture. Sharp tracks were formed by both axenically and bacterially grown Ax-3 cells. Therefore, these differences are a constitutive function of the axenic genotype, and are not affected by growth conditions. The basis of the difference in cell–particle interaction between wild-type and axenic strains is not known, although it is suggestive of differences in cell adhesiveness or patterns or membrane flow during locomotion.

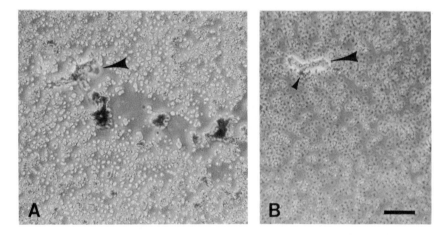

FIG. 5.   The process of track formation by wild-type and axenic cells. Wild-type (NC4) and axenic (Ax-3) cells were observed and photographed using a water immersion lens as they moved across a coverslip coated with gold particles. (A) an Ax-3 cell; (B) an NC4 cell. In both cases, the cell is indicated by a broad arrow. The large, dark objects in (A) are masses of gold particles that the Ax-3 cell has collected and then sloughed. The masses were dropped at ~15-minute intervals. The discarded masses do not adhere to the substratum and thus wash away if the coverslip is handled, leaving a clear track. In (B) the small arrow shows a tiny mass of gold particles that has just been left by the NC4 cell. The size of this mass is typical of those sloughed by wild-type cells. Note that the gold particles on the substratum are visible through the body of the cell. Wild-type cells moved over most gold particles without disturbing them, leaving indistinct tracks. Bar = 20 μm.

III. Discussion

A. Areas for Future Research

1. THE AXENIC MUTATIONS

A great deal of additional information is needed to understand the axenic mutations and their effects. The first step must be to determine which linkage groups carry mutations needed for axenic growth. As described here, current data indicate that linkage group III is essential, and linkage group II, though not essential, enhances the rate of growth. Linkage groups I, IV, and VII appear not to be involved (although the data for VII are preliminary), and the effect of linkage groups V and VI has not yet been tested.

The next step should be to determine, to the degree of resolution possible in *Dictyostelium*, whether single mutations are responsible for the effects of the axenic linkage groups II and III. So far, the axenic information on linkage group III is presumed to reside in a single mutation, *axeB*. However, there are reported to be two mutations on linkage group II that affect axenic growth, *axeA* and *axeC*. As discussed in Section II on methods, Williams *et al.* (1974b) originally reported that linkage group II was required for axenic growth. Gingold and Ashworth (1974) mapped an axenic mutation (now termed *axeC*) to the *whiA–acrA* interval of linkage group II (i.e., centromere-distal to *whiA*). In a later study, North and Williams (1978) isolated a slow-growing recombinant that had obtained only the segment of linkage group II centromere-proximal to *whiA* from the axenic parent. Believing that linkage group II was essential for axenic growth, these workers concluded that a second mutation (now termed *axeA*) must be present centromere-proximal to *whiA*. However, our data indicating that an axenic linkage group II is not required for axenic growth suggest that there may be only a single mutation involved. This can be explicitly tested using recombinants obtained from crosses for mapping the axenic mutations.

Once the individual mutations affecting axenic growth have been identified, their effects can be independently examined. The appropriate strains for such studies will be present among the segregants from the mapping crosses. This type of analysis is essential for understanding how the axenic mutations enable the cell to shift from a phagocytotic to pinocytotic feeding mode.

2. PROTEINS INDUCED DURING AXENIC GROWTH

When available, segregants carrying partial complements of the axenic mutations should be analyzed with respect to the synthesis of certain proteins that are induced both during axenic growth and during early development. These proteins include several lysosomal enzymes (Ashworth and Quance, 1972; Burns *et al.*, 1981) and the lectins, discoidin I and II (Simpson *et al.*, 1974; Frazier *et al.*, 1975). The functional significance of the axenic induction of these proteins is not understood, although it has been interpreted as representing part of the wild-type developmental response, perhaps resulting from marginal growth conditions (Burns *et al.*, 1981).

We have examined the synthesis of discoidin I in strains that carry an axenic linkage group III and a nonaxenic linkage group II. As described in this chapter, such strains grow, although slowly, in axenic medium. Preliminary data indiciate that such strains do not produce discoidin I. The implication of this finding is that the induction of this protein is actually not a response to poor growth conditions, but is instead a consequence of a mutation present on the axenic linkage group II. It remains to be determined whether or not the mutation responsible for discoidin I synthesis during axenic growth is the same mutation that enhances the rate of growth. Similar studies are needed for the axenically induced lysosomal enzymes. It is possible that these enzymes and the discoidins play functional roles in the adaptation of cells to pinocytotic growth.

3. MEMBRANE DYNAMICS

The distinctive interactions of wild-type and axenic cells with gold particles suggest that the axenic mutations may cause changes in membrane composition or membrane-cycling patterns. This is an important area for future analysis. In addition to biochemical analyses of plasma membrane composition as a function of axenic genotype and growth conditions, the path of membrane components internalized during endocytosis should be followed. Membrane turnover in axenically grown Ax-2 cells during phagocytosis and pinocytosis has been studied in detail by de Chastellier and co-workers (1983) using enzymatic labeling and electron microscopy. Similar studies of bacterially grown axenic and wild-type cells are needed.

Another potenitally informative approach is to monitor the binding and capping of selected multivalent ligands by wild-type and axenic cells. Appropriate methods have been described by Condeelis (1979) and Carboni and Condeelis (1985), who examined capping in axenically grown Ax-3 cells challenged with concanavalin A. Lectins and antibodies to cell surface components are appropriate probes for possible membrane changes accompanying the axenic genotype and/or axenic growth.

B. Summary

The axenic mutations of *Dictyostelium* offer a unique opportunity to explore the relationship between phagocytosis and pinocytosis and to examine the mechanism by which a cell shifts from one mode of feeding to the other. The axenic mutations also provide a means of exploring the relationships between endocytosis and other forms of cell motility. This chapter has described the known mutations that affect axenic growth, methods for culturing wild-type and axenic cells and measuring their growth, and methods for monitoring the effects of the axenic mutations on endocytosis and cell movement. The importance has been emphasized of distinguishing effects of the axenic genotype that are expressed constitutively (i.e., during growth on either bacteria or liquid medium) from those that are a function of axenic growth conditions. The methods described in this chapter, applied to wild-type cells and to cells carrying a full complement of the axenic mutations, have shown that the axenic mutations have constitutive effects on cell–substratum interactions, and inducible effects on cell locomotion and pinocytosis.

ACKNOWLEDGMENTS

We would like to thank Dr. John Sternfeld and Richard Birchman for their contributions to the studies described here. This work was supported by grants from the National Institutes of Health (GM11301 and GM29723).

REFERENCES

Abercrombie, M., Heaysman, J. E. M., and Pegrum, S. (1970). *Exp. Cell Res.* **62,** 389–398.
Adams, C. J., Maurey, K. M., and Storrie, B. (1982). *J. Cell Biol.* **93,** 632–637.
Albrecht-Buehler, G. (1977). *Cell* **11,** 395–404.
Albrecht-Buehler, G., and Goldman, R. D. (1976). *Exp. Cell Res.* **97,** 329–339.
Ashworth, J. M., and Quance, J. (1972). *Biochem. J.* **126,** 601–608.
Bergmann, J. E., Kupfer, A., and Singer, S. J. (1983). *Proc. Natl. Acad. Sci. U.S.A.* **80,** 1367–1371.
Besterman, J. M., and Low, R. B. (1983). *Biochem. J.* **210,** 1–13.
Besterman, J. M., Airhart, J. A., Woodworth, R. C., and Low, R. B. (1981). *J. Cell Biol.* **91,** 716–727.
Besterman, J. M., Airhart, J. A., Low, R. B., and Rannels, D. E. (1983). *J. Cell Biol.* **96,** 1586–1591.
Biswas, S., Kayman, S. C., and Clarke, M. (1984). *Mol. Cell. Biol.* **4,** 1035–1041.
Bretscher, M. S. (1983). *Proc. Natl. Acad. Sci. U.S.A.* **80,** 454–458.
Bretscher, M. S. (1984). *Science* **224,** 681–686.
Brown, M. S., Anderson, R. G. W., and Goldstein, J. L. (1983). *Cell* **32,** 663–667.
Burns, R. A., Livi, G. P., and Dimond, R. L. (1981). *Dev. Biol.* **84,** 407–416.
Carboni, J. M., and Condeelis, J. S. (1985). *J. Cell Biol.* **100,** 1884–1893.
Clarke, M. (1978). *In* "Cell Reproduction" (E. R. Dirksen, D. M. Prescott, and C. F. Fox, eds.), pp. 621–629. Academic Press, New York.
Condeelis, J. (1979). *J. Cell Biol.* **80,** 751–758.
Daukas, G., Lauffenburger, D. A., and Zigmond, S. (1983). *J. Cell Biol.* **96,** 1642–1650.
De Chastellier, C., Ryter, A., and Thilo, L. (1983). *Eur. J. Cell Biol.* **30,** 233–243.
Duffy, K. T. I., and Vogel, G. (1984). *J. Gen. Microbiol.* **130,** 2071–2077.

Farquhar, M. G. (1983). *In* "Methods in Enzymology" (S. Fleischer and B. Fleischer, eds.), Vol. 98, pp. 1–13. Academic Press, New York.

Favard-Sereno, C., Ludosky, M., and Ryter, A. (1981). *J. Cell Sci.* **51**, 63–84.

Frazier, W. A., Rosen, S. D., Reitherman, R. W., and Barondes, S. H. (1975). *J. Biol. Chem.* **250**, 7714–7721.

Gingold, E. B., and Ashworth, J. M. (1974). *J. Gen. Microbiol.* **84**, 70–78.

Glynn, P. J. (1981). *Cytobios* **30**, 153–166.

Harris, A., and Dunn, G. (1972). *Exp. Cell Res.* **73**, 519–523.

Hellio, R., and Ryter, A. (1980). *J. Cell Sci.* **41**, 89–104.

Hopkins, C. R. (1985). *Cell* **40**, 199–208.

Kayman, S. C., and Clarke, M. (1983). *J. Cell Biol.* **97**, 1001–1010.

Kayman, S. C., Reichel, M., and Clarke, M. (1982). *J. Cell Biol.* **92**, 705–711.

Lee, K.-C. (1972). *J. Gen. Microbiol.* **72**, 457–471.

Loomis, W. F. (1975). "*Dictyostelium discoideum*, a Developmental System." Academic Press, New York.

Maeda, Y. (1983). *J. Gen. Microbiol.* **129**, 2467–2473.

Morrissey, J. H., Wheeler, S., and Loomis, W. F. (1980). *Genetics* **96**, 115–123.

North, M. J. (1983). *J. Gen. Microbiol.* **129**, 1381–1386.

North, M. J., and Williams, K. L. (1978). *J. Gen. Microbiol.* **107**, 223–230.

Prem Das, O., and Henderson, E. J. (1983). *Biochim. Biophys. Acta* **736**, 45–56.

Rossomando, E. F., Jahngen, E. G., Varnum, B., and Soll, D. R. (1981). *J. Cell Biol.* **91**, 227–231.

Ryter, A., and de Chastellier, C. (1977). *J. Cell Biol.* **75**, 200–217.

Ryter, A., and Hellio, R. (1980). *J. Cell Sci.* **41**, 75–88.

Schwarz, H., and Thilo, L. (1983). *Eur. J. Cell Biol.* **31**, 212–219.

Simpson, D. L., Rosen, S. D., and Barondes, S. H. (1974). *Biochemistry* **13**, 3487–3493.

Steinman, R. M., Silver, J. M., and Cohn, Z. A. (1978). *In* "Transport of Macromolecules in Cellular Systems" (S. C. Silverstein, ed.), pp. 167–179. Dahlem Konferenzen, Berlin.

Steinman, R. M., Mellman, I. S., Muller, W. A., and Cohn, Z. A. (1983). *J. Cell Biol.* **96**, 1–27.

Takeuchi, I., Ishida, S., and Amagai, A. (1983). *Plant Cell Physiol.* **24**, 395–402.

Thilo, L., and Vogel, G. (1980). *Proc. Natl. Acad. Sci. U.S.A.* **77**, 1015–1019.

Tietze, C., Schlesinger, P., and Stahl, P. (1982). *J. Cell Biol.* **92**, 417–424.

Tolleshaug, H., Chindemi, P. A., and Regoeczi, E. (1981). *J. Biol. Chem.* **256**, 6526–6528.

Turner, R., North, M. J., and Harwood, J. M. (1979). *Biochem. J.* **180**, 119–127.

Vogel, G. (1983). *In* "Methods in Enzymology" (S. Fleischer and B. Fleischer, eds.), Vol. 98, pp. 421–430. Academic Press, New York.

Vogel, G., Thilo, L., Schwarz, H., and Steinhart, R. (1980). *J. Cell Biol.* **86**, 456–465.

Welker, D. L., and Williams, K. L. (1980). *FEMS Microbiol. Lett.* **9**, 179–183.

Williams, K. L., Kessin, R. H., and Newell, P. C. (1947a). *Nature (London)* **247**, 142–143.

Williams, K. L., Kessin, R. H., and Newell, P. C. (1974b). *J. Gen. Microbiol.* **84**, 59–69.

Yamamoto, A., Maeda, Y., and Takeuchi, I. (1981). *Protoplasma* **108**, 55–69.

Part III.   Cell Motility

The first chapter in this section is by Fechheimer and Taylor and deals with use of controlled sonication to introduce exogenous molecules into the cytoplasm of *Dictyostelium discoideum*. They describe utilization of this technique to localize a 30,000-Da actin-binding protein which they previously purified and characterized from this organism. Their chapter emphasizes the generality and possible widespread usage of this technique, and they point out its use not only in normalized immunofluorescence microscopy, but also in the measurement of intracellular pH. Condeelis and his co-workers then describe electron microscope immunolabeling of cytoskeletal proteins, which are concentrated in a cortical actin matrix. They describe procedures for fixation of amoebae, suggestions on preparation of antibodies, preparation of colloidal gold probes, optimum staining procedures, and use of myosin fragments for localizing actin filaments. The next chapter, by A. Spudich, provides one method for the isolation of the cortical actin matrix using Triton X-100 to lyse the cells. The effects of calcium ion on the integrity of the isolated matrix is described, as well as a rapid method for obtaining an actin preparation from the cells in reasonable yield.

A totally different actin isolation scheme was described by Uyemura *et al.* (*J. Biol. Chem.* **253**, 9088–9096, 1978) and was used for the basis of studies being carried out by Stone *et al.* and described in Chapter 12. Neutron diffraction can provide critical information about the structure and function of protein complexes, and organisms that can provide highly deuterated, purified proteins are greatly needed. Stone *et al.* give the details of how to obtain such deuterated proteins from *D. discoideum,* using actin as an example.

In Chapter 13, Rubenstein *et al.* describe procedures for preparing [^{35}S]methionine-labeled actin containing acetylated methionine at its amino terminus, as well as methods for studying the processing of that amino terminus.

The final two chapters of this part are concerned with tubulin structure and function. White and Katz describe biochemical and genetic approaches to microtubule function. Included in their chapter are methods for analyzing tubulin on two-dimensional gels, immunofluorescent staining of amoebae, a buffer condition using Nonidet P-40 isolation of microtubule-containing cytoskeletons, and methods for obtaining potential microtubule mutants. The last chapter of this section by Roos concerns approaches to study mitosis. Suitable strains and maintenance of cultures are described, as are methods for live observations of mitosis and staining of fixed cells.

Chapter 9

Introduction of Exogenous Molecules into the Cytoplasm of Dictyostelium discoideum Amoebae by Controlled Sonication

MARCUS FECHHEIMER

Department of Zoology
University of Georgia
Athens, Georgia 30602

D. L. TAYLOR

Center for Fluorescence Research in the Biomedical Sciences
Carnegie-Mellon University
Pittsburgh, Pennsylvania 15213

I. Introduction

Use of the cellular slime mold *D. discoideum* for studies of cell–cell interaction, development, differential gene expression, and cell movements including division, phagocytosis, locomotion, and chemotaxis is described in other chapters in this volume. We describe a method for introduction of exogenous macromolecules into the cytoplasm of living *D. discoideum* amoebae. The most general purpose of the method is to extend molecular studies of cellular constituents from the test tube to the living cell, and to provide a link between the disciplines of cell biology and biochemistry. Thus, the biochemical characteristics of macromolecules predicted from their behavior in dilute solution *in vitro* can be examined intracellularly, and the functions of specific macromolecules hypothesized from their structure and molecular interactions *in vitro* can be critically tested. The distributions of macromolecules in living cells may be observed, and the intracellular concentrations of potential secondary messenger molecules such as free calcium ions and protons that may modulate the structure and activity of cellular constituents may be examined. These measurements can be correlated both temporally and spatially with ongoing cellular activities including development and movement in order to elucidate the molecular mechanisms mediating and regulating such processes.

A number of methods for introducing macromolecules into living cells have been described (reviewed by Celis, 1984). It is difficult to microinject *D. discoideum,* and only very small numbers of cells can be microinjected in a short period of time. Therefore, we sought an alternate method for introducing exogenous molecules into the cytosol of these cells. Reports of polyethylene glycol-mediated fusion of *D. discoideum* amoebae with red blood cells (Kuhn and Parish, 1981), and electric field-induced fusion of *D. discoideum* amoebae with one another (Neumann *et al.,* 1980) indicate potential for use of these methods to introduce exogenous macromolecules into the cytosol of these cells. We have not investigated either of these techniques. Efforts to adapt the scrape-loading technique (McNeil *et al.,* 1984) to *D. discoideum* were unsuccessful, apparently due to a lack of sufficiently tight adhesion of the cells to any of the substrates employed. The mechanism by which scrape loading is believed to work is a transient mechanical shock to the plasma membrane when substrate attachment points are dislodged. Therefore, we investigated other types of mechanical shock that could be applied to cells in suspension. In the present chapter, we summarize our efforts to introduce exogenous molecules into the cytosol of living *D. discoideum* amoebae by controlled sonication (Fechheimer *et al.,* 1986), and describe some present and potential applications of this technique for experimental manipulation and analysis of the molecular mechanisms of biological processes.

II. Sonication Loading

A. Method of Sonication Loading

Dictyostelium discoideum (strain NC-4) were cultured on nutrient agar plates with *Escherichia coli* strain B/r as a food source (Bonner, 1947). Cells were washed from the plates in Bonner's salt solution, and washed five times in 15 ml of 17 mM Sorenson's Na$^+$,K$^+$-phosphate buffer containing 1 mM CaCl$_2$, pH 6.1 (PC buffer). Washed cells (10^8) were suspended in 1 ml of PC buffer in a flat-bottomed cylindrical vial 14 mm in diameter at room temperature. A 1-ml aliquot of 20 mg/ml fluorescein-labeled dextran (Sigma Chemical Co.; dye/dextran ratio of 1.1:1; MW 40,000) in PC buffer was then added, and the solution was mixed with a small stir bar at the bottom of the vial. Immediately after addition of the dextran, this solution was sonicated with stirring at room temperature for 2 seconds at power setting 1 using a Branson sonifier (model 200) with a two-step microtip (cat. no. 101-063-102). The solution was immediately diluted into 10 ml of ice-cold PC buffer, and the cells were washed six times in PC buffer by centrifugation (three times at 4°C and three times at room temperature). Loading of fluorescein-dextran was assessed either by fluorescence microscopy or flow cytometry.

B. Practical Notes for Potential Users of the Technique

Modifications of this basic protocol are devised as required. To establish the procedure after switching sonicators or cell types, it is necessary to recalibrate the power and time of sonication according to the following strategy. Select a power setting, sonicate for various periods of time, and measure cell viability using 0.04% trypan blue and a hemocytometer. Curves of viability as a function of time are obtained for each power setting. Loading of fluorescein-dextran can then be assessed at a few points on the viability curve. Optimal loading and live-cell recovery are obtained in our hands at approximately 50–70% viability. The live-cell recovery after washing is typically 40%, and the viability is near 100%.

The time and power of sonication, the position of the sonifier probe in the vial, and the volume of the solution are critical variables that must be carefully controlled. The temperature (4°–20°C), cell density (1 million to 100 million cells/ml), strain of *D. discoideum*, and extracellular free calcium ion concentration (1 mM calcium or 1 mM EGTA) have small or undetectable effects. The ionic strength, osmolarity, and nutrients present in the solution during loading can affect the technique. The presence of 10 mg/ml glucose is now used routinely, since it may facilitate recovery of the cells from the insult of loading. In addition, the intracellular pH measured in cells loaded with fluorescein-labeled dextran is ~0.5 units more alkaline if glucose is present during loading (see below). The addition of 30 mM NaCl improves the loading of fluorescein-labeled bovine serum albumin (BSA; see below).

The fluorescein-labeled dextran may be purchased from Sigma Chemical Co. (cat. no. FD-40) and is recommended for use as a probe during characterization of the method. The method has also been used to introduce DNA, fluorescein-labeled protein (see below), and a dye molecule (lucifer yellow) into living cells. Extensive characterization of the dependence on molecular weight, poly-electrolyte character, and chemical composition of the exogenous molecule to be loaded has not yet been completed.

It is important in preliminary experiments to verify that the washing protocol employed is sufficient to remove extracellular material, and that dilution and washing are sufficiently rapid to avoid pinocytosis. This is readily accomplished by mixing the probe with the cells, and washing with omission of the sonication

step. Fluorescein-labeled dextran introduced into *D. discoideum* by this procedure is virtually all intracellular, and not in the medium. In addition, the
fluorescein-labeled dextran is not simply adsorbed to the cell surface as demonstrated by lack of a significant change in fluorescence (assessed by flow cytometry) in response to a change in external pH from 6.0 to 8.0 (data not shown).

Quantitative evaluation of the intracellular compartment to which the exogenous molecules have been introduced is essential for interpretation of experiments performed with any loading method, but is technically difficult to perform
unequivocally. If the mechanism of sonication loading is a transient disruption of
the plasma membrane, then exogenous molecules will be delivered exclusively
to the cytosol. The diffuse distribution in living cells of rhodamine-labeled
dextran following loading by this method supports this interpretation. It is important to use a fluorophore that lacks pH sensitivity for qualitative evaluation of
intracellular distribution. Since fluorescence emitted by fluorescein is decreased
at low pH, this fluorophore is less readily detected if present in acidic compartments. Methods for quantitative evaluation of the subcellular compartment to
which exogenous macromolecules are delivered are under investigation.

There are several characteristics of the macromolecule loaded into the cytoplasm that should be addressed. These include toxicity, presence of contaminating molecules, and mobility in the cytoplasm. Some molecules might be toxic to
cells at certain concentrations, and this can be determined in some basic biological assays (see below). Toxicity could be due to small concentrations of contaminants, and this can be determined by further purification steps. The mobility
of the labeled macromolecule can be assessed by fluorescence recovery after
photobleaching (Jacobson and Wojcieszyn, 1984). Knowing the mobility is
important in order to determine the interaction of the labeled macromolecule with
cytoplasmic components (Luby-Phelps *et al.*, 1985). For example, low molecular weight dextrans have 100% mobility in cytoplasm of 3T3 cells (Luby-
Phelps *et al.*, 1985). Therefore, using dextrans as carriers for environment-
sensitive indicators ensures the maximum interaction with the ionic environment.
In contrast, the use of a variety of free dyes such as quin 2 and fluorescein suffers
from the fact that a fraction of the dye may bind to cytoplasmic components
(Taylor *et al.*, 1986).

C.   Characterization of Loaded Cells

Cells loaded in the presence of 10 mg/ml of fluorescein-labeled dextran, 17
mM phosphate and 1 mM CaCl$_2$, pH 6.1, have been characterized extensively.
One cell in ten in these populations is loaded with fluorescein-dextran as assessed
by flow cytometry. Calibration of the fluorescence intensity to the number of
molecules of fluorescein per loaded cell was performed using fixed neutrophils
labeled with fluorescein isothiocyanate as previously described (McNeil *et al.*,

1984). The cells that are loaded are quite heterogeneous, containing between 100,000 and 10 million molecules of dextran per cell.

The most important aspect of characterization of cells loaded by the sonication procedure was to determine whether the viability of the loaded cells had been compromised. The health of the loaded cells was examined in four ways. First, the morphology, adherence, and locomotion of cells loaded with fluorescein-labeled dextran were indistinguishable from those of unsonicated cells. Second, the sonicated cell populations developed upon removal of nutrients. Sonication-loaded cell populations formed as many mature fruiting bodies as did unsonicated cells when equal numbers of viable cells were induced to differentiate. Moreover, fluorescein-dextran-containing cells were observed in aggregation streams. Third, cells loaded with fluorescein-labeled dextran and subsequently induced to differentiate developed responsiveness to cAMP as a chemotactic factor. Living cells allowed to develop for 7 hours on minimal agar medium migrate toward the source of cAMP in a Zigmond chemotaxis chamber (Fig. 1). Fourth, cells loaded with fluorescein-labeled dextran have the capacity to endocytose rhodamine-labeled microspheres as observed by fluorescence microscopy. Flow-cytometric analysis of cells loaded with fluorescein-labeled dextran and then challenged to endocytose rhodamine-labeled microspheres indicated that the labeled cells were able to endocytose as well as unsonicated cells.

Other methods for characterization of cells loaded by the sonication-loading procedure should be devised for the specific molecules introduced and the biological processes under investigation.

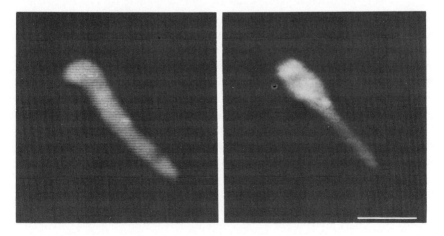

FIG. 1.   Chemotaxis of sonication-loaded *D. discoideum* amoebae in a gradient of cAMP. A single living amoeba loaded with fluorescein-labeled dextran is migrating toward the source of cAMP in a Zigmond chemotaxis chamber. The two images were recorded sequentially with an elasped time of ~1 minute. Bar = 10 μm.

III. Applications of Sonication Loading

A. Normalized Immunofluorescence Microscopy

A fundamental limitation of immunofluorescence microscopy is the difficulty of distinguishing contributions to the distribution of fluorescence intensity from the concentration of the antigenic species, the thickness of the cell, and the distribution of intracellular organelles (Wang et al., 1982; Brier et al., 1983; Taylor et al., 1984; Amato et al., 1983). It is important to determine whether an antigenic species is homogeneously distributed in the space to which it has access, or whether the molecule is "concentrated" in a specific region of the cytoplasm. This distinction is crucial for understanding the mechanisms regulating the organized and integrated responses of cells. The regulation could conceivably occur either by regulation of structure and activity of homogeneously distributed molecules, or by regulation of their distribution, or both. This problem has been discussed in detail previously (Taylor and Fechheimer, 1982). This problem of interpretation is schematically illustrated in Fig. 2, in which a cell is viewed either from above as in the light microscope, or from the side along the substrate.

The contributions of cell thickness and organelle distribution to the immunofluorescence images may be estimated by comparison of the immunofluorescence image to the image of a soluble molecule present in uniform concentration in the cytoplasm of the same cell. Fluorescein-labeled BSA has been employed as the soluble molecule in development of this technique for D. discoideum. Fluorescein-labeled BSA was introduced into living cells by the sonication-loading technique, and fixed during preparation for immunofluorescence microscopy. The distribution of the D. discoideum 30-kDa actin-binding protein (Fechheimer and Taylor, 1984) was observed in the same cells by staining with an affinity-purified antibody reactive with the 30-kDa protein and a rhodamine-labeled goat anti-rabbit IgG.

Ten million D. discoideum amoebae (strain NC-4) were suspended in 2 ml of 17 mM phosphate buffer, pH 6.1, containing 1 mM calcium chloride, 10 mg/ml glucose, 30 mM NaCl, and 42 mg/ml of FITC-BSA (dye/protein ratio of 4), sonicated at power level 1 for 2 seconds using a Branson model 200 sonicator, diluted into 10 ml ice-cold buffer, washed four times in buffer, plated onto glass slides, and allowed to attach for 5 minutes. The cells were then fixed for 20 minutes with 3.7% formaldehyde in 17 mM phosphate buffer containing 1 mM EGTA, pH 7.1, rinsed in the same solution without formaldehyde, extracted with acetone at −20°C for 2 minutes, air-dried, treated with 17 mM phosphate, 150 mM NaCl, 1 mM EGTA, pH 7.1, containing 10 mg/ml BSA, stained with 10 μg/ml affinity-purified anti-30K IgG or normal rabbit IgG for 45 minutes at room temperature, washed, stained with rhodamine-labeled goat IgG reactive

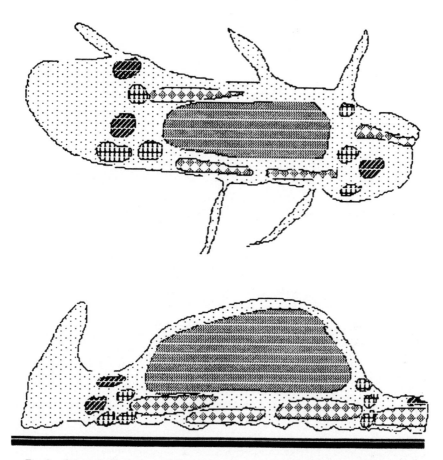

FIG. 2. Schematic illustrations of a locomoting amoeba viewed either from above (top), or along the plane of the substrate (bottom). The potential effects of cell thickness and organelle distribution on the apparent localization of a cytoplasmic constituent by immunofluorescence microscopy are illustrated.

with rabbit IgG (Cappel Laboratories; 1:500 dilution of stock), and washed. All antibodies were clarified in a Beckman airfuge (23 psi for 15 minutes) on the day of the experiment, and applied to the cells in 17 mM phosphate buffer, pH 7.1, containing 150 mM NaCl, 1 mM EGTA, and 1 mg/ml BSA. Fluorescence images were recorded on Kodak Tri-X film (ASA 400).

A cell containing fluorescein-labeled BSA fixed in the cytosol (Fig. 3, top) is shown. All of the cells in the field contain the 30-kDa protein (Fig. 3, bottom). The results indicate that the 30-kDa protein is selectively incorporated into filopodia, since these structures are not detected in the fluorescein-BSA image probably due to the very small path length of the filopodia.

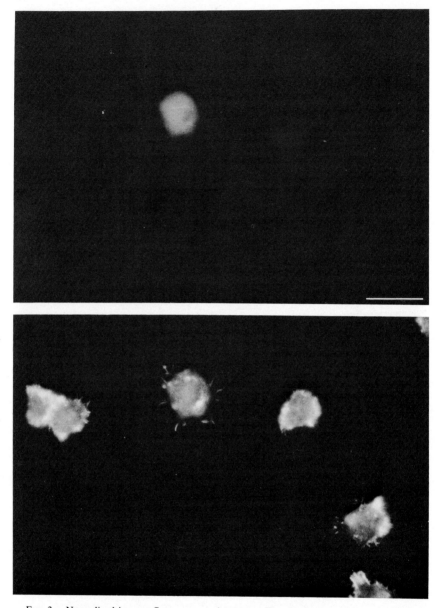

Fɪɢ. 3. Normalized immunofluorescence microscopy. The distributions of fluorescein-labeled bovine serum albumin (top) and of the *D. discoideum* 30,000-Da actin-binding protein stained with rhodamine by an indirect-immunofluorescence technique (bottom) are illustrated. Bar = 10 μm.

The example shown was selected to illustrate the general utility of this method for studies of the distributions of proteins in *D. discoideum* by immunofluorescence microscopy. The selection of the ideal probe molecule is conceivably important, since the assumption of ideal behavior in the cytosol has not been verified in *D. discoideum*. Fluorescein-labeled dextrans have been employed in some experiments, but are not well fixed during preparation for immunofluorescence microscopy. However, the much higher mobility of dextrans in cytoplasm warrants further efforts to fix the dextrans adequately (Luby-Phelps *et al.*, 1985). The low molecular weight dye lucifer yellow also holds promise as a test dye, and is held in amoebae with aldehyde fixation. Quantitative treatment of the results by digital image processing will allow an added level of refinement to the analysis.

B. Measurement of Intracellular pH in *D. discoideum*

Our initial interest in intracellular pH (pH_i) was stimulated by the observation that gelation and contraction of a soluble extract of *D. discoideum* amoebae was regulated *in vitro* by the pH (Taylor *et al.*, 1977; Condeelis and Taylor, 1977). It has subsequently been shown that the activity of an actin crosslinking protein present in this motile extract is also regulated by pH in the physiological range (Fechheimer *et al.*, 1982; Condeelis and Vahey, 1982). Spatially resolved measurements of intracellular pH have not yet defined consistent differences in the cytoplasmic pH of anterior, posterior, or central regions of the giant amoeba *C. carolinensis* (Heiple and Taylor, 1980), or in cultured fibroblasts (Tanasugarn *et al.*, 1984). However, the possibility that ion transport molecules are asymmetrically distributed or activated in membranes encourages us to investigate temporal and spatial changes in parameters such as pH and pCa.

A role for pH_i in regulation of the pathway of cell differentiation of *D. discoideum* has recently been proposed. Addition of a weak base such as ammonia favors differentiation into spore cells, and a weak acid such as acetate favors stalk cell differentiation (Gross *et al.*, 1983). In addition, both ammonia and extracellular pH regulate slug migration, orientation, and the transition to formation of a mature fruiting body (Williams *et al.*, 1984; Bonner *et al.*, 1985). Although weak base and weak acid are expected to alkalinize and acidify the cytoplasm, respectively, no measurements of the actual changes in pH_i were reported by these authors. Finally, a transient (15-minute) alkalinization to a pH of ~7.0 at 2 hours after initiation of development has been reported, and suggested to be required for normal development (Jamieson *et al.*, 1984).

The resting pH_i of the cellular slime mold *D. discoideum* has been estimated to be 6.2 using fluorescein diacetate (Jamieson *et al.*, 1984), 6.48 using ^{31}P-NMR (Jentoft and Town, 1985), 6.7 using fluorescent dextran loaded into the cytoplasm (Fechheimer *et al.*, 1986; see below), and 7.20 to 7.45 using a "null point" method (Aerts *et al.*, 1985).

The pH sensitivity of fluorescein has been used as an optical probe of cellular pH by many investigators (see Heiple and Taylor, 1982, for a review). The physical bases of the pH sensitivity are discussed elsewhere (Heiple and Taylor, 1982; Martin and Lundquist, 1975).

The pH of single cells was estimated using flow cytometry to measure the fluorescence ratio (fluorescein/rhodamine) in cells loaded using the sonication technique with a mixture of the two fluorochrome-labeled dextrans. A ratio of fluorescein/rhodamine can be used, since fluorescein is pH-sensitive while rhodamine is not. Therefore, a two-laser measurement in a flow cytometer permits measurements of pH. The fluorescence ratios were calibrated *in situ* with the flow cytometer after treatment of the cells with either weak acid (40 mM potassium acetate) or weak base (40 mM ammonium chloride) in the presence of 0.1 mM amiloride and 20 mM Tris-maleate buffer to clamp the internal pH at known values. Control experiments demonstrated that the fluorescence ratios had attained a stable value under the conditions employed. If longer times or higher concentrations of potassium acetate were used, the cell morphology became abnormal (assessed by light microscopy), and the mean rhodamine fluorescence of the loaded cells decreased (assessed by flow cytometry).

Dictyostelium discoideum amoebae loaded in the presence of 17 mM Na$^+$,/ K$^+$-phosphate, 1 mM CaCl$_2$, pH 6.1, had an intracellular pH of $\sim$5.9 ($n = 8$). If the cells were loaded and studied in the same buffer supplemented with 10 mg/ml glucose and 0.5 mg/ml of each essential amino acid (Marin, 1976), the intracellular pH was $\sim$6.7 ($n = 13$). Removal of glucose and amino acids after loading resulted in an increase in intracellular pH of 0.2 units to a pH of $\sim$6.9 ($n = 15$).

The accuracy and precision of measurements of pH$_i$ may be affected by a number of experimental variables, including (1) the method used for measurement of pH$_i$ (spectrofluorometry, NMR, weak acid–base equilibration; see Nuccitelli and Deamer, 1982, for a review); (2) the chemical form of fluorescein used in spectrofluorometric methods (e.g., fluorescein diacetate, 6-carboxyfluorescein diacetate, fluorescein-dextran); (3) the method used to introduce the probe into the cells; (4) the intracellular distribution of the probe; (5) the method used for calibration of the measured signal with respect to intracellular pH; (6) the strain of *D. discoideum* employed; (7) the state of differentiation of the cells; (8) the chemical composition and pH of the solution in which the cells are suspended. At present, it is not possible to ascribe the discrepancies among measurements reported by different laboratories to specific experimental variables. Experiments to determine whether systematic errors may be introduced by selection of one or more of the commonly employed experimental methodologies are in progress.

C. Future Applications

The ability to load macromolecules into large numbers of *D. discoideum* by using mechanical shock methods opens up a whole new dimension of research with these cells. We have presented two examples of the uses for this approach utilizing the sonication method of mechanical shock loading. Mechanical shock loading is potentially very valuable, since reasonable yields of viable loaded cells can be produced while avoiding the presence of exogenous membranes or chemicals that promote membrane fusion. Variations of the present sonication method are feasible based on the concept of producing a transient mechanical destabilization of the plasma membrane. These variations are now under investigation.

A variety of experimental manipulations and/or characterization of *D. discoideum* are now possible. The ability to normalize the fluorescence in immunofluorescence studies will permit a more critical definition of the local concentration of selected antigens. Physiological parameters in addition to pH such as pCa can be analyzed using either aequorin luminescence or one of the new fluorescent indicators of calcium. One particularly exciting possibility is the use of fluorogenic substrates for selected enzymes (Waggoner, 1986). Fluorescent analogs of selected cellular macromolecules can also be loaded into cells in order to quantify the distribution and molecular activity in living cells (Taylor *et al.*, 1984, 1986). Complementary to the use of fluorescent analogs is the application of specific molecular inhibitors such as antibodies. Incorporation of antibodies could be used to inactivate cellular constituents selectively. Finally, DNA-mediated cellular transformation and/or selected modification of protein synthesis with exogenous mRNA (Izant and Weintraub, 1985) could be used to manipulate cellular and developmental characteristics.

The main analytical tools will include flow cytometry to analyze cell populations on a cell-by-cell basis and quantitative fluorescence microscopy to define the temporal and spatial changes in selected properties of cells. These approaches coupled to molecular biological manipulations and biochemical analyses will strengthen this general approach even further. Access to the cytoplasm of living cells should yield fundamental information in the next few years.

REFERENCES

Amato, P. A., Unanue, E. R., and Taylor, D. L. (1983). *J. Cell Biol.* **96**, 750–761.

Bonner, J. T. (1947). *J. Exp. Zool.* **106**, 1–26.

Bonner, J. T., Hay, A., John, D.G., and Suthers, H. B. (1985). *J. Embryol. Exp. Morphol.* **87**, 207–213.

Brier, J., Fechheimer, M., Swanson, J., and Taylor, D. L. (1983). *J. Cell Biol.* **97**, 178–185.

Celis, J. E. (1984). *Biochem. J.* **223**, 281–291.

Condeelis, J. S., and Taylor, D. L. (1977). *J. Cell Biol.* **74**, 901–927.

Condeelis, J. S., and Vahey, M. (1982). *J. Cell Biol.* **94**, 466–471.

Fechheimer, M., and Taylor, D. L. (1984). *J. Biol. Chem.* **259**, 4514–4520.

190   MARCUS FECHHEIMER AND D. L. TAYLOR

Fechheimer, M., Brier, J., Rockwell, M., Luna, E. J., and Taylor, D. L. (1982). *Cell Motil.* **2,** 287–308.
Fechheimer, M., Denny, C., Murphy, F. R., and Taylor, D. L. (1986). *Eur. J. Cell Biol.* **40,** 242–247.
Gross, J. D., Bradbury, J., Kay, R. A., and Peacey, M. J. (1983). *Nature (London)* **303,** 244–245.
Heiple, J. M., and Taylor, D. L. (1980). *J. Cell Biol.* **86,** 885–890.
Heiple, J. M., and Taylor, D. L. (1982). *In* "Intracellular pH: Its Measurement, Regulation, and Utilization in Cellular Functions" (R. Nuccitelli and D. W. Deamer, eds.), pp. 21–54. Liss, New York.
Izant, J. G., and Weintraub, J. (1985). *Science* **229,** 345–352.
Jacobson, K., and Wojcieszyn, J. (1984). *Proc. Natl. Acad. Sci. U.S.A.* **81,** 6747–6751.
Jamieson, G. A., Frazier, W. A., and Schlessinger, P. H. (1984). *J. Cell Biol.* **99,** 1883–1887.
Jentoft, J. E., and Town, C. D. (1985). *J. Cell Biol.* **101,** 778–784.
Kuhn, H., and Parish, R. W. (1981). *Exp. Cell Res.* **131,** 89–96.
Luby-Phelps, K., Lanni, F., and Taylor, D. L. (1985). *J. Cell Biol.* **101,** 1245–1256.
McNeil, P. L., Murphy, R. F., Lanni, F., and Taylor, D. L. (1984). *J. Cell Biol.* **98,** 1556–1564.
Marin, F. T. (1976). *Dev. Biol.* **48,** 110–117.
Martin, M., and Lundquist, L. (1975). *J. Lumin.* **10,** 381–390.
Neumann, E., Gerisch, G., and Opatz, K. (1980). *Naturwissenschaften* **67,** 414–415.
Nuccitelli, R., and Deamer, D. W., eds. (1982). "Intracellular pH: Its Measurement, Regulation, and Utilization in Cellular Functions." Liss, New York.
Tanasugarn, L., McNeil, P., Reynolds, G. T., and Taylor, D. L. (1984). *J. Cell Biol.* **98,** 717–724.
Taylor, D. L., and Fechheimer, M. (1982). *In* "Calmodulin and Intracellular Calcium Receptors" (S. Kakiuchi, H. Hidaka, and A. R. Means, eds.). Plenum, New York.
Taylor, D. L., Condeelis, J. S., and Rhodes, J. A. (1977). *Proc. Conf. Cell Shape Surf. Architect.* pp. 581–603.
Taylor, D. L., Amato, P. A., Luby-Phelps, K., and McNeil, P. (1984). *Trends Biochem. Sci.* **9,** 88–91.
Taylor, D. L., Amato, P. A., McNeil, P., Luby-Phelps, K., and Tanasugarn, L. (1986). *In* "Applications of Fluorescence in the Biomedical Sciences" (D. L. Taylor, A. Waggoner, R. Murphy, F. Lanni, and R. Birge, eds.). Liss, New York.
Waggoner, A. (1986). *In* "Applications of Fluorescence in the Biomedical Sciences" (D. L. Taylor, A. Waggoner, R. Murphy, F. Lanni, and R. Birge, eds.). Liss, New York.
Wang, Y.-L., Heiple, J. M., and Taylor, D. L. (1982). *Methods Cell Biol.* **25,** 1–11.
Williams, G. B., Elder, E. M., and Sussman, M. (1984). *Dev. Biol.* **105,** 377–388.

Chapter 10

Ultrastructural Localization of Cytoskeletal Proteins in Dictyostelium Amoebae

J. CONDEELIS, S. OGIHARA,[1] H. BENNETT,[2] J. CARBONI,[3] AND A. HALL

Department of Anatomy and Structural Biology
Albert Einstein College of Medicine
Bronx, New York 10461

I. Introduction

Dictyostelium is an outstanding system in which to study the molecular mechanisms of signal transduction and pattern formation during morphogenetic cell movements. As discussed in the other chapters of this volume, much of this popularity stems from the ease with which physiological, genetic, biochemical, and cytochemical studies can be performed on well-defined cells at specific morphogenetic stages.

Dictyostelium amoebae contain one of the best-studied actin cytoskeletons, where a large number of actin-binding proteins that regulate actin assembly and function have been discovered. However, since most of these are structural proteins which lack enzymatic activity, it is difficult to assign *in vivo* functions based on *in vitro* properties alone. In this situation immunocytochemistry has become an essential tool to test hypotheses of *in vivo* function.

In this chapter we describe immunocytochemical and actin decoration techniques for use with *Dictyostelium* amoebae at the electron microscope level of

[1]Present address: Department of Biology, College of General Education, Osaka University, Toyonaka, Osaka 560, Japan.

[2]Present address: Whitehead Institute for Biomedical Research, Cambridge, Massachusetts 02142.

[3]Present address: Department of Biology, Yale University, New Haven, Connecticut 06511.

resolution that are useful for investigating the function of actin-binding proteins *in vivo* during morphogenetic cell movements.

II. Fixation of Amoebae for Electron Microscopy

Amoeboid cells of *Dictyostelium discoideum* NC-4 or Ax-3 change shape and locomote very rapidly as compared with typical vertebrate cultured cells. Slowly moving cells can be fixed with good results in glutaraldehyde or mixtures of glutaraldehyde and formaldehyde, but rapidly moving *Dictyostelium* amoebae will recoil during fixation with aldehydes, resulting in extensive shape change and organelle relocation. In order to increase the speed of fixation and resulting fidelity of preservation, a number of fixation recipes have been developed for *Dictyostelium* amoebae for use with different electron microscopic techniques.

A. Scanning Electron Microscopy

Changes in cell shape induced by aldehyde fixation are most easily observed in the scanning electron microscope. Glutaraldehyde or formaldehyde (1–2%) or mixtures of these induce extensive ruffling in amoebae. Screening for fixatives that preserve original cell shape can be done quickly in the light microscope by viewing cells that are attached to a substrate during perfusion with fixative. It is clear from such studies that OsO_4 is far superior to the aldehydes in speed of fixation and preservation of cell shape. Therefore, we have adopted OsO_4 as the primary fixative for scanning microscopy.

However, addition of OsO_4 before aldehyde fixation causes destruction of antigenicity and loss of fine detail in preparations viewed in thin section, so it is recommended that OsO_4 be used as primary fixative for scanning electron microscopy where preservation of cell shape and surface detail only are required.

The fixation procedures are as follows:

1. Cells are plated on ethanol-washed glass or plastic at 22°C in HPDF (10mM KCl, 20 mM NaPO$_4$, 0.35 mM CaCl$_2$, 1 mM MgSO$_4$, pH 6.35).
2. Buffer covering the cells is replaced with 1% OsO_4 in HPDF, pH 6.0, and then immediately (i.e., within 5 seconds) replaced with 2% glutaraldehyde in CaCl$_2$-free HPDF, pH 7.0.
3. Fixation in glutaraldehyde is continued for 2 hours. One wash with the glutaraldehyde fixative is recommended during this time to remove traces of OsO_4.
4. Cells are washed with HPDF three times for 5 minutes each (3 × 5 min) and then with distilled H_2O (3 × 5 min) to remove all traces of salt.

5. Cells are dehydrated in 50, 70, 90, and 100% ethanol (absolute) for 2 × 5 min each. Use an unopened bottle of ethanol to avoid water contamination in the last step, which will cause damage to cells during critical-point drying.
6. Critical point-dry in CO_2 and sputter-coat with gold–platinum using standard techniques.

B. High-Resolution Electron Microscopy of Thin Sections

Fixation of amoebae for high-resolution electron microscopy is much more demanding, particularly if the preparation will be used for immunocytochemistry. In this section procedures not involving immunocytochemistry are described:

1. To achieve rapid fixation and retention of fine detail, cells are fixed with glutaraldehyde containing variable amounts of Triton X-100. Use of glutaraldehyde alone will result in loss of structural fidelity due to cytoplasmic rearrangements occurring during the slow penetration of this fixative. Triton allows rapid penetration, but extraction of cytoplasmic components occurs in proportion to the detergent concentration. The concentration used varies depending on the degree of extraction desired. We usually use between 0.01 and 0.1%. Triton can also be added prior to glutaraldehyde if extensive extraction is desired as described below.

Cells are plated as described above and immersed in 2% glutaraldehyde, Triton X-100, and 4% polyethylene glycol 6000 in 20 mM NaCl, 10 mM PIPES, 2 mM MgSO$_4$, 10 mM EGTA, pH 6.8, for 30 seconds at 22°C.

2. Cells are rinsed gently once in buffer to remove detergent and then placed into 1% glutaraldehyde in step 1 buffer without detergent for 30 minutes. Tannic acid (0.2%) can be included in the fixative at this step to improve contrast and preservation of structure, but the solution must be kept above 10°C or tannic acid will precipitate.

3. Wash 2 × 5 min each in 40 mM KPO$_4$, pH 6.0, at room temperature, and once in the same on ice.

4. Fix in 0.5% OsO$_4$ in 40 mM KPO$_4$ at pH 6.0 on ice for 3 minutes.

5. Wash 2 × 5 min each with ice-cold H$_2$O.

6. Stain with 1% uranyl acetate at 4°C for 15 minutes, followed by 15 minutes in 1% uranyl acetate in 25% ethanol at room temperature.

7. Dehydrate at room temperature 2 × 10 min each in 50, 75, 100, and 100% ethanol (absolute).

We have encountered difficulty in obtaining uniform infiltration of plastic resins into amoebae, probably due to the extensive glycocalyx associated with plasma membrane. This problem can be solved by using low-viscosity resins and protracted infiltration times. We have had good success with the following:

8. Immerse dehydrated cells in a 1 : 1 mixture of propylene oxide and absolute ethanol 2 × 10 min each.

9. Transfer to 100% propylene oxide 2 × 10 min each.

10. Immerse cells in 90% epon–araldite and 10% propylene oxide overnight. This permits complete infiltration, but now care must be taken to remove all traces of propylene oxide from the cells.

11. Replace resin with fresh 100% epon–araldite and degas under vacuum for 24 hours.

12. Replace resin with fresh 100% epon–araldite and cure in an oven at 37°C overnight and then at 60°C for at least 30 hours.

The epon–araldite mixture contains 10.75 g of araldite, 10.75 g epon 812, 28.5 g dodecenyl succinic anhydride, and 0.28 ml Tris-dimethylaminomethyl phenol.

If cells are to be embedded on plastic surfaces, propylene oxide cannot be used, since it dissolves plastic. Substitute propylene oxide with hydroxypropyl methacrylate (HPMA) for this purpose (Wolosewick and Porter, 1979), beginning at step 8 as follows:

8. Immerse dehydrated cells in 90% HPMA and 10% ethanol, 3 × 15 min each.

9. Transfer to 100% HPMA, 3 × 15 min.

10. Immerse in 2 : 1 HPMA/epon–araldite and then 1 : 1 for 45 minutes each. Then place in 90% epon–araldite and 10% HPMA overnight.

11. Replace resin with fresh 100% epon–araldite at least 3 × 1 hr, and cure as above.

C. High-Voltage Electron Microscopy of Whole-Mounted Amoebae

In order to view cells directly by high-voltage electron microscopy (HVEM), cells must be spread on formvar and carbon-coated grids that have been treated with 0.1% polylysine. Cells can be monitored directly on the grids by phase-contrast microscopy during fixation.

1. Fixation is done as described in Section II,B, step 1, above. However, by making full use of the resolution and depth of field of the HVEM, it is possible to avoid extraction of cells with detergent. We usually fix cells in 2% glutaraldehyde in 20 mM NaCl, 10 mM PIPES, 2 mM MgSO$_4$, 10 mM EGTA, pH 6.8, for 20 minutes at 22°C in the absence of detergent. In such preparations most of the cells will have changed shape and even contracted during fixation. However, with HVEM it is possible quickly to survey a large number of cells and choose

those for further study whose morphology is least altered by the fixative. If rapid fixation is essential, low concentrations of detergent can be included in the fixative such as 0.05% saponin or 0.01% Triton X-100.

2. Cells are washed and fixed with 1% OsO4 for 5 minutes as in Section II,B, step 3, above.

3. Cells are washed in distilled water, dehydrated, and critical point-dried as described in Section II, A, above.

4. Grids are coated with carbon and stored desiccated over silica gel.

III. Preparation of Antibodies

In general, *Dictyostelium* and its component parts are extremely antigenic in rabbits. This simplifies considerably the production of high-titer antisera but is also a source of nonspecific immunocytochemical reactions. In this section we outline some of the procedures used to avoid the nonspecificity problem.

A. Choice and Handling of Rabbits

1. Screen rabbits for anti-*Dictyostelium* antibodies before inoculation. Serum obtained from most New Zealand white rabbits that have been bred for laboratory use will stain Western blots of one-dimensional SDS–PAGE of vegetative amoebae. The extent of reaction varies from a few minor bands, in which case the rabbit can be used for antibody production, to intense reactions involving most of the prominent polypeptides that are resolved by one-dimensional SDS–PAGE. In this case it is best not to use the animal for antibody production. Why preimmune rabbit serum contains varying titers of anti-*Dictyostelium* antibodies is not clear, other than that *Dictyostelium* is a free-living organism that produces airborne spores, so that even laboratory rabbits may have encountered *Dictyostelium* antigens prior to inoculation.

2. The contractile proteins with which we have worked are very antigenic (Condeelis *et al.,* 1981; Carboni and Condeelis, 1985). This can be advantageous in that relatively small amounts of protein are needed per inoculum and few boosts are required to generate high titers. This latter point can be troublesome, since trace contaminants in the inoculum can generate substantial titers of antibody if numerous boosts are performed. In general, we boost only once following this schedule:

    a. Bleed preimmune serum and screen by Western blotting 14 days before the first inoculation.

    b. Inoculate with 100–500 μg of protein per rabbit.

c. Boost once 2–4 weeks following step b.
d. Bleed 7–10 days after each boost.

B. Suppression of Nonspecific Reactions by Affinity Purification and Preabsorption

Immune serum is routinely affinity-purified on columns of Sepharose 4B that have covalently attached antigen. This is very important in view of the ever-present anti-*Dictyostelium* antibodies in whole preimmune serum.

A more insidious source of nonspecific reactions is anti-*Dictyostelium* antibodies in commercial preparations of whole-IgG fractions containing antirabbit antibodies. These preparations are used as secondary antibodies in staining of Western blots and for immunofluorescence and immunogold procedures. We have found this to be a problem regardless of the source of anti-rabbit IgG (i.e., goat, rat, donkey, etc.). This necessitates affinity purification of anti-rabbit IgG on columns of Sepharose 4B with covalently attached rabbit IgG. These are commercially available or can be prepared easily from CNBr-activated Sepharose 4B and rabbit IgG (Sigma). If large amounts of affinity-purified anti-rabbit IgG are not required, the whole-IgG fraction can be preabsorbed with fixed *Dictyostelium* amoebae before use for immunocytochemistry as follows:

1. Cells (8×10^7) are resuspended in 10 ml HPDF, and then an equal volume of 7.4% formaldehyde in HPDF is added.
2. Cells are fixed in suspension for 10 minutes with occasional, gentle agitation and collected at 700 g for 2 minutes.
3. The pellet is washed in 20 ml of distilled water twice and resuspended in 10 ml of −20°C acetone for 15 minutes with occasional gentle agitation.
4. The pellet is resuspended and washed once in H_2O and then in 20 mM Tris-HCl, pH 8.0, 0.9% NaCl, 1% BSA, 0.02% NaN_3 (TBS).
5. The total-IgG fraction containing antirabbit antibodies is diluted to 100 µg/ml in TBS and then mixed 1 : 1 with the cell pellet. The slurry is rotated slowly for 2 hours at room temperature.
6. Cells are pelleted in a microfuge and the supernatant containing the preabsorbed IgG fraction is filtered through a 0.22-µm filter and diluted and stored in TBS at −70°C.

IV.   Preparation of Colloidal Gold Probes

There are a number of advantages to using colloidal gold as the marker for study of the actin cytoskeleton:

1. Particles can be prepared in a range of sizes, making double or even triple labeling possible.
2. Gold probes are easily prepared and can be customized to recognize a variety of molecular targets.
3. Molecules that are bound to gold particles retain most of their biological activity.
4. Gold probes are stable for months and can be stored for longer periods if frozen in 20% glycerol.
5. Gold probes are very electron-dense and can be recognized even in preparations that have high contrast, a particularly important point when labeling the actin cytoskeleton, which must be well contrasted for viewing in conventional thin sections.

Colloidal gold probes are commercially available from a number of sources. However, due to the nonspecific reaction problems inherent with *Dicyostelium*, we prefer to prepare our own reagents with purified IgG of well-defined specificity. This also gives to us the flexibility of having custom-made gold probes available for double labeling, fluorescence microscopy, and direct labeling, resulting in higher resolution labeling than available with indirect methods. We follow the methods detailed by DeMey and Moeremans (1986) for preparation and storage of 4- to 7-nm gold probes prepared by the white phosphorus technique.

V. Optimum Staining for Preembedding Immunocytochemistry: Balancing Fixation, Probe Penetration, and Antigenicity

After specificity of antibody reaction, preservation of structure, good probe penetration, and retention of antigenicity are the main goals in high-resolution immunocytochemistry. Unfortunately, these latter three are very often mutually exclusive and this is aggravated with preembedding immunocytochemistry. Fixation that is adequate to preserve fine structure is often too harsh to preserve antigenicity or allow free penetration of the probe into the cytoplasm. Light fixation that preserves antigenicity is often inadequate to preserve structure. The best results are obtained when a balance is achieved among all three, which is seldom optimum for any single goal.

Semiquantitative methods that can be used to measure the extent to which this balance is accomplished are extremely helpful and are outlined below. Our experience with these methods is summarized in Table I.

TABLE I

Laboratory Experience with Semiquantitative Methods of Evaluating the Balance among Fixation, Probe Penetration, and Antigenicity

Treatment	Results[a]
Fixation (room temperature)	
No fixation	A, C
0.5% Glutaraldehyde, 6–10 minutes	Satisfactory
1.0% Glutaraldehyde, 10 minutes	B
2% Glutaraldehyde, 10 minutes	B
20 mM Ethylacetimide plus 3% paraformaldehyde, 8 minutes, and then 0.1% glutaraldehyde, 1 hour	A, C
1% Ethyldimethylamino propyl carbodiimide plus 0.3% glutaraldehyde, 7 minutes	A, C
Staining and washing time	
Primary antibody	
Staining: 15, 30, 60 minutes	B
3–5 hours	Satisfactory
Washing: 30, 60 minutes	A
3–5 hours	Satisfactory
Primary antibody–gold conjugate	
Staining overnight	Satisfactory
Washing 3–5 hours	Satisfactory
Secondary antibody–gold conjugate	
Staining: 15 minutes, 2 hours	B
Overnight	Satisfactory
Washing: 30, 60 minutes	A
3–5 hours	Satisfactory
Concentrations of antibodies	
Primary antibody	
3–10 μg/ml	Satisfactory
20 μg/ml	A
Secondary antibody–gold conjugate	
OD 520 nm = 0.5–1.0	Satisfactory
OD 520 nm = 1.5–2.0	A
Primary antibody–gold conjugate	
OD 520 nm = 0.5–1.0	Satisfactory
Effects of some other factors during staining and washing	
H3-P buffers and PEMS[b]	Satisfactory
TBS buffer	A, C
PBS buffer	C
No saponin	B
0.02% Saponin	Satisfactory
5 mM MgSO$_4$	No effect
Second tritonization after fixation	No effect

[a] A, High background; B, poor penetration or loss of antigenicity; C, poor preservation of fine structure.
[b] See text for buffer recipes.

TABLE II

RETENTION OF ANTIGENICITY

Condition	Antigenicity retention (%)			
	Actin[a]	Myosin	95K	120K
Unfixed	100	100	100	100
Fixed A[b]	19	39[c]	21	96
Fixed B	109	—	—	—
Fixed C	104	—	—	—

[a] All antigens were isolated from *Dictyostelium* amoebae.

[b] Fixation conditions: A, 0.5% glutaraldehyde, 8 minutes; B, 1% ethyldimethylamino propyl carbodiimide, 0.3% glutaraldehyde, 7 minutes; C, 20 mM ethylacetimide, 3% paraformaldehyde, 8 minutes, 0.1% glutaraldehyde, 1 hour.

[c] Fixed for 6 minutes according to condition A.

A. Retention of Antigenicity

A number of fixation procedures have been reported that preserve a high degree of antigenicity *in situ*. Two of the best known are the ethylacetimide–paraformaldehyde method of Geiger *et al.* (1981) and the ethyldimethylamino propyl carbodiimide method of Willingham (1980). Both of these methods are superior to glutaraldehyde in preservation of antigenicity (Table II). However, in *Dictyostelium* amoebae, they result in higher background staining and inferior preservation of structure. We have found that with proper choice of fixative concentration and time it is possible to use glutaraldehyde to achieve superior structural preservation with acceptable loss of antigenicity. Choice of the proper concentration and time for fixation is determined as follows:

1. Antigens are fixed individually in solution with varying concentrations of fixative for varying times at the appropriate temperature.
2. The reaction is stopped with 1 mg/ml NaBH$_4$ for 20 minutes.
3. Fixed antigens are quantitatively blotted (5 ng/well) onto nitrocellulose paper (NC) using a microfiltration apparatus (Bio Rad) in 20 mM Tris-HCl, 0.9% NaCl, pH 8.2 (TBS), at room temperature.
4. Wells are washed with TBS, TBS containing 500 mM NaCl, and finally TBS again.
5. NC is removed from the microfiltration apparatus, blocked with 5% BSA for 30 minutes at 45°C, and stained with primary antibody followed by ^{125}I-labeled goat anti-rabbit IgG using standard Western blotting procedures.
6. Wells are cut out and γ-counted or the entire NC is radioautographed and

quantitated by densitometry. The amount of ^{125}I associated with each well is a reproducible measure of the retention of antigenicity.

As shown in Table II, retention varies with the antigen and fixative used. *Dictyostelium* actin and 95K retain little antigenicity in glutaraldehyde, while conditions can be found for myosin and 120K where retention is acceptable.

B. Measurement of Probe Penetration

Fixation with multivalent fixatives such as glutaraldehyde results in good preservation of fine structure for electron microscopy. However, crosslinking of cytoplasm containing cytoskeletal structures decreases its penetrability to high molecular weight probes. This is particularly evident with dense probes like colloidal gold when staining antigens contained in closely packed filament networks and bundles. This can result in the absence of staining in structures that contain high concentrations of antigen.

In order to assess the efficiency of penetration of a probe, we stain amoebae with antiactin followed by goat anti-rabbit IgG–gold as described below. Since actin is known to compose microfilaments *in situ,* the density of gold label along microfilaments is a measure of the degree of penetration of the probe when comparing different regions of the cytoplasm. This has been useful in determining regimens for permeabilization of cells with detergent and optimum probe diameter.

We have found that the smaller colloidal gold probes give denser and more reproducible staining. Several methods have been described recently for production of 3- and 5-nm-diameter gold particles (Demey and Moeremans, 1986). We routinely use 5-nm particles for staining *Dictyostelium* amoebae. Particles in the 7- to 10-nm size range give lower densities of stain, while those in the > 15-nm range are not usable in our hands for staining of actin cytoskeletons in amoebae using preembedding techniques.

We have also found that extensive permeabilization is necessary to obtain uniform staining of whole-mounted amoebae with colloidal-gold probes. In general, protocols for decorating actin with myosin S1 (i.e., using glycerol or saponin) will not yield good results with gold probes. Of course, extensive permeabilization disrupts fine structure, so it is necessary to balance optimum staining against optimum fixation of fine structure.

C. Optimum Fixation

Assessment of optimum fixation is the most subjective of the measurements that must be made to design protocols for high-resolution immunocytochemistry. We do this by comparing the best fixations, which are achieved as described in

Section II, with preparations in which fixation, staining, and antigenicity have been compromised. In general, the preservation of structure in preparations that have been processed for immunocytochemistry (Sections VI and VII) is always inferior to that obtained with the fixation conditions described in Section II.

VI. Preembedding Immunocytochemistry with Gold Probes

In this section we describe fixation, permeabilization, and staining procedures that have been adopted for preembedding immunocytochemistry with colloidal gold probes on spread *Dictyostelium* amoebae and on isolated cortices prepared from amoebae in suspension. In general the results obtained with spread cells are more variable, perhaps reflecting the multiple conformational states of the cytoskeleton during spreading and locomotion. Results obtained with these procedures and isolated cortices are more reproducible. These procedures were designed around the results summarized in Table I.

A. Isolated Cortices

1. Cortices are isolated as described by Bennett and Condeelis (1984). Cortices (10^9/ml) are incubated in an equal volume of 1 mM EGTA, 2.5 mM PIPES, 2.5 mM MgSO$_4$, 10 mM KCl, 20 mM potassium phosphate, pH 7.0 (H3-P buffer) containing 1 mM PMSF, 10 μg/ml chymostatin, 400 units/ml trasylol, and 1 mg/ml phalloidin for 3 minutes on ice.

2. A 16-ml portion of 0.5% glutaraldehyde in H3-P is added per milliliter of cortices in step 1 and fixed for a total of 6–8 minutes, depending on the antigen (including spin time), at room temperature.

3. Cortices are pelleted at 700 g for 2 minutes and resuspended for 30 minutes in 15 ml of ice-cold 1 mg/ml NaBH$_4$ freshly made in H3-P without PIPES.

4. Cortices are pelleted at 700 g for 2 minutes and resuspended for 30 minutes in 3 ml of blocking buffer composed of PEMS (0.9% NaCl, 10 mM sodium phosphate, pH 7.0, 1 mM EGTA, 5 mM MgSO$_4$, 0.02% saponin) containing 1% BSA, 1% heat-inactivated fetal calf serum (FCS), 0.02% NaN$_3$, 10 μg/ml chymostatin, 400 units/ml trasylol, 12 μg/ml phalloidin, and 10 mM Tris-Cl, pH 7.0.

5. Cortices are pelleted at 700 g for 2 minutes and resuspended at 10^9/ml in 1–10 μg/ml of primary antibody in PEMS containing 1% BSA, 1% FCS, and 0.02% NaN$_3$. If the primary antibody is conjugated to the gold probe (direct-staining method), incubate overnight on ice. If an indirect-staining

method is being used, incubate here for 3–5 hours on ice. Incubations are done on a rotator for gentle agitation.

6. Cortices are washed by centrifugation (2600 g × 2 min) in 15 ml PEMS containing 0.1% BSA. These washing steps are crucial to achieve low background staining. For indirect-staining methods we do at least 5 × 0.5 hr washes with agitation to remove primary antibody. If you are using a direct-staining method, skip to step 8 below.

7. Cortices (10^9) are resuspended in 1 ml of secondary goat anti-rabbit IgG–gold and stained overnight on ice as described in step 5. A concentration of gold probe between 0.5 and 5 OD units at 520 nm is used.

8. Extensive washing is necessary to remove the gold reagent. We do at least 5 × 1 hr washes with agitation. Cortices are collected at 2600 g × 2 min each time.

9. Fix 10^9 cortices in 15 ml of 2% glutaraldehyde, 0.9% NaCl, 10 mM sodium phosphate, pH 7.0 (PBS), for 30 minutes at room temperature.

10. Postfix in 0.5% OsO_4 in PBS at pH 6.0 for 3 minutes on ice and follow, beginning at step 5, the protocol in Section II,B.

B. Spread Cells

1. All steps are done in 35-mm plastic Petri dishes. Cells are plated and allowed to locomote. Cells are fixed in 1% glutaraldehyde and 0.1% Triton X-100 for 60 seconds at room temperature as described in Section II,B, step 1.

2. The fixative is exchanged several times with 0.5% glutaraldehyde in the same buffer but without Triton X-100, and fixation is allowed to continue for 6–8 minutes as above, depending on the antigen to be stained.

3. Cells are flooded with ice-cold 1 mg/ml $NaBH_4$ in H3-P for 30 minutes, and subsequent steps are done as described above in Section VI,A, beginning with step 4, except with volumes adjusted accordingly and all exchanges by perfusion.

C. Control Experiments

Control experiments which we commonly employ for immunocytochemistry using gold probes are as follows:

1. Preabsorption of the primary antibody with purified antigen. This is useful for both direct and indirect labeling procedures. It is most convenient to perform preabsorption with antigen that is covalently attached to beads.

2. Labeling the same or equivalent preparations for two different antigens. Labeling of the same structure, regardless of probe specificity, is suspect.

3. For indirect methods, deletion of the primary antibody is a simple but effective control for probe specificity.

Due to the widespread occurrence of anti-*Dictyostelium* antibodies in preimmune serum, it is not possible to use preimmune IgG in place of primary antibody as a control with methods so sensitive as immunogold labeling.

D. Artifacts

Methods of assessing the most common artifacts inherent in these procedures have been outlined in Section V. However, there are a number of insidious artifacts that are more difficult to recognize.

Use of the indirect-staining procedures where a primary rabbit IgG is reacted with a secondary goat anti-rabbit IgG–gold probe gives excellent signal/noise ratios of 10- to 100-fold depending on the antigen. However, IgG is multivalent, and we have found that IgG-induced crosslinking of cytoskeletal structures can lead to artifactual changes in fine structure in lightly fixed cells. This problem is encountered most frequently with antigens that tend to associate with lateral arrays of filaments *in situ* such as 95K (*Dictyostelium* α-actinin). Staining for 95K or actin with the indirect procedure can lead to aggregation of filaments *in situ*. To circumvent this problem we use primary staining for such antigens. This decreases the valency of the gold conjugate and greatly reduces aggregation artifact. Unfortunately, it also reduces the signal/noise ratio (Fig. 1).

Another artifact is the tight binding of gold probes to specific structures that are unrelated to the antigen. Such binding does not appear to result from the presence of nonspecific antibodies in the reagents but rather from as yet uncharacterized high-affinity interactions of probes with specific cell structures. Binding of IgG–gold conjugates to hemocyanin and protein A–gold to *Dictyostelium* membranes are two artifacts that we have encountered. In general it is best to try a different probe if intense staining occurs in association with a specific cell structure. We usually stain equivalent preparations for more than one antigen using the same gold probe in an indirect method to check for artifacts of this nature.

VII. Ultrastructural Localization with Myosin Fragments

Decoration of actin filaments with S1 or HMM is a powerful tool, since it supplies information concerning the identity of filaments, preferred direction of polymerization, and direction of sliding during myosin-mediated motility. The permeabilization and incubation steps needed to admit S1 or HMM to the cytoplasm are usually less disruptive than those used for gold probes, resulting in

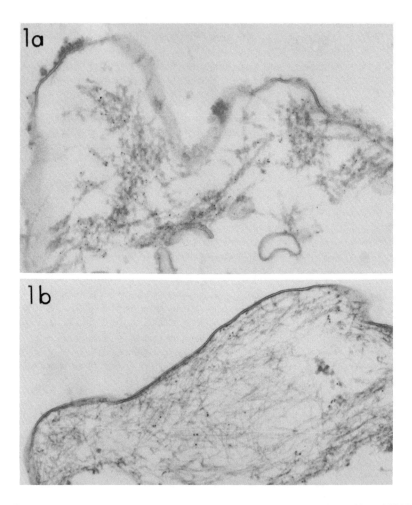

FIG. 1.   A region of the cortex of an amoeboid cell. (a) The cell was stained with anti-95 kDa followed by goat anti-rabbit IgG conjugated to gold. Such indirect-staining procedures result in aggregation of the 95 kDa-associated microfilaments in lightly fixed cells. (b) Similar preparations stained with anti-95 kDa conjugated directly to gold. The microfilaments are not aggregated in preparations stained by the direct procedure. ×60,000.

better preservation of fine structure. We prepare cells for labeling with S1 or HMM by any of the three methods described in this section.

A. Saponin Method

1. Cells in suspension or on plates are immersed in 0.1% glutaraldehyde containing 0.1% saponin for 5 minutes in DB (10 mM EGTA, 2 mM MgSO$_4$, 10 mM PIPES, 20 mM NaCl, pH 6.8, 10 μg/ml chymostatin).

2. Cells are washed at least 3 × 5 min in DB, and then 2–3 mg/ml of S1 or HMM (freshly prepared and clarified) in DB containing 0.1% saponin and 1 mM dithiothreitol (DTT) is added for 30 minutes at room temperature.

3. Cells are washed at least 3 × 5 min in DB and fixed and embedded as in Section II,B, beginning at step 2.

B. Glycerol Method

1. Cells in suspension or on plates are immersed in 50% glycerol in DB for 2.5 hours on ice.

2. Cells are sequentially transferred to 25, 10, and 0% glycerol in DB for 2.5 hours each on ice.

3. HMM or S1 is added to a final concentration of 2–3 mg/ml in DB containing 1 mM DTT and 0.02% NaN$_3$ for 24 hours on ice or 30 minutes at room temperature.

4. Cells are washed 3 × 5 min in DB, fixed, and embedded as above.

C. Triton X-100 Method

1. Cells in suspension or on plates are lysed for 5 seconds with 0.2% Triton X-100 in DB containing 400 units/ml trasylol and 10 μg/ml chymostatin.

2. Cells are washed in DB and incubated with 2–3 mg/ml of S1 or HMM in DB for 30 minutes at room temperature and then washed, fixed, and embedded as above.

Cells which have been challenged with Con A in suspension are lysed as above in 1 mM EGTA, 5 mM Tris, and 20 mM potassium phosphate, pH 7.6, and the resulting cortices are isolated and stained with S1 as described previously (Bennett and Condeelis, 1984).

D. Artifacts

Cell structure is best preserved in samples prepared with the saponin method, acceptable with the glycerol method, and most extracted with the Triton X-100 method. The efficiency of decoration of actin filaments, particularly in dense networks in the cell cortex, is the inverse of the above.

The most insidious artifact obtained with these decoration methods is the S1-induced disruption of microfilament–microfilament and microfilament–membrane contacts *in situ* (Bennett and Condeelis, 1984). These results indicate that S1 decoration should be used with caution when information about the precise location of microfilaments in the cell and their attachment to membranes is required.

VIII. Concluding Remarks

Techniques for localizing structural proteins *in situ* at high resolution in *Dictyostelium,* where cell behavior and morphogenetic stage can be defined with precision, will be extremely valuable in defining the role of the cytoskeleton in signal transduction and morphogenetic cell movements. In addition, these techniques will play a central role in establishing the *in vivo* functions of the structural components of the cytoskeleton.

In general, we have obtained valuable results with all of the techniques described above. Studies using the procedures for scanning and high-voltage electron microscopy of whole-mounted cells (Wolosewick and Condeelis, 1985) and transmission electron microscopy of thin-sectioned cells without (Bennett and Condeelis, 1986) and with decoration of actin filaments by myosin S1 (Condeelis, 1981; Detmers *et al.,* 1983; Bennett and Condeelis, 1984) have been published.

The results of our studies with IgG–gold reagents have been reported in preliminary form (Ogihara and Condeelis, 1983; Bennett and Condeelis, 1985) and summarized in detail elsewhere (Ogihara *et al.,* 1986).

Finally, this chapter contains only those techniques that we have used for high-resolution immunocytochemistry of *Dictyostelium* amoebae using preembedding techniques. This represents only a small fraction of the methods that are currently available and potentially useful with *Dictyostelium.* Further information about these can be found in a number of recent reviews (DeMey and Moeremans 1986; Polak and Varndell, 1984; Wolosewick, 1986). In addition, the procedures outlined here for staining with gold probes could probably be used, with minor modification, with a number of other labels including ferritin, peroxidase, and avidin–biotin based methods.

Acknowledgments

The authors acknowledge Dorothy Budica, Chris Hubertus, and Lorraine Letterese for preparation of the manuscript and grants from NIH (GM25813 and 5 T32 Ca09475-01) and the Hirschl trust.

References

Bennett, H., and Condeelis, J. (1984). *J. Cell Biol.* **99,** 1434–1400.
Bennett, H., and Condeelis, J. (1985). *J. Cell Biol.* **101,** 285a.
Bennett, H., and Condeelis, J. (1986). *J. Cell Sci.,* **83,** 61–76.
Carboni, J., and Condeelis, J. (1985). *J. Cell Biol.* **100,** 1884–1893.
Condeelis, J. (1981). *In* Int. Cell Biol. 1980–81 pp. 306–320.
Condeelis, J., Salisbury, J., and Fujiwara, K. (1981). *Nature (London)* **292,** 161–163.
DeMey, J., and Moeremans, M. (1986). *In* "Advanced Techniques in Biological Microscopy" (G. V. Koehler, ed.), Vol. III. Springer-Verlag, Berlin.

Detmers, P., Goodenough, U., and Condeelis, J. (1983). *J. Cell Biol.* **97**, 522–628.
Geiger, B., Dutton, A., Tokuyasu, K., and Singer, S. (1981). *J. Cell Biol.* **91**, 614–628.
Ogihara, S., and Condeelis, J. (1983). *J. Cell Biol.* **97**, 270a.
Ogihara, S., Carboni, J., and Condeelis, J. (1986). *Proc. 11th Taniguchi Int. Symp.*
Polak, J., and Varndell, I. (1984). "Immunolabelling for Electron Microscopy." Elsevier, New York.
Willingham, M. (1980). *Histochem. J.* **12**, 419–434.
Wolosewick, J. (1986). *Pfefferkorn Conf. Sci. Biol. Specimen Prep., 4th.*
Wolosewick, J., and Condeelis, J. (1986). *J. Cell Biochem.* **30**, 227–243.
Wolosewick, J., and Porter, K. (1979). *In* "Practical Tissue Culture Applications" (K. Maramorosch and H. Hirumi, eds.), pp. 59–85. Academic Press, New York.

Chapter 11

Isolation of the Actin Cytoskeleton from Amoeboid Cells of Dictyostelium

ANNAMMA SPUDICH

Department of Cell Biology
Stanford University School of Medicine
Stanford, California 94305

I. Introduction

Amoeboid cells of *Dictyostelium discoideum* have up to 50% of their total cellular actin organized into a cortical actin matrix (Giffard *et al.*, 1983). This actin matrix has been visualized by electron microscopy as a network of actin filaments in close apposition to the plasma membrane. In preparations where the cells are partially extracted, actin filaments are occasionally seen to be attached to the plasma membrane (Giffard *et al.*, 1983). This cortical actin matrix is often referred to as the cytoskeleton of these cells.

The cortical actin matrix represents a very dynamic region of the cell, since it is involved in extensions of filopodia and in cell locomotions in response to different environmental stimuli. Since this cytoskeleton may show quantitative and qualitative changes as the cell responds to stimuli that result in motile events (see, for example, Berlot *et al.*, 1987), it is of interest to isolate the cortical actin matrix in biochemical quantities. A number of proteins that interact with actin and modulate the organizational states of actin *in vitro* have now been described in *Dictyostelium* (for review, see Spudich and Spudich, 1982; Schleicher *et al.*, 1984). Localization of these actin-modulating proteins in cytoskeletons isolated from cells in different motile states and identification of any modifications of these proteins in response to appropriate stimuli would provide information on how these proteins may function *in vivo*.

The cytoskeleton isolated from vegetative cells contains as much as 50% of the

209

actin of the cell (Giffard *et al.*, 1983) and can be used as the starting material for a *Dictyostelium* actin preparation. The actin from the cytoskeleton can be depolymerized and released into a high-speed supernatant to yield a highly enriched actin fraction (>70% of the total protein is actin). This preparation can be further purified by ion exchange chromatography on DEAE–cellulose to yield highly purified *Dictyostelium* actin (Uyemura *et al.*, 1978). The cytoskeleton can also be used to isolate actin and actin-associated proteins by affinity chromatography using immobilized DNase I (Hosoya *et al.*, 1982). Actin as well as actin-associated proteins are retained by high-affinity binding to DNase I. Actin can be removed from the DNase I after the associated proteins are eluted with 0.6 *M* KI. The actin can be eluted with 3 *M* guanidine hydrochloride. This reagent, however, denatures the eluted actin, whereas elution with 10 *M* formamide yields actin which is polymerization-competent (Zechel, 1980).

II. Growth and Harvesting of Cells

Dictyostelium cells are grown as described previously (Spudich, 1982; see Sussman, Chap. 2, this volume) and harvested in log phase ($OD_{660} = 0.6$–0.7). Cells are collected by centrifugation at 400 *g* for 10 minutes and washed by resuspending the cell pellet in a 200-fold excess of 20 m*M* triethanolamine (pH 8.0) and 0.1 *M* KCl at 22°C and pelleting them again at 400 *g*. The washed cell pellet is resuspended into a 50-fold excess of a buffer containing 0.1 *M* MES [2-(*N*-morpholino)ethane sulfonic acid], pH 6.85, 2.5 m*M* EGTA, 10 m*M* TAME (*p*-tosyl arginine methyl ester), and 1 m*M* MgCl$_2$ at 22°C and pelleted as described before. The last centrifugation is done using preweighed centrifuge bottles so that the weight of the cell pellet can be estimated.

III. Preparation of Cytoskeletons by Detergent Lysis

Cytoskeletons are prepared by the following procedure. The washed cell pellet is lysed in the MES wash buffer described above, modified by the addition of 0.5 m*M* PMSF (phenylmethylsulfonyl fluoride), 20 m*M* sodium bisulfite, 0.5 m*M* ATP, and 0.5% Triton X-100 (lysis buffer with Triton). The cell pellet is evenly resuspended in 10 volumes of cold lysis buffer per gram of cell pellet. The resuspended cells are lysed by vortexing for 30 seconds in 50-ml plastic centrifuge tubes. For cell suspensions ≤50 ml, lysis is best accomplished by vortexing 10-ml aliquots at a vortex setting of 4 for ~30 seconds. For larger volumes, a Waring blender can be used at low speed using a rheostat such that good mixing can be produced without causing disruption of the cytoskeletons. Cells can also

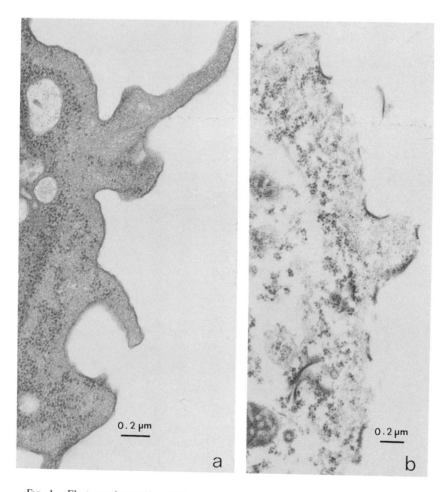

FIG. 1. Electron micrographs of thin sections of (a) the cortical region of an amoeboid cell grown in suspension, and (b) a similar cell extracted with 0.5% Triton X-100 in lysis buffer. The cortical actin matrix is clearly visible underneath the plasma membrane.

be lysed by stirring the cell suspension with a magnetic stirrer for ~10 minutes in ice.

The lysed-cell suspension (lysate) is then centrifuged at 400 g for 10 minutes to collect the cytoskeleton-containing pellet. The loose pellet (P-1) is collected by carefully aspirating the supernatant (S-1). The pellet is then resuspended gently into lysis buffer without Triton X-100 (half the volume used for the initial lysis), using a Pasteur pipet with its narrow tip broken off to avoid shearing the cytoskeletons. The suspension is centrifuged at 400 g for 10 minutes, and the supernatant is removed carefully to yield the cytoskeletal pellet. This pellet contains some nuclear remnants and some detergent-insoluble fragments of

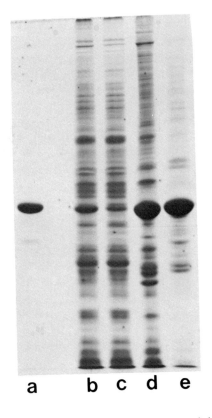

FIG. 2. SDS–PAGE pattern of the cytoskeleton preparation and the high-speed supernatant fraction prepared from the cytoskeleton. (a) Purified muscle actin standard (3 μg); (b) total *Dictyostelium* lysate (12 μg); (c) first low-speed supernatant (12 μg); (d) cytoskeleton preparation (12 μg); and (e) soluble actin fraction prepared from the cytoskeleton (12 μg). Significant enrichment of actin in the isolated cytoskeleton compared to the initial lysate is readily apparent.

membrane in addition to the cortical actin matrix (Fig. 1). SDS–polyacrylamide gel patterns show that the cytoskeletal pellet (P-2) is highly enriched for actin (Fig. 2).

A. Considerations That Affect the Isolation of the Cytoskeleton

The concentration of free calcium in the lysis buffer is an important factor that regulates the amount of actin that sediments with P-1. Using our lysis conditions (0.1 M MES, pH 6.85), we found that an EGTA concentration of 2.5 mM is necessary to stabilize the actin cytoskeleton into a low-speed sedimentable form (Giffard *et al.*, 1983). At lower EGTA concentrations the amount of actin in the low-speed sedimentable cytoskeleton was significantly reduced.

Another factor that influences the amount of actin obtained in the cytoskeletal preparation is the ratio of the weight of cell pellet to volume of lysis buffer. A ratio of 1 g of cell pellet to 10 volumes of lysis buffer is optimal for maximum yield of actin and for giving reproducible SDS-gel patterns of the cytoskeletal preparation. When the cell pellet is lysed in a more concentrated suspension, the resulting P-1 has an SDS-gel pattern more similar to that of the whole-cell lysate, suggesting that the cells are incompletely extracted. When cells are lysed in a pellet/lysis buffer ratio $>1 : 10$, the actin yield is significantly reduced (Giffard *et al.*, 1983).

B. Purification of Actin from the Cytoskeleton Preparation

Approximately 80% of the actin associated with the cytoskeleton can be dissociated by depolymerization at low ionic strength. The actin in the soluble fraction (S-3) obtained after centrifugation is purified severalfold compared to the cytoskeletal preparation. The majority of the nonactin proteins of the cytoskeletal preparation stay with the pellet, leaving actin as the predominant protein in the high-speed supernatant (Fig. 2).

To depolymerize and to solubilize the actin, all procedures are carried out at $0°–4°C$. The cytoskeletal pellet is resuspended by homogenization with a glass–Teflon homogenizer into a volume of depolymerizing buffer equaling twice the weight in grams of the starting cell pellet. The depolymerizing buffer is a low ionic strength buffer containing 10 mM triethanolamine, pH 8.0, 0.5 mM ATP, 0.5 mM dithiothreitol (DTT), and 20 μM MgCl$_2$ (G buffer). The resuspended cytoskeleton is then dialyzed against 100 volumes of G buffer, with a change to fresh G buffer after 8 hours. After dialysis, the preparation is centrifuged at 100,000 g for 2 hours to recover the solubilized actin.

C. Quantitation of Actin in the Cytoskeleton Preparation and in the High-Speed Supernatant

Using purified muscle actin standards, the yield of actin in the cytoskeletal preparation and in the high-speed supernatant can be calculated. SDS gels are scanned using a Transidyne gel scanner, and the weight of the cut-out peaks are compared to those of known amounts of purified muscle actin. One gram (wet wt) of cells yields ~1.4 mg of actin in the cytoskeletal preparation (P-2) and 1 mg of actin in the high-speed supernatant fraction (S-3). The high-speed supernatant actin represents ~5% of the total protein and 30–40% of the total actin of the cell.

IV. Concluding Remarks

The method described here for the isolation of the actin cytoskeleton of *Dictyostelium* derives from work by Spudich and Spudich (1979) using sea urchin eggs. Those studies involved the isolation of a low-speed sedimentable actin fraction from unfertilized and fertilized sea urchin eggs to study changes in assembly states of the actin upon fertilization. A number of similar procedures for isolation of cytoskeletal preparations from *Dictyostelium* have been developed by other investigators to study cytoskeleton-associated proteins. Condeelis *et al.* (Chap. 10, this volume) describe the use of isolated cortices for ultrastructural localization of proteins with gold probes. A detergent-insoluble membrane fragment associated with the filamentous actin matrix isolated from Con A-treated (Condeelis, 1979) *Dictyostelium* cells has been used by Luna *et al.* (1984) to identify a group of integral membrane proteins that have the capacity to bind actin. Tillinghast and Newell (1984) have identified a pool of folate receptors on vegetative cells that are associated with the detergent-insoluble cytoskeleton. The association of the receptor with the cytoskeleton is maintained for up to 12 hours of development. Berlot *et al.* (1987) have shown that the amount of myosin in a Triton-resistant low-speed pellet of developed *Dictyostelium* cells increases significantly as the cell responds to the chemoattractant cyclic AMP. They have also shown that the kinase activity responsible for a chemoattractant-elicited increase in myosin heavy-chain phosphorylation is in the detergent-insoluble pellet. Data such as these and others in the literature suggest that changes in the composition and organization of the isolated cytoskeleton could reflect physiological activities of the cell. The isolated cytoskeleton should be useful as a starting material for studying the modulation of proteins that are regulated during motile events.

ACKNOWLEDGMENT

This work was supported by a grant from NIH (GM27510) to James A. Spudich.

REFERENCES

Berlot, C. H., Devreotes, P. N., and Spudich, J. A. (1987). *J. Biol. Chem.*, in press.
Condeelis, J. (1979). *J. Cell Biol.* **80**, 751–758.
Giffard, R. G., Spudich, J. A., and Spudich, A. (1983). *J. Muscle Res. Cell Motil* **4**, 115–131.
Hosoya, H., Mabuchi, I., and Sakai, H. (1982). *J. Biochem.* **92**, 1853–1862.
Luna, E. J., Goodloe-Holland, C. M., and Ingalls, H. M. (1984). *J. Cell Biol.* **99**, 58–70.
Schleicher, M., Gerisch, G., and Isenberg, G. (1984). *EMBO J.* **3**, 2095–2100.
Spudich, A., and Spudich, J. A. (1979). *J. Cell Biol.* **82**, 212–226.
Spudich, J. A. (1982). *Methods Cell Biol.* **25**, 359–364.
Spudich, J. A., and Spudich, A. (1982). *In* "The Development of Dictyostelium discoideum" (W. F. Loomis, ed.), pp. 169–194. Academic Press, New York.
Tillinghast, H. S., Jr., and Newell, P. C. (1984). *FEBS Lett.* **176**, 325–330.
Uyemura, D. G., Brown, S. S., and Spudich, J. A. (1978). *J. Biol. Chem.* **253**, 9088–9096.
Zechel, K. (1980). *Eur. J. Biochem.* **110**, 343–348.

Chapter 12

Preparation of Deuterated Actin from Dictyostelium discoideum

DEBORAH B. STONE, PAUL M. G. CURMI,
AND ROBERT A. MENDELSON

Cardiovascular Research Institute and
Department of Biochemistry and Biophysics
University of California, San Francisco
San Francisco, California 94143

I. Introduction

A. Role of Deuterated Proteins in Structural Studies

Specific deuteration is an excellent tool for labeling proteins for certain structural studies. In neutron-scattering experiments, deuteration of individual components of macromolecular assemblies allows information to be obtained on the structure of individual subunits within the assembly as well as on the three-dimensional organization of the assembly (Engelman and Moore, 1975; Schoenborn, 1975). Unlike hydrogen, deuterium scatters neutrons in a manner similar to that of the other elements commonly found in biological molecules. By altering the H_2O/D_2O ratio in the solvent, the scattering properties of the solvent can be matched to those of deuterated or protonated components of a hybrid assembly, making them effectively invisible to the neutron beam; data can then be collected solely from the nonmatched components. Information on the structure of individual components within the assembly can be obtained from the radius of gyration and chord distribution. By measuring distances between pairs of similarly labeled subunits, information on the three-dimensional organization of the assembly can be obtained. These methods have been used extensively in structural studies of the bacterial ribosome (Moore *et al.*, 1975, 1984; May *et al.*,

215

1984), DNA-dependent RNA polymerase (Stöckel *et al.*, 1979), and histone complexes (Carlson, 1984).

Proton magnetic resonance studies of individual proteins have also been assisted by deuteration. For a given magnetic field strength, deuteron and proton resonances occur at very different frequencies. Consequently, selective deuteration of amino acid residues simplifies the proton magnetic resonance spectrum of a protein and facilitates identification of the remaining proton resonance lines (Crespi and Katz, 1969; Markley, 1972).

B. Production of Deuterated Organisms and Proteins

Approximately 75% of the hydrogen atoms in proteins are bound to carbon atoms in the protein backbone and side chains. These hydrogen atoms are nonexchangeable and can be replaced by deuterium only through synthetic means. Although a number of organisms have been grown in highly or fully deuterated form (Katz and Crespi, 1966, 1970), deuterated proteins have been conveniently prepared only from algae or bacteria. Autotrophic algae which are photosynthetic utilize CO_2 as their sole carbon source. They can be adapted to grow on pure D_2O and entirely inorganic nutrients, yielding fully deuterated proteins. However, their slow rate of growth and the special equipment required for their culture (Katz and Crespi, 1970) make algae difficult organisms to culture on a large scale. Bacteria are easily cultured in pure D_2O but require deuterated-carbon compounds in order to achieve full deuteration. Deuterated-carbon compounds may be supplied as glucose or succinic acid (Moore, 1979), or as extracts of deuterated algae (Crespi and Katz, 1972). The most advanced eukaryote to be cultured in a fully deuterated state is the protozoa, *Euglena gracilis*. This required both deuterated glucose and deuterated algal extracts and was achieved only after gradual adaptation to progressively higher concentrations of D_2O over a period of several months (Mandeville *et al.*, 1964). As the complexity of organisms increases, they become more fastidious in their nutritional requirements and also tolerate progressively lower levels of D_2O (Katz and Crespi, 1970).

Fortunately, for the purposes of neutron diffraction studies, it is frequently preferable to achieve only a high level (50–70%) of deuteration rather than full deuteration (Moore, 1979; May *et al.*, 1984). In bacteria this can be achieved by cultivation in medium containing protonated-carbon compounds and D_2O as the only source of deuterium (Moore, 1979). In order to study the interesting macromolecular complexes found only in higher organisms, the problem becomes one of choosing an organism which can be reliably and conveniently grown to a high level of deuteration in a relatively rapid and economic manner. We have found that *Dictyostelium discoideum* is an excellent organism for such studies. Since *Dictyostelium* readily uses bacteria as a food source, high levels of deuteration can be obtained by cultivation on deuterated *Escherichia coli* in the presence of easily tolerated levels of D_2O. This approach avoids the need for time-consuming adaptation, special cultivation equipment, and costly deuterated-carbon compounds. In this chapter we present the methods we have devised for the production of deuterated *Dictyostelium*. We also describe our method for isolating actin from deuterated *Dictyostelium* as an example for the preparation of deuterated proteins from a eukaryote.

II. Production of Deuterated *E. coli*

A flow diagram for the production of deuterated *E. coli* and *Dictyostelium* is given in Fig. 1. Deuterated *E. coli* are obtained by a minor modification of Moore's procedure (1979) for the production of partially deuterated bacteria. The culture medium is based on the inorganic salts medium, M56, of Wiesmeyer and Cohn (1960), supplemented with glucose, and contains (per 10 liters) 2.04 g $MgCl_2 \cdot 6H_2O$, 19.8 g $(NH_4)_2SO_4$, 40.8 g KH_2PO_4, 130.1 g K_2HPO_4, 15 g NaCl, and 40 g glucose. The exchangeable hydrogen in the medium components is exchanged for deuterium by dissolving the salts for 10 liters of medium in a minimum of 560 ml D_2O at 45°C and removing most of the D_2O by rotary evaporation at 45°C. The exchanged salts are then dissolved in the desired concentration of D_2O (obtained from Cambridge Isotope Labs). This yields a pH of ~7.1 with recycled D_2O (which is generally acidic). The pH should be 7.0–7.2 as measured on an ordinary pH electrode [the distinction between pH and pD

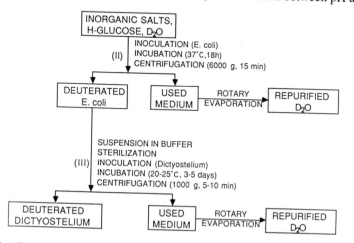

FIG. 1.   Flow diagram for the production of deuterated *E. coli* and *Dictyostelium*. For complete explanation of the protocol, refer to Sections II and III of the text.

(Salomaa *et al.*, 1964) is ignored] and may be adjusted with KOD or DCl (obtained from Sigma) or by altering the ratio of mono- and dibasic phosphate, maintaining a total concentration of 0.1 M phosphate buffer. The medium is then sterilized by passage through a 0.22-μm filter. Prefiltration through a 0.45-μm filter greatly facilitates the sterilization step. The sterile medium is distributed to sterile, dry 2.8-liter Fernbach or 4-liter Erlenmeyer flasks equipped with screw caps; the medium should occupy no more than 20% of the volume of the flask.

Stock cultures of *E. coli* B are maintained on SM agar plates (Sussman, Chap. 2, this volume). A starter culture of 100–200 ml deuterated medium is inoculated from a 2- to 3-day-old agar plate culture and grown overnight to an OD_{550} of ~2. Inoculation of the remaining medium is carried out by adding 1 ml of starter culture per 100 ml medium. The culture flasks are tightly closed (to prevent dilution of the D_2O by water vapor) and incubated overnight in an orbital shaker at 35°–37°C and 180–250 rpm. Under these conditions the *E. coli* doubling time is ~2 hours compared with ~45 minutes in the corresponding H_2O medium.

The deuterated *E. coli* ($OD_{550} \geq 3$) are harvested by centrifugation at 6000 g for 15 minutes. This is carried out at 20°–25°C to prevent H_2O condensation, and a semisterile technique is employed to minimize contamination of the *E. coli*. The pelleted cells are washed in ~100 ml sterile 0.016 M potassium phosphate buffer (prepared in H_2O), pH 6, and collected in sterile 50-ml conical tubes by centrifugation in a bench-top clinical centrifuge. Following removal of the supernatant wash, the *E. coli* pellets are frozen in a dry ice–isopropanol bath and stored at −80°C. This procedure yields 4–6 g deuterated *E. coli* per liter of medium.

D_2O is recovered from the spent medium by rotary evaporation at 60°C as described by Moore (1979). The deuterium enrichment of the recycled D_2O is measured on a xylene–bromobenzene gradient prepared in a sealed-off 25-ml buret (see Moore, 1979). Typically, 99% of the D_2O volume is recovered with a decrease in deuterium enrichment of a few tenths of a percent.

III. Production of Deuterated *Dictyostelium*

Amoebae of *D. discoideum* are grown on the deuterated *E. coli* in liquid culture using the method described by Hohl and Raper (1963) (see Fig. 1). Frozen *E. coli* are thawed and suspended evenly in 0.016 M potassium phosphate buffer (pH 6.2) containing a maximum of 47% D_2O. A concentration of 25 g *E. coli* per liter gives ~10^{10} cells/ml, which is the concentration of bacteria found by Hohl and Raper (1963) to provide optimal growth. Culture flasks having loose caps or sponge plugs are filled to one-tenth of their volume with the *E. coli*–phosphate mixture and autoclaved for 15 minutes.

Stock cultures of *D. discoideum* (Ax-3), maintained on HL5 medium (Sussman, Chap. 2, this volume), are used to inoculate the *E. coli*–phosphate mixture at 10^5 cells/ml. (Inoculation is made with protonated cells, since the inoculum will be $\leq 1\%$ of the harvested cells.) The volume of HL5 containing the required number of cells for a given flask is transferred to a sterile centrifuge tube and centrifuged at 2000 *g* for 5 minutes. After the supernatant is decanted, the pelleted cells are suspended in a few milliliters of the autoclaved and cooled bacterial suspension and added to the remainder of the flask. The inoculated flasks are incubated for 3–5 days at 20°–25°C on an orbital shaker set at 250 rpm. Growth is followed by cell counts on a hemacytometer. The generation time is usually between 10 and 15 hours but can vary considerably with the cell line and the factors outlined below. Saturation occurs at $\sim 10^7$ cells/ml and is accompanied by a clearing of the medium.

Under these conditions the rate of growth of *Dictyostelium* is sensitive to a number of factors including autoclave time, aeration, and D_2O concentration. Hohl and Raper (1963) observed a marked increase in *Dictyostelium* generation

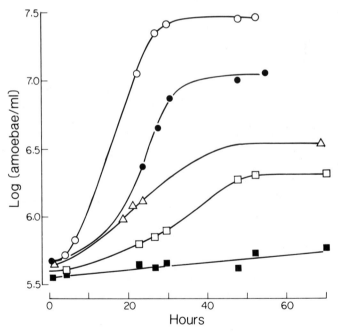

FIG. 2.   Growth of *Dictyostelium discoideum* as a function of D_2O concentration. Cultures containing protonated *E. coli* in the presence of 0% (○), 47% (●), 62.5% (□), and 75% (■) D_2O were inoculated with amoebae grown on HL5 medium. A second culture containing protonated *E. coli* in the presence of 62.5% D_2O (△) was inoculated with amoebae grown for 52 hours in the presence of 62.5% D_2O (□).

time with increasing sterilization time; the minimum time required for adequate sterilization should be used. Good aeration is promoted by loose-fitting screw caps or (preferably) sponge plugs and vigorous shaking. When the D_2O concentration is <50% such vigorous shaking in open flasks does not cause a significant decrease in the deuterium enrichment of the buffer. The effect of D_2O on the rate of growth of *Dictyostelium* on protonated *E. coli* is shown in Fig. 2. Increasing levels of D_2O lengthen the generation time and lag phase and reduce the cell density at saturation. In 75% D_2O, growth is almost completely inhibited. In 62.5% D_2O the rate of growth, lag phase, and cell density at saturation are all improved upon recultivation in the same medium, suggesting that *Dictyostelium* can adapt to higher concentrations of D_2O.

The amoebae are harvested in late log or early saturation ($\sim 10^7$ cells/ml) by centrifugation at 1000 g for 5–10 minutes. The supernatant is carefully decanted and recycled by rotary evaporation. The pelleted amoebae are washed twice in several volumes of 10 mM triethanolamine (prepared in H_2O), pH 8.0, to remove any remaining *E. coli,* and collected by centrifugation. The yield is ~ 10 g of packed cells per liter of culture medium.

IV. Measurement of the Extent of Deuteration

The extent of incorporation of deuterium into nonexchangeable carbon-bonded positions can be determined by nuclear magnetic resonance (NMR) (Moore, 1977, 1979). Ideally this measurement should be carried out on the purified protein of interest, but this can be costly. Alternatively, a good estimate of deuterium incorporation can be obtained using a sample of whole-cell protein which is free of nucleic acid. (The extent of deuterium incorporation into protein and nucleic acid can differ significantly.) The following procedure was developed as an economical method for measuring deuterium incorporation into protein as a function of growth conditions. It utilizes hot trichloroacetic acid treatment to remove nucleic acid from the protein (Schneider, 1957).

One gram of washed amoebae (obtained from a 100-ml culture) is suspended in two volumes of lysis buffer containing 10 mM triethanolamine, pH 8.0, 40 mM sodium pyrophosphate, 0.4 mM dithiothreitol, and 30% (w/v) sucrose. The cell suspension is sonicated on ice with a probe-type sonicator until 99% lysis is achieved. The lysed cells are centrifuged for 75 minutes at 150,000 g to remove unbroken cells and cell debris. The supernatant is brought to 5% trichloroacetic acid and incubated at 90°C for 3 minutes. The precipitated protein is collected by centrifugation for 10 minutes at 27,000 g and washed by suspending in 3 ml cold 5% trichloroacetic acid and collecting by centrifugation. The washed pellet is suspended in 2 ml of a solution containing 50 mM NH_4HCO_3, 0.5 mM dithiothreitol, 1 mM NaN_3, and 0.5 mM phenylmethylsulfonyl fluoride.

At this point the fraction of nucleic acid remaining in each sample can be determined from the ratio of the absorbance at 280 nm to that at 260 nm ($R_{280/260}$) (Layne, 1957). For this measurement 20 μl of each sample is solubilized with 5 μl 1 N KOH and diluted to 1.0 ml with solvent. The absorbance is measured at 260, 280, 340, and 400 nm, the readings at 340 and 400 nm being used to correct the absorbance at 260 and 280 nm for light scattering. It is desirable to achieve a value for $R_{280/260}$ of ≥1, indicating that ≤3% nucleic acid remains. More extreme treatment with trichloroacetic acid leads to significant breakdown of the protein (as detected by polyacrylamide gel electrophoresis).

The trichloroacetic acid-precipitated protein is dialyzed overnight against 5 mM NH$_4$HCO$_3$, 0.5 mM dithiothreitol, and 1 mM NaN$_3$ to ensure neutralization of the acid, then against several changes of H$_2$O for 24 hours. The samples are then freeze-dried and may be stored in a desiccator at -20°C. One gram of amoebae usually yields >20 mg freeze-dried protein. Prior to the NMR measurement the exchangeable protons in the samples are exchanged for deuterium. Each sample is suspended twice in several milliliters of 99.8% D$_2$O by vortexing and sonication and then freeze-dried. The exchanged samples are stored in a desiccator at room temperature to prevent contamination by water vapor.

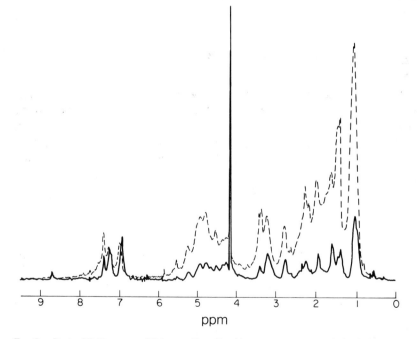

FIG. 3.   Proton NMR spectra of *Dictyostelium discoideum* supernatant protein in *d*-trifluoroacetic acid. Amoebae were grown on deuterated (~84%) *E. coli* in the presence of 47% D$_2$O (solid line) or protonated *E. coli* in the presence of H$_2$O (broken line). The sharp line occurring at 4.2 ppm is a solvent contaminant.

Proton NMR spectra are collected essentially as described by Moore (1977, 1979). We currently use a 500-MHz spectrometer operating at 25°C with 90-degree pulses and a 6-second repetition rate. Five-milligram samples are dissolved in 0.5 ml *d*-trifluoroacetic acid immediately prior to use, and 64 scans are accumulated for each sample. Typical spectra of protonated and deuterated samples are shown in Fig. 3. The deuterated amoebae used for Fig. 3 were grown on *E. coli* which had been cultured in 98% D_2O. According to Moore (1977), this should produce close to 84% deuterated *E. coli* protein. From the shape of the spectra it is evident that the proton signal of the deuterated sample is greatly reduced in the nonaromatic region but relatively unchanged in the aromatic region. The proton content of the deuterated sample can be determined from the ratio of the integrals of the proton signals of the deuterated and protonated samples. For the data in Fig. 3 this value is 26%, indicating that the extent of incorporation of deuterium into nonexchangeable carbon-bonded positions in the *Dictyostelium* protein is 74%. This approaches the highest level of deuteration obtainable using D_2O as the only source of deuterium in the *E. coli* cultivation. Reduction of the D_2O concentration in the *Dictyostelium* culture medium to 25% or 0% does not significantly change the extent of deuteration of the *Dictyostelium* protein but does influence the shape of the NMR spectrum.

V.   Purification of Deuterated Actin

Deuterated actin can be extracted from deuterated *Dictyostelium* and purified by the procedure of Uyemura *et al.* (1978), with modifications. A flow diagram of the procedure is given in Fig. 4. All solutions used for extraction and purification can be prepared in H_2O, as only exchangeable sites will lose deuterium and these are redeuterated by adding D_2O to the purified protein. The procedure described below uses 50–75 g of washed amoebae which are obtained from 5 liters of culture medium as described above.

A.   Control of Proteolysis

We have observed that *Dictyostelium* organisms grown on bacteria have greater protease activity than amoebae cultured in HL5 medium. However, proteolysis can be effectively controlled by the addition of inhibitors. A convenient method for testing the effectiveness of protease inhibitors is to incubate them with lysed (protonated) cells (or other fraction of interest) to which 1–2 mg/ml (skeletal) actin has been added. After a suitable period of incubation the concentration of undegraded actin is measured by sodium dodecyl sulfate (SDS)–polyacrylamide gel electrophoresis. Using this method, we have found chymo-

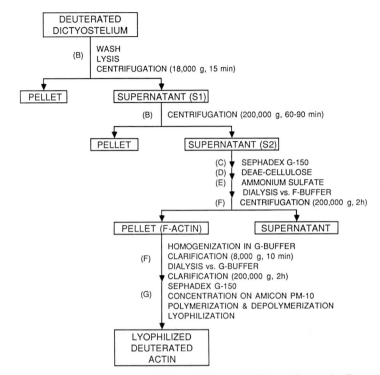

Fɪɢ. 4.   Flow diagram for the purification of deuterated *Dictyostelium* actin. For complete explanation of the protocol, refer to Section V of the text.

statin, leupeptin, and tosyl lysine chloromethyl ketone (TLCK) (all obtained from Sigma) to be the most effective inhibitors of protease activity in unfractionated, lysed cells. Currently chymostatin and leupeptin are added to the lysis buffer at a concentration of 10 mg/liter and are present in the remainder of the preparation at 2 mg/liter. TLCK is 50 mg/liter in the lysis buffer and 10 mg/liter in the remainder of the preparation. G Buffer contains all three inhibitors at a concentration of 0.5 mg/liter.

B.   Cell Lysis and Centrifugation

Washed amoebae are suspended in two volumes of lysis buffer containing 10 mM triethanolamine, pH 8.0, 40 mM sodium pyrophosphate, 30% (w/v) sucrose, 0.4 mM dithiothreitol, 3 mM NaN$_3$, and protease inhibitors. Lysis is carried out on ice by hand homogenization or sonication. Lysis is monitored by cell counts and is terminated when 10% of the original cells remain intact. The lysate is centrifuged at 18,000 g for 15 minutes; the resulting supernatant (S1) is further centrifuged at 200,000 *g* for 60–90 minutes. The high-speed supernatant (S2) contains ~80% of the actin present in whole cells.

C. Gel Filtration Chromatography I

The high-speed supernatant (S2) is applied to a 2.4-liter Sephadex G-150 column (5.4 × 105 cm) which has been completely equilibrated with 10 mM triethanolamine, pH 8.0, 20 mM sodium pyrophosphate, 50 μM ATP, 50 μM MgCl$_2$, 0.4 mM dithiothreitol, 1 mM NaN$_3$, 10% (w/v) sucrose, and protease inhibitors. Elution is carried out with the same buffer using a 30-cm head. Fractions of 20 ml each are collected and analyzed for total protein content by the procedure of Bradford (1976), and actin content by SDS–polyacrylamide gel electrophoresis on 8.5% mini-gels using the discontinuous-buffer system of Laemmli (1970) as described by Ames (1974). The gel from a typical preparation

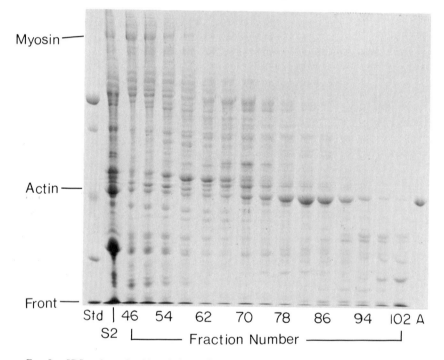

Fig. 5. SDS–polyacrylamide gel electrophoretogram of fractions from initial Sephadex G-150 column. The high-speed supernatant (S2) obtained from 50 g deuterated amoebae was applied to a 2.4-liter Sephadex G-150 column. 1.25 μl of S2 and 2.5 μl of every fourth fraction commencing with the void volume was loaded on an 8.5% SDS–polyacrylamide mini-gel. The standard (Std) contained 0.5 μg each phosphorylase B (subunit MW=92,500), bovine serum albumin (66,200), ovalbumin (45,000), and carbonic anhydrase (31,000). The actin marker (A) contained 0.5 μg protonated *Dictyostelium* actin prepared by the method of Uyemura *et al.* (1978).

is shown in Fig. 5. The actin-containing fractions appearing after all of the myosin has eluted from the column (fractions 74–90 in Fig. 5) are pooled.

D. Ion Exchange Chromatography

The pooled fractions from the gel filtration column are mixed with an equal volume of 10 mM triethanolamine, pH 8.0, 50 μM ATP, 50 μM MgCl$_2$, 0.4 mM dithiothreitol, and protease inhibitors and applied at a flow rate of 40 ml/hr (10-cm head) to a 60-ml DEAE–cellulose column (2.7 × 11 cm) which has been pre-equilibrated with 1 liter of DEAE buffer (10 mM triethanolamine, pH 8.0, 5 mM sodium pyrophosphate, 20 mM KCl, 50 μM ATP, 50 μM MgCl$_2$, 0.4 mM dithiothreitol, and protease inhibitors). After washing with one or two column volumes of DEAE buffer, the column is eluted with a linear gradient (450 ml) of 0.02–1.0 M KCl in DEAE buffer. Fractions of 6 ml each are collected and analyzed by SDS–polyacrylamide gel electrophoresis (see Uyemura *et al.*, 1978, Fig. 4). Actin elutes between approximately 150 and 250 mM KCl.

E. Ammonium Sulfate Backwash

The pooled actin from the DEAE–cellulose chromatography is brought to 50 mM triethanolamine, pH 8.0, and 65% (NH$_4$)$_2$SO$_4$ [0.39 g solid (NH$_4$)$_2$SO$_4$/ml]. After stirring for 20 minutes at 4°C, the mixture is centrifuged at 27,000 g for 20 minutes. The pellet is suspended in 10 ml backwash buffer (10 mM tri-ethanolamine, pH 8.0, 50 mM KCl, 100 μM ATP, 100 μM MgCl$_2$, 0.5 mM dithiothreitol, 1.5 mM NaN$_3$, and protease inhibitors) supplemented with (NH$_4$)$_2$SO$_4$ to 35% of saturation. The suspension is homogenized gently and stirred for 20 minutes at 4°C, then centrifuged at 27,000 g for 20 minutes. The pellet is homogenized into 10 ml backwash buffer and dialyzed overnight against 200 ml of the same solution.

F. Sedimentation and Depolymerization

The dialyzed actin is centrifuged for 2 hours at 200,000 g. The pellet of F-actin is homogenized into 10 ml G buffer (2 mM imidazole, pH 7.0, 100 μM ATP, 50 μM CaCl$_2$, 0.1 mM dithiothreitol, 1 mM NaN$_3$, and protease inhibitors) and immediately centrifuged at 8000 g for 10 minutes to remove a small amount of contaminating material. The supernatant is dialyzed for 36 hours against G buffer containing protease inhibitors and clarified by centrifugation for 2 hours at 200,000 g.

G. Gel Filtration Chromatography II

In order to improve the polymerizability of the actin, the clarified G-actin is further purified by gel filtration on a 2.5 × 50 cm column of Sephadex G-150

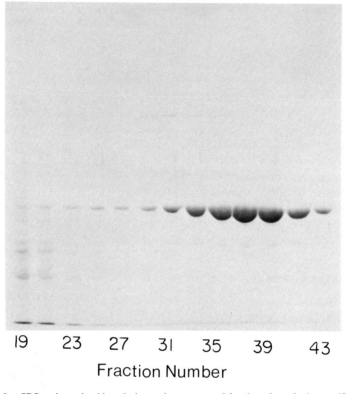

19   23   27   31   35   39   43

Fraction Number

FIG. 6.   SDS–polyacrylamide gel electrophoretogram of fractions from further purification on Sephadex G-150. 2.5 μl of every second fraction commencing with void volume fraction 19 was loaded on an 8.5% SDS–polyacrylamide mini-gel.

equilibrated in G buffer. The protein elutes in two peaks (Fig. 6). The first peak (fractions 19–23) appears at the void volume of the column and contains complexes of actin and actin-binding proteins. The second peak (fractions 31–43), containing primarily pure monomeric actin, is concentrated on an Amicon PM-10 membrane and polymerized with 5 mM MgCl$_2$ and 100 mM KCl. The F-actin is pelleted and depolymerized as described in Section V,F, then lyophilized by the method of Barden and dos Remedios (1984), and stored at −20°C.

H.   Characterization of Deuterated Actin

The protein concentration is determined by absorbance at 290 nm ($A^{1\%} = 6.3$ cm^{-1}). Approximately 10–25 mg actin is obtained from 50 g amoebae. The purity of the actin is analyzed by SDS–polyacrylamide gel electrophoresis on 8.5% mini-gels (Fig. 7). The mobility of the purified deuterated actin (lane 7) is similar to that of protonated *Dictyostelium* actin (lane 8). A 10-fold heavier loading of the purified deuterated actin (lane 9) reveals small amounts of contaminants having apparent molecular weights of 34,000–39,000. These may be proteolytic fragments of actin.

The polymerizability of the actin may be assessed in small samples of the protein by ultracentrifugation (see, for example, Cooper and Pollard, 1982) or falling-ball viscometry (Pollard and Cooper, 1982). Full characterization of the deuterated actin remains to be completed, but it appears very similar to protonated actin in its ability to assemble into filaments and to interact with myosin heads to produce "arrowheaded" complexes.

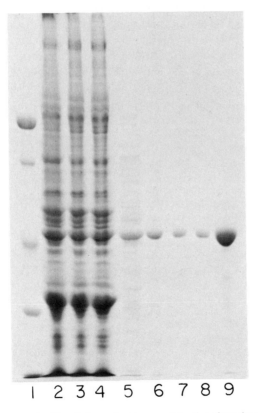

Fig. 7.   SDS–polyacrylamide gel electrophoretogram of fractions from the purification of deuterated *Dictyostelium* actin. (1) Standard containing 0.5 μg each phosphorylase B (subunit MW = 92,500), bovine serum albumin (66,200), ovalbumin (45,000), and carbonic anhydrase (31,000); (2) whole cells suspended in lysis buffer; (3) low-speed supernatant (S1); (4) high-speed supernatant (S2); (5) pooled Sephadex G-150 fractions (74–90); (6) pooled DEAE–cellulose fractions; (7) 0.2 μg purified deuterated actin (the amount of each fraction loaded on lanes 2–7 of the gel is proportional to the total volume of each fraction); (8) 0.2 μg protonated *Dictyostelium* actin prepared by the method of Uyemura *et al.* (1978); (9) 2.0 μg purified deuterated actin.

VI. Concluding Remarks

By growing *D. discoideum* on partially deuterated ($\leq$84%) *E. coli* in the presence of $\leq$47% D_2O, it is possible with only minor modification of existing procedures to extract and purify a highly deuterated ($\leq$74%) actin. This extent of deuteration is near optimal for structural studies of interest. One liter of M56 medium prepared in D_2O yields 5 g of deuterated *E. coli*, which will support the growth of 2 g of deuterated *Dictyostelium*. This weight of amoebae can yield as much as 1 mg deuterated actin. More than 95% of the D_2O used in culturing the *E. coli* and *Dictyostelium* may be recovered and used again.

The production of deuterated *Dictyostelium* represents a major advance in the supply of deuterated eukaryotic macromolecules for structural studies. The procedure outlined here is both cost- and time-efficient. D_2O is the only source of deuterium which must be purchased, and this is recycled. Both *E. coli* and *Dictyostelium* readily grow on deuterated medium without need for adaptation and with generation times which are only about twice those on comparable protonated medium. With appropriate modification this procedure should be capable of producing a variety of deuterated macromolecules from *Dictyostelium*.

ACKNOWLEDGMENTS

Some of this work was carried out in the laboratory of James A. Spudich in the Department of Cell Biology, Stanford University School of Medicine. We are indebted to Dr. Spudich and his co-workers for their assistance in the development of this procedure. We would also like to acknowledge the technical assistance of Harold Hayes. Kam C. Chan, an American Heart Association summer student, participated in the study of protease inhibitors. This work was supported by grants from NIH (HL 16683 to Robert A. Mendelson and GM 25240 to James A. Spudich); P. C. was supported by fellowships from MDA, AHA, and Postgraduate Medical Foundation, University of Sydney, Australia.

REFERENCES

Ames, G. F.-L. (1974). *J. Biol. Chem.* **249,** 634–644.
Barden, J. A., and dos Remedios, C. G. (1984). *J. Biochem. (Tokyo)* **96,** 913–921.
Bradford, M. M. (1976). *Anal. Biochem.* **72,** 248–254.
Carlson, R. D. (1984). *In* "Neutrons in Biology" (B. P. Schoenborn, ed.), pp. 47–72. Plenum, New York.
Cooper, J. A., and Pollard, T. D. (1982). *In* "Methods in Enzymology" (D. W. Frederiksen and L. W. Cunningham, eds.), Vol. 85, pp. 182–210. Academic Press, New York.
Crespi, H. L., and Katz, J. J. (1969). *Nature (London)* **224,** 560–562.
Crespi, H. L., and Katz, J. J. (1972). *In* "Methods in Enzymology" (C. H. W. Hirs and S. N. Timasheff, eds.), Vol. 26, pp. 627–637. Academic Press, New York.
Engelman, D. M., and Moore, P. B. (1975). *Annu. Rev. Biophys. Bioeng.* **4,** 219–241.
Hohl, H. R., and Raper, K. B. (1963). *J. Bacteriol.* **85,** 191–198.
Katz, J. J., and Crespi, H. L. (1966). *Science* **151,** 1187–1194.

Katz, J. J., and Crespi, H. L. (1970). *In* "Isotope Effects in Chemical Reactions" (C. J. Collins and N. S. Bowman, eds.), pp. 286–363. Van Nostrand-Reinhold, Princeton, New Jersey.

Laemmli, U. K. (1970). *Nature (London)* **227**, 680–685.

Layne, E. (1957). *In* "Methods in Enzymology" (S. P. Colowick and N. O. Kaplan, eds.), Vol. 3, pp. 447–454. Academic Press, New York.

Mandeville, S. E., Crespi, H. L., and Katz, J. J. (1964). *Science* **146**, 769–770.

Markley, J. L. (1972). *In* "Methods in Enzymology" (C. H. W. Hirs and S. N. Timasheff, eds.), Vol. 26, pp. 605–627. Academic Press, New York.

May, R. P., Stuhrmann, H. B., and Nierhaus, K. H. (1984). *In* "Neutrons in Biology" (B. P. Schoenborn, ed.), pp. 25–45. Plenum, New York.

Moore, P. B. (1977). *Anal. Biochem.* **82**, 101–108.

Moore, P. B. (1979). *In* "Methods in Enzymology" (K. Moldave and L. Grossman, eds.), Vol. 59, pp. 639–655. Academic Press, New York.

Moore, P. B., Engelman, D. M., and Schoenborn, B. P. (1975). *J. Mol. Biol.* **91**, 101–120.

Moore, P. B., Engelman, D. M., Langer, J. A., Ramakrishnan, V. R., Schindler, D. G., Schoenborn, B. P., Sillers, I-Y., and Yabuki, S. (1984). *In* "Neutrons in Biology" (B. P. Schoenborn, ed.), pp. 73–91. Plenum, New York.

Pollard, T. D., and Cooper, J. A. (1982). *In* "Methods in Enzymology" (D. W. Frederiksen and L. W. Cunningham, eds.), Vol. 85, pp. 211–233. Academic Press, New York.

Salomaa, P., Schaleger, L. L., and Long, F. A. (1964). *J. Am. Chem. Soc.* **86**, 1–7.

Schneider, W. C. (1957). *In* "Methods in Enzymology" (S. P. Colowick and N. O. Kaplan, eds.), Vol. 3, pp. 680–684. Academic Press, New York.

Schoenborn, B. P. (1975). *Brookhaven Symp. Biol.* **27**, I-10–I-17.

Stöckel, P., May, R., Strell, I., Cejka, Z., Hoppe, W., Heumann, H., Zellig, W., Crespi, H. L., Katz, J. J., and Ibel, K. (1979). *J. Appl. Crystallogr.* **12**, 176–185.

Uyemura, D. G., Brown, S. S., and Spudich, J. A. (1978). *J. Biol. Chem.* **253**, 9088–9096.

Wiesmeyer, H., and Cohn, M. (1960). *Biochim. Biophys. Acta* **39**, 417–426.

Chapter 13

Amino-Terminal Processing of Dictyostelium discoideum Actin

PETER A. RUBENSTEIN, K. L. REDMAN, L. R. SOLOMON, AND D. J. MARTIN

Department of Biochemistry
The University of Iowa
College of Medicine
Iowa City, Iowa 52242

I. Introduction

For the past few years our laboratory has been concerned with defining the NH_2-terminal processing of actin, one of the most abundant cytosolic proteins. Actins are unique in that they undergo a specific posttranslational removal of an N-Ac amino acid from the amino terminus of a completed polypeptide leading to the formation of the mature form of this protein.

Actins can be divided into two classes based on their gene structures. Class I actins are those whose genes code for an NH_2-terminal Met-Asp(Glu); examples are yeast (Ng and Abelson, 1980), *Dictyostelium* (Vandekerckhove and Weber, 1980; Firtel *et al.*, 1979), and β and γ nonmuscle actins (Gunning *et al.*, 1983). These are initially synthesized as complete polypeptides beginning with Ac-Met. Ac-Met is then removed by a specific protease, and the resulting NH_2-terminal Asp(Glu) is acetylated to yield the mature form of the protein. If acetylation of Met is prevented, then the free Met is stable (Rubenstein *et al.*, 1981; Redman and Rubenstein, 1981; Rubenstein and Martin, 1983a). This result is not unexpected, since the eukaryotic NH_2-terminal methionine aminopeptidase, which cleaves Met as a free amino acid early in translation, works only if the second amino acid in the nascent chain is small and neutral (Burstein and Schechter, 1978; Tsunasawa *et al.*, 1985).

Examples of class II actins which also begin with Ac-Asp(Glu) include mammalian and avian skeletal (Hamada *et al.,* 1982), cardiac (Gunning *et al.,* 1983; Hamada *et al.,* 1982), and smooth-muscle isoactins (Strauch and Rubenstein, 1984), *Drosophila* (Fyrberg *et al.,* 1981) and sea urchin actins (Cooper and Crain, 1982). Since their genes code for polypeptides with Met-Cys-Asp(Glu) amino termini, two amino acids must be removed. Under normal circumstances the Met is probably removed as a free amino acid early in translation, leading to a completed protein with an NH$_2$-terminal Ac-Cys-Asp(Glu). Ac-Cys is removed from the completed chain, leaving an NH$_2$-terminal Asp(Glu). which is acetylated to give the mature form of the polypeptide (Rubenstein and Martin, 1983b). The class I and class II pathways are shown in Fig. 1.

Dictyostelium discoideum actin is a molecule particularly well suited for studying actin processing *in vitro*. The organism produces essentially a single actin in large amounts, and both the actin amino acid sequence and the gene sequence have been determined (Vandekerckhove and Weber, 1980; Firtel *et al.,* 1979; Uyemura *et al.,* 1978). *Dictyostelium discoideum* actin mRNA can be

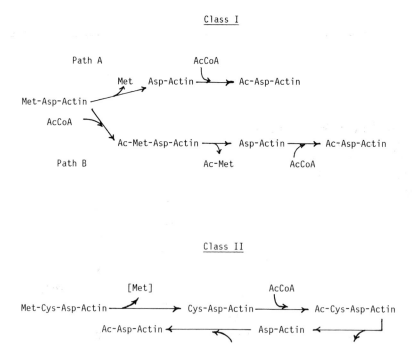

Fig. 1. Actin-processing pathways. Class I and II refer to the two groups of actins involved. Path A under class I is a *hypothetical* pathway by which actin would be processed if it lost its initiator Met early in translation, as a free amino acid, as happens with most proteins. Path B is the route actually followed.

easily obtained as a component of total-cell RNA, and it is very active in directing actin biosynthesis in an mRNA-dependent rabbit reticulocyte lysate system (Rubenstein et al., 1981).

In this chapter, we describe two assays for studying the removal of the Ac-amino acid from the amino terminus of preactin, procedures for making the actin precursor, preparation of reagents necessary for the assay, and a procedure for partial purification of the processing enzyme from rat liver. We also include a summary of prospective approaches to further investigation of this processing system.

II. Procedures for Preparing [^{35}S]Methionine-Labeled $Dictyostelium$ Actin Containing NH$_2$-Terminal Ac-Met

The overall goal of this section is to describe a method for producing [^{35}S]methionine-labeled $Dictyostelium$ actin precursor by translating actin mRNA in a rabbit reticulocyte lysate.

A. Preparation of $Dictyostelium$ RNA Containing Translatable Actin mRNA

This procedure is based on the method of Chirgwin et al. (1979). All glass-ware should be baked overnight at 100°C, and water should be treated with 0.1% (w/v) diethylpyrocarbonate for 1 hour at 60°C and subsequently autoclaved for 30 minutes to eliminate RNase contamination. One liter of Ax-3 axenic cells grown in shake flasks to early stationary phase are collected by centrifugation. Stationary-phase cells are used because they have a higher percentage of actin mRNA than vegetative cells (Tuchman et al., 1974). The cells are suspended in 9 volumes of a solution containing, per 100 ml, 60 g of guanidinium thiocyanate (Fluka), 0.5 g of N-lauryl-sarcosine (Sarkosyl), 5.0 ml of 1 M lithium citrate, pH 7, and 0.5 ml of 2-mercaptoethanol, and the suspension is homogenized immedi-ately for 1 minute using a Polytron (Brinkman). The resulting solution, 8.75 ml, is layered on 4.75 ml of a 5.7 M CsCl solution containing 0.1 M disodium EDTA in a 13.5-ml polyurethane quick-seal tube and centrifuged at 25°C for 5 hours in a Beckman Ti50 rotor at 45,000 rpm. The pelleted RNA is dissolved in 1–2 ml of water, 4 M NaCl is added (50 μl/ml) followed by 2.5 volumes of absolute ethanol. The suspension is stored overnight at −20°C. The precipitate is har-vested by centrifugation for 15 minutes at 10,000 g. The pellet is lyophilized to dryness and the residue dissolved in water, after which it is stored at −20°C. The yield per liter of cells is ~125 mg of RNA. Further purification of the mRNA by oligo(dT)-cellulose chromatography is unnecessary for these experiments.

B. Synthesis of Preactin for Use as a Processing Substrate

Our previous studies (Rubenstein *et al.*, 1981) have shown that when actin is synthesized in a reticulocyte lysate system, it accumulates in the unprocessed form, presumably because the hemin that is required for translation to occur inhibits the processing reaction (Redman, 1985). Complete rabbit reticulocyte lysate kits can be purchased from Amersham or New England Nuclear. The basic reaction mixture used for synthesis of *D. discoideum* actin contains 35 µl of mRNA-dependent rabbit reticulocyte lysate supplemented with 19 unlabeled amino acids excluding methionine, 30 µCi of L-[^{35}S]methionine (1000 Ci/ mmol), and the requisite energy mix (Pelham and Jackson, 1976). The reaction is initiated by the addition of 40 µg of *D. discoideum* RNA and allowed to proceed for 1 hour at 25°C. At this time, cycloheximide, 100 µg/ml, and methionine, 1 m*M*, are added to stop translation. An aliquot is used directly for the processing assay, or the entire translation mixture can be stored at −100°C. The actin made in this way begins with Ac-Met. If inhibition of the acetylation is desired, the acetylation inhibition system, described later, should be used prior to addition of the mRNA.

III.   Processing of *D. discoideum* Actin

This section describes methodology for following the removal of Ac-Met from the NH$_2$ terminus of the *Dictyostelium* actin precursor.

A.   Inhibition of Actin Acetylation

As seen in Fig. 1, the removal of the preactin NH$_2$-terminal Ac-amino acid is followed by acetylation of the processed actin to yield the mature form of the protein. This reacetylation occurs readily in a reticulocyte lysate and must be inhibited if processing is to be conveniently measured by the assays to be described later.

In 1977, Palmiter reported a method for inhibiting protein acetylation *in vitro* by converting endogeneous acetyl-CoA to citrate with citrate synthase and oxaloacetate. Using this method, however, we never could achieve >50% inhibition of acetylation. To improve the degree of acetylation inhibition achievable, we synthesized a nonreactive analog of acetyl-CoA, *S*-acetonyl-CoA, to serve as a potential competitive inhibitor of the protein acetyltransferase (Rubenstein and Dryer, 1980). If we first treated the reticulocyte lysate according to Palmiter and then added *S*-acetonyl-CoA, 85–95% inhibition of acetylation could be achieved (Redman and Rubenstein, 1981).

B.   Synthesis of S-Acetonyl CoA

As a reagent, monobromoacetone is synthesized according to Catch *et al.* (1948). The product is collected by distillation at 50 mm between 63.5° and 64°C as a water-clear liquid and stored in the dark at $-20°C$.

Coenzyme A (200 μmol) is dissolved in 15 ml of freshly degassed ice-cold water with magnetic stirring. Dithiothreitol (20 μmol) is added to ensure complete reduction of the coenzyme. The pH is brought to 8.0–8.2 with NaOH. Alkylation is performed by addition of monobromoacetone (250 μmol) dissolved in 5 ml of 95% ethanol just prior to use. Disappearance of -SH groups is monitored spectrophotometrically at 412 nm after adding 5 μl of the reaction mixture to 1 ml of 0.1 mM DTNB [5,5'-dithiobis(2-nitrobenzoic acid)] adjusted to pH 8.5 with 50 mM Tris buffer. The reaction with each aliquot of bromoacetone is usually complete in ≤1 minute. Solvent and excess bromoacetone are removed by lyophilization, and the product, S-acetonyl-CoA, is dissolved in 1.5 ml of water and chromatographed on a column (1.5 × 45 cm) of Sephadex G-15 equilibrated with water. Eluant fractions containing the highest concentration of material absorbing at 260 nm are pooled, and the pooled fractions are lyophilized to dryness and stored at $-20°C$. The material is also stable as a frozen aqueous solution. Acetonyl-CoA has an E^{260}_{mM} of 15.4 in water.

To prevent acetylation in a given experiment *in vitro*, the following protocol is followed. To the desired solution, before initiation of the reaction being studied, is added 1 mM of freshly made up oxaloacetic acid and 35 units/ml of porcine citrate synthase. The enzyme is made as a stock solution by centrifuging an $(NH_4)_2SO_4$ suspension of crystalline enzyme (Sigma) in a Beckman Airfuge for 5 minutes, carefully removing the supernatant solution, and dissolving the pellet in distilled water to produce the desired concentration. The reaction mixture is incubated for 7 minutes at 25°C, after which 75 μM S-acetonyl-CoA is added. Protein acetylation, at this point, is effectively inhibited. If, *prior* to the beginning of the actin translation step, the acetylation inhibition protocol is introduced, actin will be synthesized with an unblocked NH_2-Met-Asp-amino terminus (Redman and Rubenstein, 1981).

C.   Preparation of Actin-Processing Enzymes

For studying processing in our laboratory, one of three preparations of enzymes is used depending on the type of experiment being performed: rabbit reticulocyte lysate, *Dictyostelium* extracts, and partially purified rat liver preparations. Each type of enzyme is described below. In all cases, the enzyme solution is brought to 100 μg/ml of cycloheximide to block further translation and treated to inhibit acetylation as previously described immediately prior to beginning the assay.

1. Rabbit Reticulocyte Lysate

Untreated lysate is a good source of processing enzyme. However, the hemin added to promote translation must be omitted, since hemin will inhibit the processing reaction with a K_i of ~15 μM (Redman, 1985).

2. *Dictyostelium discoideum* Extract

We have recently been able to detect the presence of the actin-processing enzyme in *Dictyostelium* extracts, the first actual demonstration of this activity in a lower eukaryote (Redman *et al.*, 1985). Vegetative *Dictyostelium* cells of strain Ax-3, grown in shake flasks, are collected by centrifugation and suspended in one volume of 20 mM Tris-HCl, pH 8.0, containing 0.5 mM EDTA and 0.5 M 2-mercaptoethanol. The cells are disrupted by sonication and the resulting suspension centrifuged at 100,000 g for 1 hour at 4°C. Following dialysis of the supernatant fraction for 1 hour against the same buffer, the extract is used immediately in a processing assay. Although this preparation contains large amounts of processing activity, initial attempts to demonstrate its presence using undiluted extracts were unsuccessful because the large amounts of general proteases present in the extract destroyed the actin. Dilution of the extract 7- to 10-fold with the buffer described above circumvented the protease problem, allowing the presence of the processing activity to be demonstrated. Further purification of the *Dictyostelium* enzyme has not been attempted.

3. Partially Purified Rat Liver Enzyme

Six fresh rat livers are chilled, immediately after removal, in a buffer containing 15 mM Tris-HCl, pH 8.0, 0.5 mM EDTA, and 0.5 mM 2-mercaptoethanol. All remaining procedures are carried out at 4°C. The livers in the above buffer (3 ml/g tissue) are homogenized in a motor-driven Potter–Elvehjem device, and the homogenate is centrifuged for 15 minutes at 150,000 g. The cloudy supernatant solution is then centrifuged for 1 hour at 100,000 g. The clear-red supernatant fraction is used as a source of crude enzyme and is stored at −100°C. This preparation can be repeatedly frozen and thawed without significant loss of activity. We have partially purified this enzyme 200-fold by ammonium sulfate fractionation, Sephadex G-150 chromatography, hydroxyapatite chromatography, and chromatography on Reactive Blue-2 Sepharose. Details will be published elsewhere. Either the crude or partially purified enzyme can be utilized for most studies.

D. Enzymatic Removal of the Preactin NH$_2$-Terminal Ac-Met

1. METHOD A—TWO-DIMENSIONAL GEL ASSAY

[^{35}S]Methionine-labeled preactin in a reticulocyte lysate translation mixture is prepared as described earlier and brought to 100 μg/ml cycloheximide and 1 mM unlabeled Met to inhibit further translation. Citrate synthase and oxaloacetate followed by S-acetonyl-CoA are added as previously described to inhibit re-acetylation of the processed material. Aliquots consisting of 5–20 μl of this translation mixture are diluted 10-fold with a pH 7.5 buffered protein solution containing actin-processing enzyme. The enzyme solution should also contain cycloheximide, unlabeled Met, and the acetylation inhibition system. Tris, phosphate, and imidazole buffers work equally well in this system, and as far as we know, there are no additional cofactors required. The reaction is carried out at 25°C for the desired period. At the appropriate time, a 5-μl aliquot of the reaction mixture is analyzed by two-dimensional gel electrophoresis by the method of O'Farrell (1975), using isoelectric focusing in a pH 5–7 gradient in the first dimension and SDS-gel electrophoresis in the second. The two-dimensional system is required, since pure actin mRNA is not used, and other proteins also synthesized in the lysate exhibit the same isoelectric point as actin. The additional positive charge resulting from the removal of the NH$_2$-terminal Ac-Met causes the actin to move to a more alkaline position than the unprocessed precursor in the first dimension. A fluorogram is made from the gel using the sodium

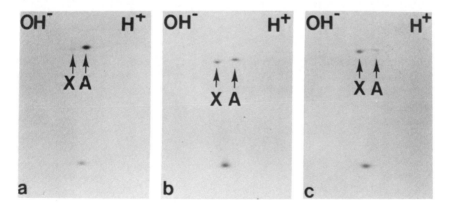

FIG. 2. Time-dependent processing of *D. discoideum* [^{35}S]methionine-labeled actin. The processing reaction and analysis by two-dimensional gel electrophoresis is described in Section II,C. Shown are autoradiograms of the actin regions of the gels. Isoelectric focusing is in the horizontal direction and SDS electrophoresis is carried out from top to bottom. **A,** The position of acetylated preactin; **X,** the position of the processed product. (a) Time 0; (b) 30 minutes; (c) 60 minutes.

salicylate method of Chamberlain (1979), and the densities in the acetylated and nonacetylated actin spots are determined by scanning densitometry. The extent of processing is calculated as the ratio of the density of the nonacetylated actin spot to the sum of the densities in the acetylated and nonacetylated actin spots. An example is shown in Fig. 2.

2. METHOD B—PEPTIDE MAPPING

The gel electrophoresis method of analysis for the actin-biosynthetic system can be misleading if the molecular structures of the various intermediates are not known. In this system, mature actin with an Ac-Asp NH_2 terminus will comigrate with preactin containing an Ac-Met-Asp NH_2 terminus. Likewise, NH_2-Met-Asp-actin will comigrate with NH_2-Asp-actin.

To establish the nature of these intermediates, it is necessary to analyze the amino-terminal actin tryptic peptide for the presence or absence of the initiating methionine. This method is based on a thin-layer electrophoresis procedure developed by Vanderkerckhove and Weber (1978) and is made possible by three characteristics of the actin NH_2-terminal tryptic peptide. First, as seen in Fig. 3, the performic acid-oxidized actin NH_2-terminal tryptic peptide has a high negative charge/mass ratio and as a result migrates at pH 6.3 far ahead of other Met-containing peptides. Second, there is a major thermolysin cleavage site between Ala_7 and Ile_8, which splits the tryptic peptide into a highly acidic N_t portion and a less acidic C_t portion. Third, the C_t portion contains a Met residue. If the protein is labeled with [^{35}S]methionine, the C_t peptide will be labeled. The N_t peptide will be labeled only if the initiator Met is still retained.

For this method, processing of actin is initiated as described in Section III,D,1, above. At the desired time, an aliquot, 0.1–1 ml, is removed and passed over a 1.5-ml DNase I actin affinity column (Worthington) previously equilibrated with 10 mM Tris-HCl, pH 8.0, containing 2.0 mM $CaCl_2$, 1.0 mM ATP,

```
                (-4)                              (-2)

                 N_t                ↓             C_t
A     Ac-ASP-GLY-GLU-ASP-VAL-GLN-ALA⌐LEU-VAL-ILE-ASP-ASN-GLY-SER-GLY-*MET-CYS-LYS

B  Ac-*MET-ASP-GLY-GLU-ASP-VAL-GLN-ALA . . . . . . . . .
```

FIG. 3. NH_2-Terminal tryptic peptide of *D. discoideum* actin. The arrow denotes the thermolysin cleavage site, and the N_t and C_t peptides produced by this enzyme. The numbers in parentheses denote the net negative charge present on each peptide at pH 6.3 following thermolysin cleavage of the performic acid-oxidized tryptic peptide. (A) The acetylated NH_2-terminal tryptic peptide of mature actin; (B) the acetylated NH_2-terminal tryptic peptide of the major actin precursor made in the reticulocyte lysate for 1 hour. The asterisk denotes the presence of [^{35}S]methionine if actin synthesis was carried out in a medium containing this labeled amino acid.

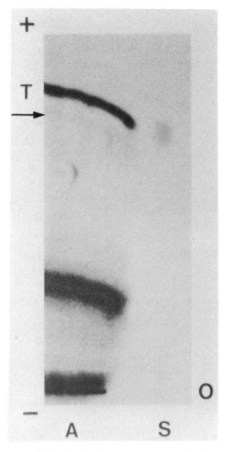

FIG. 4.   Labeled tryptic peptides from *D. discoideum* actin synthesized in a rabbit reticulocyte lysate with [^{35}S]methionine. The peptides were generated and analyzed by thin-layer electrophoresis at pH 6.3 as described in Section II,C. Shown is an autoradiogram of the electrophoretogram. **O,** origin; **A,** actin lane; **S,** position of a standard actin NH$_2$-terminal tryptic peptide. **T,** position of the NH$_2$-terminal tryptic peptide from the preactin; arrow denotes the position where this peptide would be detected following processing.

and 10% (v/v) formamide (Zechel, 1980). The column is washed with 0.4 *M* NH$_4$Cl in the same buffer, and the actin is eluted with 3 *M* guanidine HCl containing 0.5 *M* sodium acetate (Lazarides and Lindberg, 1975). Following exhaustive dialysis, the actin is lyophilized to dryness and dissolved in 0.1 ml of 97% formic acid. To this mixture is added 0.1 ml of performic acid, and oxidation is allowed to proceed for 30 minutes at 0°C. The reaction solution is diluted 10-fold with water and lyophilized to dryness.

The oxidized actin is taken up in 0.5 ml of 0.1 *M* NH$_4$HCO$_3$, and 15 μg tosyl

FIG. 5. Thermolysin digestion of the [^{35}S]methionine-labeled preactin NH$_2$-terminal tryptic peptide. **O,** Origin; C$_t$ and N$_t$ mark the peptides described in Fig. 3; arrow marks the position of orange G following thin-layer electrophoresis at pH 6.3. The peptides were generated as described in Section III,D.

phenyl ethyl chloromethyl ketone-treated trypsin (Worthington) is added. Following digestion for 2 hours at room temperature, the NH_4HCO_3 is removed by lyophilization, the residue is dissolved in water, and the resulting solution again lyophilized to dryness.

The tryptic peptides are dissolved in 20 μl of the pH 6.3 electrophoresis buffer and subjected to thin-layer electrophoresis on commercial 100-μm cellulose thin-layer plates (Eastman) at pH 6.3 in pyridine–acetic acid–H_2O (25 : 1 : 225) at 400 V using orange G as a marker dye. An autoradiogram of the electrophoresis plate is then made. The labeled NH_2-terminal actin tryptic peptide with an Ac-Met migrates just behind the orange G. Removal of the Ac-Met results in a significant retardation in the rate of migration of the NH_2-terminal tryptic peptide. A typical electrophoretogram is shown in Fig. 4.

The analysis described so far is sufficient if the structures of the NH_2-terminal tryptic peptides are known. If one wishes to determine whether a particular tryptic peptide is derived from the actin amino terminus and still contains the initiator Met, the following procedure is used. The tryptic peptide in question is eluted from the cellulose layer with water and lyophilized to dryness. The residue is taken up in 20 μl of 10 mM $NaHCO_3$ containing 2 mM $CaCl_2$ and digested with 1.5 μg of thermolysin for 2.5 hours at 30°C. The digest is applied directly to a thin layer of cellulose and subjected to electrophoresis at pH 6.3. When applying the solution to the thin-layer plate, a drying fan should not be used, since this often results in the appearance of doublets on the electrophoretogram. If the original tryptic peptide was derived from actin, a labeled C_t peptide (Fig. 3) will be observed at about 0.35 times the distance traveled by the orange G. This peptide is diagnostic for nonmuscle actins in general. If the initiator Met still remains at the NH_2 terminus of the actin amino-terminal tryptic peptide, a labeled Nt peptide (Fig. 5) will also be observed at a distance ~1.3 times that traveled by orange G. Scanning of the autoradiogram provides a means of quantitation for these peptides.

IV. Processing Specificity Studies

The specificity requirements necessary for the actin-processing enzyme to work must be quite stringent, since a whole series of short model peptides including the acetylated 18-amino acid preactin tryptic peptide are not active as substrates. To gain more information about the specificity requirements for processing, it would be convenient to have a series of different actins with various amino acid substitutions at sites of interest in the protein. Nature conveniently provides a limited number of these, and we have recently studied the processing of *Acanthamoeba* actin *in vitro*. This actin has been found to begin with Ac-Gly-Asp (Vandekerckhove *et al.*, 1984). The Ac-Gly is not removed, contrary to what would have been expected on the basis of our studies with class II actins (Strauch and Rubenstein, 1984; Rubenstein and Martin, 1983a; Solomon and Rubenstein, 1985). We wished to determine whether Ac-Gly was not removed because it was not recognized by the processing enzyme or because the organism lacked the enzyme. Our results show that neither the enzyme from rat liver nor *Dictyostelium* was capable of processing this actin *in vitro,* indicating that Ac-Gly is not recognized as a substrate by the enzyme. Furthermore, we found no evidence for the presence of the enzyme in *Acanthamoeba* (Redman *et al.,* 1985).

Comparisons of actins from various sources in these structure–function studies are often difficult, since the actins will often differ from one another at many positions. It would be more desirable for these studies to introduce one variant residue at a time into the actin and then examine its effect. This strategy might best be accomplished by carrying out site-directed mutagenesis (Zoller and Smith, 1982) on the *Dictyostelium* actin 8 gene, since it contains no introns (McKeown and Firtel, 1981). The resulting mutant actin-coding sequence could be utilized to generate actin mRNA *in vitro* (Krieg and Melton, 1984) for use in a reticulocyte lysate system. The resulting actins could then be analyzed for the ability to be processed as described in this chapter.

V.   Conclusion

The methods described above allow one to study the nature of the unusual posttranslational modifications occurring at the NH_2 terminus of virtually all actins. It is likely this modification is important for actin function, since it essentially has been conserved throughout evolution. Understanding how this processing occurs may be extremely important in elucidating the manner in which actins combine with actin-binding proteins such as myosin and various actin accessory proteins, such as depactin, which have been shown to bind to actin at or near its amino terminus (Sutoh, 1982; Sutoh and Mabuchi, 1984).

ACKNOWLEDGMENTS

This work was supported by grants to P.A.R. from the American Heart Association (81-868) and the National Institutes of Health (GM-33689). P.A.R. is an established investigator of the American Heart Association.

REFERENCES

Burstein, Y., and Schechter, I. (1978). *Biochemistry* **17,** 2392–2400.
Catch, J. R., Elliott, D. F., Hey, D. H., and Jones, E. R. (1948). *J. Chem. Soc.* 272–275.

Chamberlain, J. P. (1979). *Anal. Biochem.* **98**, 132–135.

Chirgwin, J. M., Przybyla, A. E., MacDonald, R. J., and Rutter, W. J. (1979). *Biochemistry* **18**, 5294–5299.

Cooper, A. D., and Crain, W. R. (1982). *Nucleic Acids Res.* **10**, 4081–4092.

Firtel, R. A., Timm, R., Kimmel, A. R., and McKeown, M. (1979). *Proc. Natl. Acad. Sci. U.S.A.* **76**, 6206–6210.

Fyrberg, E. A., Bond, B. J., Hershey, N. D., Mixter, K. S., and Davidson, N. (1981). *Cell* **24**, 107–116.

Gunning, P., Ponte, P., Okayama, H., Engel, J., Blau, H., and Kedes, L. (1983). *Mol. Cell. Biol.* **3**, 787–795.

Hamada, H., Petrino, M. G., and Kakunaga, T. (1982). *Proc. Natl. Acad. Sci. U.S.A.* **79**, 5901–5905.

Konauska, M. M., Padgett, R. A., and Sharp, P. A. (1984). *Cell* **38**, 731–736.

Krieg, P. A., and Melton, D. A. (1984). *Nucleic Acids Res.* **12**, 7057–7070.

Kvainev, A. N., Maniatis, T., Ruskin, B., and Green, M. R. (1984). *Cell* **38**, 993–1005.

Lazarides E., and Lindberg, U. (1975). *Proc. Natl. Acad. Sci. U.S.A.* **71**, 4742–4746.

McKeown, M., and Firtel, R. A. (1981). Cold Spring Harbor Symp. Quant. Biol. **46**, 495–505.

Maniatis, T., Fritsch, E. F., and Sambrook, J. (1982). "Molecular Cloning—A Laboratory Manual," p. 135. Cold Spring Harbor Press, Cold Spring Harbor, New York.

Ng, R., and Abelson, J. (1980). *Proc. Natl. Acad. Sci. U.S.A.* **77**, 3912–3916.

O'Farrell, P. H. (1975). *J. Biol. Chem.* **250**, 4007–4021.

Palmiter, R. D. (1977). *J. Biol. Chem.* **252**, 8781–8753.

Pelham H. R. B., and Jackson, R. J. (1976). *Eur. J. Biochem.* **67**, 247–256.

Redman, K. L. (1985). Ph.D. thesis, University of Iowa.

Redman, K. L., and Rubenstein, P. A. (1981). *J. Biol. Chem.* **256**, 13226–13229.

Redman, K. L., Martin, D. J., Korn, E. D., and Rubenstein, P. A., (1985). *J. Biol. Chem.* **260**, 14857–14861.

Rubenstein, P. A., and Dryer, R. D. (1980). *J. Biol. Chem.* **255**, 7858–7862.

Rubenstein, P. A., and Martin, D. (1983a). *J. Biol. Chem.* **258**, 3961–3966.

Rubenstein, P. A., and Martin, D. J. (1983b). *J. Biol. Chem.* **258**, 11354–11360.

Rubenstein, P. A., and Smith, P., Duechler, J., and Redman, K. (1981). *J. Biol. Chem.* **256**, 8149–8155.

Solomon, L. R., and Rubenstein, P. A. (1985). *J. Biol. Chem.*, **260**, 7659–7664.

Strauch, A. R., and Rubenstein, P. A. (1984). *J. Biol. Chem.* **259**, 7224–7229.

Sutoh, K. (1982). *Biochemistry* **21**, 3654–3661.

Sutoh, K., and Mabuchi, I. (1984). *Biochemistry* **23**, 6757–6761.

Tuchman, J., Alton, T., and Lodish, H. J. (1974). *Dev. Biol.* **40**, 116–128.

Tsunasawa, S., Stewart, J. W., and Sherman, F. (1985). *J. Biol. Chem.* **260**, 5382–5391.

Uyemura, D. G., Brown, S. S., and Spudich, J. A. (1978). *J. Biol. Chem.* **253**, 9088–9096

Vandekerckhove, J., and Weber, K. (1978). *J. Mol. Biol.* **126**, 783–802.

Vandekerckhove, J., and Weber, K. (1980). *Nature (London)* **284**, 475–477.

Vandekerckhove, J., Lal, A. A., and Korn, E. D. (1984). *J. Mol. Biol.* **172**, 141–147.

Zechel, K. (1980). *Eur. J. Biochem.* **110**, 343–348.

Zoller, M. J., and Smith, M. (1982). *Nucleic Acids Res.* **10**, 6487–6500.

Chapter 14

Biochemical and Genetic Approaches to Microtubule Function in Dictyostelium discoideum

EILEEN WHITE

Cold Spring Harbor Laboratory
Cold Spring Harbor, New York 11724

EUGENE R. KATZ

Department of Microbiology
State University of New York at Stony Brook
Stony Brook, New York 11794

I. Introduction

The cellular slime mold *Dictyostelium discoideum* represents an excellent system for studying the genetic and biochemical aspects of eukaryotic development (reviewed in this volume; Loomis, 1975, 1982). It is our intention to determine the involvement of the cytoskeleton, specifically microtubules, in important cellular processes during growth and development using *Dictyostelium* as a model system.

Microtubules are a ubiquitous component of eukaryotic cells. α- and β-tubulin, the major protein components of microtubules, are highly conserved (Luduena and Woodward, 1973; Ponstingl *et al.*, 1981; Valenzuela *et al.*, 1981) and function by assembling into the microtubule system of the cytoplasm and of the mitotic spindle (reviewed in Roberts and Hyams, 1979; Dustin, 1984). Although microtubules of the mitotic spindle are known to play a role in the separation of the daughter chromosomes at mitosis, the function of the cytoplasmic micro-

METHODS IN CELL BIOLOGY, VOL. 28

tubule network is less defined. Besides mitosis, processes occurring in the *Dictyostelium* life cycle which could be related to microtubule function include maintainance of cell shape, cell motility, modulation of the cell surface for cell–cell recognition and chemotaxis, and differentiation.

Virtually nothing is known about the function of microtubules in *Dictyostelium*. Our approach has been to probe microtubule function by examining mutants which contain genetic alterations in microtubule proteins and ask which growth and developmental processes are affected by the mutations. In order to accomplish this, genetic and biochemical methods were developed for analyzing both tubulin and the function of microtubules during the *Dictyostelium* life cycle (White *et al.*, 1983; Katz *et al.*, 1982; White, 1983): (1) A two-dimensional (2D) gel system capable of resolving single charge differences in *Dictyostelium* polypeptides was developed. Tubulin from the wild-type *Dictyostelium* was identified, and the gel system was used to screen potential microtubule mutants for electrophoretically altered microtubule proteins (Section II,A). (2) Microtubules from *Dictyostelium* amoebae were localized by indirect immunofluorescence. The structure and nature of the microtubule system was defined in both growth and development (Section II,B). (3) Conditions were established for preserving of *Dictyostelium* microtubules *in vitro*. This allowed the microtubules to be isolated and analyzed biochemically (Section II,C). (4) Methods for isolating mutants with potentially altered microtubule proteins were devised. Once mutants were obtained, tubulin and the microtubule system from them was examined and the mutant phenotypes characterized. A mutant possessing aberrations in cytoplasmic and mitotic microtubule systems is described (Section II,D).

II. Methods

A. Biochemical Analysis of *Dictyostelium* Tubulin

1. TWO-DIMENSIONAL GEL SYSTEM

We have tried numerous procedures and have obtained the best results using a combination of two methods, those of O'Farrell (1975) and Garrels (1979), with additional modifications (White *et al.*, 1983). Although there are many parameters which influence the outcome of a 2D gel, we have found that the preparation of *Dictyostelium* protein samples is the most critical aspect for obtaining satisfactory results.

a. Sample Preparation. For 2D polyacrylamide gels, 5×10^6 to 1×10^7 amoebae of strain Ax-3 (Loomis, 1971), harvested from suspension culture in HL5 medium (Watts and Ashworth, 1970), were dissolved in 60 µl of 3% SDS,

100 m*M* dithiothreitol, and immediately heated in a boiling-water bath for 45 seconds. The cooled sample was treated with DNase and RNase by adding 30 μl of a stock solution (0.5 *M* Tris, pH 7.0, 50 m*M* $MgCl_2$, 1 mg/ml DNase, 500 μg/ml RNase A), and incubated for 1 minute at room temperature. The sample was then lyophilized to complete dryness and dissolved in 180 μl isoelectric-focusing sample buffer [9.95 *M* urea, 10% Nonidet P-40 (NP-40), 2% ampholytes 5-7, 100 m*M* dithiothreitol (Garrels, 1979)]. Prior to electrophoresis, samples were warmed to 37°C and clarified by centrifugation for 2 minutes in an Eppendorf microfuge.

b. First-Dimension Isoelectric Focusing. Isoelectric-focusing tube gels 2 mm in diameter were made as described by O'Farrell (1975) and contained the following ampholytes: 3-10, 4-9, 2-4, 5-7 (Serva, Garden City Park, NY), at a ratio of 35 : 35 : 12 : 1, respectively. Gels were prefocused by raising the voltage from 200 V to 600 V over ~2 hours. After the pH gradient was established, samples (5–30 μl) were loaded and electrophoresis was carried out at 600 V for 16 hours, after which the voltage was raised to 800 V for 3 additional hours.

c. Second-Dimension SDS–Polyacrylamide Gel Electrophoresis. Isoelectric-focusing gels were extruded and equilibrated with gentle agitation in 0.375 *M* Tris, pH 8.8, 3% SDS, 50 m*M* dithiothreitol for 10 minutes, and loaded directly onto second-dimensional 10% acrylamide slab gels (Laemmli, 1970) 16 cm in height and $\frac{1}{16}$ in. thick. The isoelectric-focusing gels were gently pushed between the electrophoresis plates of the slab gels with a Teflon spatula until both gels were in direct contact (Garrels, 1979). Electrophoresis was carried out at 35–50 mA constant current, and the dye front usually took 3 hours to reach the bottom of the gel, after which the gels were either stained or blotted to DBM paper.

2. STAINING METHODS

a. Coomassie Blue. Second-dimension slab gels were stained in 50% TCA and 0.1% Coomassie brilliant blue R-250 for 20 minutes, and destained in acetic acid–methanol–H_2O (2 : 1 : 17) for several hours. Destaining was completed by soaking the gels in H_2O overnight.

b. Silver Nitrate. The most sentitive and reproducible results were obtained using the procedure of Wray *et al.* (1981) with minor modifications. Directly after electrophoresis, second-dimension slab gels were placed in 50% methanol and shaken slowly overnight. The following day, gels were washed four times over a 2-hour period with distilled H_2O to remove the methanol and stained with silver: 0.8 g of silver nitrate was dissolved in 4 ml of H_2O and added by drop to 21 ml of 0.36% NaOH, 1.4 ml 14.8 *M* NH_4OH, with constant stirring. The staining solution was brought up to 100 ml with distilled H_2O and used to stain one gel for ~12 minutes, after staining, gels were washed with distilled H_2O again (four times for a total of 45 minutes) and placed in a large volume of developer (2.5 ml 1% citric acid, 250 µl 37% formaldehyde, 500 ml distilled H_2O). With gentle rocking, spots generally appeared in several minutes but development was not complete for 15–20 minutes. Development of the gels was stopped and preserved in 50% methanol until they could be photographed.

3. IDENTIFICATION OF *Dictyostelium* TUBULIN BY WESTERN TRANSFER ANALYSIS

a. Electrophoretic Transfer of Polypeptides to Diazo Paper. Activation of the ABM paper to the DBM form was carried out according to the method of Alwine *et al.* (1979). Briefly, ABM paper (Schleicher and Schuell, Keene, NH) was washed once in cold 1.2 *M* HCl and, while on ice in a hood with 67.5 ml of cold 1.2 *M* HCl, 1.82 ml of a 10 mg/ml solution of $NaNO_2$ was added with constant shaking. After 30 minutes, the paper was washed twice rapidly with distilled H_2O, twice with cold transfer buffer [25 m*M* phosphate buffer, pH 6.5 (Symington *et al.*, 1981)], and placed in a sandwich with the gel. Two-dimensional slab gels had previously been equilibrated in transfer buffer and the gel–DBM–filter paper sandwich was supported on Scotch Brite pads and clamped tightly between stiff plastic grids (Towbin *et al.*, 1979). Electrophoretic transfer of proteins from 2D gels to the DBM paper was in 25 m*M* phosphate buffer at 30 V (1.3 A) for 4 hours at 4°C (Symington *et al.*, 1981).

b. Immunochemical Localization of Tubulin on 2D Gels. DBM paper blots were incubated with antitubulin antibodies in Tris–EDTA–NaCl–NP-40 (Symington *et al.*, 1981), followed by iodinated protein A. Transfers were autoradiographed at −70°C with Dupont Cronex Lightning-Plus intensifying screens.

The α and β subunits of tubulin from mammalian brain are acidic polypeptides with a molecular weight of 55,000. Both subunits migrate, in this and in other gel systems, slightly above and to the right (more acidic) of actin. *Dictyostelium* tubulin has the same molecular weight as brain tubulin but is much more basic, migrating to the left of actin (Fig. 1; White *et al.*, 1983). This is a highly unusual result, since, to date, tubulins from a wide variety of organisms have been shown to comigrate with brain tubulin on 2D gels. In addition, the migratory positions of *Dictyostelium* α- and β-tubulin are reversed with respect to those of brain tubulin. The α-tubulin species from brain migrates slower than β, but in *Dictyostelium* it is the α-tubulin which migrates faster.

Despite the highly conserved nature of tubulins, it is clear that there may be biochemical as well as functional differences (see below) between *Dictyostelium* microtubules and brain microtubules. Whether or not this reflects differences in the primary structure of the tubulin proteins remains to be determined.

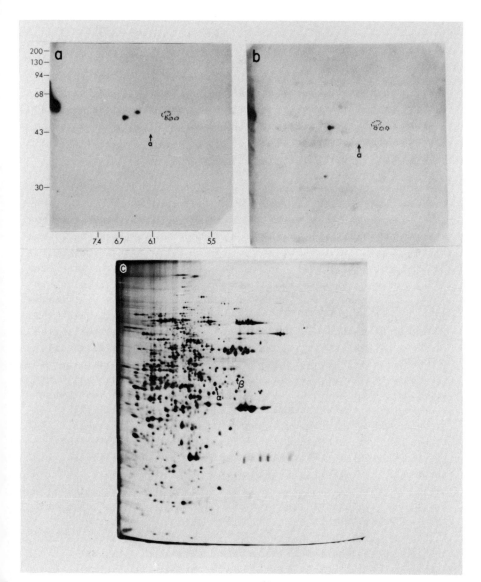

FIG. 1. Identification of *Dictyostelium* tubulin on 2D gels. Unlabeled slime mold proteins, resolved by 2D polyacrylamide gel electrophoresis were blotted to diazo paper (a,b) or stained with silver (c). Tubulins were identified by incubating the paper with antitubulin antibodies and iodinated protein A. (a) Anti-sea urchin egg α- and β-tubulin antibody; (b) anti-yeast α-tubulin antibody YL1/2. The migration position of actin (arrow) and chick brain tubulin (circles) are indicated. Purified chick brain tubulin was run in the second dimension only on the left side of (a) and (b) as an internal control. Molecular weights ($\times 10^{-3}$) are marked on the left and pH on the bottom of the gel for reference.

B. Indirect Immunofluorescence of Amoebae

1. Whole Amoebae

The microtubules in amoebae are extremely sensitive to different methods of fixation, and, in fact, many common methods used to preserve microtubules in mammalian cells do not work for *Dictyostelium*. The following, developed by the Cappucinelli laboratory (Unger *et al.*, 1979), is the most satisfactory procedure, of the many we have tried.

Amoebae at a density of 2×10^6/ml in HL5 medium were attached to glass coverslips by spreading the slips with 200 μl of the culture in a humid chamber. After 30 minutes the amoebae have sufficiently attached and spread out onto the glass. Amoebae on coverslips are fixed with 3.7% formaldehyde in HL5 medium at room temperature for 10 minutes. After fixation it is very important to wash the coverslips for 30 minutes in four changes of PBS (150 mM NaCl, 20 mM Na$^+$,K$^+$-phosphate, pH 7.4). The amoebae can then be fixed and permeabilized with methanol (9 minutes), then acetone (6 minutes), both at $-20°$C, and air-dried. After rehydration in PBS, the coverslips were incubated with antitubulin antibodies diluted in PBS, 1% BSA, for 1 hour at 37°C. The coverslips were washed for 20 minutes with four changes of PBS and then incubated with the FITC- or rhodamine-conjugated second antibodies (Miles-Yeda, Rehovot, Israel) diluted 1 : 200 in PBS, 1% BSA. After washing again for 20 minutes in PBS, coverslips were rinsed in distilled H$_2$O, and mounted in Aquamount (Lerner Laboratories, Stamford, CT). Slides often did not keep well and were therefore photographed the same day they were made, whenever possible. Slides were examined with a Zeiss Photomicroscope III equipped with epifluorescence optics using a 100× oil-immersion objective and a UG5 exciter filter. Black and white photographs were taken at ASA 1600 with Tri-X film, and the film was push-processed.

The cytoplasmic microtubule array in *Dictyostelium* consists of numerous microtubules which radiate outward from a single microtubule-organizing center (NAB) attached to the nucleus, to the cell peripheri (Fig. 2; White *et al.*, 1983; Unger *et al.*, 1979; Kuriyama *et al.*, 1982; Roos *et al.*, 1984; Omura and Fukui, 1985; Kitanishi-Yumura *et al.*, 1983). This arrangement of microtubules persists in early aggregation, with the microtubules eventually aligning with the long axis of the cell during streaming (Yumura and Fukui, 1983; White and Farsi, unpublished). At the onset of mitosis this cytoplasmic microtubule system depolymerizes and an intranuclear mitotic spindle forms (Roos, 1975; Cappuccinelli *et al.*, 1981, 1982; Roos *et al.*, 1984; McIntosh *et al.*, 1985). At the completion of mitosis the spindle depolymerizes and the cytoplasmic array is reestablished.

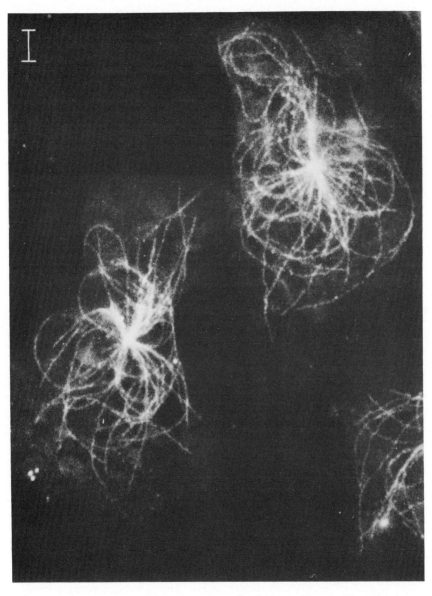

FIG. 2. Indirect immunofluorescence of *Dictyostelium* microtubules using antitubulin antibodies. Amoebae were attached to coverslips, fixed, and stained with the anti-yeast tubulin antibody YOL/34, diluted 1 : 400, followed by the corresponding FITC-conjugated second antibody. × 1200; bar = 1 μm.

2. Antibodies Which Recognize *Dictyostelium* Microtubules

We have screened a number of antitubulin antibodies for the ability to cross-react with *Dictyostelium* microtubules and have found that despite the highly conserved nature of tubulin, many antibodies do not cross-react (White *et al.*, 1983). The antibodies we have tested which do cross-react are the monoclonal antibodies directed against a tubulin from yeast, YOL/34, YL1/2 (Kilmartin *et al.*, 1982), a polyclonal antibody directed against sea urchin egg tubulin from Dr. Keigi Fujiwara (Harvard Medical School), monoclonal antibodies DM_1A, directed against chick brain α-tubulin, and MT_2, directed against mouse brain β-tubulin (Blose *et al.*, 1984). A polyclonal antibody directed against mammalian brain and another against *Tetrahymena* axonemal tubulin (Van der Water *et al.*, 1982) do not cross-react, where monoclonal antibody DM_1B (Blose *et al.*, 1984) directed against chick brain β-tubulin recognizes only the microtubule-organizing centers from amoebae (Farsi and Katz, unpublished). Similar results demonstrating a lack of cross-reaction of antitubulin antibodies with *Dictyostelium* microtubules has been reported by Roos *et al.* (1984).

C.   Preservation and Isolation of *Dictyostelium* Microtubules

1.   Cytoskeleton Extraction Buffers

Classic buffers used for extracting cells and preserving microtubules from a variety of sources (for an example, see Osborn and Weber, 1977) do not preserve microtubules from *Dictyostelium*. High detergent concentrations, low ionic strength, and protease inhibitors are necessary to preserve amoebal microtubules in an intact cytoskeleton and are included in HEMS buffer: 50 mM HEPES, 2 mM EGTA, 5 mM Mg acetate, 10% sucrose, at pH 7.4, 50 μg/ml leupeptin, and 2% cemulsol NPT-12 or NP-40 (White *et al.*, 1983).

2.   *In Situ* Extraction and Indirect Immunofluorescence

Coverslips with adhering amoebae were drained, placed in a 30 × 15 mm plastic Petri dish, and extracted with 1 ml of HEMS buffer plus a nonionic detergent, for 1 minute. The extraction buffer was then removed and replaced with 1 ml of HEMS buffer without detergent. The cytoskeletons were fixed on the coverslips by immersing them in methanol at $-10°C$ for 6 minutes. As with whole amoebae, microtubules in cytoskeletons are very sensitive to the method of fixation and are preserved best in methanol, which does not work well for unextracted amoebae. After fixation, the coverslips were rehydrated in PBS for 30 seconds and stained by the method described for whole amoebae.

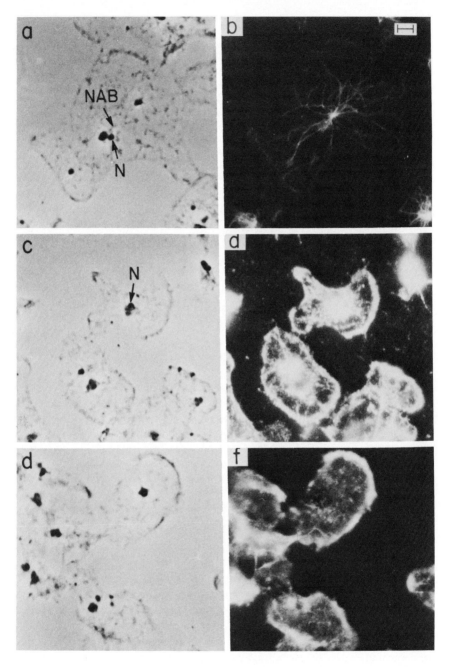

FIG. 3. The distribution of tubulin, myosin, and actin in isolated *Dictyostelium* cytoskeletons. Amoebae of Ax-3 growing in suspension culture were plated on coverslips, extracted with HEMS buffer plus detergent, fixed, and stained for indirect immunofluorescence with one of the following monoclonal antibodies: (b) antitubulin, (d) antimyosin, (f) antiactin. (a), (c), and (e) are phase-contrast micrographs of the corresponding fields. Nuclei (N) and nuclear-associated body (NAB) are marked. × 1200; bar = 1 μm.

The optimum conditions for isolation of the amoebal microtubule system preserve the arrangement of microtubules found in intact amoebae (Fig. 3; White *et al.*, 1983). Microtubules prepared in this way are stable in HEMS buffer for at least 30 minutes and, unlike mammalian brain microtubules, amoebal microtubules are stable in the presence of calcium and cold temperatures (4°C) (White *et al.*, 1983).

When cytoskeletons are stained with monoclonal antibodies directed against *Dictyostelium* myosin and actin (a generous gift of Dr. Margaret Clarke, Albert Einstein College of Medicine, Bronx, NY), there is staining of the cell periphery and pseudopodia (Fig. 3c–f). This cortical actomyosin probably corresponds to the cortical actin (Clarke *et al.*, 1975; Eckert and Lazarides, 1978; Rubino *et al.*, 1984), myosin (Rubino *et al.*, 1984), calmodulin (Bazari and Clarke, 1982), and actin-binding protein (Condeelis *et al.*, 1981; Fechheimer *et al.*, 1982) filament network of whole cells. A cortical membrane cytoskeleton and actin matrix have been isolated from whole amoebae (Luna *et al.*, 1981; Greenberg-Giffard *et al.*, 1983), indicating that the cell cortex is the location of the contractile machinery of *Dictyostelium* amoebae. Microtubules span the distance between this cortical actomyosin contractile system and the nucleus.

3. PREPARATIVE ISOLATION OF CYTOSKELETONS IN SUSPENSION

Amoebae in log phase of growth were harvested at room temperature in an International Clinical tabletop centrifuge at 300 g for 1.5 minutes. A pellet of 5 × 10^7 cells was vortexed and then extracted for 1 minute in 2.5 ml of HEMS buffer and detergent with gentle swirling, and the resulting cytoskeletons were harvested by centrifugation at 900 g for 1 minute in the tabletop centrifuge. The resulting cytoskeletal pellet was then processed for electrophoresis by the method described for whole amoebae.

To assess the degree of preservation of microtubules, cytoskeletons prepared in suspension could be monitored by immunofluorescence. A cytoskeletal pellet was resuspended in a small volume of HEMS buffer and fixed in suspension by adding a much larger volume of HEMS buffer containing 3.7% formaldehyde. Fixed cytoskeletons were harvested and washed four times with PBS by centrifugation in a tabletop centrifuge. The pellet was taken up in PBS–1% BSA, and the suspension was spread on coverslips and air-dried. The coverslips were then placed in cold methanol (9 minutes), then acetone (6 minutes), and stained by the method described for whole amoebae (not shown).

When 2D gels of total *Dictyostelium* polypeptides and isolated cytoskeletons are compared, the cytoskeletal fraction appears to be enriched for a small subset of cellular proteins (Fig. 4). Among the major components are actin, α- and β-tubulin, and a 200K protein which comigrates with myosin (White *et al.*, 1983). Other cytoskeletal proteins of unknown identity are the 28K, 50K, 60K, and groups of polypeptides of 68K and 94K in molecular weight (Fig. 4). The β-

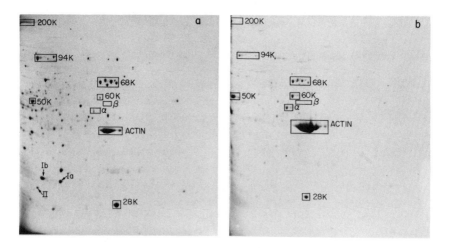

FIG. 4. Cytoskeletal proteins of *Dictyostelium*. Two-dimensional gel of (a) total amoebal proteins, and (b) cytoskeletal proteins of detergent-extracted amoebae, of strain Ax-3 growing in liquid medium. The proteins were visualized by staining with Coomassie blue. The most prominent cytoskeletal proteins are indicated by numbers corresponding to their approximate molecular weight ($\times$ 10^{-3}) in SDS gels. The presence of discoidins Ia, Ib, and II is marked.

tubulin is not well represented because of the large amount of actin focusing in the same region.

4. ATTEMPTS TO PURIFY TUBULIN

Methods for purifying brain tubulin as well as tubulin from other sources have relied on the ability of microtubules to undergo reversible cycles of assembly and disassembly *in vitro*. High tubulin concentrations are required for tubulin self-assembly. In *Dictyostelium*, however, the tubulin concentration in crude extracts is low relative to that of brain extracts, and this has so far prevented the isolation of tubulin and microtubules by this method. The purification of *Dictyostelium* tubulin has also been hampered by high protease levels and a protein in amoebal extracts which irreversibly inhibits tubulin assembly (Weinert *et al.*, 1982). Partially purifying tubulin by biochemical fractionation prior to attempting self-assembly methods was successful in purifying yeast (Kilmartin, 1981) and plant tubulin (Morejohn and Fosket, 1982), and a similar approach may be necessary for *Dictyostelium*.

D. Methods for Obtaining Potential Microtubule Mutants

Mutations conferring resistance to toxic drugs are often related to the primary site of action of the drug. Many mitotic inhibitors interact directly with tubulin, and mutants selected for resistance to mitotic inhibitors have been found to have alterations in the structural genes for α-and β-tubulin (Sheir-Neiss *et al.*, 1978; Cabral *et al.*, 1980, 1981; Keates *et al.*, 1981; Neff *et al.*, 1983; Burland *et al.*, 1984). However, no tubulin mutants, or mutants with any microtubule defects, have been isolated in *Dictyostelium* where cell biology, genetics, and development are so well represented. We have undertaken the task of selecting mutants resistant to mitotic inhibitors, some of which should have alterations in the structural genes for tubulin or other cytoskeletal components. But first, it was necessary to identify effective mitotic inhibitors in *Dictyostelium*.

1. Effective Mitotic Inhibitors

Many potential mitotic inhibitors were screened for (1) inhibition of amoebal growth at low concentrations (<25 μg/ml), (2) efficient block of mitosis by Giemsa staining, and (3) a specific effect on *Dictyostelium* microtubules (i.e., depolymerization) by indirect immunofluorescence. Drugs which did meet these criteria are CIPC [isopropyl *N*-(3-chlorophenyl)carbamate] (White *et al.*, 1981; Katz *et al.*, 1982; White, 1983), EPC (ethyl *N*-phenylcarbamate) (Kitanishi *et al.*, 1984), NOC (Nocodozole), and TBZ (thiabendazole) (Cappuccinelli *et al.*, 1979; Welker and Williams, 1980; White, 1983; Katz *et al.*, 1982). Drugs which are either ineffective or marginally effective are colchicine, vinblastine (Cappuccinelli and Ashworth, 1976), Benomyl (Welker and Williams, 1980), IPC, griseofulvin (Cappuccinelli and Ashworth, 1976), taxol and maytansine (White and Katz, unpublished experiments).

Stock solutions of CIPC (Sigma) and TBZ (a generous gift from Merck, Sharp and Dohme, Rahway, NJ) were made by dissolving them in dimethylformamide at a concentration of 20 mg/ml. NOC (Aldrich Chemical Co.) was disolved at 5 mg/ml in dimethyl sulfoxide. Stock solutions of the drugs were diluted into HL5 or SM agar medium at the desired concentration (5–10 μg/ml).

2. Selection of Mutants

CIPC, NOC, and TBZ were used to select resistant mutants in the axenic strain Ax-3. Selection for CIPC-resistant mutants, for example, was accomplished by plating spores of freshly cloned and nonmutagenized Ax-3 on SM agar plates containing 10 μg/ml CIPC that were seeded with the bacterial food source. Plates were then incubated for 1 week at 22°C. Resistant mutants arose as plaques with a frequency of 1.7×10^{-5} and were further purified by a second passage on CIPC plates. The resulting CIPC-resistant mutants were clonally purified and streak-tested for growth temperature sensitivity at 27°C (White, 1983). A similar strategy was employed with NOC and TBZ, and a preliminary characterization of some of the mutants is reported in Katz *et al.* (1982) and White (1983).

III. Summary

Methods have been developed for analyzing tubulin and microtubules from the cellular slime mold *D. discoideum*. α- and β-tubulin have been identified on high-resolution 2D gels, and microtubules have been isolated in cytoskeleton preparations from amoebae (White *et al.*, 1983). These studies have revealed properties unique to *Dictyostelium* tubulin. Amoebal microtubules can be visualized by indirect immunofluorescence, which has aided in the identification of inhibitors which specifically depolymerize microtubules and block amoebae in mitosis. The mitotic inhibitors CIPC, NOC, and TBZ have been used to select resistant mutants which are currently the subjects of biochemical, morphological, and genetic analysis (Katz *et al.*, 1982; White, 1983).

One mitotic inhibitor-resistant mutant, CIPC 6, was found to be temperature-sensitive for growth at 27°C as well as CIPC-resistant. At the restrictive temperature amoebae from this mutant are deficient in the passage through mitosis. After incubation for 12 hours at the restrictive temperature, 20% of the CIPC 6 amoebae displayed condensed chromosomes, compared to 2% at the permissive temperature, as determined by Giemsa staining. Examination of the microtubules of this mutant by indirect immunofluorescence showed abnormal spindle microtubule formation at the restrictive temperature, which is the likely cause of the mitotic arrest (White, 1983). Cytoplasmic microtubules were also disrupted in nonmitotic amoebae of CIPC 6 at 27°C. This temperature-sensitive loss of microtubule function suggested the possibility that tubulin from CIPC 6 might be altered. When tubulin from CIPC 6 was examined on 2D gels, no reproducible electrophoretic change was observed from that of the wild type.

Through further characterization of mitotic inhibitor-resistant mutants like CIPC 6, more mitotic or microtubule mutants will be identified. Among these mutants, some should contain electrophoretically altered tubulin, microtubule-associated proteins, or components of the amoebal cytoskeleton. Possessing *Dictyostelium* mutants with known biochemical alterations in cytoskeletal proteins should reveal significant information regarding the function of these proteins in eukaryotic growth and development.

REFERENCES

Alwine, J. C., Kemp, D. J., Parker, B. A., Reiser, J., Renart, J., Stark, G. R., and Wahl, G. M. (1979). *In* "Methods in Enzymology" (R. Wu, ed.), Vol. 68, pp. 220–242. Academic Press, New York.

Bazari, W. L., and Clarke, M. (1982). *Cell Motil.* **2**, 471–482.

Blose, S. H., Meltzer, D. L., and Feramisco, J. R. (1984). *J. Cell Biol.* **98**, 847–858.

Burland, T. G., Schedl, T., Gull, K., and Dove, W. F. (1984). *Genetics* **108**, 123–141.

Cabral, F., Sobel, M. E., and Gottesman, M. M. (1980). *Cell* **20**, 29–36.

Cabral, F., Abraham, I., and Gottesman, M. M. (1981). *Proc. Natl. Acad. Sci. U.S.A.* **78,** 4388–4391.

Cappuccinelli, P., and Ashworth, J. M. (1976). *Exp. Cell Res.* **103,** 387–393.

Cappuccinelli, P., Fighetti, M., and Rubino, S. (1979). *FEMS Microbiol. Lett.* **5,** 25–27.

Cappuccinelli, P., Unger, E., and Rubino, S. (1981). *J. Gen. Microbiol.* **124,** 207–211.

Cappuccinelli, P., Rubino, S., Fighetti, M., and Unger, E. (1982). *In* "Microtubules in Microorganisms" (P. Cappuccinelli and N. R. Morris, eds.), pp. 71–98. Dekker, New York.

Clarke, M., Schatten, G., Mazia, D., and Spudich, J. A. (1975). *Proc. Natl. Acad. Sci. U.S.A.* **72,** 1758–1762.

Condeelis, J., Salisbury, J., and Fujiwara, K. (1981). *Nature (London)* **292,** 161–163.

Dustin, P. (1984). "Microtubules," 2nd Ed. Springer-Verlag, Berlin.

Eckert, B., and Lazarides, E. (1978). *J. Cell Biol.* **77,** 714–721.

Fechheimer, M., Brier, J., Rockwell, M., Luna, E. J., and Taylor D. L. (1982). *Cell Motil.* **2,** 287–308.

Garrels, J. I. (1979). *J. Biol. Chem.* **254,** 7961–7977.

Greenberg-Giffard, R., Spudich, J. A., and Spudich, A. (1983). *J. Muscle Res. Cell Motil.* **4,** 115–131.

Katz, E. R., Scandella, D., White, E., Cole, M. L., and Kasbekar, D. (1982). *In* "Microtubules in Microorganisms" (P. Cappuccinelli and N. R. Morris, eds.), pp. 109–128. Dekker, New York.

Keates, R. A., Sarangi, F., and Ling, V. (1981). *Proc. Natl. Acad. Sci. U.S.A.* **78,** 5638–5642.

Kilmartin, J. V. (1981). *Biochemistry* **20,** 3629–3633.

Kilmartin, J. V., Wright, B., and Milstein, C. (1982). *J. Cell Biol.* **93,** 576–582.

Kitanishi, T., Shibaoka, H., and Fukui, Y. (1984). *Protoplasma* **120,** 185–196.

Kitanishi-Yumura, T., Blose, S. H., and Fukui, Y. (1985). *Protoplasma* **127,** 133–146.

Kuriyama, R., Sato, C., Fukui, Y., and Nishibayashi, S. (1982). *Cell Motil.* **2,** 257–272.

Laemmli, U. K. (1970). *Nature (London)* **227,** 680–685.

Loomis, W. F. (1971). *Exp. Cell Res.* **64,** 484–486.

Loomis, W. F. (1975). "*Dictyostelium discoideum.* A Developmental System." Academic Press, New York.

Loomis, W. F. (1982). "The Development of *Dictyostelium Discoideum.*" Academic Press, New York.

Luduena, R. F., and Woodward, D. O. (1973). *Proc. Natl. Acad. Sci. U.S.A.* **70,** 3594–3598.

Luna, J. E., Fowler, V., Swanson, J., Branton, D., and Taylor, D. L. (1981). *J. Cell Biol.* **88,** 396–409.

McIntosh, J. R., Roose, U.-P., Neighbors, B., and McDonald, K. L. (1985). *J. Cell Sci.* **75,** 93–129.

Morejohn, L. C., and Fosket, D. E. (1982). *Nature (London)* **297,** 426–428.

Neff, N. F., Thomas, J. H., Grisafi, P., and Bostein, D. (1983). *Cell* **33,** 211–219.

O'Farrell, P. H. (1975). *J. Biol. Chem.* **250,** 4007–4021.

Omura, F., and Fukui, Y. (1985). *Protoplasma* **127,** 212–221.

Osborn, M., and Weber, K. (1977). *Cell* **12,** 561–571.

Ponstingl, H., Krauhs, E., Little, M., and Kempf, T. (1981). *Proc. Natl. Acad. Sci. U.S.A.* **78,** 2757–2761.

Roberts, K., and Hyams, J. S. (1979). "Microtubules." Academic Press, London.

Roos, U.-P. (1975). *J. Cell Sci.* **18,** 315–326.

Roos, U.-P., DeBrabender, M., and De Mey, J. (1984). *Exp. Cell Res.* **151,** 183–193.

Rubino, S., Fighetti, M., Unger, E., and Cappuccinelli, P. (1984). *J. Cell Biol.* **98,** 382–390.

Sheir-Neiss, G., Lai, M. H., and Morris, N. R. (1978). *Cell* **15,** 639–647.

Symington, J., Green, M., and Brackmann, K. (1981). *Proc. Natl. Acad. Sci. U.S.A.* **78** 177–181.

Towbin, H., Staehelin, T., and Gordon, J. (1979). *Proc. Natl. Acad. Sci. U.S.A.* **76,** 4350–4354.

Unger, E., Rubino, S., Weinert, T., and Cappuccinelli, P. (1979). *FEMS Microbiol. Lett.* **6,** 317–320.

Valenzuela, P., Quiroga, M., Zaldivar, J., Rutter, W. J., Kirschner, M. W., and Cleveland, D. W. (1981). *Nature (London)* **289,** 650–655.

Van De Water, L., Guttman, S., Gorovsky, M., and Olmsted, J. B. (1982). *Methods Cell Biol.* **24,** 79–96.

Watts, D. J., and Ashworth, J. M. (1970). *Biochem. J.* **119,** 171–174.

Weinert, T. A., Cappuccinelli, P., and Wiche, G. (1982). *Biochemistry* **21,** 782–787.

Welker, D. L. and Williams, K. L. (1980). *J. Gen Microbiol.* **116,** 397–407.

White, E. (1983). Ph.D. thesis, SUNY at Stony Brook.

White, E., Scandella, D., and Katz, E. R. (1981). *Dev. Genet.* **2,** 99–111.

White, E., Tolbert, E., and Katz, E. R. (1983). *J. Cell Biol.* **97,** 1011–1019.

Wray, W., Boulikas, T., Wray, V. P., and Hancock, R. (1981). *Anal. Biochem.* **118,** 197–203.

Yumura, S., and Fukui, Y. (1983). *J. Cell Biol.* **96,** 857–865.

Chapter 15

Probing the Mechanisms of Mitosis with Dictyostelium discoideum

URS-PETER ROOS

Institut für Pflanzenbiologie
Universität Zürich
CH-8008 Zürich, Switzerland

I. Introduction

The descriptions, in admirable detail, of mitosis and cell division in living and fixed plant and animal cells and tissues over a hundred years ago (Flemming 1882; Strasburger, 1880) stood at the beginning of a long road the end of which we have not yet reached. More than once since that time the solution of the "enigma mitosis" seemed to be just around the corner. Over the decades, and especially since phase contrast, differential interference contrast (DIC), and electron microscopy have become routine techniques, innumerable descriptive accounts and much structural data have accumulated, so that today we know a lot about the time course of mitosis and spindle structure in many lower and higher eukaryotic cells. Our knowledge about the physiology, biochemistry, and molecular biology of the mitotic apparatus is considerably more meager. In particular, we are still ignorant about the motors that make chromosomes move and spindles elongate. The reader will find much food for thought in the reviews by Forer (1982), Fuge (1977), Inoué (1981b), McIntosh (1979), Nicklas (1971), Petzelt (1979), Pickett-Heaps *et al.* (1982), and in a monograph edited by Zimmerman and Forer (1981).

Research on mitosis begins with the choice of a suitable organism, and *Dictyostelium discoideum* is an attractive candidate. Strains with a stable, low chromosome number are available, they are easy to grow in mass culture for biochemical analysis, the cell cycle is short (reviews: Loomis, 1975; Newell, 1982; Bonner, 1967), and log-phase cultures contain a suitable proportion of amoebae in mitosis for observations on selected individual cells (Roos and Camenzind, 1981; Roos *et al.,* 1984; Sussman and Sussman, 1962). Furthermore, mitosis in *D. discoideum* has many structural and functional features of higher eukaryotic cells, but its spindle is considerably less complex (Moens, 1976; Roos *et al.,* 1984; McIntosh *et al.,* 1985). The small nucleus and consequently small mitotic spindle are minor disadvantages that can be overcome with a judicious use or adaptation of the increasingly sophisticated equipment and techniques.

To date, mitosis in *D. discoideum* has been studied cytologically by light and electron microscopy, hence the emphasis on appropriate methods in the following text. Drawing from the experience of other investigators with other organisms, I will also point out approaches not yet tried with *D. discoideum,* but for which this cellular slime mold may be equally suitable.

II. Cultures

A. Suitable Strains

Strain NC-4 (cf. Bonner, 1967) of *D. discoideum* was the most widely used before the introduction of axenic strains (cf. Loomis, 1975; Sussman, this volume, Chapter 2). It is a stable haploid strain with seven chromosomes (Sussman and Sussman, 1962). Strain NC-4 is grown nonaxenically with a bacterium, preferably *Escherichia coli* or *Aerobacter aerogenes,* as food source (Bonner, 1967). Among the strains of *E. coli,* the most frequently used has been B/r (Loomis, 1975), but in our laboratory we always use strain B. Whatever the chosen bacterial strain, it should not usually produce long filaments, as the amoebae phagocytize these less easily. The presence of bacteria in liquid cultures is a drawback that makes certain biochemical experiments difficult, but for many other purposes the bacteria can be reasonably well separated from the amoebae by a low-speed centrifugation. Maeda (1983) has succeeded in growing strain NC-4 axenically, but the cells were smaller and the generation time was much longer than in nonaxenic culture.

In spite of their other advantages, axenic strains are of doubtful use for studies on mitosis. Their cell cycle is at least twice as long as that of NC-4 (Weijer *et al.,* 1984; Zada-Hames and Ashworth, 1977), 20–30% of the amoebae in axenic culture are multinucleate, and almost 30% of the nuclei have an aberrant number of chromosomes (Zada-Hames, 1977; Zada-Hames and Ashworth, 1977). Mitotic aberrations are a likely consequence, and these are hardly desirable conditions for investigating the fundamental mechanisms of nuclear division.

B. Maintenance Cultures

We maintain strain NC-4 on agar plates with SM/2 medium. This is Bonner's medium (1967), but with nutrients and salts at half the concentrations. Fruiting bodies are more abundantly formed on this than on the full-strength medium (H. R. Hohl, personal communication), and it is also very suitable for *E. coli*.

If during inoculation of agar plates the spores of *D. discoideum* are evenly spread over the entire area covered by the bacterial inoculum, fruiting bodies will form uniformly in contrast to a point inoculation. Cultures are incubated at 21°–23°C to maturation of the fruiting bodies (3–5 days), after which they can be stored for $\geq$3 months at ~5°C. Separate plates inoculated with *E. coli* alone are transferred to the cold when the bacterial lawn is creamy.

Desiccating spores (Laine *et al.*, 1975; Loomis, 1975; Spudich, 1982) is a more convenient method for long-term storage than agar cultures. It is also a safeguard in case agar cultures are lost or become contaminated. In our laboratory, K. Iselin (personal communication) has developed a technique for freezing spores. Difco skim milk (17% in distilled water) and 20% glycerol are autoclaved separately for 10 minutes and mixed 1:1 just before use. Spores are suspended in the mixture, transferred to cryoampules, and frozen in liquid nitrogen. Thawing is done by swirling the ampule in a water bath at 20°C.

C. Cultures for Studies on Mitosis

We routinely use cultures in liquid medium which are in log phase, grown for approximately 24–32 hours at 21°–23°C. Thus, flasks inoculated one morning are ready for use the next morning, and one has a full day ahead for experiments or preparations. As inoculum we use spores from agar cultures <10 days old, for the rate of germination declines with increasing age.

The formula for the liquid medium is slightly modified from Sussman (1961): 0.5 g yeast extract, 5.0 g peptone, 5.0 g dextrose, 0.64 g $Na_2HPO_4 \cdot 2H_2O$ (in the original: 1.5 g $K_2HPO_4 \cdot 12H_2O$), 1.45 g KH_2PO_4, 0.5 g $MgSO_4 \cdot 7H_2O$, distilled water to make 1000 ml, final pH ~6.5.

Normally, we inoculate 100 ml of liquid medium contained in a 500-ml Erlenmeyer flask (cf. Roos and Camenzind, 1981), but when fewer amoebae are needed, 40 ml in a 200-ml Erlenmeyer flask will suffice. Agitation and inocula are the critical factors determining the development of a culture. The volume of medium in a flask must be just right to allow sufficient sloshing on the reciprocal shaker at ~90 cycles/min. It is also imperative to continue agitating flasks intermittently once they have been removed from the shaker for experimentation, otherwise the amoebae will die instantly.

A large bacterial inoculum will promote a rapid increase in the bacterial population, with a concomitant pH depression that slows down multiplication of the amoebae. Too big a spore inoculum leads to a premature depletion of bacteria, with accompanying clumping of the amoebae.

We do not routinely count spores of the inoculum, because there are so many other imponderables (age of medium, germination rate, etc.) that affect the development of a culture. We simply scrape fruiting bodies from an agar plate with a loop and tease the tangle apart in the flask. When one ultimately needs a pure population of amoebae, as for biochemical studies, separation of spores from stems of sorocarps is desirable. This is most easily done by shaking the crude inoculum in sterile water and passing the suspension through a plug of sterile gauze in a sterile test tube.

An important step is to check the state of a liquid culture with the light microscope before it is used further. We place a flat drop of the culture on a microscope slide and examine it with a low-power objective (e.g., $16\times$) in phase contrast or, preferably, DIC. We then visually estimate the relative cell number and we check whether the amoebae are mostly solitary or partly clumped and whether there are dividing amoebae in the sample (see below). Clumping, unless extensive, is no cause for concern. Clumped amoebae will disperse when placed on a coverslip or slide, and in suspension the clumps can be broken by repeated passages through a sterile pipet.

The mitotic index (MI) in log-phase cultures, as determined from cells fixed on coverslips (see Section IV), is $\sim 2\%$ (Roos and Cattelan, 1981). It can be increased with chemical agents blocking mitosis, but colchicine is only partially effective even at very high concentrations (Cappuccinelli and Ashworth, 1976; Rubino et al., 1982; Zada-Hames, 1977), and colcemid (Roos, 1982), the vinca alkaloids, and griseofulvin (Cappuccinelli and Ashworth, 1976; Rubino et al., 1982) are similarly ineffective. The most potent inhibitors of mitosis in cellular slime molds are benzimidazole derivatives (Welker and Williams, 1980), notably nocodazole (Cappuccinelli et al., 1979; Roos, 1980; Roos and Cattelan, 1981, 1983).

We routinely use nocodazole to accumulate amoebae in mitosis. Applied for 1 hour at a concentration of $1.65 \times 10^{-5}\,M$, it raises the MI to $\sim 17\%$ (Roos and Cattelan, unpublished) and its effect is readily reversible. Prolonged exposure to nocodazole raises the MI to as much as 60%, but at the expense of mitotic aberrations and incomplete recovery of the cells from the treatment (Roos and Cattelan, 1981, 1983).

A stock solution of nocodazole is prepared in dimethyl sulfoxide (DMSO) at a concentration of 1 mg/ml. Storing smaller aliquots frozen for one-time use avoids repeated thawing and freezing and guarantees a fresh solution every time. The stock solution is added directly to the liquid culture, but this must be done drop by drop under continuous shaking to ensure complete solubility of the compound in the medium (M. De Brabander, personal communication).

For recovery from nocodazole, amoebae are collected by low-speed centrifugation and resuspended in conditioned medium, prepared from an untreated parallel culture from which the amoebae have been removed by centrifugation. A wave of mitoses follows within 15–30 minutes, but synchrony is not strict, for nocodazole arrests cells in all the stages of nuclear division. Most dividing amoebae enter mitosis during the treatment and will therefore form a metaphase spindle during recovery. From observations on live cells, on spindles stained by indirect immunofluorescence with an antibody against tubulin, and on amoebae stained with 4', 6-diamidino-2-phenylindole (DAPI), we have no indication that mitosis is not normal after recovery from a 1-hour exposure to $1.65 \times 10^{-5} M$ nocodazole (Roos and Cattelan, 1983).

A method to synchronize cell division in axenic strains without a chemical inhibitor has been used by Soll et al. (1976) and Weijer et al. (1984). It is based on the principle that starving amoebae in stationary-phase cultures are arrested in late G_2 of the cell cycle. Upon dilution into fresh medium the cells undergo division synchronously. Whether this synchrony is tight enough for studies on mitosis has not been investigated cytologically. With strain NC-4 a similar mechanism may be operating, for serendipitous observations showed that numerous amoebae from a clumped culture in postlog phase divide almost immediately when allowed to spread on a microscope slide or coverslip. For the accumulation of a large quantity of mitotic amoebae this possibility of synchronization clearly deserves further attention.

Mitotic or cell cycle mutants, as now exist for *Aspergillus nidulans* (Bergen et al., 1984; Morris et al., 1982; Oakley and Morris, 1983), *Saccharomyces cerevisiae* (e.g., King and Hyams, 1983), and *Schizosaccharomyces pombe* (Toda et al., 1983; Umesono et al., 1983), are another attractive possibility for synchronization and the accumulation of mitotic cells. In *D. discoideum*, many temperature-sensitive mutants for growth have been isolated (Loomis, Chap. 3, this volume; Newell, 1982), some of which might in fact be mitotic mutants (B. M. Coukell, personal communication). Even if this proved not to be the case, true mitotic mutants should be obtainable, too.

III.   Live Observations of Mitosis

A.   Observation Chambers

Various miniature chambers have been designed for the light-microscopic observation of living cells (e.g., Heunert, 1961, 1971). G. Gerisch (1963, and personal communication) used a chamber allowing observations on individual *Dictyostelium* amoebae sandwiched between a block of agar and a coverslip.

We have found it preferable to observe amoebae in an unconstrained situation, i.e., adhering to a coverslip in a thin layer of liquid medium (Fig. 1). The chambers are not thicker than an ordinary slide, so optimal Köhler illumination can be achieved and the lower cover glass allows condenser immersion. A tiny amount of a nontoxic antifog should be evenly spread on the inside of the lower coverslip, for droplets of condensing water will otherwise drastically diminish image quality.

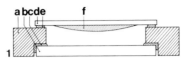

Fig. 1.   Schematic cross section of a culture chamber for live observations of mitosis. a, Aluminum slide; b, circular coverslip, 0.4 mm thick; c, paraffin sealing; d, square or rectangular coverslip, e.g., 24 × 24 mm or 25 × 35 mm, 0.13–0.18 mm thick; e, Vaseline sealing; f, liquid nutrient medium—the amoebae adhere to coverslip d. From Roos and Camenzind (1981), with permission.

Whereas the lower coverslip of a chamber is semipermanently mounted with paraffin, the upper coverslip is disposable and a new one is prepared prior to every observation session. For this purpose, humid chambers are prepared from Petri dishes containing a moist filter paper on top of which a piece of glass rod bent into a V or Z shape is placed. An ordinary glass slide is put on top of the rod, as support for one or several coverslips large enough to cover the hole in the aluminum chamber.

Sterility may be desired, but it is not required for preparations that are kept only for a few hours or less. We normally just rub the coverslips clean with a paper towel, then seed each with a drop or two from a log-phase culture in liquid medium and close the Petri dish to let the amoebae settle and attach in the humid atmosphere. This will take only 5–10 minutes, but there is no harm in keeping them as long as 1–2 hours. It is very important to keep the drop of culture rather flat, for otherwise the amoebae in the center will die, presumably because of lack of oxygen. Cell density on a coverslip can be adjusted at this stage by the size of the drop of culture. Mitotic amoebae are scarce when cells are too densely seeded. Cell density, degree of flattening, and general state of the amoebae can be easily checked as described above (Section II,C) when a coverslip is placed on a glass slide on the stage of an upright microscope, but care must be taken lest the preparation begins to dry. We check routinely to make sure we only use good coverslip preparations for live observations, experiments, or fixation and embedding (see Section VI).

Excess medium is then withdrawn with a Pasteur pipet held vertically to the coverslip until only a very thin film of liquid remains that affects the optical quality of the preparation only minimally. To prevent drying, the coverslip is then quickly inverted on top of the aluminum chamber, the hole of which has been ringed with Vaseline. The medium on the coverslip should not touch the rim of the hole and the Vaseline should not smear into the chamber. The coverslip is pressed down so that the Vaseline seals it airtight. No liquid nutrient medium must ever fill the space between the two cover glasses, otherwise the amoebae will die instantly. In a well-sealed chamber the amoebae stay alive for several days, and we have occasionally found fruiting bodies inside.

Mitotic amoebae can be found immediately after a chamber is sealed, but if

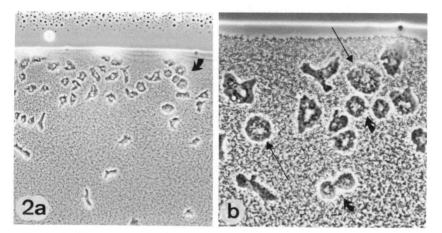

Fig. 2.   (a) Phase-contrast micrograph at low power of live amoebae at the periphery of the layer of liquid medium in a chamber as in Fig. 1. The arrow points to two amoebae in mitosis. (b) Higher magnification, showing two amoebae in mitosis (slender arrows) and two in cytokinesis (thick arrows).

one waits until many of them have moved, together with bacteria, to the periphery of the film of medium (cf. Roos and Camenzind, 1981), there are numerous mitotic amoebae, many of them flatter than toward the center of the preparation (Fig. 2). Amoebae in the early stages of mitosis are easily recognized: they are stationary, more or less circular, and they lack the prominent nucleolus of interphase cells.

B.   Light Microscopy

A heat-absorption filter must be used in the path of light to avoid cell damage or artifacts. The condenser should always be immersed for best resolution.

The axis of mitotic spindles is more or less parallel to the plane of the coverslip, hence also to the optical plane, especially in flat amoebae. Details such as the spindle pole bodies are occasionally visible, but otherwise only the outline of the nucleus is recognizable during mitosis, for the chromosomes of *D. discoideum* are optically hidden in the dispersed nucleolus (Roos and Camenzind, 1981). When weighing the pros and cons of phase contrast versus DIC, one has to consider the possibility of obtaining optical sections and the disadvantages of superimposition of objects in different optical planes (cf. McIntosh, 1982).

No technique of image recording equals the resolving power of the human eye, therefore some details will always be lost. Because of its good resolution compared to its convenient dimensions, 35-mm film is the material of choice. Whenever the light is sufficient we use Kodak technical pan film 2415, which has an extremely good resolution and can be developed for various degrees of contrast (consult the appropriate technical bulletins). An electronic flash is indispensable for still pictures of fast events, and it also allows one to follow the cells at a low light level that further reduces the possibility of heat damage.

For recording movements in real time or time lapse, video microscopy is superseding 16-mm film in most cases. Excellent results can be obtained with relatively inexpensive equipment (De Brabander et al., 1985; Inoué, 1981a, 1986). Combined with digital image processing, video microscopy has opened exciting new avenues for research on motile events even below the limit of resolution of the light microscope (e.g., Allen, 1985; Allen et al., 1985). For *Dictyostelium* this offers the first opportunity to also study chromosome movement in live cells. Preliminary observations (Roos, unpublished) demonstrated that chromosomes are visible in a digitally processed image obtained with video-enhanced contrast (AVEC-DIC, see Allen et al., 1981).

Another approach that now becomes realizable is to follow chromosomes in live cells vitally stained with DNA-specific dyes, such as DAPI (cf. Lin et al., 1977). Normal epifluorescence with UV excitation kills amoebae instantly (Roos, unpublished). A way out of this dilemma is to irradiate the cells with innocuous intensities of UV light and to intensify video-electronically the resulting weak emission signal. The technology is now available and such observations will no doubt be carried out in the very near future.

Similarly, it should be possible to follow individual components of the mitotic apparatus, e.g., the spindle, or even different classes of spindle microtubules (MTs), and the spindle pole bodies (SPBs), after marking them with fluorescently labeled antibodies or spindle molecules microinjected into living amoebae (fluorescent analog cytochemistry; Wang et al., 1982; Salmon et al., 1984).

IV. DNA-Specific Dyes for Fluorescence Microscopy

Fluorescent dyes specific for DNA, like DAPI, Hoechst 33258 (Latt and Wohlleb, 1975), and mithramycin (Ward et al., 1965), are excellent chromosome stains. In *D. discoideum* we have obtained weak staining with mithramycin, but we use DAPI for many purposes, as a counterstain in immunofluorescence preparations (Roos et al., 1984), or as a quick stain to check the effects of mitotic inhibitors (Roos and Cattelan, 1981). DAPI reveals very fine details of interphase nuclei, of chromosome configurations, ingested bacteria, and mitochondria (cf. Roos, 1982).

Fixation with buffered glutaraldehyde gives superior preservation of cell components stained by DAPI, but the problem of autofluorescence (Collins and Goldsmith, 1981) must be kept in mind. To facilitate stain penetration we prefer a permeabilizing prefixation (see Sections V,B and V,D). Staining is sufficient after 10–15 minutes in a solution of 0.1 μg/ml in the fixation buffer or in distilled water. Amoebae can be mounted in the staining solution, in buffer, in distilled water, in 90% glycerin, or in gelvatol (Rodriguez and Deinhardt, 1960), the choice depending on the desired durability of the preparations. An agent retarding fading of fluorescence should be added to the mounting medium. We have had better results with *p*-phenylenediamine (Johnson and De Nogueira Araujo, 1981), than with *n*-propyl gallate (Giloh and Sedat, 1982) or DABCO (Langanger *et al.*, 1983; but see also Böck *et al.*, 1985).

V. Immunocytochemistry of the Mitotic Spindle

A. Suitable Antibodies and Markers

MTs are a major component of the mitotic spindle of *D. discoideum* (McIntosh *et al.*, 1985; Moens, 1976; Roos, 1980) which can be revealed for light microscopy by indirect immunofluorescence (Cappuccinelli *et al.*, 1981; Roos *et al.*, 1984). Polyclonal and monoclonal antibodies against tubulins of different origin or against proteins associated with MTs (microtubule-associated proteins, MAPs) have been produced in many laboratories, and some are available commercially. As is true for other lower eukaryotes (e.g., Lemoine *et al.*, 1984), the MTs of *D. discoideum* do not bind certain antibodies raised against brain tubulin from mammals, but the monoclonal antibody YL 1/2 of Kilmartin *et al.* (1982), raised against tubulin isolated from the yeast *Saccharomyces uvarum*, gives excellent results (Roos *et al.*, 1984; see also White *et al.*, 1983).

Antibodies coupled to colloidal gold have become a powerful tool in immunocytochemistry (De Mey, 1983; Slot and Geuze, 1984) and have virtually superseded the peroxidase–antiperoxidase technique (Sternberger, 1974) for combined light and electron microscopy. MTs of mitotic spindles can be revealed clearly (e.g., De Mey *et al.*, 1982), but one must take into account that cells must be extracted considerably with nonionic detergent to allow the larger gold granules to reach the antigen sites.

Selective immunocytochemistry of specific spindle components has become possible with antibodies or sera reacting with kinetochores (e.g., Moroi *et al.*, 1980), the spindle poles (Calarco-Gillam *et al.*, 1983), or both (Vandre *et al.*, 1984). Such probes have proved useful for studies on the fate of these structures during the cell cycle and mitosis (Brenner *et al.*, 1981), but also for analyses of spindle function (Cox *et al.*, 1983) and for the molecular dissection of the kinetochore (Palmer and Margolis, 1985; Sullivan *et al.*, 1985). We could certainly learn much from *D. discoideum* in this respect, but little exploration has been done. Our own attempts to localize kinetochores, SPBs, and calmodulin have been unsuccessful. There will no doubt be further advances as more antibodies or sera are tested.

B. Fixation and Permeabilization of Amoebae

The following account is mainly based on our experience with antibodies against tubulin, but it certainly has a greater range of applications. We have tested many different regimens of fixation and permeabilization (e.g., Roos *et al.*, 1984), checking each for preservation of cell structure, especially MTs, by transmission electron microscopy of ultrathin sections. Kitanashi *et al.* (1984) have used a "sandwich" technique for immunofluorescence of MTs, which yields flat cells, but in many cases the necessary additional handling and artificial flattening prior to fixation may be inconvenient.

Formaldehyde as a fixative makes amoebae permeable, but it preserves cell shape poorly at the light-microscopic level and it does not preserve MTs at the ultrastructural level. Fixation with acetone and methanol at −20°C (cf. Osborn and Weber, 1982; Weber and Osborn, 1979) also renders cells permeable but detaches many amoebae from the coverslip and preserves them poorly. Glutaraldehyde preserves cell structure, especially MTs, superiorly (Weber and Osborn, 1979). Nevertheless, if first attempts with an antibody are negative, one is well advised to use other fixation regimens to assess the influence of fixation on the outcome.

We found that the fixing action of glutaraldehyde must be counteracted by a fairly aggressive permeabilizing agent to prevent shrinking of the amoebae. Our standard protocol (listed below) is therefore based on a delicate balance between the lysing effect of Triton X-100 and the fixing effect of glutaraldehyde. Autofluorescence induced by glutaraldehyde (see Section IV) should be checked with a preparation fixed identically but not incubated with any antibody.

C. Antibody Staining, Embedding, and Microscopy

Treatment with sodium borohydride is necessary with glutaraldehyde-fixed cells to reduce excess aldehyde groups that could bind antibodies nonspecifically (Weber and Osborn 1979). Normal goat serum prevents nonspecific binding of the second antibody. Nevertheless, this type of nonspecific binding should also be checked with a control from which the first antibody is omitted.

Embedding is best in a water-soluble semipermanent or permanent mounting medium, such as 90% glycerol or gelvatol (Rodriguez and Deinhardt, 1960). If *p*-phenylenediamine is added to the mounting medium, the preparations can be viewed repeatedly over several weeks with minimal loss of fluorescence intensity. Sodium azide and sodium iodide may be even superior antifading reagents, for according to Böck et al. (1985), they not only retard bleaching, but also enhance the fluorescence intensity of FITC.

We have occasionally observed autofluorescence of chromatin and chromosomes with the FITC filters and noticed that this was after viewing the preparations with the DAPI filters, therefore presumably induced by the UV light (cf. Collins and Goldsmith, 1981).

D. Protocol for Tubulin Immunofluorescence

All steps are carried out at room temperature.

1. Seed amoebae onto a coverslip (see Section III,A) cleaned for 5 minutes in a mixture of equal parts of ether and methanol for better adherence of amoebae during fixation (cf. Roos et al., 1986). Withdraw medium and put on a cushion of 1% glutaraldehyde in PHEMO containing 0.2% Triton X-100. PHEMO is the MT-stabilizing buffer of Schliwa et al. (1982), with 10% DMSO added. Fix and permeabilize for 2 minutes.

2. Replace first solution with 1% glutaraldehyde in PHEMO, fix for 10 minutes.

3. Rinse for 10 minutes each, once with PHEMO, twice with Tris-buffered saline (TBS, 0.5 mg/ml).

4. Treat with fresh solution of $NaBH_4$ in TBS (10 mg/20 ml) for 10 minutes; repeat with fresh solution.

5. Rinse three times with TBS, 10 minutes each.

6. Incubate for 30 minutes with 5% normal goat serum (NGS) in TBS, 75 μl per coverslip.

7. Rinse for 10 minutes with TBS.

8. Incubate for 2 hours or overnight (in a humid box) with the first antibody, e.g., rat monoclonal antitubulin YL 1/2 (Kilmartin et al., 1982) diluted 1:800 in TBS containing 1% NGS, 75 μl per coverslip.

9. Rinse three times with TBS, 10 minutes each, on a rocking table.

10. Dilute FITC-labeled goat anti-rat IgG 1:40 or 1:80 in TBS, incubate each coverslip for 0.5 to 1 hour with 75 μl in the humid box.

11. Rinse as in step 9.

12. Counterstain for 10 minutes with a fresh aqueous solution of DAPI (0.1 μg/ml).

13. Rinse twice with distilled water, 10 minutes each.

14. Embed in gelvatol or 90% glycerin containing p-phenylenediamine (see Section IV).

Rhodamine- or TRITC-labeled phallotoxin (Wulf et al., 1979) can be added to the solution of the second antibody, or, alternatively, the preparations are incubated with this peptide specific for F-actin in an additional step between steps 11 and 12, for the simultaneous localization of MTs and F-actin (cf. Roos et al., 1986).

VI. Combined Light and Electron Microscopy of Mitotic Amoebae

The combined light and electron microscopy of individual mitotic amoebae can be approached in two ways: (1) selection after fixation and embedding on the coverslip, or (2) preselection of live amoebae and fixation on the microscope stage.

A. Fixation and Embedding *in Situ*

1. Coverslips with adhering amoebae, cleaned, seeded, and checked as described (Sections II,C and V,D), are transferred to shallow plastic caps and fixed.

2. Coverslips are provided with landmarks, such as scratches made with a diamond-tipped marker, which will help in relocating the cells. The coverslips are cleaned and seeded as described. A perfusion chamber (Roos et al., 1986) is fashioned from a glass slide onto which two double strips of parafilm are applied as supports for the coverslip, which is inverted over them. The edges of the coverslip are sealed with valap (one part each of Vaseline, lanolin, and paraffin; cf. Molè-Bajer and Bajer, 1968), except for an opening in the middle of each of the two small sides. Amoebae can then be observed and recorded live until the desired stage, when they are fixed by perfusion of fixing solution through the chamber. Fixing solution poured with a syringe drop by drop at one of the openings in the seal will rush through the chamber compartment by capillary action. Further passage of fixative can be effected by holding a filter paper to the opposite opening. Recording the area containing the amoebae with a low-power objective after perfusion will greatly facilitate relocation of cells. A video recording is obviously advantageous, because it can be played back instantly. The coverslip is removed and processed further in a plastic cap, cell side up.

Many regimens of fixation for transmission electron microscopy of *D. discoideum* amoebae have been used by various investigators over the years (cf. Loomis, 1975; Sameshima, 1985). We obtained good fixation with glutaraldehyde and osmium tetroxide in Sørensen's phosphate buffer (McIntosh et al., 1985; Roos, 1980; Roos and Camenzind, 1981). Because glutaraldehyde causes shrinkage even when added directly to a growing culture, we have recently also used a fixation regimen as for immunofluorescence (cf. Section V,D). Postfixation with osmium tetroxide, staining en bloc with uranyl acetate, and infiltration with the embedding medium are carried out as described (McIntosh *et al.*, 1985; Roos and Camenzind, 1981).

As the last step, the coverslips are inverted over the wells (depth = 0.7 mm) of a homemade silicone rubber mold (cf. Chang, 1971) filled with embedding medium. The back of the coverslip is carefully scraped and wiped as clean of resin as possible. After polymerization, the coverslips are removed from the chips by immersion in liquid nitrogen.

B. Selection and Relocation of Individual Amoebae

The plastic chips are mounted cell side up in special aluminum slides (Fig. 3). Immersion oil must fill the space between the chip and the supporting glass, and a layer of immersion oil applied to the upper face of the chip further improves the optical quality even for dry objectives. These preparations, too, are no thicker than an ordinary one, and conditions for correct Köhler illumination are maintained. Relocated or selected amoebae are photographed with phase contrast and scribed with a diamond-tipped marker fastened to the turret in place of one of the objectives.

The chip is carefully cleansed of immersion oil with ethanol, the back side also with acetone (careful: it softens or partially dissolves some epoxy resins). It is then pressed onto a double sticky tape on a piece of cardboard and the scribed area is cut out. We use a sharp-edged, hollow bit (inner diameter 2 or 4 mm) on a

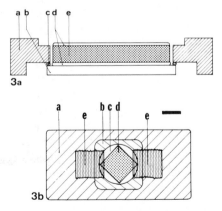

FIG. 3. Aluminum slide for mounting a plastic chip with embedded amoebae. (a) Schematic cross section. a, Aluminum slide; b, square coverslip, 0.4 mm thick; c, glue sealing; d, immersion oil; e, plastic chip, mounted cell side up. (b) Top view, drawn to scale (bar = 1 cm). a, Aluminum slide; b, recessed portion of slide; c, bottom coverslip; d, plastic chip; e, adhesive tape fixing the chip in position.

drill press. This way we avoid having to cut a chip into several pieces and can use it again for selecting other amoebae. The small disk is glued to a plastic stub for sectioning. Because the axis of the mitotic spindle was more or less parallel to the plane of the coverslip, sections parallel to the surface of such a block will yield longitudinal or slightly oblique sections of the spindle (Fig. 4).

If cross sections of the spindle are desired, one's choice is limited to amoebae whose spindle axis is approximately perpendicular to the direction of one of the stage movements of the light microscope. The diamond marker is then used to scratch a line as close as possible to the selected amoeba (Fig. 5). Additional markings are helpful for identification of the mitotic amoeba in ultrathin sections.

An adjustable wedge placed between the slide holder of the microscope stage and the aluminum slide would make it possible to vary the angle between the stage drive and the spindle axis, so that some amoebae whose spindle is not perpendicular could be included.

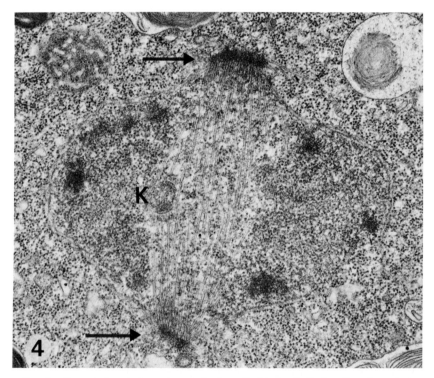

Fig. 4. Transmission electron micrograph of an ultrathin longitudinal section of a metaphase nucleus. The two spindle pole bodies (arrows) are prominent, and a pair of kinetochores (K) is visible to the left of the central spindle.

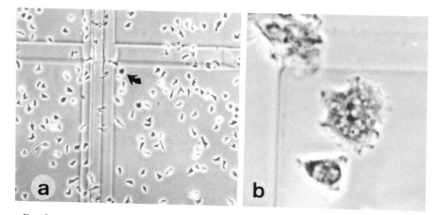

FIG. 5.   Amoeba embedded in a plastic chip marked for cross sections of the mitotic spindle. (a) The horizontal line marks the boundary for trimming the block down to the telophase cell (arrow). The vertical scratch facilitates relocation of the amoeba in sections. (b) The horizontal line is perpendicular to the spindle axis of the telophase nucleus.

A rectangular piece with the marked amoeba is excised from the plastic chip and glued to the side of a square supporting stub, the cells facing the stub, and the portion with the amoeba protruding. The protruding portion is reinforced with a drop or two of resin. Epoxy glue is not suitable for this, because it is porous and its sectioning properties are poor. The block so obtained is trimmed down to the line near the amoebae, which saves much time before one starts with serial ultrathin sections.

VII.   Summary and Outlook

Mitosis in *D. discoideum* has been studied in live amoebae (Roos and Camenzind, 1981), by immunofluorescence (Cappuccinelli *et al.*, 1981; Roos *et al.*, 1984), and by electron microscopy (Moens, 1976; McIntosh *et al.*, 1985). From the live observations we learned that the spindle elongates by a factor of three or more during anaphase and telophase at a velocity of 4 μm/min, which is faster than the rate of chromosome movement in many cells. Clearly, these facts must be explained by any model of mitosis. The immunofluorescence studies provided insight into the overall mode of formation of the MT spindle as the interphase complex of MTs breaks down, and the reverse process at the end of nuclear division.

Electron microscopy has revealed that the spindle consists of the ensemble of kinetochore MTs that link chromosomes directly to the SPBs, plus the central spindle, which is made up essentially of two sets of interdigitating pole MTs. As the spindle elongates, the number of MTs decreases while the remaining MTs elongate by addition of subunits at their distal, overlapping ends. Concomitantly, the order between interdigitating MTs from opposite poles increases. These results are compatible with the idea that the spindle elongates by MT sliding and elongation.

Unresolved details of spindle formation and structure in *D. discoideum* are the genesis of the SPBs from the single, highly structured microtubule-organizing center of interphase amoebae (cf. Roos, 1980), and the precise relationship between the ends of spindle MTs and the SPBs. For a test of models of mitosis, the polarity of the spindle MTs must be known (cf. Haimo, 1982; Heidemann and Euteneuer, 1982). Our current work addresses these questions, as well as that of whether our common fixation procedures for electron microscopy are adequate, or if cryofixation and freeze-substitution (Heath and Rethoret, 1982; Howard, 1981) are required for the preservation of structures as labile as spindle MTs.

The future of mitosis research belongs to cells that are accessible to many different methodological approaches: cytological, biochemical, molecular, and genetic. Yeasts and *Aspergillus* have a head start on the latter two approaches (Blackburn and Szostak, 1984; Morris *et al.*, 1982), but as organisms with a cell wall they are less well suited for cytological and biochemical analysis (see, however, Kilmartin and Adams, 1984). Especially with mutants one would like to be able to check the phenotype quickly by light microscopy, rather than by tedious electron microscopy. Here lies one of the advantages of *D. discoideum* as a "naked" cell. Now that the genetic approach has become feasible (Loomis, Chap. 3, this volume) and the biochemical analysis of microtubule proteins has been worked out (White *et al.*, 1983), there are many places this modest organism can go.

Acknowledgments

I am grateful to Dr. J. De Mey for his competent instruction in immunofluorescence, to Mrs. H. Cattelan for help with the figures, and to Prof. H. R. Hohl for encouragement and support. The original work done in this laboratory and in collaboration was also supported in part by funds from the Swiss National Science Foundation and the Jubiläumsstiftung of the University of Zurich.

References

Allen, R. D. (1985). *Annu. Rev. Biophys. Biophys. Chem.* **14**, 265–290.
Allen, R. D., Allen, N. S., and Travis, J. L. (1981). *Cell Motil.* **1**, 291–302.
Allen, R. D., Weiss, D. G., Hayden, J. H., Brown, D. T., Fujiwake, H., and Simpson, M. (1985). *J. Cell Biol.* **100**, 1736–1752.
Bergen, L. G., Upshall, A., and Morris, N. R. (1984). *J. Bacteriol.* **159**, 114–119.

Blackburn, E. H., and Szostak, J. W. (1984). *Annu. Rev. Biochem.* **53**, 163–194.
Böck, G., Hilchenbach, M. Schauenstein, K., and Wick, G. (1985). *J. Histochem. Cytochem.* **33**, 699–705.
Bonner, J. T. (1967). "The Cellular Slime Molds," 2nd Ed. Princeton Univ. Press, Princeton, New Jersey.
Brenner, S., Pepper, D., Berns, M. W., Tan, E., and Brinkley, B. R. (1981). *J. Cell Biol.* **91**, 95–102.
Calarco-Gillam, P. D., Siebert, M. C., Hubble, R., Mitchison, T., and Kirschner, M. (1983). *Cell* **35**, 621–629.
Cappuccinelli, P., and Ashworth, J. M. (1976). *Exp. Cell Res.* **103**, 387–393.
Cappuccinelli, P., Fighetti, M., and Rubino, S. (1979). *FEMS Microbiol. Lett.* **5**, 25–27.
Cappuccinelli, P., Unger, E., and Rubino, S. (1981). *J. Gen. Microbiol.* **124**, 207–211.
Chang, J. P. (1971). *J. Ultrastruct. Res.* **37**, 370–377.
Collins, J. S., and Goldsmith, T. H. (1981). *J. Histochem. Cytochem.* **29**, 411–414.
Cox, J. V., Schenk, E. A., and Olmsted, J. B. (1983). *Cell* **35**, 331–339.
De Brabander, M., Geuens, G., Nuydens, R., Moeremans, M., and De Mey, J. (1985). *Cytobios* **43**, 273–283.
De Mey, J. (1983). *In* "Immunocytochemistry: Applications in Pathology and Biology" (J. M. Polak and S. Van Noorden, eds.), pp. 82–113. Wright-PSG, Boston.
De Mey, J., Lambert, A. M., Bajer, A. S., Moeremans, M., and De Brabander, M. (1982). *Proc. Natl. Acad. Sci. U.S.A.* **79**, 1898–1902.
Flemming, W. (1882). "Zellsubstanz, Kern und Zelltheilung." Vogel, Leipzig.
Forer, A. (1982). *In* "The Cytoskeleton in Plant Growth and Development" (C. W. Lloyd, ed.), pp. 185–201. Academic Press, New York.
Fuge, H. (1977). *Int. Rev. Cytol. Suppl.* **6**, 1–58.
Gerisch, G. (1963). "Entwicklung von *Dictyostelium*." 16mm-film C 876. Institut für den Wissenschaftlichen Film, Göttingen, FRG.
Giloh, H., and Sedat, J. W. (1982). *Science* **217**, 1252–1255.
Haimo, L. T. (1982). *Methods Cell Biol.* **24**, 189–206.
Heath, I. B., and Rethoret, K. (1982). *Eur. J. Cell Biol.* **28**, 180–189.
Heidemann, S. R., and Euteneuer, U. (1982). *Methods Cell Biol.* **24**, 207–216.
Heunert, H. H. (1961). *Res. Film* **4**, 59–65.
Heunert, H. H. (1971). *Zeiss Inf.* **41-101.1-d**, 3–14.
Howard, R. J. (1981). *J. Cell Sci.* **48**, 89–103.
Inoué, S. (1981a). *J. Cell Biol.* **89**, 346–356.
Inoué, S. (1981b). *J. Cell Biol.* **91**, 131s–147s.
Inoué, S. (1986). "Video Microscopy." Plenum Press, New York.
Johnson, G. D., and De Nogueira Araujo, G. M. (1981). *J. Immunol. Methods* **43**, 349–350.
Kilmartin, J. V., and Adams, A. E. M. (1984). *J. Cell Biol.* **98**, 922–933.
Kilmartin, J. V., Wright, B., and Milstein, C. (1982). *J. Cell Biol.* **93**, 576–582.
King, S. M., and Hyams, J. S. (1983). *Eur. J. Cell Biol.* **29**, 121–125.
Kitanishi, T., Shibaoka, H., and Fukui, Y. (1984). *Protoplasma* **120**, 185–196.
Laine, J., Roxby, N., and Coukell, M. B. (1975). *Can. J. Microbiol.* **21**, 959–962.
Langanger, G., De Mey, J., and Adam, H. (1983). *Mikroskopie (Vienna)* **40**, 237–241.
Latt, S. A., and Wohlleb, J. C. (1975). *Chromosoma* **52**, 297–316.
Lemoine, A., Mir, L., and Wright, M. (1984). *Protoplasma* **120**, 43–50.
Lin, M. S., Comings, D. E., and Alfi, O. S. (1977). *Chromosoma* **60**, 15–25.
Loomis, W. F. (1975). "*Dictyostelium discoideum*. A Developmental System." Academic Press, New York.
McIntosh, J. R. (1979). *In* "Microtubules" (K. Roberts and J. S. Hyams, eds.), pp. 428–441. Academic Press, New York.

McIntosh, J. R. (1982). *Methods Cell Biol.* **25**, 33–56.
McIntosh, J. R., Roos, U.-P., Neighbors, B., and McDonald, K. L. (1985). *J. Cell Sci.* **75**, 93–129.
Maeda, Y. (1983). *J. Gen. Microbiol.* **129**, 2467–2473.
Moens, P. B. (1976). *J. Cell Biol.* **68**, 113–122.
Molè-Bajer, J., and Bajer, A. (1968). *Cellule* **67**, 257–265.
Moroi, Y., Peebles, C., Fritzler, M., Steigerwald, J., and Tan, E. M. (1980). *Proc. Natl. Acad. Sci. U.S.A.* **77**, 1627–1631.
Morris, N. R., Kirsch, D. R., and Oakley, B. R. (1982). *Methods Cell Biol.* **25**, 107–130.
Newell, P. C. (1982). *In* "The Development of *Dictyostelium discoideum*" (W. F. Loomis, ed.), pp. 35–70. Academic Press, New York.
Nicklas, R. B. (1971). *Adv. Cell Biol.* **2**, 225–297.
Oakley, B. R., and Morris, N. R. (1983). *J. Cell Biol.* **96**, 1155–1158.
Osborn, M., and Weber, K. (1982). *Methods Cell Biol.* **24**, 97–132.
Palmer, D. K., and Margolis, R. L. (1985). *Mol. Cell. Biol.* **5**, 173–186.
Petzelt, C. (1979). *Int. Rev. Cytol.* **60**, 53–92.
Pickett-Heaps, J. D., Tippit, D. H., and Porter, K. R. (1982). *Cell* **29**, 729–744.
Rodriguez, J., and Deinhardt, F. (1960). *Virology* **12**, 316–317.
Roos, U.-P. (1980). *In* "Microtubules and Microtubule Inhibitors" (M. De Brabander and J. De Mey, eds.), pp. 385–398. Elsevier/North-Holland, New York.
Roos, U.-P. (1982). *In* "Microtubules in Microorganisms" (P. Cappuccinelli and N. R. Morris, eds.), pp. 51–69. Dekker, New York.
Roos, U.-P., and Camenzind, R. (1981). *Eur. J. Cell Biol.* **25**, 248–257.
Roos, U.-P., and Cattelan, H. (1981). *Experientia* **37**, 658 (Abstr.).
Roos, U.-P., and Cattelan, H. (1983). *Abstr. EMBO Workshop Mitosis, Univ. College, London.*
Roos, U.-P., De Brabander, M., and De Mey, J. (1984). *Exp. Cell Res.* **151**, 183–193.
Roos, U.-P, De Brabander, M., and Nuydens, R. (1986). *Cell Motil. Cytoskel.* **6**, 176–185.
Rubino, S., Unger, E., Fogu, G., and Cappuccinelli, P. (1982). *Z. Allg. Mikrobiol.* **22**, 127–131.
Salmon, E. D., Leslie, R. J., Saxton, W. M., Karow, M. L., and McIntosh, J. R. (1984). *J. Cell Biol.* **99**, 2165–2174.
Sameshima, M. (1985). *Exp. Cell Res.* **156**, 341–350.
Schliwa, M., van Blerkom, J., and Pryzwansky, K. B. (1982). *Cold Spring Harbor Symp. Quant. Biol.* **46**, 51–67.
Slot, J. W., and Geuze, H. J. (1984). *In* "Immunolabelling for Electron Microscopy" (J. M. Polak and I. M. Varndell, eds.), pp. 129–142. Elsevier, New York.
Soll, D. R., Yarger, J., and Mirick, M. (1976). *J. Cell Sci.* **20**, 513–523.
Spudich, J. A. (1982). *Methods Cell Biol.* **25**, 359–364.
Sternberger, L. A. (1974). "Immunocytochemistry." Prentice-Hall, New York.
Strasburger, E. (1880). "Zellbildung und Zelltheilung," 3rd Ed. Fischer, Jena.
Sullivan, K. F., Cleveland, D. W., Saunders, W., Rothfield, N., and Earnshaw, W. S. (1985). *Abstr. Int. Symp. Microtubules Microtubule Inhibitors, 3rd, Beerse, Belgium.*
Sussman, M. (1961). *J. Gen. Microbiol.* **25**, 375–378.
Sussman, M., and Sussman, R. R. (1962). *J. Gen. Microbiol.* **28**, 417–429.
Toda, T., Umesono, K., Hirata, A., and Yanagida, M. (1983). *J. Mol. Biol.* **168**, 251–270.
Umesono, K., Toda, T., Hayashi, S., and Yanagida, M. (1983). *J. Mol. Biol.* **168**, 271–284.
Vandre, D. D., Davis, F. M., Rao, P. N., and Borisy, G. G. (1984). *Proc. Natl. Acad. Sci. U.S.A.* **81**, 4439–4443.
Wang, Y.-L., Heiple, J. M., and Taylor, D. L. (1982). *Methods Cell Biol.* **25**, 1–11.
Ward, D. C., Reich, E., and Goldberg, I. H. (1965). *Science* **149**, 1259–1263.
Weber, K., and Osborn, M. (1979). *In* "Microtubules" (K. Roberts and J. S. Hyams, eds.), pp. 279–313. Academic Press, New York.

Weijer, C. J., Duschl, G., and David, C. N. (1984). *J. Cell Sci.* **70,** 111–131.

Welker, D. L., and Williams, K. L. (1980). *J. Gen. Microbiol.* **116,** 397–407.

White, E., Tolbert, E. M., and Katz, E. R. (1983). *J. Cell Biol.* **97,** 1011–1019.

Wulf, E., Deboben, A., Bautz, F. A., Faulstich, H., and Wieland, T. (1979). *Proc. Natl. Acad. Sci. U.S.A.* **76,** 4498–4502.

Zada-Hames, I. M. (1977). *J. Gen. Microbiol.* **99,** 201–208.

Zada-Hames, I. M., and Ashworth, J. M. (1977). *In* "Development and Differentiation in the Cellular Slime Moulds" (P. Cappuccinelli and J. M. Ashworth, eds.), pp. 69–78. Elsevier/North Holland, New York.

Zimmerman, A. M., and Forer, A. (1981). "Mitosis/Cytokinesis." Academic Press, New York.

Part IV. Chemotaxis

The next layer of cell biological complexity, chemotaxis, naturally follows the section on cell motility. Part IV begins with a chapter by Konijn and Van Haastert, who describe how to measure chemotaxis. Several different assays are described, as well as methods for giving localized pulses of chemoattractants, time-lapse cinematography, and other procedures pertinent to the study of the chemotactic response. Devreotes *et al.* next describe the cyclic AMP signaling system and methods to purify and study the cAMP receptor. Their chapter includes further information on growth and development of cells, how to monitor spontaneous oscillations, several chemotaxis assays, ways of measuring intracellular and secreted cAMP, and how to assay for adenylate cyclase. They go on to describe binding assays for cAMP and its analogs, receptor photoaffinity labeling, receptor modification, and purification procedures for the cAMP receptor. Their chapter concludes with a section on characterization of transmembrane signaling mutants. In Chapter 18, Berlot describes methods for examining changes in myosin phosphorylation during the chemotactic response. Her chapter includes conditions for labeling cells *in vivo* with [^{32}P]orthophosphate, conditions that eliminate phosphatase and protease activities in extracts, and methods to isolate phosphorylated protein rapidly by immunoprecipitation. The last chapter in Part IV is by Fukui *et al.*, who describe detailed procedures for an agar-overlay technique for immunofluorescence. This technique provides higher resolution than others that have been employed, so that individual myosin thick filaments and their dynamics during the chemotactic response have been able to be measured.

Chapter 16

Measurement of Chemotaxis in Dictyostelium

THEO M. KONIJN AND PETER J. M. VAN HAASTERT

Cell Biology and Morphogenesis Unit
Zoological Laboratory
2311 GP Leiden, The Netherlands

I. Introduction

Chemotaxis is the oriented response and movement of cells or organisms to gradients of chemicals in their environments. The response is positive when cells move to higher concentrations of chemicals such as nutrients or attractants secreted by neighboring cells; a negative response means turning away from higher concentrations of components like toxic substances or repellents. Bacteria and true slime molds are attracted to a wide variety of nutrients such as sugars and amino acids. The cellular slime molds only respond to specific attractants which lead them to their food source or, in case of starvation, to social partners to form cell aggregates.

For many years research on chemotaxis was limited to the demonstration that an extract or compound was attractant or repellent. Lately a plausible sequence of events from ligand–chemoreceptor interaction to the final directed movement start to emerge. The early steps of signal transfer have become an exciting field of research. Here, neurotransmitter, hormonal, and chemotactic signals seem to converge. Crucial elements in the transfer of all these signals from the outside to the inside of the cell surface are appropriate receptors, guanine nucleotide regulatory proteins, calcium, phosphoinositol, protein kinase C, and the nucleotide cyclases. Probably during early evolution, cells only could survive by adequate responses to environmental signals, of which the transduction mechanism through the cell surface has been largely preserved. Therefore, the elucidation of

283

the molecular mechanism of chemotaxis may contribute to the explanation of signal transduction in general.

To investigate the transfer of chemotactic signals one has to measure the response to chemoattractants. A shift from observing cell behavior to exploring the mechanism of signal transduction required more sophisticated chemotactic assays. Measurement of simple qualitative positive or negative responses was supplemented with semiquantitative observation of chemotaxis. Several assays have been developed; some have already been applied to other cells, and others are specifically designed to measure a chemotactic response in the slime molds, as will be shown in the next part of this chapter.

II.   Chemotactic Assays

Among the chemotactic assays, different division lines may be drawn. Assays may be divided according to a response to a direct gradient of attractant or exposure of cells to an equally distributed attractant. Another division depends on the measurement of chemotaxis in the unicellular or in the multicellular stage. Normally chemotaxis is measured by exposure of cell populations to the attractant; single cells, however, can also be activated by a chemotactic compound.

A.   Early Assays

Most assays for the measurement of chemotaxis in the cellular slime molds depend on a direct detection of a gradient of attractant by the amoebae.

In the wake of Pfeffer's success in demonstrating chemotaxis in a variety of organisms (1884), abortive trials were made early in this century by exposing *Dictyostelium* cells to gradients of sugars or malic acid. However, in the early 1940s Runyon (1942) demonstrated chemotaxis in the cellular slime molds. He deposited differently aged amoebae at both sides of a cellophane membrane. When the older cells aggregated, an attractant secreted by these cells induced the younger cells to form aggregation patterns coinciding with the aggregates at the other side of the membrane. He concluded that a chemotactic substance diffused through the permeable membrane.

Runyon's observations were confirmed by Bonner (1947), who used a variety of simple but convincing approaches to show that chemotaxis controls cell aggregation in the cellular slime molds. He used two coverslips placed side by side under water. An aggregation center near the edge of one coverslip attracted sensitive amoebae from the other coverslip, even when both coverslips were 20–30 μm apart. When water flowed past an aggregate, only downstream amoebae moved to the center, while the upstream amoebae were not attracted. Apparently a chemical signal secreted by the center induced in both cases responding cells to migrate to the source of the attractant. Francis (1965) made use of Bonner's underwater-flow technique. He refined the flow test by regulating the depth of the water and the flow rate; the density of the downstream tail of the aggregate (upstream amoebae moved at random) changed according to the size of the center and the sensitivity of the responding cells. Francis was able to replace the center by a continuous injection of concentrated washwater of sensitive amoebae.

Shaffer realized that a flow test disturbs existing diffusion gradients (1956). He therefore sandwiched his cells between a block of agar and a glass surface and carried out his chemotaxis experiments in a moist chamber. Washwater from slugs repeatedly transferred to one side of the agar block induced the sandwiched amoebae to respond positively only to this side. Washwater was quickly inactivated except when it was filtered through a cellophane membrane; apparently an inactivating enzyme, later identified as a phosphodiesterase (Chang, 1968), destroyed the chemotactic compound present in the washwater and later identified as cyclic AMP (cAMP; Konijn et al., 1967). The sandwich assay, like the flow assay, is not suitable for a quantitative approach.

B.   Boyden Assay

Boyden (1962) introduced the micropore-filter method to assay chemotaxis in leukocytes. The Boyden chamber consists of two compartments which are separated by a micropore filter. A suspension of leukocytes occupies the upper compartment, and the chemotactic substance has been dissolved in the lower compartment. Chemotactic molecules diffuse through the filter from below and cells on the upper surface migrate, to reach the highest attractant concentration, into the filter of which the pore size is smaller than the cell diameter. The optimal pore size for *D. discoideum* cells is 1.2 μm (Bonner et al., 1971). Steepness of the gradient, though gradually decaying, induces the cells to move to the lower surface of the filter. The number of cells penetrating through the filter within a certain time period is a measure of the activity of the attractant. The count of cells at the lower site of the filter is facilitated after fixation and staining. The assay is rather elaborate and time-consuming. Testing various concentrations of attractants requires a number of chambers. An advantage is its constant humidity, which avoids different degrees of dryness of the agar surface as observed in some other assays.

C.   Small-Population Assay

The small-population assay has been dealt with previously (Konijn, 1970, 1975). Suspended cells are placed as a small drop on a hydrophobic agar surface of a specific rigidity. The agar surface becomes hydrophobic after repeated washing with deionized water (Ennis and Sussman, 1958). Such agar is water-repellent after boiling and gelation, even when salts have been replenished (Konijn and Raper, 1961). Within the boundaries of the drop, the agar is hydrophilic and cells move freely as on plain nonnutrient agar. They do not cross the drop margins except at agar concentrations of ≤0.3% (w/v) when cells move into the agar and outside the boundaries of the drop. At a slightly higher rigidity of the agar surface, amoebae only leave the drop when attracted by an outside source such as a cell aggregate (Fig. 1), dense bacterial population, or active extract. Attracting and responding drops have a diameter of ~0.6 mm, obtained by placement of 0.1-μl drops on the hydrophobic agar surface. The responding drops contain 500–1000 cells. The strength of the chemotactic activity exerted by the attracting drop can be measured semiquantitatively by increasing the distance between attracting and responding drops and determining the distance at which 50% of the responding drops react positively. A drawback of this approach is the sensitivity of the assay to the rigidity of the agar. A slightly higher or lower stiffness of the agar affects the ease with which cells cross the margins of the drop. To eliminate this variable, the rigidity of the agar surface should be increased by using higher agar concentrations. If a weight of 40–80 g, depending on the responding species, is needed to push a microscope slide through the agar surface, then chemotactically activated cells do not cross the boundaries of the drop. Instead, they press against the side closest to the attracting drop (Fig. 2). With at least twice as many cells pressed against this margin than occur on the opposite side, the response is considered to be positive. *Dictyostelium dis-*

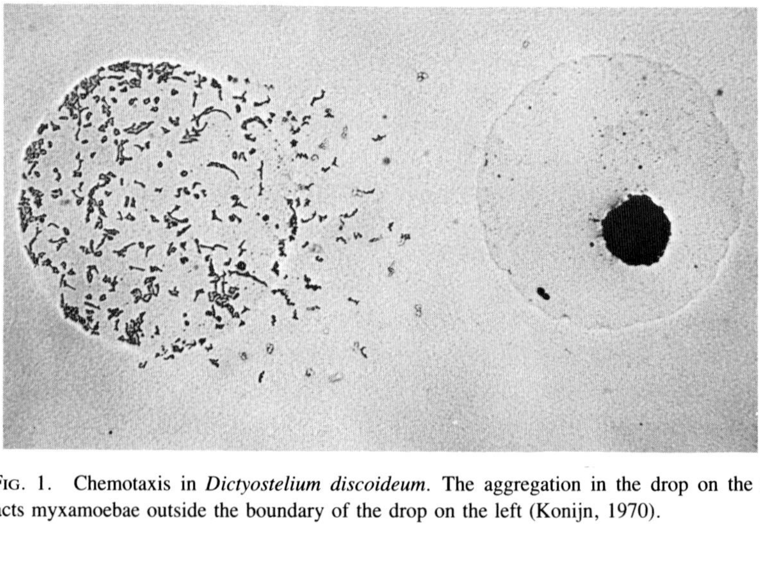

FIG. 1. Chemotaxis in *Dictyostelium discoideum*. The aggregation in the drop on the right attracts myxamoebae outside the boundary of the drop on the left (Konijn, 1970).

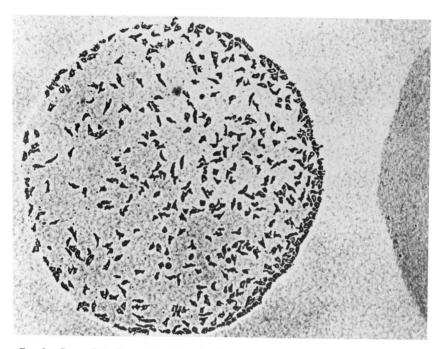

FIG. 2. Drops (0.1 μl) containing 3 × 10⁻¹³ g cyclic AMP were deposited three times at 5-minute intervals to the right of the responding drop (Konijn, 1970).

coideum cells react this way on hydrophobic agar with a surface rigidity of 40 g. Most other species require a surface rigidity of 80 g. Details on the measurement of the rigidity of the agar surface are given elsewhere (Konijn and Raper, 1961). More simply, one freeze-dries agar washed with deionized water, pours some plates with different agar concentrations, and uses for future experiments the agar concentration which results in the desired stiffness to keep the cells inside the drop. Also, on this more rigid agar the strength of the attractant can be measured by increasing the distance between attracting and responding drops. After gelation the agar should be stored in the refrigerator for 1 day before being used, and dishes >1 week old should be discarded.

The experiment is facilitated by drawing four to six lines on the bottom of the Petri dish with hydrophobic agar and placing 20–30 drops on each line while observing the deposition through a low-magnification binocular microscope. Normally experimenters need no more than 1 hour to learn the technique to place 100–150 equally sized drops, ~0.1 μl each, on the agar within a couple of minutes by use of a micropipet, hand-drawn from a Pasteur pipet. Fast deposition of drops leaves the Petri dishes uncovered for only a short period and avoids large variations in cell density. One pipet-filling should not be used for more than two rows of drops.

Instead of increasing the distance between attracting and responding drops, one also may keep a constant distance of 200–300 μm between both drops and vary the concentration of the attractant by serial dilutions. Particularly for testing the chemotactic activity of analogs of known attractants, serial dilution of the test substances is an obvious choice.

Summarizing, the small-population assay gives reproducible results, is sensitive to low concentrations of the attractants (10^{-8}–10^{-9} M), allows several determinations within a short time period (positive response of sensitive amoebae to attractants and their analogs can be observed within half an hour), and, if quantitative results are required, two- to threefold differences in the concentration of the attractant can be shown this way. For smaller differences this assay is unsuitable.

D. Cellophane Square Assay

The cellophane square and the small-population assay are most frequently used for the measurement of chemotactic activity. A significant difference between the cellophane square assay (Bonner et al., 1966) and the previously mentioned assays is the equal distribution of the attractant through the agar medium. The attractant is mixed with liquid agar and, after gelation, squares of cellophane on which amoebae have settled will be placed on the agar surface. The cells cross the edge of the cellophane membrane (Fig. 3), and the distance between one edge and the four amoebae which moved furthest away from it is measured after 1, 2, and 4 hours. This distance is compared to the outward movement of cells on a cellophane membrane which has been placed on plain agar. Also, this test allows semiquantitative measurement of chemotactic activity by serial dilution. More active attractants allow cells to move further away from the edge of the cellophane square. Supposedly the cells themselves produce enzymes which inactivate locally the chemotactically active substance in the agar, creating a gradient which would orient the amoebae away from the cellophane square. This gradient therefore has not been produced by the experimenter but by the cells themselves.

The advantage of this indirect assay is its simplicity. It is less suitable if only minute quantities of the attractant to be tested are available and only one constant concentration of a specific attractant can be tested in one Petri dish. Also, the time period of exposure to the active substance is longer than in some other assays. A modification of the cellophane square test is the deposition of the amoebae directly on the cAMP-containing agar (Soll and Mitchell, 1982). Drops of a dense cell suspension 300 μm in diameter are placed on 2% nonnutrient agar containing 10^{-7} M cAMP. The diameter of the drop is measured just after

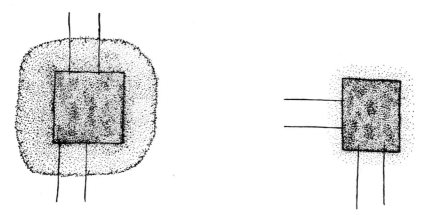

FIG. 3. Drawings of two photographs of the cellophane square assay. The control is at the right, and an experimental with active-rate substance is shown on the left. The central cellophane square is ~4 mm long (Bonner *et al.*, 1966).

plating and after 2 hours. The increase of the diameter of the drop depends on the developmental stage of the amoebae.

E.   Well Assay

Another simple chemotaxis test is the use of a small agar dish with a well in the center (Wallace and Frazier, 1979; Nandini-Kishore and Frazier, 1981). A 3-ml aliquot of a 2% agar solution is poured into a dish of 35 mm diameter. A well of 5 or 7 mm diameter is punched in the center and filled with a chemoattractant solution. Then, 1 -μl drops containing sensitive amoebae are placed on the agar at different distances from the well. After 3 hours all drops containing cells are photographed. Also, this assay becomes semiquantitative by varying the distance between the drop and the well and the concentration of the attractant. Similar to the cellophane square assay, only one attractant concentration can be tested in one dish, and the cells are exposed to the attractants for a long time before the chemotactic effect is registered.

III.   Localized Pulses of Chemoattractants and Iontophoresis

The small-population assay and iontophoresis allow cells to be stimulated repeatedly while the cellophane square and Boyden (1962) assays expose cells to a constant or a gradually less steep gradient of attractant. Application of pulses of cAMP to sensitive amoebae allows an imitation of the natural pulses of cAMP secretion. Cells of the more advanced species of *Dictyostelium* do not release attractant constantly; the interval varies with the species, the developmental stage, and the temperature. Cells of *D. discoideum* start to excrete synchronized pulses of cAMP ~4 hours before aggregation with an interval of 10 minutes. At aggregation time the interval between two pulses is reduced to 2–3 minutes, except at lower temperatures, at which more time elapses before the next pulse of cAMP occurs. Other species may require more time between two pulses of cAMP (Konijn, 1972). In the case of *D. rosarium,* a 20-minute interval between two pulses has been measured at 22°C.

Natural and artificial pulses of cAMP enhance differentiation in these species (see Devreotes, 1983). Application of these pulses by iontophoresis, instead of using small populations, has the advantage of introduction of the cAMP-containing microelectrode inside the field of amoebae. This way the attractant concentration to which the cells have been exposed can be determined more accurately and rapidly (Robertson *et al.,* 1972). Fluorescein, in solution a negatively charged ion with similar ionic weight to cAMP, may be added as a marker. The iontophoretically applied pulse of cAMP was delivered within 2 seconds to an area with a 10-μm diameter and contained 10^{10}–10^{12} molecules of attractant. The diffusion coefficient of cAMP in the agar was 0.97×10^{-5} cm^2/5 min (Cohen *et al.,* 1975). The interval between two pulses has been chosen to be shorter than the natural time interval to entrain any spontaneously occurring aggregation centers. Time-lapse films showed that 2- to 4-hour starving amoebae were attracted chemotactically within a region ~100 μm in diameter without relay of the signal. After 6 hours cells were attracted and relayed the signal to the more peripherally situated cells. After 2 more hours, streams were formed and pointed directly to the microelectrode. Gerisch *et al.* (1975) released cAMP pulses by iontophoresis near single cells and demonstrated chemotactically induced pseudopod formation within a few seconds. If the same cell was stimulated from the opposite side within 10 seconds, extrusion of a pseudopod could also be observed at that side within an interval of 5 seconds. Apparently, the refractory period only lasts some seconds (Alcantara and Monk, 1974). Enhancement of diffusion of chemoattractant out of the micropipet by iontophoresis is not a necessity. Gerisch and co-workers (Gerisch *et al.,* 1985; Gerisch and Keller, 1981) pulsed chemoattractants near amoebae or granulocytes directly from a micropipet clamped in a micromanipulator. The movement of the cells was observed through an inverted phase-contrast microscope. Swanson and Taylor (1982) stimulated different parts of the surface of single cells chemotactically by allowing the attractant to diffuse from a microneedle. They observed a polarity in the responsiveness of the cell surface to the attractant. The microneedle was fixed in a micromanipulator, and the cell movements were monitored and recorded with a Vidicon TV camera or on Pan-X 35-mm film.

The most simple way of pulsing is the periodic application of small drops containing attractant near amoeboid populations on a hydrophobic agar surface.

This small-population assay with repeated application of aqueous bacterial extract was the key for the purification and identification of the acrasin of *D. discoideum* as cAMP (Konijn *et al.*, 1967). Pulsatile deposition of analogs of cAMP and folic acid allowed a detailed study of the relationship between structure and chemotactic activity in *D. discoideum*. This led to the construction of a model of the cAMP–chemoreceptor interaction in which the anticonformation of the cyclic nucleotide binds to the receptor by specific hydrogen bonds (Mato *et al.*, 1978; Van Haastert and Kien, 1983). Also, the possible interactions between folic acid and the different types of folate-binding sites have been monitored this way (De Wit *et al.*, 1985).

IV.   Time-Lapse Cinematography

One of the most fascinating sights during the development of *Dictyostelium* is the pulsatile movement of cells to the center of aggregation. That this coming together of aggregating cells did not happen gradually was demonstrated in the 1930s by Arndt (1937), who made the first time-lapse film of *Dictyostelium*. The oscillations depend on cell density and time after harvesting (Gingle, 1976). Below a critical density, cells enter the aggregate solely by chemotaxis. At higher cell density the threshold signal concentration for relay is apparently reached early. Halfway between the interphase between the vegetative and aggregative stages time-lapse films reveal fast waves spreading over the entire amoeboid population. Later these pulses are constrained to local regions originating from a small group of cells in the center. The interval between two pulses declines from ~10 minutes a couple of hours before aggregation to 2 minutes during aggregation. Chemotaxis and relay are coupled processes; the chemoattractant bound to the ligand induces two processes, one leading to directed movement and the other to the secretion of more cAMP. The synchronization of the cAMP secretion results in the concentric waves of inward-moving cells chemotactically responding to waves of cAMP relayed outward. The oscillation frequency depends on temperature. The oscillation interval rises from 20 minutes at 8°C to 5 minutes at 22°C with a linear Arrhenius plot (Nanjundiah *et al.*, 1976). The increase in oscillation frequency corresponded to an increase in the mean speed of amoeboid movement.

Compared to Arndt's hand-cranked film, taking time-lapse film has become a rather simple operation. Chemotaxis mostly has been observed on an agar surface. The directed movement of the cells can be recorded using an inverted or noninverted phase-contrast microscope. Drying out of the agar is prevented by covering the bottom of the Petri dish with a plastic disk having a hole of the diameter of the objective. An electrically heated metal coil around the objective prevents fogging of the lens. Warm-water tubing around the objective also prevents fogging. Where the focus distance is sufficiently long, the movie film can be taken with the Petri dish turned upside down. Triocular microscopes can be connected directly with the movie camera. In case of a monocular microscope, a side viewfinder is placed between the microscope and the movie camera to focus the object. The film used in the camera is selected according to the desired contrast and sensitivity. High-contrast film is suitable to demonstrate the movement of chromosomes. Panatomic film exposes the different shades of gray and should be used when the movements of organelles within the cell are investigated during chemotaxis. Panchromatic films are usually the right choice for chemotactic movement of cells.

Since chemotaxis is affected by temperature (Konijn, 1965), a low light intensity should be used to illuminate the field of amoebae. Superspeed film requires low light intensities, and heat produced by light is reduced even more by placement of a water cell containing a 6% $CuSO_4 \cdot 5H_2O$ solution between the light source and the condenser; the copper solution reduces the light intensity by about one-third. With longer time intervals intermittent light avoids heat. Intermittent light is specifically useful when high light intensities are needed such as with high magnification. Under such conditions heat-absorbent glass between light source and condenser is insufficient. Irregular lighting of the frames may be due to opening of the camera shutter while the light intensity is not yet maximal. An experienced eye may judge the required light intensity needed; a photo voltmeter, however, is more reliable.

The movie camera is connected with a time-lapse movie control. A time interval of 10 seconds between each two frames reveals the wavelike movements of the cells when the film is projected with 16 frames per second. Single frames may be selected to be used in publications. To assure registration of the magnification, the first frames of a film are taken of a reticule. Chemotactic movement is mostly filmed with a 100–150 times magnification. If after developing and drying of the film several spots are visible during projection, then drops of water may have dried on the film. Addition of Photo Flo to the water prevents these spots. Refresh the developer in time—especially when it has been standing for some months or when it becomes yellow. Registration of the following information is necessary for duplication or publication of the results: object, kind of movie film, light intensity, exposure time, time interval between two frames, magnification, kind of developer, and developing times.

V.   Chemotactic Response to Spatial and Temporal Gradients

The outcome of measurement of the type of gradient a cell responds to depends largely on the applied technique. The type of gradient needed for chemotaxis in bacteria is clear: they respond to temporal stimulation by attractants

(Macnab and Koshland, 1972; Brown and Berg, 1974). Bacteria compare concentrations over distances substantially greater than their own length and move with a lower frequency of tumbling at higher concentrations of attractant and higher frequency of tumbling at regions of lower attractant concentration.

Amoebae and leukocytes migrate much slower. If they compare changes in the attractant concentration as a function of time, then a substantial period of time would pass before they moved a distance equal to their own length. Bonner (1947) proposed, therefore, that amoebae respond to a spatial concentration gradient measured over their total length. Evidence for a spatial gradient was obtained with the small-population assay (Mato *et al.*, 1975). The chemotactic threshold distances were measured after drops with different concentrations of cAMP were placed at various distances from amoebal drops. A linear regression analysis of a double logarithmic plot of distance against concentration gave a straight line with a slope value of 1/4.25, which is closer to the theoretical value of a spatial gradient (slope = 1/4) than of a temporal gradient (slope = 1/5). The concentration difference between front and back at which cells would still respond chemotactically was calculated to be 1%. A similar shallow gradient has been given for a spatial gradient response of leukocytes (Zigmond, 1974), based on a different assay. Futrelle (1982) applied a moving-micropipet source of cAMP to attract amoebae. The movements of some cells were registered and computer analysis led to the conclusion that they responded to a spatial gradient.

Another possible mechanism is a chemotactic response of amoebae to a temporal gradient of attractant. Optical recording of light scattering in cell suspensions favors a response to a temporal gradient (Gerisch *et al.*, 1974). Also in a cell suspension chemoattractants are expected to affect pseudopodial activity. cAMP injection into the cell suspension (Fig. 4) shows one response peak in vegetative cells and two response peaks in aggregation-competent cells. One of these peaks is supposed to be related to the chemotactic response of the cells to a temporal gradient. This gradient would be detected by extending and retracting filopodia of a cell migrating on a solid surface. Vicker and co-workers (1984) designed a gradient chamber (Fig. 5) in which evenly spread cells were exposed to predeveloped spatial gradients. Their results indicate that spatial gradients effect chemokinesis without adaptation and do not therefore effect true chemotactic movement. Such chemokinesis alone possibly causes a transient attraction or pseudochemotaxis (Lapidus, 1980).

The classical approach to study the spatial integration of the chemical signal with a variety of assays may, whatever clever design, not lead to a definite answer. Oriented movement of amoebae and leukocytes still is an unsolved problem where spatial or temporal gradients, or a combination of both could be the responsible mechanism (Gerisch, 1982; Zigmond *et al.*, 1981). Adaptation to the signal is possibly involved (Van Haastert, 1983). More information on the molecular mechanisms of the response may be needed to come to a more definite answer.

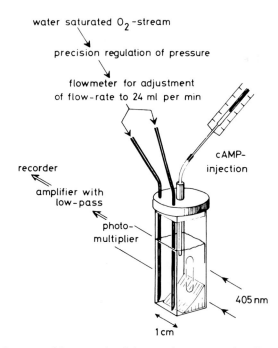

FIG. 4.   Equipment used for measuring light-scattering responses in cell suspensions. Suspensions of 2×10^8 cells/ml are aerated and simultaneously stirred through two needles by a carefully equilibrated O_2 stream. Noise produced by bubbling is suppressed by a low-pass filter in the recording pathway of optical density. A sloping surface of paraffin at the bottom of the optical cuvette prevents sedimentation of cells in the corners not directly seized by the oxygen bubbles. cAMP and other nucleotides are injected in a small volume of solute by a microsyringe (Gerisch *et al.*, 1974).

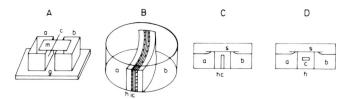

FIG. 5.   Gradient chambers: (A) The penetrable-microfilter migration pad (m) spans a gap (g) between two wells (a and b) containing BBSS with or without cAMP. Once the cell-laden carrier strip (c) is in place, the chambers are incubated in 100% humidity. (B) Two carrier filter pads (c) are juxtaposed between either two migration pads (not shown) and/or two impenetrable border pads (i). The assembly is pressed between two 2-mm slabs (h) of 1% agarose or two layers of three micropore filters (450 μm thick) and bowed, then it is sealed across a 35-mm-diameter Petri dish forming a source and a sink compartment (a and b). (C) A side view of a 5-mm-wide barrier (h) of 1% agarose in BBSS, which spans a plastic tissue culture Petri dish (35 mm diameter). A cell suspension is drawn into the ~100-μm-wide slit (c), and the cells settle and crawl on the dish under the agarose. A cover glass (s) levels the source and sink (a and b) at the block and improves its optical quality. (D) Cells were drawn into the narrow (<1 mm), low chamber (c) within the 3-mm-wide barrier (h), as shown in (C) (Vicker *et al.*, 1984).

VI. Negative Chemotaxis

Both negative and positive chemotaxis were discovered last century (Engelmann, 1881; Pfeffer, 1888). Mainly the latter process has been thoroughly studied, especially during the last decades. The response to repellents of amoebae has been sporadically investigated. Cells may move away from sources of abiotic and biotic repellents, the latter secreted by bacteria or amoebae. Samuel (1961) showed negative chemotaxis by placement of a cellophane square close to a drop of amoebae and turning the square 90 degrees after a few cells migrated on the cellophane square. Cells on the square, again exposed to secreted products from the amoebal drop, turned right-angled in the opposite direction. The problem of negative chemotaxis was approached in different ways by Keating and Bonner (1977). They placed amoebae on the division line of two agar layers, one deep and the other shallow. The cells on the shallow agar migrated much farther than their neighbors on the deep agar because of the accumulation of repellents in the thin agar layer. Also, if two cell populations were placed side by side, fewer cells moved between the populations than at the opposite sides. The repellent passed a dialysis membrane.

The small-population assay allows measurement of negative chemotaxis in a semiquantitative way and could be used for testing purified fractions of repellents (Kakebeeke, *et al.*, 1979). Only vegetative cells responded negatively to repellents, which are secreted by both vegetative and aggregation-competent cells. Vegetative cells repelling each other cover larger territories with less competition for prey. There is more than one repellent, and their concentration is lowered by inactivating enzymes.

VII. Chemotaxis and Relay in the Multicellular Stages

Recently compelling evidence has been presented that cellular slime mold species use cAMP pulses to regulate morphogenetic movement during fruiting body formation (Schaap *et al.*, 1984; Schaap and Spek, 1984; Schaap, 1985; Schaap and Wang, 1985; Schaap and Van Driel, 1985). Similar to the pulsations during aggregation of the more advanced cAMP-responsive species, wavelike motions occur in the tip of all species just before culmination, which proceeds by the chemotactic attraction of cells to the center of wave propagation. cAMP receptors and phosphodiesterase activity appear on the cell surface just before culmination begins. Two techniques that were necessary to observe chemotactic movement and pulsations were time-lapse cinematography as described in Section IV and aggregates in small populations on a hydrophobic agar surface. Pulses of cAMP were applied close to the drops containing preculmination structures. In the presence of cAMP the preculmination structures disintegrated while the attractants typical to induce aggregation, such as folic acid in *D. minutum*, were without effect. In the primitive and in the more advanced cellular slime molds, oscillatory cAMP secretion and relay contribute to the regulation of cell movement during culmination. Durston *et al.* (1976), using time-lapse cinematography, also showed pulsatile movements in culminating *D. discoideum* structures. A disadvantage of time-lapse films of later stages is the difficulty of distinguishing individual cells. This problem may be partly overcome by staining cells with neutral red. The vital dye appears in large intracellular vacuoles of cells that typically occupy the front and rear guard of the slug and can be tracked in time-lapse films (Durston and Vork, 1979). The stained vacuole is an autophagic vacuole that is specific to the prestalk cells and rear guard cells in mature slugs. A subpopulation of the prespore cells also contain neutral red-containing vacuoles. If stained prestalk cells and unstained prespore cells are mixed, the stained cells accumulate via directional movement to the tip. Prestalk and prespore cells sort out via directional movements of the prestalk cells to a cAMP source (Matsukuma and Durston, 1979), which can be shown on time-lapse films.

VIII. Conclusions

The first chemotaxis assays for cellular slime molds were meant to prove that amoebae move chemotactically to the center of an aggregate. These qualitative assays were gradually succeeded by semiquantitative assays when chemoattractants were isolated, purified, characterized, and sometimes identified. The further development of assays will be directed to the elucidation of cell behavior in gradients of chemoattractants. We already see a considerable extension of automation of data acquisition (e.g., Futrelle *et al.*, 1982). This automation allows new assays to be performed which focus both on the cell population and on individual cells in chemotactic gradients with defined temporal and spatial components (Vicker *et al.*, 1984; Varnum and Soll, 1984; Varnum *et al.*, 1985).

REFERENCES

Alcantara, F., and Monk, M. (1974). *J. Gen. Microbiol.* **85**, 321–334.
Arndt, A. (1937). *Wilhelm Roux Arch.* **136**, 681–744.
Bonner, J. T. (1947). *J. Exp. Zool.* **106**, 1–26.
Bonner, J. T., Kelso, A. P., and Gillmor, R. G. (1966). *Biol. Bull.* **130**, 28–42.
Bonner, J. T., Hirshfield, M. F., and Hall, E. M. (1971). *Exp. Cell Res.* **68**, 61–64.
Boyden, S. (1962). *J. Exp. Med.* **115**, 453–466.
Brown, D. A., and Berg, H. C. (1974). *Proc. Natl. Acad. Sci. U.S.A.* **71**, 1388–1392.

Chang, Y. Y. (1968). *Science* **160**, 57–59.
Cohen, M. H., Drage, D. J., and Robertson, A. (1975). *Biophys. J.* **15**, 753–764.
Devreotes, P. N. (1983). *Adv. Cyclic Nucleotide Res.* **15**, 55–96.
De Wit, R. J. W., Bulgakov, R., Pinas, J. E., and Konijn, T. M. (1985). *Biochim. Biophys. Acta* **814**, 214–226.
Durston, A. J., and Vork, F. (1979). *J. Cell Sci.* **36**, 261–279.
Durston, A. J., Cohen, M. H., Drage, D. J., Ptel, M. J., Robertson, A., and Wonio, D. (1976). *Dev. Biol.* **52**, 173–180.
Englemann, T. W. (1881). *Pflüger's Arch. ges. Physiol.* **26**, 537.
Ennis, H., and Sussman, M. (1958). *Proc. Natl. Acad. Sci. U.S.A.* **44**, 401–411.
Francis, D. (1965). *Dev. Biol.* **12**, 329–346.
Futrelle, R. P. (1982). *J. Cell. Biochem.* **18**, 197–212.
Futrelle, R. P., Traut, J., and McKee, W. G. (1982). *J. Cell Biol.* **92**, 807–821.
Gerisch, G. (1982). *Annu. Rev. Physiol.* **44**, 535–552.
Gerisch, G., and Keller, H. U. (1981). *J. Cell Sci.* **52**, 1–10.
Gerisch, G., Malchow, D., and Hess, B. (1974). *In* "Biochemistry of Sensory Functions" (L. Jaenicke, ed.), pp. 279–298. Springer-Verlag, Berlin.
Gerisch, G., Malchow, D., Huesgen, A., Nanjundiah, V., Roos, W., and Wick, U. (1975). *In* "Developmental Biology" (D. McMahon and C. F. Fox, eds.), Vol. 2, pp. 76–88. Benjamin, New York.
Gerisch, G., Weinhart, U., Bertholdt, G., Claviez, M., and Stadler, J. (1985). *J. Cell Sci.* **73**, 49–68.
Gingle, A. R. (1976). *J. Cell Sci.* **20**, 1–20.
Kakebeeke, P. I. J., and De Wit, R. J. W., Kohtz, S. D., and Konijn, T. M. (1979). *Exp. Cell Res.* **124**, 429–433.
Keating, M. T., and Bonner, J. T. (1977). *J. Bacteriol.* **130**, 144–147.
Konijn, T. M. (1965). *Dev. Biol.* **12**, 487–497.
Konijn, T. M. (1970). *Experientia* **26**, 367–369.
Konijn, T. M. (1972). *Adv. Cyclic Nucleotide Res.* **1**, 17–31.
Konijn, T. M. (1975). *In* "Primitive Sensory and Communication Systems" (M. J. Carlile, ed.), pp. 101–152. Academic Press, London.
Konijn, T. M., and Raper, K. B. (1961). *Dev. Biol.* **3**, 725–756.
Konijn, T. M., Van de Meene, J. G. C., Bonner, J. T., and Barkley, D. S. (1967). *Proc. Natl. Acad. Sci. U.S.A.* **58**, 1152–1154.
Lapidus, I. R. (1980). *J. Theor. Biol.* **86**, 91–103.
Macnab, R. M., and Koshland, D. E. (1972). *Proc. Natl. Acad. Sci. U.S.A.* **69**, 2509–2512.
Mato, J. M., Losada, A., Nanjundiah, V., and Konijn, T. M. (1975). *Proc. Natl. Acad. Sci. U.S.A.* **72**, 4991–4993.
Mato, J. M., Jastorff, B., Morr, M., and Konijn, T. M. (1978). *Biochim. Biophys. Acta* **544**, 309–314.
Matsukuma, S., and Durston, A. J. (1979). *J. Embryol. Exp. Morphol.* **50**, 243–251.
Nandini-Kishore, S. G., and Frazier, W. A. (1981). *Proc. Natl. Acad. Sci. U.S.A.* **78**, 7299–7303.
Nanjundiah, V., Hara, K., and Konijn, T. M. (1976). *Nature (London)* **260**, 705.
Pfeffer, W. (1884). *Untersuch. Bot. Inst. Tübingen* **1**, 363–482.
Pfeffer, W. (1888). *Untersuch. Bot. Inst. Tübingen* **2**, 582.
Robertson, A., Drage, D. J., and Cohen, M. H. (1972). *Science* **175**, 333–335.
Runyon, E. H. (1942). *Collect. Net.* **17**, 88.
Samuel, E. W. (1961). *Dev. Biol.* **3**, 317–336.
Schaap, P. (1985). *Differentiation* **28**, 205–208.
Schaap, P., and Spek, W. (1984). *Differentiation* **27**, 83–87.

Schaap, P., and Van Driel, R. (1985). *Exp. Cell Res.*, **159**, 388–398.
Schaap, P., and Wang, M. (1985). *Cell Differ.* **16**, 29–33.
Schaap, P., Konijn, T. M., and Van Haastert, P. J. M. (1984). *Proc. Natl. Acad. Sci. U.S.A.* **81**, 2122–2126.
Shaffer, B. M. (1956). *J. Exp. Biol.* **33**, 645–657.
Soll, D. R., and Mitchell, L. H. (1982). *Dev. Biol.* **91**, 183–190.
Swanson, J. A., and Taylor, D. L. (1982). *Cell* **28**, 225–232.
Van Haastert, P. J. M. (1983). *J. Cell Biol.* **96**, 1559–1565.
Van Haastert, P. J. M., and Kien, E. (1983). *J. Biol. Chem.* **258**, 9636–9642.
Varnum, B., and Soll, D. R. (1984). *J. Cell Biol.* **99**, 1151–1155.
Varnum, B., Edwards, K. B., and Soll, D. R. (1985). *J. Cell Biol.* **101**, 1–5.
Vicker, M. G., Schill, W., and Drescher, K. (1984). *J. Cell Biol.* **98**, 2204–2214.
Wallace, L. J., and Frazier, W. A. (1979). *Proc. Natl. Acad. Sci. U.S.A.* **76**, 4250–4254.
Zigmond, S. H. (1974). *Nature (London)* **249**, 450–452.
Zigmond, S. H., Levitsky, H. L., and Kreel, B. J. (1981). *J. Cell Biol.* **89**, 585–592.

Chapter 17

Transmembrane Signaling in Dictyostelium

PETER DEVREOTES, DONNA FONTANA, PETER KLEIN,
JANE SHERRING, AND ANNE THEIBERT

Department of Biological Chemistry
The Johns Hopkins University
School of Medicine
Baltimore, Maryland 21205

I. Cell–Cell Communication in Development

Dictyostelium is becoming increasingly recognized as a genetically and biochemically accessible system for studies of transmembrane signaling and cell–cell communication in development. Early in development, an intercellular communication system appears which coordinates the highly organized aggregation of 1 million cells. The elegant wave patterns which appear during aggregation are organized by cyclic AMP (Fig. 1A, B). Cyclic AMP (cAMP) acts extracellularly as a cell–cell signaling molecule and chemoattractant. At aggregation centers, cAMP levels spontaneously oscillate and each peak initiates one of the propagated cAMP waves. The leading edge of each passing cAMP wave provides a gradient that orients the chemotactically sensitive cells toward the aggregation center. Cells move up the gradient for several minutes until the peak of the wave reaches the position of the cell. The cells then move randomly until the next wave elicits another coordinated movement step (Fig. 1C).

The two responses essential to this coordinated aggregation are chemotaxis and cAMP signaling (i.e., the synthesis and secretion of cAMP in response to extracellular cAMP). Both are mediated by a cell surface receptor and provide excellent models for transmembrane signaling in eukaryotic cells (Devreotes, 1982). Most of the known components of these response systems are under tight developmental regulation. The periodic cAMP signaling, in turn, regulates the

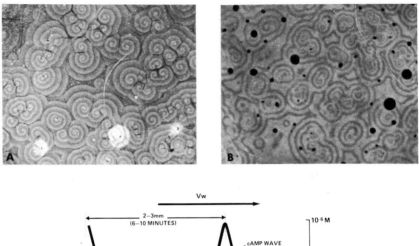

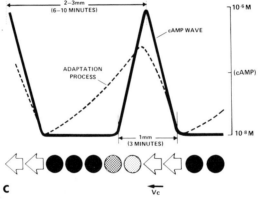

Fɪɢ. 1.　Chemotaxis and cAMP signaling mediate aggregation in *Dictyostelium*. (A) Dark-field photograph of aggregating cells at 5 hours in the developmental program. Territories containing ~1 million cells are about 1–2 cm in diameter. (B) Print of fluorographic image of cAMP waves within monolayer of aggregating cells. Waves were detected by a solid-phase isotope dilution technique referred to in text. (C) Dynamics of signal relay and chemotaxis. The heavy line representing the cAMP concentration is drawn from analyses of scans of the fluorographic images of the cAMP waves. Symbols in the lower part of the diagram represent a single radial line of cells: open arrows, cells moving toward the center; shaded circles, randomly oriented cells, arrow vectors, the speed and direction of motion of the cAMP wave (V_w = 300 μm/min) and the moving cells (V_c = 20 μm–min).

expression of these components as well as a host of other developmentally controlled gene products (Chisholm *et al.*, 1985). This chapter focuses on the properties of surface receptor-mediated responses. Current techniques to control cellular sensitivity and monitor chemotaxis, cAMP signaling, and adenylate cyclase activation are outlined. Receptor-binding assays are reviewed, and a protocol for receptor purification is offered. Finally, the assays are brought together in a flow sheet for characterization of transmembrane signaling mutants.

II. Growth and Development

A. Standard Growth and Development

Ax-3 cells of *Dictyostelium discodieum* are grown in 400 ml of HL5 medium (Watts and Ashworth, 1970) to a density of 5×10^6/ml. To initiate development, cells are centrifuged (600 g, 5 minutes), resuspended in an equal volume of development buffer (DB) (5 mM Na$_2$HPO$_4$ 5 mM NaH$_2$PO$_4$, 2 mM MgSO$_4$, 200 μM CaCl$_2$; Marin and Rothman, 1980), centrifuged again, resuspended at 2 $\times 10^7$/ml in DB, and shaken at 110–130 rpm in an Erlenmeyer flask. The optimal flask/cell suspension ratio is 10:1. Ratios of 15:1 and 4:1 have a deleterious effect on the expression of early developmental markers. Spontaneous oscillations (see Section III) begin at 3.5 hours and continue through 7 hours. Early developmental markers, such as adenylate cyclase and surface cAMP receptors, reach peak expression (at least 12-fold above growth levels) within 5–6 hours. Cell lines which do not meet these criteria are discarded, and a new clone is selected from frozen stocks.

NC-4 cells are grown in association with *Enterobacter aerogenes* to a density of 5×10^8 per 100-mm Petri dish. To initiate development, cells are freed of bacteria with three washes in DB, resuspended, and shaken as for Ax-3 cells. With NC-4 cells, pulsing with cAMP at 50 nM final concentration (80-μl pulse delivered in 5 seconds with Gilson Minipuls pump) at 10-minute intervals achieves maximum levels of cAMP-binding activity after 10–12 hours.

The duration of development of NC-4 cells can be extended by developing the cells on agar (10 ml of 1.5% agar in DB per 100-mm Petri dish) at 7°C (Alcantara and Monk, 1974). Then, 15 ml of washed cells at 5×10^6/ml are allowed to adhere for 15 minutes. The clear fluid is pipeted off and the top is left off for ~20 minutes until the surface is dried to a "mat" finish. After incubation at 7°C for 15 hours, patterns of wave-directed cell motion should be readily visible.

B. Large-Scale Growth and Development

Ax-3 cells (30 liters) are grown in fifteen 4-liter Erlenmeyer flasks, each containing 2 liters of HL5 medium. Cells are shaken at 200 rpm for 3 days to a final density of $5–8 \times 10^6$/ml. Cells are then collected either by standard centrifugation at 600 g for 6 minutes or by continuous-flow centrifugation. The entire 30 liters can be centrifuged in one Sharples T-I continuous-flow centrifuge, which has a capacity for $>1.5 \times 10^{11}$ cells. The interior of the cylindrical rotor is fitted with a sheet of plastic which covers the wall of the rotor exactly (without overlap). This plastic sheet facilitates the later removal of the cell pellet. The rotor is adjusted to 16,000 rpm, and then cells are allowed to flow in by gravity at a flow rate of 2.2 liters/min. The cell pellets from either method are then washed by resuspension in an equal volume of DB and recentrifugation, resuspended at $2-4 \times 10^7$/ml and shaken at 160 rpm. Large Ax-3 cultures must be pulsed with cAMP for optimal development (50 nM cAMP final concentration delivered every 6 minutes from hours 1 to 5 after initiation of development).

III.   Control of Cellular Sensitivity and Applied cAMP Stimuli

Figure 2 illustrates the interaction of the components of the cAMP signaling system. Also indicated are inhibitors or treatments which block interactions at specific steps. An increase in the occupancy of the surface cAMP receptors leads to activation of adenylate cyclase. This results in a transient increase in intracellular cAMP, which is secreted at a rate proportional to the intracellular cAMP level. The secreted cAMP binds to surface receptors, creating a positive-feedback loop (dashed arrow). After several minutes, the rate of cAMP synthesis is attenuated by an adaptation process (likely due to receptor modification), and the cAMP present is degraded by extracellular and cell surface phosphodiesterases (PDE). Deadaptation then ensues and the cyclic begins again, yielding oscillations in cAMP signaling with a period of 6–11 minutes (Devreotes, 1982).

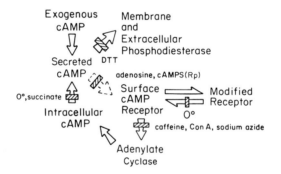

FIG. 2.   Interaction of components of the signaling response. Increases in surface receptor occupancy lead to rapid activation of adenylate cyclase and less rapid increases in the fraction of modified receptors, which may attenuate the activation. The modification is reversed when cAMP is removed. Intracellular cAMP and the rate of cAMP secretion increase in parallel. Secreted cAMP binds to surface receptors, creating a positive-feedback loop (dashed arrow), or is degraded by cell surface and secreted phosphodiesterases. Binding is inhibited by adenosine 3′,5′-phosphorothioate-R_p stereoisomer [cAMPS(R_p)] ($K_i = 30$ μM; Van Haastert and Kien, 1984) and adenosine ($K_i = 300$ μM) or 2′-*O*-methyladenosine ($K_i = 30$ μM; Theibert and Devreotes, 1984). Receptor-mediated activation of the adenylate cyclase is inhibited by caffeine (IC$_{50}$ = 0.2 mM; Brenner and Thoms, 1984), Con A (IC$_{50}$ = 10 μg/ml; Fontana and Devreotes, 1984), and sodium azide (IC$_{50}$ = 10 μM; Dinauer *et al.*, 1980). Receptor demodification is inhibited at 0°C (Devreotes and Sherring, 1985). cAMP secretion is inhibited at 0°C (Van Haastert, 1984a) and by sodium succinate (IC$_{50}$ = 3 mM; Williams *et al.*, 1984). Extracellular phosphodiesterases are inhibited by DTT ($K_i = 500$ μM; Henderson, 1975).

In order to assess ͺ ·operly the response to an exogenous stimulus, the cells must be in a basal state prior to its application and the effects of both the positive-feedback loop and extracellular phosphodiesterase must be minimized. Strategies to monitor and control cellular sensitivity and apply defined stimuli make use of the inhibitors and perfusion techniques to interrupt specific steps in the cAMP signaling system.

A.   Monitoring Spontaneous Oscillations

Spontaneous oscillations in cAMP signaling can be monitored by recording changes in light scattering which occur concurrently. Using the light-scattering trace as a guide, cells can be sampled or stimulated at different phases of the oscillatory cycle. Previously, cells were stirred by bubbling in a cuvette, but this precludes the use of large volumes (Gerisch and Hess, 1974). Currently, large volumes are monitored by shaking the cells in a beaker from which a small portion is continuously pumped through a flowthrough cuvette and returned (Fig. 3). A simple apparatus can be built using a Gilson Minipuls peristaltic pump and a 12 × 75 mm disposable glass test tube as the cuvette. Tygon tubing (i.d. = i.d. = $\frac{1}{16}$ in., o.d. = $\frac{1}{8}$ in.) and pump tubing (Rainin cat no. 39-628) serve as the connecting lines. The ends of the tygon tubing are passed through a two-holed rubber stopper (#00) and capillary pipets (100 μl, Gold Seal Glassware) are pushed ~1.5 cm into the ends of the tubing. The tubing is withdrawn from the stopper until the ends of the capillary pipets, covered by the tygon tubing, seal

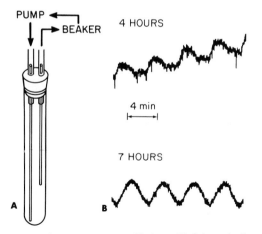

FIG. 3.   Apparatus to monitor spontaneous oscillations. (A) Schematic diagram showing details of the flowthrough cuvette. (B) Spontaneous oscillations were recorded from 60 ml of cell suspension from 3.5 to 7.5 hours of development. Examples at 4 and 7 hours are shown.

tightly into the holes of the stopper. The capillary pipets are cut so that the *inlet* is near to and the *outlet* is 2.5 cm from the bottom of the tube. The length of the connecting tubing should be as short as possible; ~30 cm works well. The volume of cell suspension in the tubing and the 2 ml in the test tube are exchanged in ~2 minutes with the pump operated at maximal speed. There is no time lag between the response and the record, since the cells in the beaker, tubing, and tube respond simultaneously. Light scattering is measured with a spectrophotometer at 425 nm. The continuous pumping throughout development appears to have no deleterious effect on the cells. In Ax-3 cells, oscillations begin at ~3.5 hours of development and have the familiar double-humped appearance observed by the standard bubbling method. The period is initially about 7–11 minutes and then decreases to ~5.5 minutes. At about 6–7 hours the pattern switches to the sinusoidal oscillations reported by Gerisch *et al.* (1979) (Fig. 3). NC-4 cells oscillate less reliably with a period of ~11 minutes.

B.   Interference with Feedback Loops

1.   CAFFEINE AND DITHIOTHREITOL

Caffeine blocks activation of the adenylate cyclase and thus cAMP production (Brenner and Thoms, 1984). Dithiothreitol (DTT) inhibits all phosphodiesterase exposed to the medium (Henderson, 1975). Thus, in the presence of these two inhibitors, cells can neither increase nor decrease the cAMP concentration. Oscillations cease and the cells attain a basal state. NC-4 (6×10^7/ml) or Ax-3 (4×10^7/ml) cells are treated with 5 mM caffeine for 30 minutes, then with 10 mM DTT. cAMP stimuli are added within a few minutes. After 20 minutes >85% of the added cAMP remains for concentrations ranging from 100 pM to 10 μM. Surface cAMP receptor modification and myosin heavy- and light-chain phosphorylation dose–response curves have been measured with this method (Devreotes and Sherring, 1985; Berlot *et al.*, 1985). An advantage of this method is that dense suspensions of cells can be controlled and sampled at numerous times during the response. However, it is difficult to remove the stimulus rapidly. This can be achieved, at least for <100 nM cAMP, by addition of 5 mM 2'-*o*-methyladenosine (Theibert and Devreotes, 1984).

2.   0°C SHAKING

Developed cells are diluted to 5×10^6/ml and shaken at 200 rpm for 20–30 minutes. This is followed by one wash in DB at 22°C. Cells are resuspended in cold DB at 2×10^7/ml and shaken at 120 rpm on ice. Prestimulus samples are taken and cells are stimulated with cAMP and 10 mM DTT. Cells can be warmed to 22°C as the stimulus is added to enhance the response, although warming itself will trigger a response within 2 minutes. This procedure has been used to monitor receptor-mediated activation of adenylate cyclase in both Ax-3 and NC-4 cells. The basis of the technique is probably that the rapid shaking interferes with the oscillations and the cells attain a basal state. Since cAMP secretion is markedly reduced at 0°C (Van Haastert, 1984a), the oscillations do not resume. This method is slightly less convenient than Method 2 but does not require inhibitors.

3. INFUSION OF cAMP

The level of the applied stimulus can be controlled by continuous infusion of cAMP so that a balance between infusion and phosphodiesterase hydrolysis is achieved (Wurster and Butz, 1983). The steady-state concentration is given by the infusion rate divided by the rate constant for hydrolysis. With Ax-2 cells at a density of 2×10^7/ml, an infusion rate of 0.5 nM/sec leads to a steady-state concentration of ~25 nM cAMP. Lower infusion rates give correspondingly lower steady-state concentrations. The concentrations achieved depend on cell density, strain, and clonal origin. This method does not require inhibitors and is recommended for studies of the effects of chronic cAMP application on gene expression. The stimulus is easily removed by stopping the infusion. The technique may not be useful for very rapid responses, since the steady state is achieved slowly ($t_{1/2}$ = ln 2/rate constant of degradation = 1 minute). The method can be combined with the use of caffeine (but not DTT) to ensure that the cells are in a basal state when the infusion begins.

4. RAPID-PERFUSION TECHNIQUES

Rapid perfusion interrupts the extracellular positive-feedback loop by removing cAMP as it is secreted (Devreotes *et al.*, 1979). Exogenous cAMP that is degraded by the phosphodiesterase is quickly renewed by the continued perfusion. The advantage of perfusion methods is the rapid, convenient exchange of perfusing solutions. The techniques are most useful for triggering the secretion of specific metabolites or direct observation of intact cells. Rapid cell shape changes in response to cAMP and folic acid stimuli have been studied with a cover glass perfusion chamber technique (Section IV). The secretion of cAMP in response to exogenous cAMP and the effects of numerous inhibitors on that response has been extensively characterized using a filter-perfusion technique (Section V). Changes in intracellular metabolites or the localization of histochemically detectable proteins can be monitored during rapid perfusion, but this requires a multichannel apparatus.

IV. Chemotaxis Assays

Three types of assays are described in this section: the small-population and micropipet assays, both of which measure true chemotaxis, and a cover glass perfusion chamber assay that monitors rapid chemoattractant-induced cell shape changes that reflect chemotaxis.

A. Small-Population Assays

1. Double-Spot Assay

Small droplets (100 nl) of cells at 5×10^6/ml are spotted on the surface of washed agar (Konijn, 1970). The excess fluid evaporates in a few minutes and the cells are confined within a circular area 0.5 mm in diameter. After 30 minutes, a similar droplet of test substance is applied at a center-to-center distance of 1 mm. Cells are observed at 10-minute intervals, and a substance is scored as positive if at least twice as many cells are pressed against the edge closest to as against the edge farthest from the test substance. Twenty populations are observed for a given concentration of test substance. Data is reported as *percentage of populations* responding. The chemotactic response in *Dictyostelium* NC-4 cells and other species of cellular slime molds has been extensively characterized using this assay (Van Haastert and Konijn, 1982; Van Haastert *et al.*, 1982a). Ax-3 cells respond more readily in the presence of 5 m*M* caffeine. The assay has yielded quantitative dose–response data for a wide variety of chemoattractants. Since a redistribution of the cells is required, the assay is not useful for monitoring responses triggered within the first few minutes of application of the chemoattractant.

2. Single-Spot Assay

In a simpler version of a small-population assay, cells are spotted on the surface of agar containing a given concentration of test substance. If the cells degrade the test substance (i.e., cellular phosphodiesterase would degrade cAMP in the agar), a gradient will form across the perimeter of the spot. The test substance is scored as positive if it stimulates the cells to spread radially. Although the chemoattractants cAMP (Konijn *et al.*, 1967), folic acid (Pan *et al.*, 1972), and glorin (Shimomura *et al.*, 1982) were identified using this assay, it requires that the cells produce an enzyme that will degrade the test substance. In addition, substances which are not chemoattractants, such as 5′-AMP, can also induce radial spreading (Van Haastert *et al.*, 1982b).

B. Micropipet Assays

1. DISTANT PIPETS

Chemotaxis can be elicited with chemoattractant-filled micropipets. Sufficient control of the stimulus is maintained by allowing it to diffuse from the pipet tip (Futrelle *et al.*, 1983). cAMP is loaded into pipets with resistances of 20–100 $M\Omega$, which have femtoliter-per-second diffusional efflux rates. Steep gradients are formed by the point sources. With 100 μM cAMP in the pipet, the concentration drops to 100 pM within 100 μm of the tip. Futrelle devised a clever method of rapidly applying and removing cAMP gradients by making and breaking the contact of the pipet with the agar surface on which the cells are supported. The initial upshift in cAMP concentration caused by application of the pipet to the agar triggers a pseudopod retraction, or "cringe" response. If the duration of gradient application is extended beyond 20 seconds, the "cringe" is followed by elongation and movement toward the pipet. The duration of chemotactic movement equals the duration of the applied gradient. The advantage of this technique is the rapid application and removal of defined gradients. The somewhat sophisticated apparatus used in the original experiments may deter investigators wishing merely to verify that a given set of cells are chemotactically sensitive.

2. CLOSE PIPETS

Micropipets brought closer to cells trigger localized pseudopod extension within a few seconds (Gerisch *et al.*, 1979; Swanson and Taylor, 1981). Again, cAMP is allowed to diffuse from the pipet and concentrations drop off steeply from the tip. Lower concentrations ($= 5$ μM) of cAMP are used, since shorter distances are involved. The majority of cells tested are initially unpolarized; pseudopod extension will occur at any point. Cells led by the pipet to establish an elongated form become polarized. The front responds immediately to slight repositioning of the pipet. The tail requires 30–90 seconds to respond (Swanson and Taylor, 1981).

C. Cover Glass Perfusion Chamber Assay

A perfusion assay allows observation of chemotactic responses on a time scale of a few seconds (Fontana *et al.*, 1984). The perfusion chamber consists of a stainless-steel block the size and thickness of a microscope slide which contains an elongated diamond-shaped hole and 26-gauge inlet and outlet needles (Fig. 4). Tygon tubing ($\frac{1}{32}$ in. i.d.) is adapted to the needles with tubing (Intermedic PE 50). A Gilson Minipuls pump draws fluid from a reservoir through the chamber

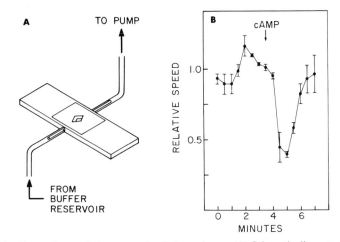

Fɪɢ. 4.   Cover glass perfusion assay of cell shape change. (A) Schematic diagram of perfusion chamber. (B) Response of cells to cAMP stimulation assessed by speed measurement. Relative speed normalized to speed prior to cAMP addition (arrow). A value of 1.0 represents a speed of ~7 μm/min. Values are average for nine cells.

and into a waste container, exchanging the solution in the chamber every 2–3 seconds. Developed NC-4 cells, taken at 5 hours from shaken cultures, are diluted to 5×10^6/ml and vigorously pipeted to disrupt aggregates. A 5-μl drop is placed on a cover glass, and cells are allowed to adhere for 3 minutes. Loose cells are washed off by gently dipping the cover glass in buffer, and the adherent cells are visible as a grayish 1-mm spot. The cover glass is inverted onto the chamber, which has been prepared by sealing the floor with a cover glass, completely filling the inlet and outlet tubing with buffer, and coating the upper surface with vacuum grease. Once the upper cover glass is sealed, the pump is started and a field is selected containing about 10–15 cells at 400×. At higher densities the cells respond poorly, and at lower densities insufficient numbers of cells are recorded. The behavior of the cells is time-lapse video-recorded.

Amoebae move randomly with a speed of about 7–10 μm/min (Fig. 4). When an appropriate stimulus is introduced, the amoebae retract their pseudopodia. After 15–30 seconds, the pseudopodia are reextended in all directions; the amoebae become immobilized and appear flattened. After 2–3 minutes, the cells resume their random motion even in the continued presence of stimulus. The resumption of motion in the presence of a continuous level of stimulus suggests that the chemotactic response adapts. Adaptation occurs at all doses of cAMP. Once cells have adapted to a low concentration, a further response can be elicited by directly increasing the stimulus. If cells are exposed to serial increments in the stimulus concentration (1 n*M*, 10 n*M*, 100 n*M*, and 1 μ*M*, each for 2 minutes with no interposed recovery periods), the cells reduce their speed to ~40% of the prestimulus value and remain at this slow speed for >10 minutes. The shape change response appears to be an expression of the chemotactic response in general. Growing amoebae, which are chemotactic to folic acid but not cAMP, show the same shape change response to folic acid but not to cAMP (Fontana *et al.*, 1984).

V.   Measurements of Intracellular and Secreted cAMP

A.   Total-cAMP Levels

For measurement of total-cAMP levels, samples are extracted in 5% trichloroacetic acid (TCA), neutralized, and cAMP level determined by isotope dilution. Anti-cAMP antibodies (Klein and Brachet, 1975; Rahmsdorf *et al.*, 1976; Pahlic and Rutherford, 1979; Abe and Yanagisawa, 1983), the regulatory subunit of protein kinase (Gerisch and Wick, 1975; Brenner, 1977, 1978), and erythrocyte ghosts (Devreotes *et al.*, 1979) have been used as binding proteins. In all cases, elimination of the signal by prior treatment of the sample with phosphodiesterase provides a useful control. Radiolabeled cAMP should be taken through the sample preparation to assess recovery. A variation of the isotope dilution assay was devised to determine the distribution of cAMP with cell monolayers (Tomchik and Devreotes, 1981). The binding reaction is carried out directly on a Millipore filter blot of the monolayer. In a fluorograph of the filter, regions of high cAMP concentration within the monolayer appear as light regions against a dark background in the fluorograph.

B.   Intracellular and Extracellular cAMP

Since the cAMP levels in *Dictyostelium* can change rapidly, when measuring intracellular and extracellular cAMP levels, proper preparation of the sample is crucial. Therefore, for an accurate measurement, the amoebae must be rapidly separated from their surrounding medium (within seconds) and the phosphodiesterase rapidly inactivated. The approaches generally used to separate the amoebae from the surrounding medium are (1) centrifugation through silicone oil, (2) retention on filters, and (3) removal of medium from amoebae attached to tissue culture dishes.

1.   CENTRIFUGATION THROUGH SILICONE OIL

Gerisch and Wick (1975) devised an assay which separates cells from medium in <2 seconds and simultaneously acidifies both the cell layer and the medium.

In a microfuge tube, a layer of silicone is placed over a glucose–TCA solution which contains glass beads. The cell suspension is layered over the oil. A plunger, consisting of a glass tube closed at one end with a Millipore filter and containing TCA, is positioned such that the filter is above the cell suspension. Centrifugation drives the amoebae through the silicone layer into the glucose–TCA solution. The plunger penetrates only through the upper layer, forcing the medium into the lumen, where it mixes with the TCA. The plunger is removed and its contents are assayed for extracellular cAMP. The remaining medium and the silicone are discarded, a more dense silicone is added, and the tube is again centrifuged. The acidified cell extract, which now lies above the silicone layer, is removed for assay of intracellular cAMP. TCA is extracted, the pH adjusted, and cAMP determined using one of the competition methods. While this method should result in accurate measurements, the amount of time which must be invested in setting up the assay is a drawback.

2.   RAPID FILTRATION

This method is thoroughly discussed by Brenner (1978). If the amoebae are developed on filters, the top filter can be placed on a Buchner funnel and rapidly washed with buffer at 22°C. The harvesting and washing process is complete in 15 seconds, and the filter is immediately immersed in TCA. Washing is necessary to remove extracellular cAMP. However, significantly reduced intracellular cAMP levels are recorded if the wash is performed with ice-cold buffer. The proposed explanation is that the cold temperature inhibits the adenylate cyclase and not the phosphodiesterases. This method is easily set up and if the filtration is done rapidly, it yields accurate results. Because a filter containing amoebae can be employed—i.e., the cells do not have to be in suspension—the cAMP concentration of *Dictyostelium* at all stages of development can be assayed. The filter technique can also be employed if the amoebae are in suspension, with the suspension being added to the filter apparatus (Rahmsdorf *et al.*, 1976).

3.   SUBMERGED ATTACHED CELLS

Other methods to separate medium from cells take advantage of the ability of submerged *Dictyostelium* to aggregate when attached to glass or plastic surfaces. Medium is rapidly removed by aspiration or perfusion and the cells are harvested in TCA. This technique was used to investigate the effect of inhibitors of aggregation on intracellular cAMP levels (Klein and Brachet, 1975), and initially to measure cAMP-stimulated cAMP secretion. Its applicability is limited, because most investigators do not develop *Dictyostelium* in this fashion.

C.   cAMP-Stimulated cAMP Synthesis

1.   FILTER-PERFUSION ASSAY

A useful method for examining cAMP levels, both intracellular and secreted, looks at relative levels of ^{3}H-cAMP. This technique employs a perfusion apparatus (Devreotes *et al.*, 1979; Fig. 5, *left*) to control the sensitivity of the cells (see Section III). Upon the initiation of development, the amoebae are shaken with bacteria previously labeled with [^{3}H]adenosine. After 3 hours, amoebae are plated on nonnutrient agar. At the onset of aggregation, cells are placed on Millipore filters (3-μm pore size) supported in the perfusion apparatus. The pump lines above the filters deliver 45-μl drops of buffer which cover the filter surface (1 cm^2). The pump lines are connected to rocker arms such that two lines are capable of servicing each filter. Rocking the arm causes a rapid switch in perfusion solution. The solution flowing past the amoebae is collected in tubes containing excess cAMP and acid. ^{3}H-cAMP is purified using the sequential dowex and alumina columns described by Salomon (1979). To determine relative changes in intracellular cAMP levels, each filter is inverted into a microtiter well containing 8% formic acid, 10^{-3} M cAMP, and trace amounts of ^{14}C-cAMP (Dinauer *et al.*, 1980). The cAMP is purified using the Salomon columns, and the ratio of ^{3}H-cAMP to ^{14}C-cAMP is a measure of the relative level of cAMP within the amoebae.

The perfusion device has been used to show that cAMP secretion increases at least 100-fold upon cAMP stimulation (Devreotes and Steck, 1979; Fig. 5, *right*). The increased synthesis and secretion of cAMP is transient, even during persistent stimulation, and a further secretion is possible only if the stimulus concentration is raised or if the stimulus is removed and the cells are allowed to recover. The adaptation process, which causes a cessation in cAMP synthesis, occurs even when the synthesis of cAMP is blocked with caffeine (Theibert and Devreotes, 1983).

The perfusion device has also been used to study the action of inhibitors (Theibert and Devreotes, 1983; Fontana and Devreotes, 1984; Sussman, 1985) and a stimulator of cAMP signaling (Devreotes, 1983). The responses are quantified by integrating under curves similar to those seen in Fig. 5. A closer examination of one of the inhibitor studies demonstrates the versatility of the perfusion technique. The lectin concanavalin A (Con A) was shown to inhibit the signaling response in a noncompetitive fashion. α-Methylmannoside prevented the inhibition by Con A (Fig. 5, *bottom*, A–C). The inhibition could also be reversed by α-methylmannoside (Fontana and Devreotes, 1984).

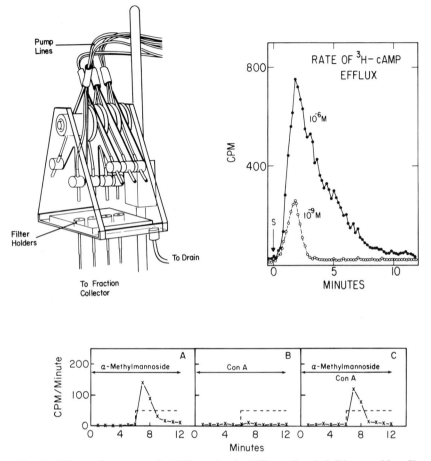

FIG. 5.   Filter-perfusion assay of cAMP-stimulated cAMP secretion. *Left:* Diagram of four-filter perfusion apparatus. *Right:* Rate of [³H]cAMP release induced by 0 to 1 n*M* (○) or 0 to 1 μ*M* (●) concentration steps. Arrow (S) represents first fraction of application of the stimulus which was held constant thereafter. *Bottom:* Stimuli of 1 μ*M* cAMP are indicated by dashed lines. Release of ³H-cAMP is plotted as (- -×- -). Arrows indicate the presence of (A) 50 μ*M* α-methylmannoside, (B) 10 μg/ml Con A, or (C) 10 μg/ml Con A plus 50 μ*M* α-methylmannoside.

2.   STIMULATION WITH 2′-deoxy-cAMP

cAMP-Stimulated cAMP secretion can also be measured by stimulation of unlabeled cells with 2′-deoxy-cAMP (Van Haastert, 1984a). This compound effectively activates the response leading to the production of cAMP. Assay of cAMP is carried out by isotope dilution using the regulatory subunit of the protein kinase as described above. Since 2′-deoxy–cAMP binds to the regulatory subunit only at very high concentrations, the applied stimulus does not interfere with the measurement of cAMP.

VI.   Assay of Adenylate Cyclase

A.   Adenylate Cyclase Assay

1.   LYSIS TECHNIQUES

Lysis techniques include freeze–thawing (Klein, 1976), sonication (Roos and Gerisch, 1976), lyophilization–homogenization (Pahlic and Rutherford, 1979), nitrogen cavitation, permeabilization by digitonin (Renart *et al.*, 1981), and filter lysis (Das and Henderson, 1983). The filter lysis technique appears to be the most useful method for assay of adenylate cyclase. Cells ($<5 \times 10^7$/ml) are loaded into syringes and rapidly pushed through two 5-μm pore size nucleopore filters. For large volumes, a syringe system with a 47-mm filter holder is used (Antlia Pneumatic Hand Pump, Schleicher & Schuell). For rapid assays, cells (100–800 μl at 4×10^7/ml) are pipeted into a 1-ml tuberculin syringe on which has been fitted a double-layer 5-μm nucleopore filter between the barrel tip and the needle (20 gauge 1.5 in.).

2.   REACTION CONDITIONS

The basic reaction mix (RM) contains the following:

10–50 mM Tris, pH 7.5–8.0
0.05–0.5 mM ATP (Sigma A2383)
α-^{32}P-ATP
1–10 mM MgCl$_2$ or MnSO$_4$
0.1–1 mM cAMP
12 mM Creatine phosphate
250 μg/ml Creatine phosphokinase

The regeneration system is required if assays are >2 minutes in duration. DTT (10 mM) is often included as an additional phosphodiesterase inhibitor; it is essential if 0.1 mM cAMP is used but is not necessary if 1 mM cAMP is present. The [α-^{32}P]ATP is synthesized by the method of Johnson and Walseth (1979). About 10^7 cpm are added per 200 μl RM. Cell lysates or membranes are incubated at 22°C in 1× RM for 0.1–60 minutes, and the reaction is stopped by the addition of an equal volume of 1% SDS, 9 mM ATP, 1 mM cAMP. The sample is diluted to 1 ml and the [^{32}P]cAMP is purified by sequential dowex and alumina chromatography (Salomon, 1979).

It is useful to measure activity in the presence of 2–10 mM MnSO$_4$. In higher eukaryotes, MnSO$_4$ stimulates activity by apparently "uncoupling" the catalytic portion of the enzyme from regulation by its associated G protein (Ross and Gilman, 1980). In *Dictyostelium*, basal activity in lysates is stimulated ~10-fold by MnSO$_4$, and activity is no longer enhanced by cAMP stimulation of intact cells prior to cell lysis. Thus, the MnSO$_4$-stimulated activity apparently reflects the unregulated adenylate cyclase activity, a very useful parameter.

B. Basal Adenylate Cyclase Activity

1. Crude Extracts

In broken cells, basal activity is stable for at least 25 minutes at 25°C. These preparations exhibit non-Michaelian kinetics with two apparent K_m values of 17 μM and 400 μM for Mg^{2+}-ATP (de Gunzburg *et al.*, 1980). The V_{max} is ~30 pmol/min/mg. Activity is inhibited by 0.1–1 mM CaCl$_2$. MnSO$_4$ stimulates activity from 2- to 10-fold and antagonizes the CaCl$_2$ inhibition (Loomis *et al.*, 1978). Vegetative cells contain a heat-stable inhibitor (Cripps and Rutherford, 1981), as well as a heat-stable activator protein which has a molecular weight of 13,000 (Devreotes, 1982).

2. Membranes, Purification, and Solubilization

Membranes are prepared by large-volume filter lysis at a density of 5 × 10^7/ml. Lysis buffer consists of 200 mM sucrose, 10 mM Tris, pH 8.5, and either 0.2 mM EDTA or 2 mM MgCl$_2$ plus 0.2 mM EGTA plus PICS (1 mg/ml phenanthroline, 250 μg/ml CBZ-phenylalanine, 100 μg/ml PMSF, 1 mg/ml benzamidine, 10 μg/ml TPCK, 10 μg/ml TLCK, 10 μg/ml TAME, 0.5 μg/ml antipain, 1 μg/ml leupeptin, 0.5 μg/ml aprotin, 0.6 μg/ml chymostatin, 0.6 μg/ml pepstatin). The lysate is diluted twofold with HEG (10 mM HEPES, pH 7.5, 1 mM EDTA, 10% glycerol) and centrifuged at 7000 rpm (Sorvall SS-34 rotor) for 20 minutes. The pellet is resuspended at a two-fold concentration in HEG. Most of the activity is recovered in the pellet at about a fourfold purification. This preparation has a K_m value of ~400 μM for Mg^{2+}-ATP; in Mn^{2+}-ATP, the V_{max} increases about fivefold and the K_m value is reduced to 40 μM. The activity can be further enriched by extraction in 12 mM CHAPS for 15 minutes at 0°C followed by centrifugation at 18,000 rpm (Sorvall SS-34) for 45 minutes. All of the activity is recovered in the detergent-extracted pellet, with a

30-fold purification increasing the V_{max} for Mn^{2+}-ATP to ~2 nmol/min/mg. A wash of 2 M NaCl and 5 mM PP$_i$ results in further purification over cells, but only 30% of the activity is recovered. This CHAPS-extracted fraction is composed of bilayer vesicles with a diameter of 0.5–2 μm. The recovery of the enzyme into the CHAPS-resistant matrix may depend on the initial membrane preparation. If the membranes are washed with 1 M NaCl prior to extraction, ~40% of the activity is reported to be solubilized (Hagman, 1985).

C. Adenylate Cyclase Activation

1. cAMP STIMULATION OF INTACT CELLS

Regulation of cAMP synthesis during the signaling response occurs at the level of adenylate cyclase, which is tightly controlled by receptor occupancy (see Fig. 2). *In vitro* activation of the *Dictyostelium* enzyme by cAMP has not yet been achieved. In fact, none of the reactions activated by cAMP in intact cells are stimulated when cAMP is added to broken cells. However, extracts of cells sampled at the peak of a cAMP signaling response and quickly lysed and assayed, synthesize cAMP at an elevated rate. Cells in DB minus $CaCl_2$ are lysed directly into an equal volume of 2× RM. Extracts of unstimulated cells synthesize cAMP at 1–6 pmol/min/mg. In extracts of stimulated cells, the initial rate is 10–100 pmol/min/mg; however, this decays to ~150% of the basal rate within 30 seconds (Roos and Gerisch, 1976; Roos *et al.*, 1977). In extracts of adapted cells (following 5–10 minutes of persistent cAMP stimulation), the rate is again 1–6 pmol/min/mg. These differences are observed only in $MgCl_2$; with $MnSO_4$ in the assay, the high rate is observed in basal, stimulated, and adapted cells. The decay of the activated state *in vitro* can be retarded by holding the extracts at 0°C or prevented by freezing (Roos *et al.*, 1977; Klein *et al.*, 1977). Activation can also be measured by lysing stimulated cells with the detergent CHAPS (10 mM) in the presence of RM. Activity in $MgCl_2$ is almost completely inhibited, but is measurable in the presence of 0.5–1 mM $MnSO_4$, which apparently is not a high enough concentration to "uncouple" activation.

2. GMPPNP STIMULATION

Even though it has been reported that *Dictyostelium* contains a GTP-binding protein similar to the G_s regulatory protein in higher eukaryotes (Leichtling *et al.*, 1981; Fontana *et al.*, 1984), and a GTP effect on cAMP binding has been observed in membranes (Van Haastert, 1984b), previous attempts to activate the adenylate cyclase with guanine nucleotides in homogenates or purified membranes have failed. It appears that the catalytic component is very rapidly uncoupled from regulation by its G protein (and receptor) by cell lysis. However, we have recently discovered a method to activate the enzyme *in vitro* with GTP and GMPPNP (Theibert and Devreotes, 1986). Cells are washed and held at 0°C to ensure basal activity, as described in Section III. Cells are lysed in the presence of 50 μM GMPPNP or lysed directly into a tube containing GMPPNP and incubated for 5 minutes at 0°C prior to assay. These conditions yield a 12- to 27-fold stimulation by GMPPNP (Table I). Addition of GMPPNP during or immediately after lysis is essential. Addition after 5 minutes is ineffective. GDP does not stimulate and, in fact, slightly inhibits basal activity. Incubation of the lysate at 0°C is required, and the peak in activity begins to plateau after 5 minutes. Longer incubations (>10 minutes) at 0°C lead to a loss in activity, except if 10% glycerol is added to the lysate, in which case the overall stimulation is reduced but is stable for at least 15 minutes at 0°C. GMPPNP activation is stable to freezing if 10% glycerol is present. cAMP (1 μM) is usually present during the cell lysis. However, we have observed large activations by GMPPNP even in the absence of exogenously added cAMP during lysis. Since a low level of cAMP could be present in the suspension, we cannot rule out that a very short *in vivo* cAMP activation occurs.

Fractionation experiments were performed to determine the subcellular distribution of the GMPPNP-activated complex. Following centrifugation in a microfuge for 3 minutes, all activity is recovered into the pellet, but there is about a two- to threefold loss of activated activity which is not recovered by adding back the supernatant. The activity in the pellet from cells lysed in the presence of GMPPNP is usually two- to sixfold higher than pellets from cells lysed in its absence. These fractionation experiments suggest that the activated system will have to be disrupted and then reconstituted to identify the individual components.

TABLE I

ADENYLATE CYCLASE ACTIVITY IN AX-3

Experiment number	Addition during Lysis (pmol/min/mg)					Fold stimulation
	None	cAMP	cAMP + GDP	GMPPNP	cAMP + GMPPNP	
1	3.7	11.2	3.5	—	31	—
2	2.8	4.8	—	22	36	—
3	8.0	—	—	—	148	18
4	4.7	—	—	—	106	22
5	2.5	—	—	—	31	12.5
6	5.8	—	—	—	157	27
7	3.0	—	—	—	119	40

TABLE II

ADENYLATE CYCLASE IN NC-4 AND MUTANTS

Addition during lysis or assay	Source of lysate (pmol/min/mg)		
	Synag 49	Synag 7	NC-4
None	3.8	2.1	2.6
	5.2	5.5	5.7
cAMP + GMPPNP	6.3	7.9	72
	7.8	9.6	183
MnSO₄	14	54	32
	14.3	46	44

3. MUTANTS AND RECONSTITUTION ASSAYS

Two mutants in the signaling system have been characterized. These mutants were obtained by nitrosoguanidine-induced mutagenesis to about a 40% kill rate. Aggregation minus clones were selected and characterized (C. Frantz, Ph.D. thesis, University of Chicago). Neither mutant synthesizes significant cAMP in response to cAMP, but each has a normal number of surface cAMP-binding sites and a normal chemotactic response, and can synergize to form spores with wild-type cells. One, designated *Synag 49,* has low basal cyclase activity in the presence of $MgCl_2$, and *in vivo*-stimulated cells show a very small (twofold) activation of cyclase activity. The activity in the presence of $MnSO_4$ is also very low (Table II). This suggests that *Synag 49* is deficient in catalytic activity. The other mutant, *Synag 7,* has low basal activity in the presence of $MgCl_2$ and shows no activation of cyclase activity in *in vivo*-stimulated cells. The activity in the presence of $MnSO_4$, however, is the same as that in wild type. This suggests that the biochemical defect in *Synag 7* lies between receptors and cyclase in the signal transduction pathway. Surprisingly, P. Van Haastert has determined that there is an *in vitro* GTP effect on the affinity of the cAMP receptor in *synag 7* (personal communication), suggesting that it may contain a G protein which can interact with receptors. Using the mutants *synag 7* and *synag 49,* we devised a reconstitution assay to investigate the mechanism of GTP-dependent stimulation.

4. SOLUBLE ACTIVATOR

Our preliminary results indicate that the 100,000 *g* supernatant from lysates of *synag 49* (or wild-type cells) contains a heat-labile, Biogel P-30-excluded component which can activate *synag 7* lysates or membranes. Up to 40-fold activations are observed (Table III). The *synag 7* lysates must be prepared in the presence of GMPPNP to reconstitute activity. Furthermore, the activator has no

TABLE III

EFFECT OF SUPERNATANTS ON *Synag 7* ACTIVITY

Experiment number	Source of supernatant (pmol/min/mg)		
	Synag 7	*Synag 49*	Ax-3
1	5.6	48.5	105
2	5.4	90	204

effect on *synag 7* activity assayed in $MnSO_4$. These data, taken together with Van Haastert's observation that *synag 7* receptor affinity is modulated by GTP, suggest that *synag 7* is not defective in G protein but in a factor required for G-protein regulation. Supernatants prepared from *synag 7* cells appear to lack the activator (Table III). The kinetics and reversibility of activation of *synag 7* extracts with the activator suggests that the mechanism is stoichiometric and reversible as opposed to catalytic. Studies are under way to determine the identity and exact function of the soluble activator and to demonstrate directly that it is missing or nonfunctional in *synag 7*. A soluble activator of GTP-dependent, hormone-stimulated cyclase has been reported in reticulocytes (Shane *et al.*, 1985). An exciting possibility is that the activator is another component of G-protein systems in general and *synag 7* represents the first mutation in this component.

VII.   cAMP Binding, Receptor Photoaffinity Labeling, and Receptor Modification

A.   cAMP-Binding Assay

In each of the binding assays that have been described, cells are incubated with ^3H-cAMP or ^{32}P-cAMP in the presence of a phosphodiesterase inhibitor such as DTT. In most of these assays, it is not possible to wash the cells after binding, because the receptor–cAMP complex dissociates rapidly (Mullens and Newell, 1978; Van Haastert and de Wit, 1984). Cells with cAMP bound can be centrifuged and the pellets counted directly without washing (Henderson, 1975); cells can be placed onto Millipore filters, allowing the buffer containing free ligand to pass through the filter; or they can be centrifuged through silicone oil, which separates cells from the aqueous phase (Klein, 1976). The oil assay can also be used to measure cAMP binding in plasma membranes (Van Haastert, 1984b).

Saturated ammonium sulfate stabilizes cAMP binding to the receptor (Van Haastert and Kien, 1983), permitting washing steps which greatly lower the nonspecific binding. The K_D for cAMP binding from this assay is 2 nM, lower than that reported using other assays. The number of sites per cell is $\sim$180,000, which is threefold greater than detected in the standard assays. Only the ammonium sulfate assay will be described in detail below, since it yields a high ratio of specific to nonspecific binding, is very reproducible, and is relatively simple.

Cells are prepared at the appropriate time(s) in development, washed once in DB, once in DB with 10 mM DTT, and then resuspended in DB–DTT at 10^8/ml at 0°C. In a 12 × 75 mm glass test tube, 100 μl of cells is added to 100 μl of ^{3}H-cAMP ($\sim$20 Ci/mmol) at 400 nM in DB–DTT. In a parallel tube, nonradioactive cAMP is added at 10^{-4} M to determine nonspecific binding. The reaction is vortexed, and, after 30 seconds, 3 ml of 98% saturated ammonium sulfate is added. After 5 minutes (to allow complete stabilization of binding), cells are centrifuged at 4000 rpm for 10 minutes (Sorvall HS-4). The supernatant is discarded, and the cells are washed again in ammonium sulfate. The pellet is then resuspended in water and counted. The nonspecific binding is typically 1–2% of the maximal binding. If binding is being measured during development, time points can be taken from a single flask of developing cells and kept at 0°C until all time points are ready for assaying.

B. Binding of cAMP Analogs

The several cAMP-binding proteins in *Dictyostelium* can be distinguished pharmacologically. Van Haastert (1983) tested 16 cAMP analogs and found a characteristic order of affinity for the surface-binding sites that correlated with the order of potency for stimulation of chemotaxis and cGMP accumulation and was distinct from the order of substrate specificity of the membrane phosphodiesterase and the affinity of the regulatory subunit of protein kinase. For example, 2'-deoxy-cAMP has a high affinity for the surface receptor but not for the regulatory subunit, whereas N^6-monobutyryl-cAMP has a high affinity for the regulatory subunit but does not bind to the surface receptor even at 0.1 mM. These differences in analog specificity have been used to distinguish the surface receptor from the regulatory subunit in photoaffinity labeling (Klein *et al.*, 1985) and in studies of the effects of exogenous cAMP on gene expression during development (Schaap and van Driel, 1985).

C. Photoaffinity Labeling

The surface cAMP receptor can be photoaffinity-labeled using 8-N$_3$-[^{32}P]cAMP, which reveals a doublet of bands at MW 40,000 (R form) and 43,000 (D form) in autoradiographs (Theibert *et al.*, 1984). Cells are prepared as for the binding assay. 8-N$_3$-[^{32}P[cAMP ($\sim$60 Ci/mmol, available from ICN) is dried from methanol solution with N$_2$ and redissolved in DB–DTT at 1.0 μM at 4°C. Binding of the photolabel to cells is done as for the cAMP-binding assay

(200 μl reaction volume), except that the cell pellet after the second ammonium sulfate wash is resuspended at a density of 10^8/ml in ammonium sulfate. A small aliquot (10–20 μl) is removed to measure noncovalent binding, and the remainder is placed on a plastic Petri dish at <8°C. The sample is irradiated with a short-wave UV lamp (254 nm) for 5–20 minutes. The cells are then lysed in receptor buffer, RB (25 mM Tris, pH 7.5, 5 mM EDTA, 5 mM DTT, with PICS as described in Section VI) at a 10:1 dilution and centrifuged at 12,000 rpm for 30 minutes (Sorvall SS-34). The membranes are then solubilized in Laemmli sample buffer in a volume equal to the original reaction volume (200 μl). Membranes from 10^7 cells must be diluted into at least 200 μl to prevent aggregation of the receptor. Once solubilized in SDS, the receptor is stable at room temperature for at least 3 days and can be stored at −20°C for several weeks. The samples are then electrophoresed on 10% PAGE (Laemmli, 1970), stained with Coomassie blue, dried, and autoradiographed. A 50-μl sample should require a 15- to 20-hour exposure.

D. Receptor Modification

A reversible shift in electrophoretic mobility from the R form to the D form has been shown to be stimulated by cAMP, with kinetics and concentration dependence that correlate closely with that of the adaptation process (Devreotes and Sherring, 1985). Furthermore, the shift to the D form is associated with a 10-fold increase in the level of phosphorylation of the receptor. When endogenous cAMP synthesis is suppressed, the R form of the receptor predominates in autoradiographs of phosphorylated or photolabeled cells. After stimulation with cAMP, the D form predominates (Fig. 6A).

In order to label preferentially either the R or D form of the receptor, aggregation stage cells are washed once in DB, resuspended at 10^8/ml at 22°C, and shaken at 150 rpm. To label the R form, cells are treated with 5 mM caffeine for 30 minutes. To label the D form, cells are treated with 10 mM DTT, followed by 10 μM cAMP for 13 minutes. Con A (100 μg/ml) is then added, since it improves recovery of receptor. After 2 minutes, cells are washed in DB–DTT, resuspended at 2 × 10^8/ml in DB–DTT, and photolabeled as described above. Frequently, the R form of the receptor appears more intense than the D form from a parallel set of cells, even when the level of noncovalent binding of the photolabel is similar, presumably due to selective degradation of the D form of the receptor.

To phosphorylate the receptor *in vivo*, cells at the aggregation stage are washed twice in 10 mM MES, pH 6.8, resuspended at 10^8/ml in 10 mM MES, and shaken at 150 rpm. Carrier-free $^{32}PO_4$ (1.0 mCi/ml of cells at 10^8/ml) is added for 30 minutes. For the R form, 5 mM caffeine is added at the same time. For the D form, DTT and cAMP are added 15 minutes after addition of $^{32}PO_4$. Con A is added and membrane pellets prepared as described above (Fig. 6A).

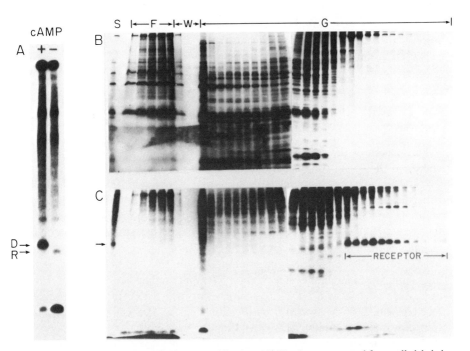

FIG. 6. Receptor phosphorylation and purification. (A) Membranes prepared from cells labeled *in vivo* with $^{32}P_i$. One set of cells (+) was treated with cAMP prior to membrane preparation. (B, C) Fractions from hydroxyapatite column analyzed on standard Laemmli gels. Coomassie stain (B) or autoradiograph (C) of same gel are shown. First lane is applied sample (S); the following lane is empty. Next five lanes show flowthrough from every other column volume during loading (F). Next three lanes are taken at widely spaced intervals during washes (W). Following lanes show every two-thirds column volume during application of phosphate gradient (G). Fractions pooled for further purification are marked "RECEPTOR" on the autoradiograph.

E. Resolving R and D Forms by Polyacrylamide Gel Electrophoresis

The resolution of the two forms of the receptor depends markedly on the concentration of bis-acrylamide in the gel. Furthermore, the mobility of both forms of the receptor relative to other proteins (e.g., actin) also varies with the concentration of bis-acrylamide. For example, using a 10% gel with 0.05% bis ("low bis"), the two forms of the receptor are widely separated (as in Fig. 6A). The slower mobility form of the receptor (D form) migrates slower than actin, while the higher mobility form (R form) migrates faster than actin. Using a 10% gel with 0.25% bis ("standard bis"), the two forms migrate more closely together and both have a slightly higher mobility than actin. On a 10% gel with 1% bis ("high bis"), the two forms have very similar mobilities but a significantly higher mobility than actin. Due to this anomalous behavior, a "high bis" gel resolves the receptor from contaminants that coelectrophorese on the "low bis" gels used in the purification described in Section VIII.

VIII.   Purification of the cAMP Surface Receptor

Purification of the surface cAMP receptor is hindered by its low abundance and the difficulty in solubilizing *Dictyostelium* membranes in mild detergents. The low abundance requires a large volume of cells (Section II), which should be collected at the peak of receptor-binding activity (Section VII). The low solubility in mild detergents (such as Triton X-100, NP-40, octyl-glucoside, lubrol-PX, and CHAPS) has precluded the use of many standard chromatographic techniques such as ion exchange, lectin, affinity, and gel permeation chromatography. The receptor can be solubilized completely in SDS at sufficiently high dilution of membranes (see Section VII), but binding activity is lost. However, once labeled with ^{32}P by *in vivo* labeling with $^{32}PO_4$ or by photoaffinity labeling with 8-N_3-[^{32}P[cAMP, purification steps in SDS can be monitored by Cerenkov counting and by SDS–PAGE/autoradiography.

The overall scheme for purification of the D form is shown below (purification of the R form is the same except that cells are pretreated with caffeine rather than cAMP):

Developed cells (aggregation stage) (100%)
| cAMP
↓
D Form of receptor
| 98% Saturated ammonium sulfate
↓
Osmotic Lysis
| Centrifugate at 12,000 *g*
↓
SDS-Solubilized membranes (30%)
| Add ^{23}P-labeled membranes as tracer
↓
Hydroxyapatite column (10%)
| Concentrate eluate
↓
Preparative SDS–PAGE (low bis) (3%)
| Elute and lyophilize
↓
Second SDS–PAGE (high bis) (1.5%)

A.   Purification Protocol

1.   PREPARATION OF MEMBRANES

One liter of aggregation-stage cells at 10^8/ml is stimulated with cAMP as described in Section VII and centrifuged in two 500-ml buckets at 2500 rpm in a Sorvall GS3 rotor. The pellets are each resuspended in 500 ml 98% saturated ammonium sulfate at $4°-10°C$ and centrifuged at 7000 rpm for 10 minutes. The pellets are resuspended in 80 ml saturated ammonium sulfate, pooled, and added to 4 liters of stirring RB (Section VII) at $0°C$. Membranes are collected at 10,000 rpm for 30 minutes (Sorvall GS3 rotor). The pellets are added to 1 liter SDT (1.5% SDS, 5 mM DTT, 25 mM Tris, pH 6.8). Once solubilized in 1.5% SDS, the receptor is stable at room temperature for at least 3 days and is also stable at $-20°C$ for at least 6 weeks.

2.   HYDROXYAPATITE COLUMNS IN SDS (Moss and Rosenblum, 1972)

Dry hydroxyapatite (BioGel HTP) is resuspended in column buffer (25 mM Tris, pH 6.8, 0.1% SDS) at 1 g/10 ml, de-fined, then adjusted to a 50% slurry. For large preparations, 50-ml columns are poured in 2.5 × 20 cm Bio-Rad econocolumns. Several columns can be run simultaneously by connecting them to a peristaltic pump with multiple lines (e.g., Gilson Minipuls). Sand is placed in the bottom of the column at 10% of the column volume, and 50 ml of column buffer is added. Then, 100 ml of the 50% slurry is added, the fines are aspirated, and the column is allowed to pack for 1–3 days.

The SDS-solubilized membranes are thawed and filtered through Whatman 3M paper in a Buchner funnel. Then, ^{32}P-labeled membranes prepared as in Section VI are added as a tracer at 1–2 ml/liter of unlabeled membranes. The sample is loaded at room temperature at a rate of 5–8 minutes per column volume with a maximum of 10 column volumes of membranes per column. The vast majority of proteins pass through the column, as shown in Fig. 6B. It is essential to avoid running the column dry, since the hydroxyapatite hardens and the columns slow or stop completely. The column is then washed in 5 column volumes of 10 mM sodium phosphate, pH 6.4, 0.1% SDS, 2 mM DTT, followed by 5 column volumes of 0.2 M sodium phosphate, pH 6.4, 0.1% SDS, 2 $M(M)$ DTT (which elutes a significant amount of contaminating protein). The column is then washed with 10 column volumes of 0.2 M sodium phosphate, pH 5.4, 2 mM DTT without SDS, to lower the concentration of SDS in the subsequent elution down to 0.05%.

A linear gradient of sodium phosphate, pH 6.4 (six column volumes of 0.2 M and six column volumes of 0.6 M) with 0.05% SDS and 2 mM DTT is applied at a flow rate of one column volume per 15 minutes. The flow rate during elution is critical to the resolution of the cAMP receptor. We found that the slower the flow rate, the later in the gradient the receptor eluted—that is, further from the bulk of eluted proteins. For example, at a flow rate of 2 minutes per column volume, the receptor begins to elute at 0.2 M with a large amount of contamination, while at 15 minutes per column volume, it elutes at 0.45–0.6 M (see Fig. 6B and C). Once the gradient is completed, another six column volumes of 0.6 M sodium phosphate, 0.05% SDS, 2 mM DTT is applied.

3. CONCENTRATION OF ELUATE

The eluate is collected in fractions of one to two column volumes, which are kept at 0°C during collection. After 1 hour, >90% of the receptor coprecipitates with the SDS. The precipitated fractions are centrifuged at 7500 g for 20 minutes and the supernatant discarded. The pellets can be stored at −20°C for several weeks. Small aliquots of each fraction (2%) are run on SDS–PAGE/auto-radiography to localize the receptor. (The initial one or two fractions containing receptor are usually excluded due to significant contamination.) The pellets containing purified receptor are pooled. (These redissolve when warmed to room temperature and yield a volume of 5% the original volume.) The pooled fraction is immediately (since the receptor is unstable at this point) reprecipitated in smaller centrifuge tubes at 0°C and centrifuged at 18,000 g, which further con-centrates the sample to 0.5% of the original volume. These pellets are diluted 1:1 with Laemmli sample buffer lacking SDS.

4. PREPARATIVE SDS–PAGE

The pooled fractions are then loaded onto a 3-mm, 10% "low bis" gel (see Section VII) with a 4% stacking gel. No more than 5 ml of the 100-fold concen-trated eluate is loaded per preparative gel. Gels are run at 50 V for ~16 hours. These fractions still contain 0.6 M phosphate and require several hours to stack. The gels are dried without fixation at 60°C on Whatman 3M paper and auto-radiographed overnight. The autoradiograph is lined up with the gel and the receptor band, which is the most intensely phosphorylated band at this step, is excised with a razor blade. The dry gel slice is cut into 2- to 3-mm pieces, which are shaken in a 20-ml vial containing 10 ml of 0.05% SDS and 0.05% am-monium bicarbonate overnight. The next day, the buffer is removed and the elution repeated. The eluates are pooled and filtered through glass wool. Typ-ically 80% of the ^{32}P-labeled protein elutes from the gel. The eluate, which is stable at room temperature for up to 4 days, is lyophilized and resuspended in 0.5 ml of water or Laemmli sample buffer without SDS. It is important not to resuspend the lyophilized protein in a smaller volume, since it will aggregate.

B. Final Purification and Bookkeeping

Purification of the receptor relies on the copurification of the receptor from the large membrane preparation with the ^{32}P-labeled band from the added tracer. Although a single band is excised from the "low bis" preparative gel, when the eluted protein is electrophoresed on a "high bis" gel (see Section VII), silver staining reveals three protein bands. Only one of the bands coelectrophoreses with the phosphorylated tracer (see Fig. 7).

The overall recovery at each step of the purification estimated by scanning the major phosphorylated band in autoradiographs is included in parentheses in the

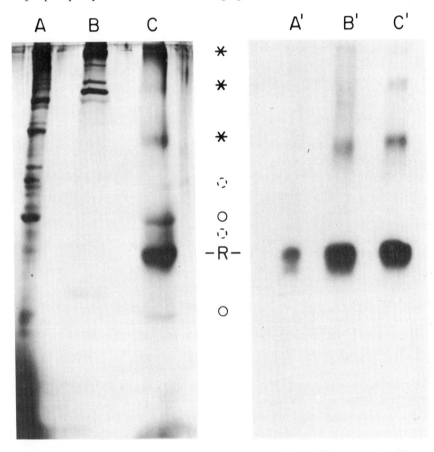

FIG. 7. Final purification and bookkeeping. "Hi bis" SDS–PAGE analysis of membranes (A,A'), hydroxyapatite column eluate (B,B'), and preparative "low bis" SDS–PAGE eluate (C,C'). Gels were silver-stained (A,B,C) or Western-blotted with specific receptor antiserum (A',B',C'). In C and C', R indicates the position of the receptor (MW = 43,000). Higher molecular weight aggregates of the receptor are indicated by *. Contaminants are indicated by circles.

flowchart above. Specific polyclonal rabbit serum was raised against the purified receptor. Figure 7B shows a Western blot of the gel in Fig. 7A. The overall recovery estimated from the Western blot analysis is 1.8%. Based on the molecular weight of the receptor, the number of sites per cell (assuming there is one site per MW 40,000 polypeptide), and these recoveries, it is expected that ~1 μg of highly purified receptor should be obtained from each liter of growing cells. About 20 to 30 μg are typically purified from each 30-liter preparation.

IX. Characterization of Transmembrane Signaling Mutants

To bring together the various assays presented in this chapter, we now show how these assays can be applied to assess quickly whether a mutant has a lesion in transmembrane signaling (Fig. 8).

A. Developmental Time Course of cAMP Binding

An initial step is to determine the time course of development relative to the parental strain. Cells are starved at 22°C in suspension in the absence or presence of cAMP pulses as described in Section II. At 2-hour intervals, 2 ml of suspension are removed and held at 0°C to be used for cAMP-binding assays. The total time period covered is 10 hours for Ax-3 strains and 16 hours for NC-4 strains. Binding is most conveniently assayed with the ammonium sulfate technique at 1 nM and 100 nM ^{3}H-cAMP. This experiment should give a rough indication of the time course of development and of the number and affinity of surface receptors at the peak (6 hours for Ax-3 and 12 hours for NC-4).

Alterations in the affinity or number of receptors should be documented by Scatchard plots. Defects in individual classes of cAMP-binding sites can be assessed by the modifications in the binding assays described by Van Haastert and de Wit (1984). In addition to receptor mutations, altered cAMP-binding affinity could also reflect G-protein mutations which can be assessed by examining the effects of guanine nucleotides on binding *in vitro*. Potential receptor defects can be further investigated by photoaffinity labeling of the receptor. A structural change might be revealed on SDS–PAGE.

If cAMP-binding sites are normal with respect to number and affinity (even if expression must be induced by pulses of cAMP), then it is suggested to proceed with alternate assays. If the timing of appearance of binding sites is altered, the timing of other early developmental markers such as the adenylate cyclase or membrane phosphodiesterase should be examined to determine the generality of the timing defect. Developmental timing alterations can be characterized in detail with the methods of Soll (1979, 1983; see also Soll, Chap. 22, this volume).

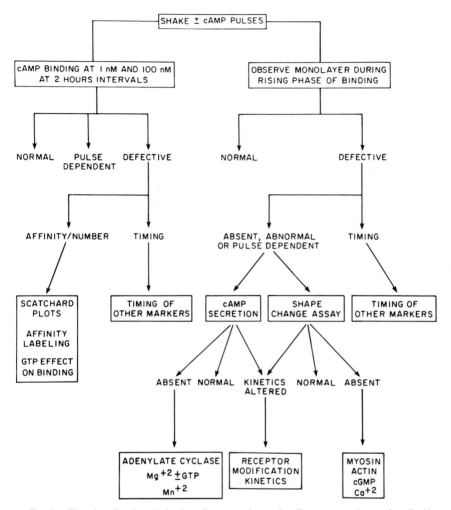

FIG. 8. Flowchart for characterization of transmembrane signaling mutants. Assays described in this chapter are enclosed in boxes. Possible results of assays are not enclosed.

B. Observation of Wave Patterns

It is also very informative to observe cells deposited in a monolayer on an agar surface as described in Section II. These can be observed at intervals or time-lapse video-recorded. In each of these tests, it is critical that parental strains are assayed in parallel, since rate of development is very condition-dependent. If wave and aggregation patterns and their timing of appearance in early development are normal, we suggest that the defect is not in transmembrane signaling. (However, we have no idea how defects in transmembrane signaling that are expressed late in development will affect phenotype. This is an important area to be defined in future research.) All of the mutants showing altered wave or aggregation patterns examined thus far have shown defects in one or more of the assays described in this chapter. If the patterns are absent or abnormal or dependent on pulsatile application of cAMP, assays of cAMP-stimulated cAMP secretion and cell shape changes should be carried out.

1.   cAMP SIGNALING

Absence of the cAMP secretion response should be pursued further with assays of adenylate cyclase *in vitro*. It is most informative to carry out three simultaneous assays in the presence of $MgCl_2$, $MgCl_2$ plus GMPPNP, and $MnSO_4$ for the mutant and parental strains in parallel. The $MnSO_4/MgCl_2$ ratio tests for the catalytic function of the adenylate cyclase, independent of its regulation. The $MgCl_2$ plus $GMPPNP/MgCl_2$ ratio tests for the function of the G protein or other components required for G-protein regulation of the adenylate cyclase such as the cytosolic activator.

2.   SHAPE CHANGE ASSAY

The cAMP-stimulated shape change should also be assessed. If this response is absent, further chemotaxis-related responses should be examined, such as cAMP-stimulated myosin phosphorylation (Berlot *et al.*, 1985), actin and myosin translocation to the detergent-insoluble matrix (McRobbie and Newell, 1983, 1984; Yumura and Fukui, 1985), cGMP accumulation (Mato *et al.*, 1977), and Ca^{2+}-influx and K^+-efflux measurements (Bumann *et al.*, 1984; Malchow, 1985).

3.   KINETICS OF RECEPTOR MODIFICATION

The assays for the cAMP-stimulated cAMP secretion and cell shape changes assess the kinetics of these responses. If the kinetics are altered, the defect may reside in the rate of receptor modification. The kinetics of receptor modification can be assessed by monitoring phosphorylation or the shift in electrophoretic mobility assayed by photoaffinity labeling or Western blot analysis.

X.   Summary and Perspectives

Dictyostelium provides a biochemically and genetically accessible system for studies of transmembrane signaling. Large quantities of cells are easily and inexpensively grown and brought to the appropriate developmental stages (Sections II and III). Transmembrane signaling mutants can be simply isolated by screening for morphological aberrations during development. The differentiated phenotype is reversible so that, once isolated, the mutants can be grown and propagated as the wild type (Section IX). A cell surface cAMP receptor has been identified and purified, and a specific antiserum has been raised (Sections VII and VIII). Evidence suggests that this receptor may regulate chemotaxis (Section IV), activation of adenylate cyclase (Sections V and VI), and gene expression. An extensive, cAMP-mediated phosphorylation of the receptor may cause adaptation of the physiological responses. The recent demonstration of *in vitro* guanylnucleotide regulation of the adenylate cyclase (Section VI) suggests that control mechanisms, similar to those found in vertebrates, may operate in this primitive eukaryotic organism. We expect that future detailed investigation of transmembrane signaling phenomena in this model system will continue to lead to discoveries that will provide important insights into mechanisms in higher organisms.

ACKNOWLEDGMENTS

P.N.D. is an American Heart Association Established Investigator. This work was supported by NIH (GM28007) and NSF (DCB-8417094) awarded to P.N.D.

REFERENCES

Abe, K., and Yanagisawa, K. (1983). *Dev. Biol.* **95,** 200–210.
Acantara, F., and Monk, M. (1974). *J. Gen. Microbiol.* **85,** 321.
Berlot, C., Spudich, J., and Devreotes, P. (1985). *Cell* **43,** 307–314.
Brenner, M. (1977). *J. Biol. Chem.* **252,** 4073–4077.
Brenner, M. (1978). *Dev. Biol.* **64,** 210–223.
Brenner, M., and Thoms, S. (1984). *Dev. Biol.* **101,** 136–146.
Bumann, J., Wurster, B., and Malchow, D. (1984). *J. Cell Biol.* **98,** 173–178.
Chisholm, R., Fontana, D., Theibert, A., Lodish, H. F., and Devreotes, P. N. (1985). *In* "Microbial Development" (Losick and Shapiro, eds.). Cold Spring Harbor Press, Cold Spring Harbor, New York.
Cripps, M., and Rutherford, C. (1981). *Exp. Cell Res.* **133,** 309–316.
Das, P., and Henderson, E. (1983). *Biochim. Biophys. Acta* **736,** 45–56.
Devreotes, P. N. (1982). *In* "The Development of Dictyostelium Discoideum" (W. Loomis, ed.). Academic Press, San Diego.
Devreotes, P. (1983). *Dev. Biol.* **95,** 154–162.
Devreotes, P., and Sherring, J. (1985). *J. Biol. Chem.* **260,** 6378–6384.
Devreotes, P., and Steck, T. (1979). *J. Cell Biol.* **80,** 300–309.
Devreotes, P., Derstine, P., and Steck, T. (1979). *J. Cell Biol.* **80,** 292–300.
Dinauer, M., McKay, S. A., and Devreotes, P. N. (1980). *J. Cell Biol.* **86,** 537.
Fontana, D., and Devreotes, P. (1984). *Dev. Biol.* **106,** 76–82.
Fontana, D., Theibert, A., Wong, T.-Y., and Devreotes, P. (1984). *In* "The Cell Surface in Cancer and Development" (M. Steinberg, ed.). Plenum, New York.
Futrelle, R., Traut, J., and McKee, W. (1983). *J. Cell Biol.* **92,** 807–821.

Gerisch, G., and Hess, B. (1974). *Proc. Natl. Acad. Sci. U.S.A.* **71**, 2118–2122.

Gerisch, G., and Wick, U. (1975). *Biochem. Biophys. Res. Commun.* **65**, 364–370.

Gerisch, G., Malchow, D., Roos, W., and Wick, U. (1979). *J. Exp. Biol.* **81**, 33–47.

Gunzburg, de, J., Veron, M., and Brachet, P. (1980). *Cell Biol. Int. Rep.* **4**, 533–539.

Hagman, J. (1985). *Cell Biol. Int. Rep.* **9**, 491–494.

Henderson, E. (1975). *J. Biol. Chem.* **250**, 4730–4735.

Johnson, R., and Walseth, T. (1979). *Adv. Cyclic Nucleotide Res.* **10**, 135–167.

Klein, C. (1976). *FEBS Lett.* **68**, 125–128.

Klein, C., and Brachet, P. (1975). *Nature (London)* **254**, 432–434.

Klein, C., Brachet, P., and Darmon, M. (1977). *FEBS Lett.* **76**, 145–147.

Klein, P., Theibert, A., Fontana, D., and Devreotes, P. (1985). *J. Biol. Chem.* **260**, 1757–1763.

Konijn, T. (1970). *Experientia* **26**, 367–369.

Konijn, T., van de Meene, J., Bonner, J., and Barkley, D. (1967). *Proc. Natl. Acad. Sci. U.S.A.* **58**, 1152–1154.

Laemmli, U. (1970). *Nature (London)* **227**, 680–685.

Leichtling, B., Coffman, D., Yaeger, E., and Rickenberg, H. (1981). *Biochem. Biophys. Res. Commun.* **102**, 1187–1195.

Loomis, W., Klein, C., and Brechet, P. (1978). *Differentiation* **12**, 83–89.

McRobbie, S., and Newell, P. (1983). *Biochem. Biophys. Res. Commun.* **115**, 351–359.

McRobbie, S., and Newell, P. (1984). *J. Cell Sci.* **69**, 139–144.

Marin, F., and Rothman, F. (1980). *J. Cell Biol.* **87**, 823–827.

Mato, J., Krens, F., Van Haastert, P., and Konijn, T. (1977). *Proc. Natl. Acad. Sci. U.S.A.* **74**, 2348–2351.

Moss, B., and Rosenblum, E. (1972). *J. Biol. Chem.* **247**, 5194–5198.

Mullens, I., and Newell, P. (1978). *Differentiation* **10**, 171–176.

Pahlic, M., and Rutherford, C. (1979). *J. Biol. Chem.* **254**, 9703–9707.

Pan, P., Hall, E., and Bonner, J. (1972). *Nature (London)* **237**, 181–182.

Rahmsdorf, H., Cailla, H., Spitz, E., Moran, M., and Rickenberg, H. (1976). *Proc. Natl. Acad. Sci. U.S.A.* **73**, 3183–3187.

Renart, M., Sebastian, J., and Mato, J. (1981). *Cell Biol. Int. Rep.* **5**, 1045–1054.

Roos, W., and Gerisch, G. (1976). *FEBS Lett.* **68**, 170–172.

Roos, W., Malchow, D., and Gerisch, G. (1977). *Cell Differ.* **6**, 229–239.

Ross, E. M., and Gilman, A. (1980). *Annu. Rev. Biochem.* **49**, 533–64.

Solomon, Y. (1979). *Adv. Cyclic Nucleotide Res.* **10**, 35–55.

Schaap, P., and Van Driel, L. (1985). *Exp. Cell Res.* **159**, 388–398.

Shane, E., Yeh, M., Feigin, A., Owens, J., and Bilezikian, J. (1985). *Endocrinology* **117**, 255–262.

Shimomura, O., Suthers, H., and Bonner, J. (1982). *Proc. Natl. Acad. Sci. U.S.A.* **79**, 7376–7379.

Soll, D. (1979). *Science* **230**, 841–891.

Soll, D. (1983). *Dev. Biol.* **95**, 73–91.

Swanson, J., and Taylor, L. (1981). *Cell* **28**, 225–232.

Theibert, A., and Devreotes, P. N. (1983). *J. Cell Biol.* **97**, 173–178.

Theibert, A., and Devreotes, P. (1984). *Dev. Biol.* **106**, 166–173.

Theibert, A., and Devreotes, P. (1986). *J. Biol. Chem.* **261**, 15121–15125.

Theibert, A., Klein, P., and Devreotes, P. N. (1984). *J. Biol. Chem.* **259**, 12318–12321.

Tomchik, K., and Devreotes, P. (1981). *Science* **212**, 443–446.

Van Haastert, P. (1983). *J. Biol. Chem.* **258**, 9643.

Van Haastert, P. (1984a). *J. Gen. Microbiol.* **130**, 2559–2564.

Van Haastert, P. (1984b). *Biochem. Biophys. Res. Commun.* **124**, 597–604.

Van Haastert, P., and Kien, E. (1983). *J. Biol. Chem.* **258**, 9636.

Van Haastert, P., and Konijn, T. (1982). *Mole. Cell Endocrinol.* **26**, 1–17.

Van Haastert, P., and de Wit. (1984). *J. Biol. Chem.* **259**, 1332–3325.
Van Haastert, P., de Wit, R., Grimjpma, Y., and Konijn, T. (1982a). *Proc. Natl. Acad. Sci. U.S.A.* **79**, 6270–6274.
Van Haastert, P., Jastorff, B., Pinas, J., and Konijn, T. (1982b). *J. Bacteriol.* **149**, 99–105.
Watts, D., and Ashworth, J. (1970). *Biochem. J.* **119**, 171–174.
Williams, G., Elder, E., and Sussman, M. (1984). *Dev. Biol.* **105**, 377–383.
Wurster, B., and Butz, U. (1983). *J. Cell Biol.* **96**, 1566–1570.
Yumura, S., and Fukui, Y. (1985). *Nature (London)* **314**, 194–196.

Chapter 18

Identification of Chemoattractant-Elicited Increases in Protein Phosphorylation

CATHERINE BERLOT

Department of Cell Biology
Stanford University School of Medicine
Stanford, California 94305

I. Introduction

One of the advantages of working with *Dictyostelium discoideum* is that it not only undergoes a reproducible series of responses upon stimulation with the chemoattractant cyclic AMP (cAMP), but it also can be grown easily in large enough quantities that even minor proteins can be purified and analyzed biochemically. Thus there is great potential for being able ultimately to dissect such events as chemotaxis and signal relay into a series of reactions triggered by the binding of cAMP to specific cell surface receptors. One way to approach this problem involves identifying the target components involved in executing the response, then isolating regulators of these components and working back toward the initial event triggered by occupied cAMP receptors. According to this approach, it is essential, first, to identify those changes that occur *in vivo* upon stimulation of cells with chemoattractant. Only then is it possible to design and interpret correctly the results of biochemical experiments with purified components.

There are numerous examples of how alterations in protein phosphorylation can affect motility, growth, and development in cells transformed by specific viruses or treated with various growth factors, hormones, or chemoattractants (for review see Rosen and Krebs, 1981). Developed amoebae stimulated with cAMP undergo a series of reactions which may be involved in triggering the cellular responses, possibly via modulation of phosphoproteins. These reactions

METHODS IN CELL BIOLOGY, VOL. 28

include the activation of both adenylate cyclase and guanylate cyclase to result in transient increases in the intracellular concentrations of both cAMP and cGMP, an increase in the rate of $^{45}Ca^{2+}$ influx, acidification of the medium, alterations in the methylation states of several proteins and phospholipids, and a small increase in the number of intracellular vesicles as seen by electron microscopy (for review see Devreotes, 1982). There is precedent to believe that several of these reactions may result in changes in protein phosphorylation. cAMP-dependent protein kinase has been identified in *Dictyostelium* (Majerfeld *et al.*, 1984), although no specific *Dictyostelium* substrates have been identified. The Ca^{2+} influx could serve to activate calmodulin, which has been isolated from *Dictyostelium* (Bazari and Clarke, 1981; Clarke *et al.*, 1980), although it has not been shown to activate any kinases. It is also possible that the Ca^{2+} influx could be regulating the activity of a *Dictyostelium* Ca^{2+}- and phospholipid-dependent protein kinase.

Dictyostelium possesses numerous phosphorylated proteins, which can make the analysis of *in vivo* phosphorylation patterns difficult. To quantitate the degree of phosphorylation of a protein it is necessary to be able to isolate *in vivo* ^{32}P-labeled protein under conditions in which background labeling is low, where the degree of phosphorylation is not modified after cell lysis, and where it is possible to look at a representative population of molecules. Hopefully the techniques described here, using *Dictyostelium* myosin as an example, will provide a good starting point in the identification of phosphorylated proteins which show changes in response to cAMP stimulation and which may play roles in mediating cellular behavior.

II.   Conditions for Labeling Amoebae with ^{32}P

Dictyostelium amoebae can be labeled with [^{32}P]orthophosphate both during log-phase growth and during starvation. External phosphate is required for growth but not for development. Therefore, it is possible to label cells in buffer containing carrier-free isotope during development, while during growth it is necessary to achieve a balance between inhibiting growth by adding too little carrier phosphate and inhibiting incorporation of phosphate into protein by adding too much carrier. Under conditions which permit satisfactory incorporation of label into growing cells, the doubling time is increased significantly, which may alter normal physiology. Thus it is preferable to label during development if the protein of interest is expressed during both growth and development.

A.   Labeling during Log-Phase Growth

To label during growth, amoebae can be grown in the defined medium of Franke and Kessin (1977) with the concentration of phosphate reduced from 5 to

0.4 mM (Kuczmarski and Spudich, 1980; see also Spudich, this volume, Chapter 1). This reduction in phosphate concentration lengthens the doubling time of the cells from 12 to 20 hours. Cells will grow in this medium for at least four doubling times. Amoebae grow poorly in medium with less phosphate. The amount of ^{32}P incorporated into a TCA-precipitable form plateaus at ~60%, and this level is attained after three or four doubling times. It is recommended that amoebae to be labeled under these conditions be grown for three or four generations in this medium to allow for equilibration of the intracellular ATP pools with ^{32}P. Then the specific radioactivity of the ATP will be known, and the degree of protein phosphorylation can be quantitated (see Section V,A). The amount of ^{32}P necessary for labeling will depend on the moles phosphate/mole protein ratio. To get an overnight exposure of an autoradiograph of an SDS gel loaded with 0.5 μg of myosin labeled under these conditions, where myosin has between 0.1 and 0.5 mole of phosphate per mole of heavy chain and per mole of 18,000-Da light chain (Kuczmarski and Spudich, 1980), ~0.03 mCi/ml of [^{32}P]orthophosphate is added to the medium.

B. Labeling during Development and Stimulation with cAMP

Amoebae can be labeled in the following phosphate-free development buffer: 20 mM MES (pH 6.8), 0.2 mM CaCl$_2$, 2 mM MgSO$_4$. Amoebae developed in this buffer begin spontaneous oscillations at ~3.5 hours as monitored by light scattering (see Devreotes et al., Chap. 17, this volume). When looking for cAMP-mediated increases in protein phosphorylation, the length of the labeling time may be critical. It is not necessary to equilibrate the ATP pool with ^{32}P because the specific radioactivity of the ATP can be measured during the response (see Section V,B). When [^{32}P]orthophosphate is added to cells developed longer than 3.5 hours, the ATP pool takes ~2 hours to equilibrate (M. Brenner, personal communication; C. Berlot, unpublished observations). However, after ~20 minutes of labeling, no increase in ATP specific radioactivity during the interval of a response (~8 minutes) is detectable. Also, the specific radioactivity of ATP does not change as a result of cAMP stimulation. Thus it is not necessary to label for long amounts of time. In fact, long labeling times may mask any cAMP-mediated increases in protein phosphorylation. In the case of myosin phosphorylation, when amoebae are developed for 3.5 hours and then labeled for 20 minutes, cAMP-stimulated phosphorylation increases are readily observed. However, when amoebae are labeled for 4 hours from the beginning of development, these increases are very difficult to see over the rather high background. Under the longer labeling conditions, only 25% of the phosphorylation sites on the myosin heavy chain are removable after a 30 minute cold chase with 10 mM phosphate. However, in amoebae labeled 3.5 hours after development was initiated, for 20 minutes or up to 2 hours, 70% of the phosphorylation sites can be removed with this cold chase. These results suggest that when amoebae are labeled early in development, some myosin phosphorylation sites are labeled which do not turn over as development proceeds. These sites can then mask the turnover of those phosphorylation sites which do respond to cAMP stimulation.

To observe whether cAMP stimulation causes any changes in protein phosphorylation, amoebae can be ^{32}P-labeled and stimulated during spontaneous oscillations or can be pretreated with caffeine, which inhibits oscillations by preventing activation of adenylate cyclase (Devreotes *et al.*, Chap. 17, this volume), and then labeled. To monitor spontaneous oscillations, light scattering is measured spectrophotometrically (Devreotes *et al.*, Chap. 17, this volume). After incubation of the suspension in ^{32}P, cells are stimulated with a saturating dose of cAMP (2×10^{-6} *M,* final concentration) ~ 1 minute prior to a spontaneous stimulus. Although myosin is a major *Dictyostelium* protein (0.5% of the total cell protein), it is very poorly labeled when developed cells are incubated with [^{32}P]orthophosphate. Under these conditions, myosin from unstimulated cells incorporates labeled phosphate to a stoichiometry of 0.05 mole of phosphate per mole of heavy chain and 0.03 mole of phosphate per mole of light chain. Therefore, a considerable amount of isotope must be used. To observe myosin phosphorylation, cells are labeled with [^{32}P]orthophosphate (0.1 mCi/ml) for 20 minutes prior to stimulation. The autoradiograph of a gel loaded with 0.5 μg of myosin yields a good overnight exposure. If amoebae are labeled in the presence of caffeine, more [^{32}P]orthophosphate must be added because caffeine reduces the amount of ^{32}P incorporated into myosin. Under these conditions, myosin from unstimulated cells incorporates labeled phosphate to a stoichiometry of 0.03 mole of phosphate per mole of heavy chain and 0.007 mole of phosphate per mole of light chain. In this case, incubation of amoebae in 0.5 mCi/ml of [^{32}P]orthophosphate for 30 minutes yields a good overnight exposure. This effect of caffeine on myosin phosphorylation may apply to other proteins whose phosphorylation levels are regulated by cAMP. During spontaneous oscillations, responses to one stimulus do not completely subside before the next stimulus appears; therefore, a true basal state may never be achieved. Because caffeine inhibits spontaneous oscillations, it is possible that the effect of caffeine on myosin phosphorylation may be merely to allow the cells to attain a true basal state before application of the stimulus.

III.   Buffer Conditions to Preserve Phosphorylated Protein

In order to determine if the protein of interest is phosphorylated and to investigate whether or not its phosphorylation levels are altered by cAMP stimulation, it is necessary to be able to isolate the protein rapidly at various times during the response under conditions where proteolysis is minimized, no phosphate is added by kinases or removed by phosphatases after cell lysis, and ^{32}P background contamination is minimized. In this section I describe the lysis buffer which was chosen to provide these conditions for the isolation of myosin from ^{32}P-labeled amoebae. In the following section I describe a method of rapid isolation by immunoprecipitation.

The following lysis buffer (LB) is diluted 1:1 with a cell suspension: 40 mM Tris-Cl (pH 7.5), 0.2% NP-40, 2 mM DTT, 10 mM EDTA, 2 mM PMSF, 2 mM TAME, 200 μM TPCK, 200 μM TLCK, 20 mM NaHSO$_3$, 100 μg/ml RNase A

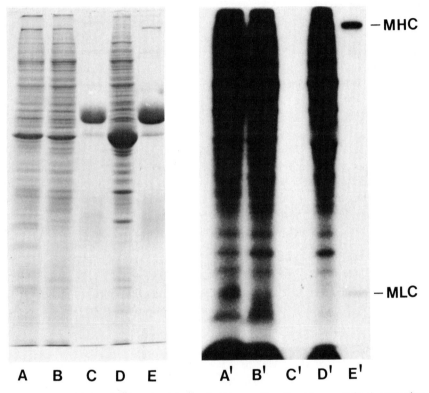

FIG. 1.   Specificity of immunoprecipitation for *Dictyostelium discoideum* myosin. A suspension of developed amoebae was treated with 5 mM caffeine for 30 minutes and then labeled with [^{32}P]orthophosphate (0.5 mCi/ml) for 30 minutes. Aliquots of the suspension were immunoprecipitated as described in the text and then subjected to SDS–polyacrylamide gel electrophoresis according to Laemmli (1970). Lanes A–E, Coomassie-stained 12% polyacrylamide gel; lanes A'–E', autoradiograph of A–E. Lane A, Total lysate; lane B, supernatant from 20,000 rpm centrifugation of lysate; lane C, final washed pellet after immunoprecipitation with preimmune serum; lane D, supernatant from immunoprecipitation with antimyosin antibodies; lane E, final washed pellet after immunoprecipitation with antimyosin antibodies. MHC, Myosin heavy chain; MLC, myosin light chain. From Berlot *et al.* (1985), Figure 1; used with permission.

(Worthington), 50 mM sodium pyrophosphate, 200 mM NaF, 2 mM ATP, and 200 mM potassium phosphate (pH 7.5). PMSF, TAME, TPCK, TLCK, NaHSO$_3$, EDTA, and pyrophosphate inhibit proteases. Myosin is stable in ice-cold lysis buffer for several hours on ice, as judged by SDS–polyacrylamide gel electrophoresis. Myosin immunoprecipitated from this buffer is recovered quantitatively and without proteolytic cleavage (Fig. 1, compare lane A, total lysate which was immediately TCA-precipitated, with lane E, myosin immunoprecipitate). Pyrophosphate and NaF prevent phosphatase activity as determined by the following criterion. Myosin labeled with ^{32}P *in vitro* with a specific heavy chain of kinase (Kuczmarski and Spudich, 1980) was incubated with a cell lysate or with lysis buffer alone as a control for 30 minutes on ice and then immunoprecipitated. The same amount of radioactivity was recovered in both cases. Pyrophosphate is the most effective inhibitor, but NaF has a slight additional inhibitory effect. Molybdate and phosphate have no effect on *Dictyostelium* myosin phosphatase activity. ATP is included in the lysis buffer to prevent addition of ^{32}P to myosin by kinases after cell lysis.

^{32}P-Labeled nucleic acids can cause serious background problems, especially in the higher molecular weight region of a gel. Most of this contamination appears to be due to RNA, because RNase A significantly reduces the background while micrococcal nuclease and DNase I have little effect. It is important to use Worthington RNase A, because the Sigma product contributes protease activity. Potassium phosphate is included in the lysis buffer as a cold chase for the unincorporated [^{32}P]orthophosphate. When cold phosphate is present from the moment of cell lysis until the end of the immunoprecipitation, then there is minimal free [^{32}P]orthophosphate present at the gel stage. Otherwise, the free ^{32}P can artifactually label proteins (such as gel markers) which are not actually phosphorylated *in vivo,* as well as present significant radioactive contamination hazards. The above lysis buffer should serve as a good buffer with which to begin the rapid purification of any phosphorylated protein from *Dictyostelium.*

IV. Rapid Isolation of *in Vivo* Phosphorylated Protein by Immunoprecipitation

To address the question of whether or not cAMP stimulation alters phosphorylation of the protein of interest, it is necessary to be able to examine protein phosphorylation in samples taken at short intervals (i.e., 10 seconds) over a period of 10 minutes. Each of these samples must be individually purified, preferably under conditions where the yield is good, to guard against examining a small subpopulation of the total which may not be representative. Immunoprecipitation with high-affinity specific antibodies is the method of choice.

Immunoprecipitation requires only a few hours, during which time myosin is stable in lysis buffer. This method does not require large amounts of starting material at each time point, and it is possible to attain close to 100% recovery, as determined by quantitation of material on SDS gels.

A.   Antibodies

IgG is purified from polyclonal serum as follows, all manipulations being performed at 4°C or on ice. Ammonium sulfate is added to 30% saturation, and the serum is centrifuged at 20,000 rpm for 20 minutes in a Sorvall SS-34 rotor. The supernatant is brought to 50% saturation in ammonium sulfate and similarly centrifuged. The 50% pellet is dissolved in 5 ml of phosphate-buffered saline with azide (PBS-N_3) and dialyzed for 24 hours against two changes of 2 liters of PBS-N_3. The dialysate is chromatographed on a 2.5 $\times$ 100 cm Sephadex G150 column. The peak IgG-containing fractions are pooled, aliquoted, and stored at -80°C. IgG concentration is determined according to A_{280} of 1.5 = 1 mg/ml.

B.   Preparation of IgG–*Staphylococcus* A Cell Mixture

Staphylococcus A cells (Pansorbin) are washed three times in immunoprecipitation (IP) buffer (IP buffer is half-strength LB minus RNase A) plus 1 mg/ml ovalbumin. Then, 50 μl of these washed cells are added to 40 μl IgG (1 mg/ml) and incubated at 4°C on a rotator for at least 30 minutes.

C.   Preparation and Immunoprecipitation of Cell Lysates

Cell suspensions (2 $\times$ 10^7 cells/ml) containing up to 5 $\times$ 10^6 cells are added to an equal volume of ice-cold lysis buffer and centrifuged for 30 minutes at 20,000 rpm in a Sorvall SS-34 rotor. The supernatants are added to preadsorbed IgG–*Staphylococcus* A cell mixtures and incubated for 30 minutes at 4°C with rotation. The samples are centrifuged in a microfuge and the pellets are resuspended in 1 ml IP buffer with 1 mg/ml ovalbumin. The pellets are washed three times and then once with IP buffer minus ovalbumin. The pellets are frozen to facilitate resuspension. The samples are resuspended in SDS sample buffer and boiled for 5 minutes. The supernatants from a microfuge spin are then loaded on a polyacrylamide gel.

D.   Comments on the Immunoprecipitation Technique

The preclearing of the antigen is important for removing particulate matter which would otherwise contribute a significant ^{32}P background problem. However, it is important to determine that the protein of interest is soluble under these conditions. In the case of myosin, 1 mM ATP is essential for solubility because of its ability to dissociate actomyosin. Also, it is possible that other proteins may not be as stable in lysis buffer as myosin. A method of preclearing the antigen which is faster than a 30 minute centrifugation involves passing cells through a filter with a pore size of 0.45 μm (Amicon Corp., Lexington, MA), which is held with a prefilter (Millipore, cat. no. AP 25 020 00) at the end of a syringe in a Swinnex filter holder, as described by Fechheimer and Cebra (1982). This method preclears the antigen instantaneously, but has the drawback that it contaminates the filter holders, which must be recycled, with ^{32}P. The preclearing centrifugation, on the other hand, can be done in disposable microfuge tubes in two-hold adapters in a Sorvall SS-34 rotor.

With immunoprecipitation it should be possible to isolate the majority of the total protein. This can be checked for by examining the supernatant from the immunoprecipitation if the protein is major enough to be distinguishable on an SDS gel from the rest of the proteins in its molecular weight region. Otherwise the amount of antigen remaining in the supernatant can be quantitated by re-precipitation with a fresh addition of IgG–*Staphylococcus* A cell complexes. Depending on the affinity of the antibody, it may be necessary to concentrate the *Dictyostelium* amoebae so that the lysate volume can be reduced in order to immunoprecipitate all of the antigen. However, amoebae cannot be concentrated much beyond a density of 4×10^7/ml, since the effectiveness of the lysis buffer as a proteolysis inhibitor decreases at higher cell densities. In order to immunoprecipitate antigens in a reproducible, quantitative manner, it is important to be sure that all of the components are properly calibrated. This involves first checking that all of the antibody added to the *Staphylococcus* A cells is bound, and then that all of the antigen added to the IgG–*Staphylococcus* A cell mixture is precipitated.

E. Demonstration of Myosin Immunoprecipitation

Figure 1, lane E, shows Coomassie stain of the *Dictyostelium* proteins immunoprecipitated by a polyclonal antimyosin IgG. The proteins seen in addition to the myosin heavy chain come from the antibody preparation or the *Staphylococcus* A cells. When myosin is immunoprecipitated from ^{32}P-labeled amoebae, the only ^{32}P-labeled proteins seen in the autoradiograph of the immunoprecipitate (Fig. 1, lane E′) comigrate with the heavy chain and the 18,000-Da light chain of myosin from *Dictyostelium*. Virtually all of the myosin is immunoprecipitated by antimyosin antibodies (Fig. 1, compare lanes A, D, and E), whereas no myosin is immunoprecipitated by the preimmune serum (lane C).

V. Quantitation of Amount of Phosphorylation

A. Vegetative Cells

After amoebae have been grown in defined medium containing [^{32}P]orthophosphate for four or more doubling times, the specific radioactivity of the intracellular ATP pools will be the same as that of the extracellular orthophosphate. Therefore the moles phosphate/mole protein ratio can be determined by comparing the radioactivity of a known amount of phosphate in defined medium with that in a known amount of immunoprecipitated protein. The protein can be quantitated by scanning gels containing immunoprecipitates and relating the peak areas to those obtained using pure protein of known amount. Coomassie-stained gels are scanned using an RFT Scanning Densitometer (Transidyne General Corp., Ann Arbor, MI) at 600 nm. The peaks are then cut out and weighed.

B. Developed Cells

When using labeling conditions in which the intracellular ATP pools are not equilibrated with ^{32}P, it is necessary to withdraw cell aliquots for determination of the specific radioactivity of ATP at the same time as the aliquots are removed for immunoprecipitation of protein.

1. METHOD FOR ISOLATING [^{32}P]ATP

Cells are lysed into an equal volume of 0.1 N perchloric acid. Lysates are kept on ice until use or can be frozen. To precipitate out the inorganic phosphate, which would otherwise present background problems, a modification of the method of Sugino and Miyoshi (1964) is used. The lysate is pelleted in the microfuge for 1 minute. To 30 μl of the supernatant is added: 30 μl of 1.54 mM potassium phosphate (pH 7.5), 100 μl of 1 N perchloric acid, 200 μl of 0.02 M ammonium molybdate, and 50 μl of 0.1 M TEA-HCl (pH 5.0). After 5 minutes on ice this mixture is pelleted in a microfuge for 5 minutes. Then, 10 μl of the supernatant is added to 150 μl of 1.54 mM potassium phosphate (pH 7.5), and various amounts of this mixture are spotted on PEI cellulose (Polygram CEL 300 PEI/UV$_{254}$, Macherey-Nagel) with 1 μl of a 10 mM ATP standard. The plate is chromatographed at 4°C in 0.5 M ammonium sulfate for the first third of the plate followed by 0.7 M ammonium sulfate. ATP spots are localized by visualization of the standards with UV light. The spots are then cut out and Cerenkov radiation is measured. Recovery of cellular ATP by this method is 75%, as determined by the use of internal standards.

2. CALCULATION OF MOLES OF PHOSPHATE PER MOLE OF PROTEIN RATIO

Gerisch et al. (1979) determined that the intracellular ATP concentration for *Dictyostelium* is ~1 mM and does not change during spontaneous oscillations. Bumann et al. (1984) estimated that the volume of a cell is 5.2×10^{-10} cm^3. Given this information, it is possible to compare the amounts of ^{32}P incorporated into known quantities of cellular ATP and of the protein of interest. Specific radioactivity of ATP is calculated by correcting for the dilutions made during the isolation of ATP to calculate cpm/cell, which is converted to cpm/intracellular volume, which is then converted to cpm/mole ATP. As described above, the amount of radioactivity in a known amount of protein is determined by scanning bands on gels and counting the radioactivity. The specific radioactivity of the protein can be converted into moles of phosphate per mole of protein by comparison with the specific radioactivity of the ATP.

It is important to emphasize that the stoichiometry of protein phosphorylation obtained by labeling developed cells for short intervals represents only those phosphorylation sites which turn over during the labeling period. Those sites which do not turn over will not be detected since they are not labeled in these experiments. As mentioned above (see Section II,B), 70% of the myosin phosphorylation sites observed when amoebae are labeled after 3.5 hours of development can be removed with a cold chase. However, only 25% of the myosin phosphorylation sites observed in amoebae labeled from the beginning of development can be removed with this cold chase. These results suggest that some phosphorylation sites are labeled early in development and have a relatively low turnover rate later in development. Therefore, although the myosin heavy chain incorporates 0.05 mole of phosphate per mole of heavy chain when amoebae are labeled after 3.5 hours of development, the total amount of phosphate on the heavy chain may be greater than this.

VI.   Summary and Conclusions

Modulation of protein phosphorylation is undoubtedly one of the mechanisms by which *D. discoideum* transduces receptor-mediated cAMP signals to bring about cellular responses such as chemotaxis and signal relay. Since it is often possible to obtain nonspecific phosphorylation *in vitro,* it is important to identify biologically relevant substrates by observing *in vivo* levels of phosphorylation in amoebae before and after cAMP stimulation.

Using the techniques described in this chapter for labeling amoebae with [^{32}P]orthophosphate and then immunoprecipitating myosin under conditions in which the *in vivo* levels of phosphorylation are preserved, it has been demonstrated (Berlot *et al.*, 1985) that phosphorylation of both the heavy chain and the 18,000-Da light chain of myosin transiently increases in response to a cAMP stimulus. Immediately before the phosphorylation increase, the heavy chain also exhibits a transient decrease in phosphorylation (Fig. 2A,B). In the presence of caffeine, which inhibits activation of adenylate cyclase, these responses are still obtained. In this case the magnitude of the responses is greater and the duration is longer (Figs. 2C,D). This pattern of myosin phosphorylation correlates with the series of shape changes induced in amoebae exposed to a temporal increase in cAMP concentration. This stimulus causes amoebae to cease random movement for ~20 seconds, a response referred to as a cringe (Futrelle *et al.*, 1982). The cells then extend pseudopods in all directions, which causes them to flatten on the substrate. After approximately 2–3 minutes, the amoebae adapt to the stimulus and resume random movement (Fontana *et al.*, 1985; Chisholm *et al.*, 1985).

Direct observation of cAMP-mediated *in vivo* changes in phosphorylation of a protein is the necessary first step in implicating this phosphorylation as one of the transducers of the signal. To determine how the protein phosphorylation is regulated, it is necessary to be able to assay for cAMP-mediated changes in phosphorylation rate *in vitro*. This enables one to ask if the response is due to changes in kinase or phosphatase activity or to some change in the availability of the protein as a substrate, as well as to open the way for isolation of modulating factors. Methods have been established to measure the rate of phosphorylation *in vitro* both of the myosin heavy chain, which appears to be associated in an insoluble form with its kinase, and of the myosin light chain (Berlot *et al.*, 1985). These methods should apply to most other phosphoproteins.

The discovery that the *in vivo* phosphorylation levels of a protein change in response to stimulation of amoebae with cAMP can be extremely useful in elucidating previously unanticipated components of the cellular response. If the biochemical functions of such a component are already known, the *in vivo* response data can suggest further biochemical experiments that may be relevant to the *in vivo* roles of the protein. For example, in the case of myosin, light chain phosphorylation activates myosin for movement along actin filaments (Griffith *et al.*, 1987), while heavy chain phosphorylation prevents assembly of myosin into thick filaments and inhibits actin-activated ATPase activity (Kuczmarski and Spudich, 1980). Since the cAMP-induced increases in heavy chain and light chain phosphorylation have seemingly opposite effects on myosin function, it becomes of interest to determine whether the cAMP-induced increases in heavy chain and light chain phosphorylation occur on the same population of myosin molecules *in vivo* and if so to determine the effects of simultaneous heavy chain and light chain phosphorylation on purified myosin. One scenario for how cAMP-induced changes in myosin phosphorylation may result in directed cell movement is that light chain phosphorylation activates motive force production

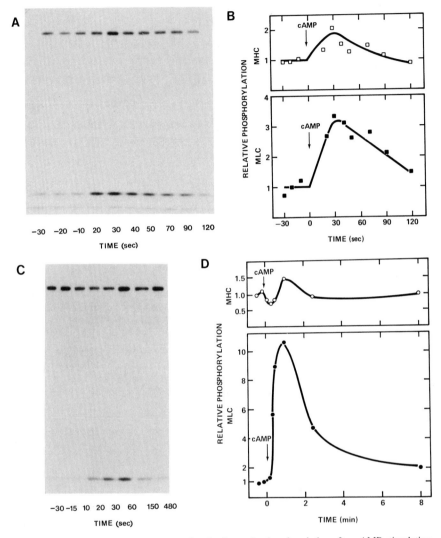

FIG. 2. Transient increase in *in vivo* levels of myosin phosphorylation after cAMP stimulation. Developed cell suspensions were labeled with [^{32}P]orthophosphate. One suspension was labeled with 0.1 mCi/ml for 20 minutes, then stimulated with $2 \times 10^{-6}\,M$ cAMP (A,B). A second suspension was pretreated with caffeine for 30 minutes, then labeled with 0.5 mCi/ml for 30 minutes, after which a stimulus of $2 \times 10^{-6}\,M$ cAMP was applied (C,D). At the indicated times, aliquots of the suspensions were taken, immunoprecipitated for myosin, and subjected to SDS–polyacrylamide gel electrophoresis. (A,C) Autoradiographs of gels of immunoprecipitated time points; (B, D) relative phosphorylation of myosin versus time after cAMP stimulation as quantitated by densitometry. From Berlot *et al.* (1985), Figure 2; used with permission.

while heavy chain phosphorylation gives direction to the movement by initiating redistribution of myosin filaments in a polarized manner via modulation of the assembly–disassembly cycle. Further analysis of the effects of phosphorylation on myosin as well as purification and characterization of additional regulatory factors should help to provide a better understanding of the mechanism by which *Dictyostelium* responds to a chemotactic signal.

ACKNOWLEDGMENTS

This work was supported by grants from the National Institutes of Health to Dr. James A. Spudich (GM 25240) and to Dr. Peter N. Devreotes (GM 28007). Ms. Berlot is a trainee of the Medical Scientist Training Program at Stanford Medical School.

REFERENCES

Bazari, W. L., and Clarke, M. (1981). *J. Biol. Chem.* **256,** 3598–3603.
Berlot, C. H., Spudich, J. A., and Devreotes, P. N. (1985). *Cell* **43,** 307–314.
Bumann, J., Wurster, B., and Malchow, D. (1984). *J. Cell Biol.* **98,** 173–178.
Chisholm, R., Fontana, D., Theibert, A., Lodish, H., and Devreotes, P. (1985). *In* "Microbial Development" (R. Losick and L. Shapiro, eds.), pp. 219–254. Cold Spring Harbor Press, Cold Spring Harbor, New York.
Clarke, M., Bazari, W. L., and Kayman, S. C. (1980). *J. Bacteriol.* **141,** 397–399.
Devreotes, P. N. (1982). *In* "The Development of Dictyostelium Discoideum" (W. Loomis, ed.), pp. 117–168. Academic Press, New York.
Fechheimer, M., and Cebra, J. J. (1982). *J. Cell Biol.* **93,** 262–268.
Fontana, D., Theibert, A., Wong, T.-Y., and Devreotes, P. (1985). *In* "The Cell Surface in Cancer and Development" (M. Steinberg, ed.). Plenum, New York.
Franke, J., and Kessin, R. (1977). *Proc. Natl. Acad. Sci. U.S.A.* **74,** 2157–2161.
Futrelle, R. P., Traut, J., and McKee, W. G. (1982). *J. Cell Biol.* **92,** 807–821.
Gerisch, G., Malchow, D., Roos, W., and Wick, U. (1979). *J. Exp. Biol.* **81,** 33–47.
Griffith, L. M., Downs, S., and Spudich, J. A. (1987). *J. Cell Biol.,* in press.
Kuczmarski, E. R., and Spudich, J. A. (1980). *Proc. Natl. Acad. Sci. U.S.A.* **77,** 7292–7296.
Laemmli, U. K. (1970). *Nature (London)* **227,** 680–685.
Majerfeld, I. H., Leichtling, B. H., Meligeni, J. A., Spitz, E., and Rickenberg, H. V. (1984). *J. Biol. Chem.* **259,** 654–661.
Rosen, O. M., and Krebs, E. G. (1981). *Cold Spring Harbor Conf. Cell Prolif.* **8.**
Sugino, Y., and Miyoshi, Y. (1964). *J. Biol. Chem.* **239,** 2360–2364.

Chapter 19

Agar-Overlay Immunofluorescence: High-Resolution Studies of Cytoskeletal Components and Their Changes during Chemotaxis

YOSHIO FUKUI,[1] SHIGEHIKO YUMURA,[2]
AND TOSHIKO K. YUMURA

Department of Biology
Faculty of Science
Osaka University
Toyonaka, Osaka 560, Japan

I. Introduction

The agar-overlay method described here overcomes two major problems faced by conventional immunofluorescence. These are (1) drastic shrinkage and/or deformation of the cells caused by formaldehyde fixation and (2) insufficient resolution because of superimposition of the fluorescence due to the round or cylindrical shape of *Dictyostelium* cells (Eckert and Lazarides, 1978; Bazari and Clarke, 1982; Rubino *et al.*, 1984). We overcame these problems by flattening the living cells with an overlay of a thin agarose sheet while conducting instantaneous fixation with cold absolute methanol. This procedure results in cells that are 1–2 μm thick and 20–30 μm wide and aids orientation of the filamentous structures in parallel to the focal plane. Using high-titer monoclonal anti-

---

[1]Present address: Department of Cell Biology and Anatomy, Northwestern University Medical School, Chicago, Illinois 60611.

[2]Present address: Department of Biology, Faculty of Science, Yamaguchi University, Yamaguchi 753, Japan.

METHODS IN CELL BIOLOGY, VOL. 28

bodies, we have described specific distributions of actin, myosin, and tubulin in *Dictyostelium* (Figs. 1 and 2) (Yumura and Fukui, 1983; Kitanishi *et al.*, 1984; Yumura *et al.*, 1984; Kitanishi-Yumura *et al.*, 1985). More recently, a minor improvement of the fixation step rewarded us with the first evidence of myosin thick filaments in this organism (Yumura and Fukui, 1985). It was also shown that the myosin filaments transiently change their distribution responding to a chemotactic stimulus in a dynamic manner (Fig. 3) (Yumura and Fukui, 1985).

"Agar-overlay" immunofluorescence seems to be generally useful for small and round vertebrate nonmuscle cells as well as those of lower organisms. This chapter describes the detailed methods especially for *Dictyostelium*.

II.   Agar-Overlay Immunofluorescence

A.   Agar-Overlay Technique

1.   PREPARATION OF THE AGAROSE SHEET

1. Dissolve 2 g of immunochemical-grade agarose [e.g., agarose M, LKB, Sweden; Dojin Agarose-II (800–600 mg/cm^2), Wako Pure Chemicals Ind., Osaka] in 50 ml of hot 15 mM Na/K-phosphate buffer (pH 6.5) to make a 2% (w/v) solution.
2. Place two strips of thin glass, ~3 × 22 mm wide, made of coverslips (grade no. 1; 0.15 mm thick) on each edge of a clean slide glass as a spacer, and then drop ~1 ml of the hot agarose solution on this slide glass.
3. Put another piece of slide glass onto the agarose, and press both edges of the slide with the fingers until the agarose gels.
4. Put this sandwich into a Petri dish filled with the buffer, and then carefully slip the upper slide glass off the agarose sheet.
5. Cut the agarose sheet using a razor blade to make a ~8 × 8 mm sheet.

2.   PREPARATION OF THE CELLS

1. Harvest the cells and resuspend them in buffer (15 mM Na/K-phosphate buffer, pH 6.5) to ~10^6 cells/ml. Cells grown on solid media are more resistant than axenically grown cells and hence are more suitable for this technique.
2. Put an aliquot of the cell suspension on a clean 18 × 18 mm coverslip, and let them become attached to the coverslip by incubating for 5 minutes.
3. Remove the excess buffer from the cell suspension, and carefully put the agarose sheet onto the cells. A razor blade can be used to pick up a piece of agarose sheet.

4. Using a Pasteur pipet and pieces of filter paper, remove excess buffer from the periphery of the agarose sheet. Next, incubate the cells in a moist chamber until they reach the appropriate stage (e.g., preaggregation stage).
5. Prior to fixation, thoroughly absorb the excess buffer from the surface of the agarose sheet using a small piece of filter paper. This step requires a very light touch and extreme care. Once the periphery dries sufficiently, the sheet will not fall off from the coverslip during the subsequent procedure.
6. Observe the cells under a phase-contrast microscope equipped with a 20× lens. Further absorption of the buffer with a small piece of filter paper generates mechanical pressure, which flattens the cells completely. In some states, small intracellular organelles (e.g., mitochondria) are clearly visible moving with a saltatory motion.

3. FIXATION PROTOCOLS

a. Methanol Fixation. Dip the coverslip instantaneously into −10°C methanol in a 100-ml beaker, and allow 5 minutes for fixation in a deep-freezer. (Do not use −20°C methanol! This temperature freezes the agarose and destroys the cell structure.) Using a staining rack (e.g., Coors, Philadelphia, PA; Ikemoto Chemical Industrial Co., Ltd., Tokyo) is recommended. This fixation is especially good for staining microtubules (Kitanishi *et al.*, 1984, 1985).

b. Formaldehyde–Methanol Fixation. Mix 2.7 ml of formalin (37%) with 97.3 ml of absolute methanol to make a 1% formalin–methanol solution. Freeze it (−10°C) and dip the sample instantaneously as described in Section II,A,3,a. Although this fixation is good for any antigen, the major advantage is in the preservation of individual myosin "thick" filaments (Yumura and Fukui, 1985) (Fig. 2c). It can also be applied for double-staining with an antibody and a DNA stain, 4′,6′-diamidino 2-phenylindole (DAPI) (Fig. 1a–c) (Kitanishi-Yumura *et al.*, 1985), or with two different antibodies (Fig. 2).

c. Two-Step Fixation. Put a small aliquot of 2% formalin in 15 mM phosphate buffer on the agarose-overlaid sample and allow 5 minutes for fixation at room temperature. Next, dip the sample into 1% formalin–methanol solution (−10°C) and allow 5 minutes for extraction. This fixation is especially good for preserving actin filaments and is also suitable for double-staining with a variety of fluorescent probes (Fig. 1d–f).

4. EXTRACTION OF SOLUBLE SUBSTANCE

Carefully transfer the samples from a fixing solution to phosphate-buffered saline (PBS; 0.15 M, pH 7.2, 3–5 min × 3). This step is also for washing out the fixatives which otherwise may interrupt specific binding of the antibody with its antigenic site.

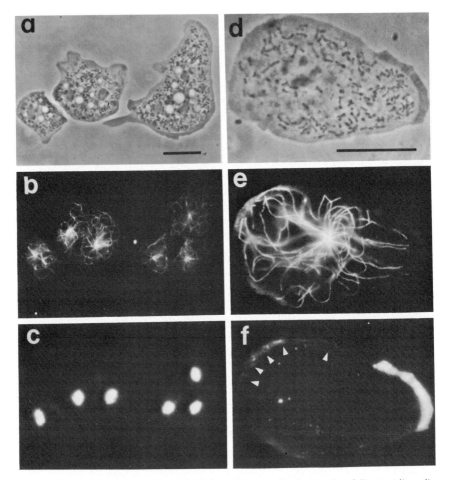

Fig. 1. Phase-contrast (a, d) and fluorescent (b, c, e, f) micrographs of *Dictyostelium discoideum* amoebae prepared by the "agar-overlay" technique. The same cells were stained by double-fluorescent staining with anti-α-tubulin (b) and DAPI (c) or anti-α-tubulin (e) and rhodamine-phalloidin (f). Note that F-actin forms an interwoven knitlike meshwork in the posterior ectoplasm as well as long cables running along with the side body (f) (arrowheads). The cells were fixed by the formaldehyde–methanol (a–c) or the two-step (d–f) fixation method. Bar = 10 μm.

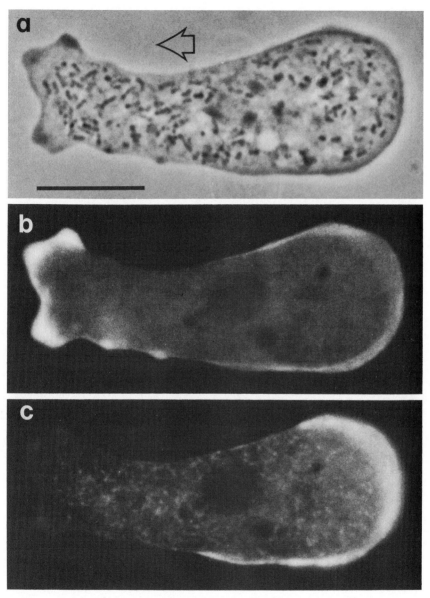

FIG. 2.   Double immunofluorescence of *Dictyostelium discoideum* amoebae. The cell was fixed by the formaldehyde–methanol fixation protocol. (a) Phase-contrast micrograph. The arrow indicates the direction of locomotion. (b) Direct immunofluorescence using TMRITC-labeled anti-*Dictyostelium* actin. (c) Indirect immunofluorescence using anti-*Dictyostelium* myosin followed by FITC-labeled goat anti-mouse IgG. Fluorescent rods in the cytoplasm represent individual myosin thick filaments (Yumura and Fukui, 1985). Bar = 10 μm.

B. Indirect Immunofluorescence

1. ANTIBODIES

Dilute the antiserum, ascites fluid, or culture medium of the hybridoma which contains monoclonal antibody with PBS containing 0.1% NaN_3 in a ratio of 1:25, 1:500, or 1:1, respectively. Test a range of two- to fivefold dilutions for the culture medium. Using well-characterized and specific antibody is essential for good results. For detailed immunological technique, see Hudson and Hay (1980).

2. PREADSORPTION OF THE SECOND ANTIBODY

1. Sediment $\sim 10^8$ cells and allow 5 minutes for extraction with cold methanol ($-10°C$).
2. Wash the lysate by centrifugation with PBS (3000 rpm, 5 min × 3).
3. Resuspend the washed lysate in 500 μl of 1:25 fluorescein isothiocyanate (FITC)- or tetramethylrhodamine isothiocyanate (TMRITC)-labeled second antibody diluted with PBS containing 0.1% NaN_3. Affinity-purified immunoglobulins are recommended (e.g., Sigma, St. Louis, MO; Cappel, Cochranville, PA; Zymed, South San Francisco, CA).
4. Incubate the mixture in a small vial or a tube on an ''end-over-end'' rotator for 1 hour at 4°C. If the result is not satisfactory, try drastic adsorption for 1 hour at 37°C.
5. Spin down the lysate using an Eppendorf-type microcentrifuge (13,000 rpm, 30 minutes 4°C).
6. Collect the preadsorbed antibody and keep it at 4°C in the dark. This antibody can be preserved for several weeks with minimal loss of the activity. For long-term preservation, it can be frozen at $-70°C$ for at least 1 year.

3. STAINING PROCEDURE

1. Apply 10–15 μl of the antibody over the agarose sheet, and incubate it for 30 minutes at 37°C in a moist chamber.
2. Wash with PBS (5 min × 3). For the sample stained with antitubulin, add 0.05% Tween-20 to PBS. This not only shortens the washing time (3 min × 3) but also results in a clearer fluorescent image.
3. Incubate with 10–15 μl of the preadsorbed second antibody labeled with fluorochrome for 30 minutes at 37°C.
4. Wash as described in step 2.

TABLE I

EXCITATION–EMISSION CHARACTERISTICS
OF FLUOROCHROMES

	Wavelength (nm)	
Fluorochrome	Excitation	Emission
FITC	475–495	520
TMRITC	510–565	580–595
NBD	460–470	510–650
DAPI	340	450

5. In a 100-ml beaker filled with distilled water, carefully peel the agarose sheet off the coverslip. The sheet will fall off when a corner is stripped off with a small spatula or forceps. The cells remain stuck to the coverslip. This step also washes out the salts which cause precipitation of the mounting medium.
6. Mount and observe under a fluorescence microscope. An epifluorescence microscope using an oil-immersion lens with large NA is recommended (Table I).

4. MOUNTING MEDIUM

1. Dissolve 20 g of polyvinyl alcohol (e.g., Gelvatol, PVA 20-30, Monsanto Polymers and Petrochemicals Co., St. Louis, MO; PVA 2000, Wako Pure Chemical Ind., Osaka) in 80 ml of PBS, and mix for 1 day with a magnetic stirrer.
2. Add 40 ml of glycerol and 1.2 ml of 10% NaN_3 to the 80 ml of PVA, and mix for 1 day (Rodrigues and Deinhardt, 1960).
3. Dissolve 100 mg of p-phenylenediamine in 10 ml of PBS (Johnson and Nogueira-Aranjo, 1981), and mix with the PVA–glycerol (step 2) on a stirrer for several hours in the dark.
4. Adjust the pH to 8.0 with 0.5 M carbonate–bicarbonate buffer (pH 9.0).
5. Distribute small aliquots (1 ml) into airtight vials and keep them frozen in a deep-freezer.
6. Each time before use, warm the vial to melt the medium, and return it to the deep-freezer after use. This medium hardens at room temperature in a few hours, and the extent of fading of FITC is almost negligible during several minutes of exposure.
7. The sample mounted with the medium can be kept for at most 1 week in a refrigerator, but the best results are obtained in 2–3 days. During pro-

longed preservation, *p*-phenylenediamine becomes brownish, and this interferes with the fluorescence. Sealing the periphery of the sample with nail enamel can help prevent oxidation.

C. Fluorescent Phallotoxins

7-Nitrobenz-2-oxa-1,3-diazonylyl (NBD)-phallacidin, rhodamine- or fluorescein-phalloidin (Molecular Probes, Inc., Junction City, OR) is a useful probe for staining *Dictyostelium* F-actin.

1. Prepare the cells by the "two-step fixation" protocol.
2. Evaporate 1 unit of the probe and redissolve it in *20* μ*l* (instead of 200 μl) of PBS.
3. Apply this solution to a coverslip either with or without the agarose sheet, and incubate the coverslips for 30 minutes at 37°C.
4. Briefly rinse with PBS and then with distilled water.
5. Mount with the medium and observe under a fluorescence microscope. The rhodamine-phalloidin has several advantages over the other probes for the following reasons: (1) rhodamine is relatively resistant to fading, and (2) a possible longer exposure results in a thicker image of the filaments on the recording film, which makes it possible to identify fine structures (Fig. 1f).

D. Photomicroscopy

For fluorescence microscopy, Tri-X (ASA 400/DIN 27, Eastman Kodak, Rochester, NY) is recommended. Tri-X can be sensitized fourfold using either Acufine or Diafine (Acufine, Inc., Chicago, IL). For phase-contrast microscopy, Panatomic-X (ASA 32/DIN16, Kodak) gives the best results. For color slides, Ektachrome (ASA 400/DIN 27, Kodak) is recommended. If necessary, color pictures can be processed by direct print by Kodak.

III. Summary

Cells that are flattened by overlaying with a thin sheet of agarose can be instantaneously fixed with freezing absolute methanol containing 1% formalin. This procedure results in good preservation of the cytoskeleton. Use of this technique ("agar-overlay immunofluorescence") clarified that (1) *Dictyostelium* myosin exists *in situ* as thick filaments (Yumura and Fukui, 1985), (2) the thick filaments are arranged in a meshwork at the posterior cortex of a polarized cell performing directed locomotion, at the constricting portion of a dividing cell

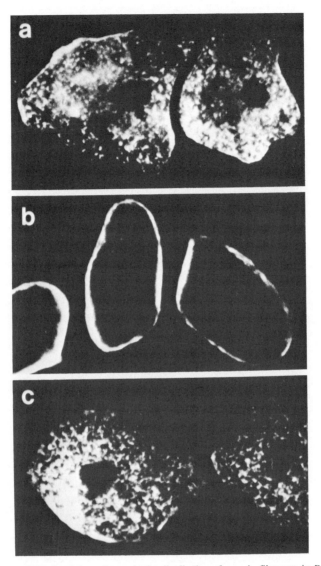

FIG. 3.   A rapid and transient change in the distribution of myosin filaments in *Dictyostelium discoideum* subsequent to the treatment with 10^{-6} *M* adenosine 3′,5′-cyclic monophosphate (Yumura and Fukui, 1985). The endoplasmic filaments disappeared and accumulated in the cortex after 2 minutes (b), and the original distribution was recovered in 3 minutes (c). The experiment was performed at 5°C.

forming a contractile ring, and at the outermost lateral periphery of a cell engaging in spiral aggregation (Fig. 1f; Yumura *et al.*, 1984; Yumura and Fukui, 1985), and (3) the distribution of thick filaments changes dramatically in response to the chemoattractant cAMP in 1 minute (Fig. 3; Yumura and Fukui, 1985). This technique can provide valuable information on the dynamic features as well as the detailed organization of cytoskeletal elements which, otherwise, cannot be visualized with sufficient resolution.

ACKNOWLEDGMENTS

We would like to express our sincere gratitude to Dr. Stephen H. Blose of Cold Spring Harbor Laboratory for his enthusiastic introduction of immunofluorescent techniques at the first AMBO workshop held at Kyoto in 1981.

REFERENCES

Bazari, W. L., and Clarke, M. (1982). *Cell Motil.* **2,** 471–482.
Blose, S. H., Meltzer, D. I., and Feramisco, J. R. (1984). *J. Cell Biol.* **98,** 847–858.
Eckert, B. S., and Lazarides, E. (1978). *J. Cell Biol.* **77,** 714–721.
Hudson, L., and Hay, F. C. (1980). "Practical Immunology," 2nd ed. Blackwell, Oxford.
Johnson, G. D., and Nogueira-Aranjo, G., de C. (1981). *J. Immunol. Methods* **43,** 349–350.
Kitanishi, T., Shibaoka, H., and Fukui, Y. (1984). *Protoplasma* **120,** 185–196.
Kitanishi-Yumura, T., Blose, S. H., and Fukui, Y. (1985). *Protoplasma* **127,** 133–146.
Rodrigues, J., and Deinhardt, F. (1960). *Virology* **12,** 316–317.
Rubino, S., Fighetti, M., Unger, E., and Cappuccinelli, P. (1984). *J. Cell Biol.* **98,** 382–390.
Yumura, S., and Fukui, Y. (1983). *J. Cell Biol.* **96,** 857–865.
Yumura, S., and Fukui, Y. (1985). *Nature (London)* **314,** 194–196.
Yumura, S., Mori, H., and Fukui, Y. (1984). *J. Cell Biol.* **99,** 894–899.

Part V. Cell Adhesion and Cell–Cell Recognition

Once *Dictyostelium* cells become competent to respond to their chemotactic signal, they typically form aggregates containing ~100,000 cells. The formation and maintenance of these aggregates require special forms of cell—cell recognition. Cell–substratum adhesion is also fundamental to cells during this stage of development as well as during other stages. In Chapter 20, Bozzaro *et al.* describe methods to study the role of cell surface glycoproteins in cell interactions. Cell–substratum adhesion, cell adhesion to bacteria, and homotypic cell–cell adhesion are all described. Both immunochemical and genetic methods are detailed. Barondes *et al.* in Chapter 21 present methods for studying the endogenous lectins called discoidins I and II, which are involved in cell–substratum adhesion and spore coat formation, respectively. They describe the purification of the discoidins and how to assay for them, purification of a discoidin-binding polysaccharide from *Dictyostelium discoideum* and of a cell surface receptor for the cell-binding site of discoidin I, methods for examining adhesion of cells to tissue culture dishes, video microscopy for observing cell migration into aggregates, and fluorescent light microscopy and electron microscopy analysis of histochemical localizations of these lectins and their glycoconjugate ligands.

Chapter 20

Cell Adhesion: Its Quantification, Assay of the Molecules Involved, and Selection of Defective Mutants in Dictyostelium and Polysphondylium

SALVATORE BOZZARO,[1] RAINER MERKL,
AND GÜNTHER GERISCH

Max-Planck-Institut für Biochemie
D-8033 Martinsried bei München
Federal Republic of Germany

I. Introduction

Cells of *Dictyostelium discoideum* adhere to substrata such as glass or plastic surfaces, to bacteria, to each other, and also to cells of related species. These various contact interactions have different biological functions and different specificities. Adhesion to substratum is a prerequisite of movement of single cells and of their chemotactic response during aggregation. Attachment of bacteria precedes their phagocytosis and is important for nutrition of *D. discoideum* cells if the bacteria are suspended rather than fixed to a substratum. Homotypic cell–cell adhesion is involved in the aggregation process and in the maintenance of multicellularity in the slug and culmination stages. In the following we describe methods applicable to either one of these contact interactions and cover also cell adhesion in a related species, *Polysphondylium pallidum*. Methods taken from the literature will be reviewed only briefly. Detailed protocols will be given only for methods that have been used in our laboratory.

---

[1]Present address: Cattedra di Biologia Generale, University of Turin, 10126 Turin, Italy.

METHODS IN CELL BIOLOGY, VOL. 28

II. Cell–Substratum Adhesion: Derivatized Polyacrylamide Gels as Chemically Defined Surfaces

An advantage of a polyacrylamide matrix as carrier for specific ligands is that *D. discoideum* cells adhere very weakly to the underivatized gels. The degree of derivatization can be modified, thus allowing variation of the number of ligands per surface area of the gel relative to the density of corresponding binding sites on the cell surface. The method is applicable to a great variety of ligands, although it has been mainly worked out for the linkage of sugars to the gel matrix (Schnaar *et al.*, 1978).

Incubation of developing *D. discoideum* cells on the surface of glucoside-derivatized gels blocks development precisely at the aggregation stage: preaggregative development, chemotaxis to cyclic AMP (cAMP), and streaming of the cells are not inhibited under these conditions (Bozzaro and Roseman, 1983b), but the formation of tight aggregates and the following appearance of "postaggregative" transcripts is prevented. Thus these gels can be used for the investigation of developmentally regulated gene expression (Bozzaro *et al.*, 1984). Selection of mutants defective in sugar-binding sites on the cell surface may provide insight into the function of these sites.

A. Preparation of Sugar-Derivatized Gels

Sugars are linked to the polyacrylamide gel matrix through an amino group on a C_6 spacer (Fig. 1). The 6-amino-hexyl-*O*-sugars required for derivatization are synthesized and purified according to Weigel *et al.* (1979). They are stable at $-20°C$ under chloroform or as a dry powder.

In the following procedure it is important to use exclusively superpure distilled or deionized water, since otherwise the cells may unspecifically stick to the gel matrix. The aminohexyl-sugar (≤ 0.2 m*M*) is dissolved in 0.5 ml of water and mixed on ice with a 1.5-fold molar excess of $Ca(OH)_2$. Subsequently a total of 1.1 molar equivalents of acryloylchloride (E. Merck, 6100 Darmstadt, FRG) diluted 10-fold with dry acetone is added in four to five aliquots at intervals of 2 minutes. The reaction mixture is stirred during each addition and every 20 seconds thereafter. After the last addition of acryloylchloride, 1 ml of a solution of 60% acrylamide and 3% *N-N'*-methylenebisacrylamide (both electrophoresis grade) in water are added at room temperature under vortexing, followed by addition of 0.1 ml of saturated $Ba(OH)_2$. After incubation at 37°C for 20 minutes, the insoluble salts are removed by centrifugation for 5 minutes at 1000 *g*. The supernatant is then deionized by mixing with MB-3 Amberlite. The resin is filled into a column to a final volume of 1.8 ml, washed with superpure water, and excess water is removed by pressing air through the column. Supernatant is

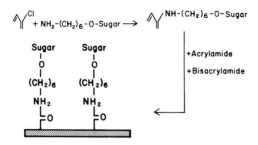

F<small>IG</small>. 1. Synthesis of sugar-derivatized polyacrylamide gels. For explanation see text. From Bozzaro and Roseman (1982).

added and stirred in the column for 60 minutes at room temperature with the resin. Then air is pressed through the column to recover the deionized solution as quantitatively as possible. To the solution, $1M$ HEPES buffer, pH 8.0, 0.1 M potassium persulfate, water, and 10% TEMED in water (v/v) are added in this sequence to adjust the acrylamide concentration to 20%, and to yield final concentrations of 50 mM HEPES, 1 mM potassium persulfate, and 0.1% TEMED. The solution is then poured between glass plates carefully cleaned with a polyphosphate-free detergent like 10% Extran MA 03 (E. Merck, 6100 Darmstadt, FRG) or with a mixture of 30 ml of 65% nitric acid in 1 liter of concentrated sulfuric acid. The glass plates are separated by a 0.2-mm plastic spacer. Following polymerization, one glass plate is removed, the gel cut into squares of 0.6 cm^2, washed with water, and subsequently with phosphate buffer, pH 6.0 (see Section IV,A), containing 60 mM NaCl. The gels can be stored in the refrigerator with 10% isopropanol added to the buffer and are washed before use at least three times for 20 minutes in the buffer to remove the isopropanol.

B. Assay of Cell Adhesion to the Sugar-Derivatized Gels

The gel pieces are blotted onto filter paper to remove excess water and placed in a circular array in a 60-mm-diameter polystyrene Petri dish (Falcon no. 1007), using a thin spatula for transfer. No air bubbles should be entrapped between the gels and the plastic surface. To avoid drying, a drop of the phosphate-NaCl buffer is pipeted on top of each gel which is removed immediately before adding the cells. The cells are washed once in the buffer and adjusted to 1 × 10^7/ml. Aliquots of 30 µl are placed on top of each gel piece. After incubation of the cells on the gels for the desired period, the Petri dish is placed on a gyratory shaker, 9 ml of phosphate-NaCl buffer is gently pipeted in the middle of the dish, and the dish is shaken for 30 seconds at 90 rpm. The buffer containing the unbound cells is aspirated, and cell adhesion assessed either microscopically or, more accurately, by assaying alanine transaminase (Section II,C).

In the phosphate-NaCl buffer *D. discoideum* cells adhere strongly to gels derivatized with glucose, maltose, cellobiose, *N*-acetylglucosamine, or mannose, but not to underivatized gels or to gels derivatized with galactose, lactose, or only with the 6-aminohexanol spacer (Bozzaro and Roseman, 1983a). If NaCl is omitted, ~30% of the cells will bind to any of these latter gels. NaCl causes the cells to round up for 15–20 minutes before they start to move on the gels. If this should be avoided, the cells can be preincubated for 20 minutes at room temperature in the phosphate-NaCl buffer.

C. Alanine Transaminase Assay for Determining Cell Numbers

Alanine transaminase is an intracellular enzyme in *D. discoideum* (Firtel and Brackenbury, 1972) and also in *P. pallidum*. Advantages of the enzyme are its convenient, sensitive assay and its stability in cell extracts. In the lysis buffer described below the decay of enzyme activity is <5% after 3 hours at room temperature, 6 hours in the refrigerator, or 6 days in the freezer.

Gels with the adhering cells are transferred into wells of plastic dishes loaded

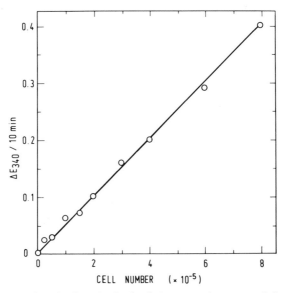

Fɪɢ. 2.   Determination of cell number by the alanine transaminase assay. Cells of *D. discoideum* strain AX2 starved for 1 hour were adjusted to 1×10^7/ml. An aliquot was diluted 10 times with 0.1 *M* Tricine buffer, pH 7.6, containing 0.2% Triton X-100. Aliquots of the lysate (0.1–0.8 ml) were transferred to 1-ml plastic cuvettes, the volume adjusted to 0.8 ml with lysis buffer, and alanine transaminase assayed by adding 0.1 ml of the mix described in the text. Cell number refers to the number of lysed cells added to the cuvette. From Bozzaro and Roseman (1982).

with 1 ml of lysis buffer containing 0.1 M Tricine-HCl buffer, pH 7.6, and 0.2% Triton X-100. Of each lysate 0.9 ml are mixed with 0.1 ml of a solution containing, per milliliter, 0.1 M Tricine-HCl buffer, pH 7.6, 44 mg L-alanine, 2 mg α-ketoglutarate, 0.8 mg NADH, and 20 units of lactic dehydrogenase. Alanine transaminase is assayed by determining the decrease of absorbance at 340 nm. Its rate is directly proportional to the number of cells in the range shown in Fig. 2. For each experiment a calibration curve should be run with a dilution series from the same batch of cells as placed on the gels.

D. Selection of Mutants Defective in Binding to Sugar-Derivatized Gels

Adhesion to derivatized gels can be used for the selection of mutants defective in certain carbohydrate-binding sites. The following protocol has been successfully applied to the selection of mutants with no or with reduced binding to N-acetylglucosamine (Bozzaro and Roseman, unpublished).

Exponentially growing AX2 cells are washed twice in 17 mM Soerensen phosphate buffer, adjusted to pH 7.0, resuspended in the buffer at a density of 1 $\times$ 10^7/ml and incubated under gentle shaking with 1 mg/ml of 1-methyl-3-nitro-1-nitrosoguanidine for 20 minutes at 23°C in the dark. The cells are washed and cultivated axenically for 3 days in four separate flasks. Dividing the mutagenized cells into aliquots immediately after mutagenesis helps to isolate independently produced mutants of the same phenotype.

Grown cells are washed and starved for 1 hour at 23°C in 17 mM Soerensen phosphate buffer, pH 6.0 (see Section IV,A), washed and preincubated for 20 minutes at room temperature in phosphate-NaCl buffer at a concentration of 8 $\times$ 10^6 cells/ml, before they are transferred onto gels. From each flask a total of 2.4 $\times$ 10^6 cells are incubated for 20 minutes on 10 pieces of glucose-derivatized gels. After addition of 9 ml of buffer the dish is swirled at 90 rpm for 1 minute. The gels with the attached cells are transferred into glass tubes containing phosphate-NaCl buffer. The cells are detached by vortexing, concentrated to 8 $\times$ 10^6/ml by centrifugation, and incubated on N-acetylglucosamine-derivatized gels for 30 minutes. Unbound cells are removed by swirling in 9 ml phosphate-NaCl buffer at 80 rpm for 20 seconds. Unbound cells are subjected to fresh N-acetylglucosamine gels, and cells remaining unbound are plated on nutrient agar plates with *E. coli* B/2. Individual clones are retested for failure of binding to N-acetylglucosamine-derivatized gels.

The selection for mutants which fail to bind to N-acetylglucosamine leads to an enrichment for morphogenetic mutants which are blocked at the loose-mound stage, suggesting a relation between sugar-binding sites and the control of late development.

III.   Cell Adhesion to Bacteria: Quantification
by Assaying Phagocytosis

A prerequisite of the phagocytosis of bacteria or other particles in shaken suspensions is their attachment to the cell surface of the phagocytes. The rate of phagocytosis is thus a measure of the adhesion of bacteria to *D. discoideum* cells. The rate of uptake of glucose-derivatized latex beads has been compared with that of underivatized beads to show that sugar-binding sites on the surface of *D. discoideum* cells are involved in the attachment of particles before they are phagocytosed (Vogel, 1983; also see Vogel, Chap. 6, this volume).

A.   Conditions of Phagocytosis Assays

For the phagocytosis assay, washed bacteria and growth-phase cells of *D. discoideum* are taken up in phosphate buffer, pH 6.0, and 0.3 ml of a suspension containing 4×10^9 bacteria/ml is mixed with 0.3 ml of a suspension of 4×10^6 *D. discoideum* cells/ml. The mixture is shaken for 45 minutes at 23°C on a gyratory shaker with 150 rpm in glass tubes of 13 mm diameter. Subsequently 4.6 ml of ice-cold phosphate buffer are added, and the *D. discoideum* cells pelleted by centrifugation at 125 *g* for 3 minutes. The pellet is washed twice with 5 ml buffer, and bacteria might be determined in the three supernatants by nephelometry. More accurate are assays in which bacteria are labeled with fluorescein isothiocyanate (FITC; Vogel *et al.*, 1980) or with ^{14}C-labeled amino acids, and the label is determined in the washed pellet. Electron microscopy of cells has shown that virtually no *E. coli* B/r or *Salmonella minnesota* R595 cells remain attached to the surface of *D. discoideum* AX2 cells after two washes. However, this may not apply to other *Dictyostelium* strains or to *Polysphondylium* species and other bacterial strains. Vogel *et al.* (1980) recommend centrifugation of the cells through an aqueous solution of 20% (w/w) polyethylene glycol 6000 to separate them from noningested bacteria.

B.   Labeling of Bacteria with ^{14}C-Labeled Amino Acids

The following procedure is similar to the one used by Chadwick *et al.* (1984). The bacteria are cultivated overnight at 37°C on a gyratory shaker in nutrient medium containing, per liter, 0.5 g sodium citrate, 1 g $(NH_4)_2SO_4$, 3 g KH_2PO_4, 7 g K_2HPO_4, 0.1 g $MgSO_4 \cdot 7H_2O$, 2.7 g glucose, and 250 μCi L-[^{14}C]lysine, which is added to the sterilized solution shortly before use. The bacteria are washed three times in phosphate buffer, pH 6.0, and adjusted to 4×10^9 bacteria/ml. They can be stored under gentle shaking in the cold for up to 1 week. The specific activity should be $1-2 \times 10^4$ cpm per 1×10^8 bacteria. The radioactivity incorporated after the phagocytosis assay into the washed cell pellet is determined by dissolving the cells in 2.5 ml of scintillant and counting in a β-counter.

IV. Homotypic Cell–Cell Adhesion in *D. discoideum*

In both currently used methods, Coulter counting and agglutinometer assays, the agglutination of suspended cells under conditions of constant shear is determined. In Coulter counting, agglutination and measurement are separated in time. In the agglutinometer, measurements can be taken on-line. None of these methods yields a measurement of "cell adhesiveness" in exact physical terms, as the viscometer assay of Curtis (1969) does. But for routine biochemical purposes these methods have their established merits.

A. Quantification by Coulter Counting

An advantage of Coulter counting is the possibility to determine the size distribution of agglutinates, whereas the agglutinometer assay yields only averages of particle sizes. However, in practice the proportion of single cells is usually determined as representing cell adhesiveness (Orr and Roseman, 1969). This requires a uniform cell population to start with, since a subpopulation of cells with reduced adhesiveness would falsify the results. Since the ratio of single cells at the end of the experiment to total cells at the beginning is determined, the possibility of cell lysis during the experiment should be taken into account. Lysis would shift the data into the same direction as aggregation of the cells.

In order to monitor the decrease in the number of single cells with a Coulter counter, it is first necessary to calibrate the counter with polystyrene or other beads of known size, and to determine the size distribution of unaggregated cells relative to these beads. The window of counted particles should be set such that debris are excluded, and that an overlap of large single cells with aggregates of smaller cells is minimized. Errors may be caused by agglutination of the cells during the Coulter counting and by the variability of cell sizes. This is particularly true for axenically grown cells of which a high proportion is multinucleate. These cells may divide into smaller ones during an experiment. Agglutination of cells during the measurement can be reduced by adding 150 mM NaCl buffered with phosphate to pH 7.2. We prefer to fix single cells and also the agglutinates formed during the experiments by adding to the samples an equal volume of 5% trichloroacetic acid (TCA; Tetsch, 1970) or glutaraldehyde (Bozzaro *et al.*, 1983).

Standardized shear forces during cell agglutination can be applied by agitating cells in test tubes with flat bottoms or in Erlenmeyer flasks. Alternatively, the cell suspension can be gently agitated in test tubes on a rotating drum (Gerisch, 1961). We routinely use 17 mM Soerensen phosphate buffer, pH 6.0, for agglutination assays, which can be replaced, if necessary, by other low ionic strength buffers of pH 6.0–7.4. A 50-fold concentrated stock solution of the Soerensen buffer contains, per liter, 99.86 g KH_2PO_4 and 17.8 g $Na_2HPO_4 \cdot 2H_2O$, and can be stored in the cold; the diluted buffer is designated "phosphate buffer, pH 6.0" throughout this paper. Other buffers which preserve cell shape, motility, and agglutinability are (1) 10mM Tricin-HCl, pH 7.0, 10 mM KCl, 5 mM NaCl; (2) barbital buffer, pH 7.3 (Beug *et al.*, 1973) containing, per liter, 0.575 g barbital, 0.375 g barbital sodium salt, 0.102 g $MgCl_2 \cdot 6H_2O$, 0.022 g $CaCl_2 \cdot 2H_2O$, and 0.806 g NaCl. A fivefold concentrated stock solution can be stored in the refrigerator.

B.   Quantification Using an Agglutinometer

In the agglutinometer devised by Born and Garrod (1968), suspended cells are stirred in a test tube and agglutination is monitored by recording turbidity. The same principle is applied in the agglutinometer of Beug and Gerisch (1972). It allows simultaneous recordings of 23 samples relative to a reference, in special cuvettes which are rotated with constant speed (Figs. 3 and 4). The sample volume is 0.2 ml. This small volume is of advantage for testing the effects of antibodies or expensive substances on cell adhesion. The recommended cell concentration is 1×10^7/ml. At this concentration no rounding up of cells due to the exhaustion of oxygen in the air space of the cuvettes has been observed after an incubation period of 1 hour.

In this instrument the fraction of unscattered light is determined. Since the apparent optical density is roughly a linear function of particle number (Beug and Gerisch, 1972), the agglutinometer responds most sensitively to differences in the average sizes of small aggregates, while sizes of aggregates containing more than five cells are only poorly distinguished from each other. To some extent the range of sensitivity can be adjusted to the strength of adhesiveness by varying the speed of rotation between 16 and 80 rpm. At higher speeds agglutinated cells may be sedimented by centrifugal forces.

The buffers described under Section IV,A can be used for the agglutinometer assay. For assaying the developmentally regulated EDTA-stable contacts, the

---

FIG. 3.   Agglutinometer for the quantification of cell adhesion. Unscattered light as a function of the agglutinate size is measured in 23 cuvettes of suspended cells and in one reference cuvette. The samples are subjected to constant shear forces (Beug and Gerisch, 1972). The measuring device consists of (A) the cuvettes filled with cell suspension, (B) a rotating carrier for 24 cuvettes, and the optical system, which is specified in (C). Dimensions are in millimeters. The cuvettes with oblong cavities are manufactured by Hellma (D-7840 Mühlheim, FRG), drawing no. 110.041.02. During rotation the inner surfaces of the cuvettes are wetted continually, preventing the adhesion of cells to the glass. The fluid motion generates the shear forces counteracting cell adhesion.

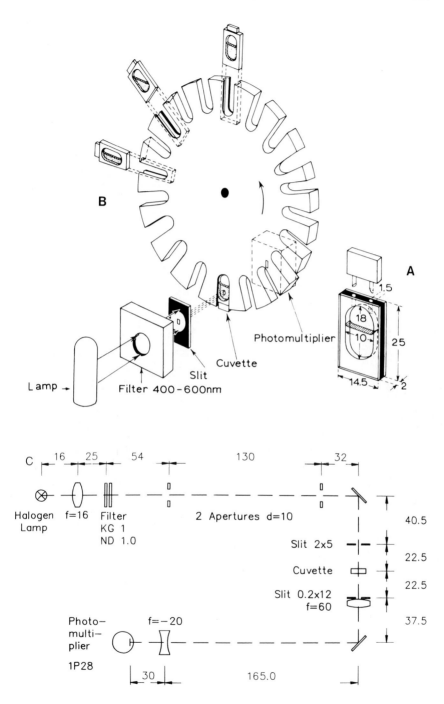

B

A

1.5

18

10   25

14.5   2

Photomultiplier

Cuvette

Slit

Lamp →

Filter 400-600nm

C   16   25   54   130   32

Halogen   f=16   Filter   2 Apertures d=10
Lamp   KG 1
                 ND 1.0

40.5

Slit 2x5

22.5

Cuvette

22.5

Slit 0.2x12
f=60

37.5

Photo-   f=-20
multi-
plier

1P28

30

165.0

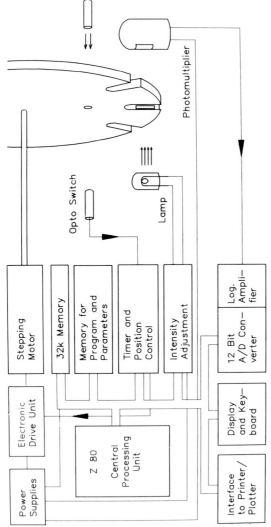

FIG. 4. Block diagram of the control unit of the agglutinometer. Controlled by a microprocessor, the system works automatically after the settings for the experiment are programmed and the cuvette carrier is filled. The agglutinometer measures the signals of all cuvettes at specified time intervals during the run. After the experiment the agglutination kinetics can be plotted. The *central processing unit* controls the system, the peripheral devices, and the data flow. The *memory* can store at least 640 cycles of measurement. The *display and keyboard* circuits control the input and output. All functions of the instrument are programmed using the keyboard. Actual data like speed or time are shown by the display. A malfunction of the essential parts of the system, like the stepping motor, position control, lamp and analog/digital (A/D) converter is indicated by a warning lamp. The *timer* generates pulses which are sent to the *electronic drive unit*. The electronic drive unit controls the motion of the stepping motor which turns the rotating carrier. The *opto switch* triggers measurement of the reference cuvette. By counting the number of pulses sent to the stepping motor after the opto switch signal, the *position control* determines when each of the other cuvettes is in the light path and ready to be measured. The circuit for *intensity adjustment* regulates the intensity of the lamp by means of pulse width modulation. The signal of the *photomultiplier* is fed through a *logarithmic amplifier* and digitized by the *A/D converter*.

phosphate buffer, pH 6.0, is supplemented with 10 mM EDTA. At this concentration EDTA almost completely suppresses the EDTA-labile adhesiveness of growth-phase cells in the AX2 of AX3 strains as well as in wild-type V12/M2. However, in the HU1628 strain, which is often used as a marker strain for the mapping of mutations, cell–cell adhesion of growth-phase cells has been found to be much less EDTA-sensitive; an EDTA concentration of 40 mM is required to dissociate these cells satisfactorily though still incompletely. Higher concentrations are unfavorable to the cells.

The agglutinometer allows recording of the rates of agglutination of single cells and of the rates of dissociation in a suspension of aggregated ones. An equilibrium is normally reached within 20–60 minutes, after which time the sizes of agglutinates become practically constant.

The agglutinometer has found its major application in the identification of cell surface molecules involved in intercellular adhesion by use of the immunological approach described in the following (Müller and Gerisch, 1978). It can be used to determine the acquisition of EDTA-stable adhesiveness as an indicator of cell development (Gerisch et al., 1975). The agglutinometer can also be applied to the quantitative assay of lectins that agglutinate Dictyostelium cells or erythrocytes (Beug and Gerisch, 1972; Bozzaro and Gerisch, 1978; Yoshida et al., 1984).

During the preaggregation and aggregation stages the cells of D. discoideum produce and release cAMP periodically. The responses to the cAMP pulses produced are accompanied by changes in cell shape and probably also in adhesiveness, which can be recorded by monitoring light scattering (Gerisch and Hess, 1974). The agglutinometer can be adapted to these recordings (Gerisch, unpublished results).

Turbidity does not provide an unequivocal measure of cell agglutination, since it is influenced not only by shape changes but also by lysis of cells. Effects of different treatments on cell shape can be compensated by normalizing turbidity in an aliquot of cells which are completely dissociated by an Fab preparation that does not influence cell shape (oscillating or otherwise fluctuating changes will not be compensated for). Samples should always be checked at the end of the experiment under the microscope to avoid misinterpretation of the data. Bacteria-grown cells will sometimes stick to dead bacteria, thus simulating EDTA-stable intercellular adhesion, and also lysed D. discoideum cells may cause unspecific clumping of intact cells.

V. Assay of the Species Specificity of Cell–Cell Adhesion

If cells of different species are mixed, they sort out either during aggregation or at later stages of development (Raper and Thom, 1941). Sorting out at the aggregation stage can be due either to different chemoattractants which are produced and recognized by these species, or to incompatibility of their cell–cell adhesion systems. In order to eliminate chemotaxis as a factor involved, two techniques have been devised:

1. Cells of one species are used as a monolayer while those of the other species are applied in suspension. After gentle shaking the number of cells of the second species (and of control cells of the first one) which are bound to the monolayer is determined (Walther *et al.*, 1973). This technique has been used for combination of *D. discoideum* and *D. purpureum* cells (Springer and Barondes, 1978).

2. Motility of the cells is inhibited by 2,4-dinitrophenol (DNP) before the cells are mixed and gently agitated in suspension, e.g., in the agglutinometer described above. Two classes of clusters, consisting predominantly of cells of either one or the other species, would indicate specificity of the adhesion systems. This technique has been applied to mixtures of *D. discoideum* and *P. pallidum* cells (Bozzaro and Gerisch, 1978; Gerisch *et al.*, 1980).

For both techniques cells have to be labeled to distinguish them. Labeling of the cells at the beginning of the experiment by fluorescent dyes has found wide application. For details we refer to the original description by Springer and Barondes (1978). For inhibiting the motility of *D. discoideum* and *P. pallidum* cells during an experiment and for labeling them at the end we use the following procedure. To a suspension of 1×10^7 cells/ml of Soerensen phosphate buffer, pH 6.0, 30–50 µl of a stock solution of 1 mM DNP is added. The stock solution is prepared by heating DNP with the phosphate buffer in a water bath. The concentration of DNP required for rounding up the cells without killing them is critical; it depends on pH and cell concentration and is about twofold higher for *D. discoideum* than for *P. pallidum*. At higher concentrations DNP acts as a fixative. At the low concentrations recommended, it completely rounds up the cells in a reversible manner. When *D. discoideum* cells are washed after 1–3 hours of incubation they will continue development.

Mixtures of DNP-treated *D. discoideum* and *P. pallidum* cells are allowed to agglutinate for 30 minutes at 16 rpm in the agglutinometer (Fig. 3) or at 90 rpm on a gyratory shaker, and are fixed immediately thereafter by adding one volume of Bouin's fixative to five volumes of the sample. The fixative consists of 15 ml of saturated aqueous picric acid solution, 5 ml of 37% formaldehyde, 1 ml acetic acid. It stabilizes the agglutinates, preserves the reactivity of species-specific carbohydrate groups with antibodies, and does not cause background fluorescence. The fixative is removed by washing in 150 mM NaCl buffered with 100 mM sodium phosphate, pH 7.2 (PBS), until the yellow color has disappeared. The material is then incubated with 1% bovine serum albumin (BSA) in PBS, before the cells are labeled with FITC or tetramethylrhodamine isothiocyanate-conjugated antibodies.

Species-specific anticarbohydrate antibodies can be raised in rabbits by immunization with entire cells that had been heated in 150 mM NaCl for 1 hour at 70°C. The heated cells are washed and injected without adjuvant in alternate days for a total of 12 days (Gerisch et al., 1969). The antisera obtained proved to be highly species-specific reagents for cell surface labeling but nevertheless required absorption with cells of other species to eliminate weak cross-reactivities.

VI.   Immunological Approaches to the Molecular Basis of Cell–Cell Adhesion

One strategy to identify molecules involved in intercellular adhesion is based on the use of univalent antibody fragments (Fab) against total membrane antigens for the blockage of intercellular adhesion. "Contact sites" are identified by an assay which includes quantification of cell adhesion in the presence and absence of Fab, using the agglutinometer shown in Fig. 3, and neutralization of the adhesion-blocking Fab by fractions of solubilized membrane antigens. One difficulty in this approach is that often carbohydrate residues are highly immunogenic, such that antibodies which are neutralized by more than a single glycoprotein will dominate the immune response. In these cases carbohydrate residues may be partially or completely removed from the antigens before immunization to increase the proportion of antiprotein antibodies.

Purified target antigens of blocking Fab can be used for the production of monoclonal antibodies against specific epitopes, and antibodies against the protein moiety of the contact sites can be applied to screening for DNA clones in an expression library (Noegel et al., 1986). Since the Fab neutralization assay is an indirect approach, any function in cell adhesion of a molecule identified as a target site of adhesion-blocking Fab requires confirmation by other methods. Mutants specifically defective in the synthesis of the entire molecule or of some of its constituents, such as carbohydrate residues, may provide final proof (Noegel et al., 1985).

A.   Preparation of Fab for Blockage of Cell Adhesion

An immunization schedule which has been successful for raising polyspecific antisera in rabbits against D. discoideum membrane fractions to preparing high-titer adhesion-blocking Fab has been published by Beug et al. (1970). An interval of 2 months between a first series of injections and the boost improves the adhesion-blocking titer of the Fab. Aluminum hydroxide as an adjuvant may give rise to Fab with a better blocking activity than Freund's adjuvant. It is nevertheless worthwhile to try both adjuvants, since the spectrum of epitopes recognized by antibodies often differs with the adjuvant employed. The immune responses of individual rabbits to complex antigens such as total membranes vary considerably. It is therefore advisable to test the adhesion-blocking activity of Fab from each animal and to pool only antisera with satisfactory titers. Between 0.5 and 2 mg Fab/ml should be enough for complete dissociation of aggregation-competent cells. Rabbit IgG and Fab are purified by conventional procedures (Beug *et al.*, 1973) and can be stored at $-20°C$ after dialysis against distilled water and lyophilization.

For the preparation of monoclonal adhesion-blocking Fab from mouse IgG, the sensitivity of this Fab to papain should be taken into account. Optimal incubation times for complete cleavage of the heavy chains with minimal loss of Fab have to be tested for each batch of IgG by SDS–polyacrylamide gel electrophoresis. In our hands, cellulose-bound papain (Sigma) has been most effective in preserving activity of the Fab produced.

B.   Assay of Adhesion-Blocking Activity and Absorption of Fab

To test the adhesion-blocking activity of Fab in the agglutinometer, phosphate buffer (pH 6.0) is employed or, if divalent cations are wanted or optimal conditions for antibody binding are needed, the barbital buffer described under Section IV,A can be used. The addition of 2 mg BSA/ml of the buffers is recommended.

To distinguish between blockage of EDTA-stable and EDTA-labile cell adhesion either growth-phase cells which form only EDTA-labile contacts are used as test cells, or aggregation-competent cells are used in the presence of 10 m*M* EDTA where only EDTA-stable contacts are formed.

The Fab can be made specific for developmentally regulated antigens, primarily the contact site A glycoprotein, by absorption with either intact growth-phase cells or a membrane fraction from these cells.

For absorption with intact cells, 2×10^8 growth-phase cells are washed, pelleted, and resuspended in a solution of 10–20 mg Fab or IgG in 1 ml of adequate buffer. After incubation on ice for 20 minutes, the cells are spun down and the procedure is repeated with the supernatant and fresh cells for another three times. Subsequently the supernatant is cleared by centrifugation for 20 minutes at 27,000 *g*.

Exhaustive absorption can be checked by indirect labeling of growth-phase and aggregation-competent cells with fluorescent antibodies; only the latter should be labeled with the absorbed antibody.

For absorption with a membrane fraction, growth-phase cells are frozen and thawed and the 27,000 *g* pellet containing plasma membranes is washed twice with buffer. Then, 10–30 mg Fab or IgG in 1 ml buffer is absorbed for 30 minutes on ice with membranes equivalent to 5×10^8 cells, corresponding to ~12 mg protein in the membrane fraction. The supernatant obtained by centrifugation for 30 minutes at 27,000 *g* is absorbed three times as described and finally centrifuged for 60 minutes at 100,000 *g*.

C. Assay of Contact Sites, the Target Sites of Adhesion-Blocking Fab

Neutralization of Fab by solubilized membrane antigens is assayed in the agglutinometer by mixing Fab with antigen fractions and adding test cells thereafter. The Fab concentration is adjusted such that the turbidity of cells with Fab alone is ~80% of the turbidity of completely dissociated cells. For determining the degree of neutralization of the adhesion-blocking activity of Fab, a dilution series of the Fab should be run in parallel with the experimental samples and a control of cells should be included to which antigen but no Fab is added.

VII. Purification of Contact Sites

A. Extraction with Detergents

Glycoproteins that neutralize adhesion-blocking Fab directed against membrane antigens of either *D. discoideum* or *P. pallidum* can be solubilized from membranes of aggregation-competent cells of these species by detergent extraction. Crude membrane fractions are prepared by centrifugation of cell homogenates for 30 minutes at 27,000 g. The pellet is taken up at a concentration of 4 mg protein/ml in either 2% sodium cholate or 1% Triton X-100 in buffers of pH 6–7.4, supplemented with 0.15 M NaCl. After 1 hour of solubilization in the cold, the 100,000 g supernatant is tested for absorption of adhesion blocking Fab. Since *D. discoideum* as well as *P. pallidum* cells tolerate maximally 0.02% sodium cholate or 0.001% Triton X-100, the extract needs to be adequately diluted for the assay of Fab neutralization.

Alternatively, detergents with a high critical micellar concentration can be partially removed by dialysis. For that purpose Triton X-100 should be replaced by another nonionic detergent. Octyl-oligooxyethylene, a mixture of octyltri-, tetra-, and pentaoxyethylenes (Bachem, 4416 Bubendorf, Switzerland), is an easily removable detergent with low denaturing activity (Rosenbusch et al., 1982). For the extraction of *D. discoideum* glycoproteins it is used at a concentration of 1%.

B. Butanol–Water Extraction

A butan-1-ol–water two-phase system provides another efficient extraction procedure for target antigens of adhesion-blocking Fab from *D. discoideum* and *P. pallidum* membranes (Huesgen and Gerisch, 1975; Bozzaro and Gerisch, 1978). It is particularly useful for the purification of membrane glycoproteins for immunological purposes, although it might denature the proteins. For that purpose a 27,000 g pellet fraction from homogenates of aggregation-competent cells or a plasma membrane fraction obtained by the dextran–polyethylene glycol method of Brunette and Till (1971) is stirred for 1 hour on ice in 10 mM Tris-HCl buffer, pH 7.0, containing 1.5 M KCl and pelleted by centrifugation for 30 minutes at 10,000 g. KCl-extracted membranes containing 10–20 mg protein are suspended in 10 ml ice-cold phosphate buffer, pH 6.0, or in 10 mM piperazine-HCl, pH 5.5 (see Section IV,A). (Raising the amount of protein to 50 mg results in higher specific activity of the contact site A glycoprotein but lower yield.) Under stirring on ice 7 ml of ice-cold butanol is added dropwise. Stirring is continued for 5 minutes before the phases are separated by centrifugation for 5 minutes at 3000 g. The water phase is extensively dialyzed against the phosphate buffer and then centrifuged for 60 minutes at 100,000 g. The clear supernatant contains ~0.1 mg protein/ml and can be concentrated 10-fold on an Amicon YM-30 filter.

C.   Purification of Contact Site Glycoproteins from *D. discoideum* and *P. pallidum*

For purification of the contact site A glycoprotein from butanol–water extracts, enrichment in a Triton X-114 phase, DEAE–cellulose chromatography in 10 mM piperazine-HCl buffer, pH 5.5, in the presence of 0.1% Triton X-100, followed by preparative SDS–polyacrylamide gel electrophoresis has been employed (Stadler *et al.*, 1982). A procedure for affinity purification is given in Section IX,A. Fractions containing the contact site A glycoprotein can be detected by spotting 5 μl to a nitrocellulose filter. The filter is dried and equilibrated in 10 mM Tris-HCl, pH 7.8, 150 mM NaCl, 0.05% Tween 20, and 0.02% sodium azide, labeled with contact site A-specific ^{125}I-labeled antibody, extensively washed in the Tween–NaCl buffer, and autoradiographed. Alternatively, the spotted areas can be cut out and the bound radioactivity counted in a γ-counter. If no contact site A-specific antibodies are available, ^{125}I-labeled wheat germ agglutinin can be used as a substitute for labeling immunoblots after SDS–polyacrylamide gel electrophoresis. The contact site A glycoprotein is the major target site of this lectin in aggregation-competent *D. discoideum* cells (Yoshida *et al.*, 1984).

The contact site 1 glycoprotein of *P. pallidum* has an apparent molecular weight of ~64,000 or 56,000, depending on the strain (Bozzaro *et al.*, 1981; Toda *et al.*, 1984a, 1987; Francis *et al.*, 1985). This glycoprotein is purified from butanol–water extract by fractionated ammonium sulfate precipitation followed by similar procedures as worked out for the contact site A glycoprotein of

D. discoideum. For the butanol–water extraction, the use of water instead of buffer is preferable (Toda *et al.*, 1984a).

VIII.   Production of Monoclonal Antibodies against Contact Sites

A.   Immunization and Screening Procedures

The following immunization schedules are indicative; they may be modified according to the quantities of antigen available. For raising monoclonal antibodies, 4-week-old female BALB/c mice are injected intraperitoneally with not more than 250 μl of a 1:1 (v/v) emulsion of antigen solution containing ~50 μg protein and complete Freund's adjuvant. An intraperitoneal boost is given after 4 weeks or later with the same amount of antigen plus incomplete Freund's adjuvant. At the fourth day thereafter the spleen cells are fused with Sp2-01 or X63 Ag 8.653 myeloma cells using conventional procedures. Alternatively, antigen is injected at amounts increasing from 10 to 50 μg protein per injection at weekly intervals, alternately adsorbed to 50–100 μl of Alugel S, or together with 50–100 μl *Bordetella pertussis* antigen suspension, or without adjuvant. At ≥4 weeks after the last injection, 30–50 μg protein adsorbed to Alugel S is injected intraperitoneally, followed on the next day by injection of the same amount of antigen without adjuvant, and by fusion of spleen cells at 2 days after this injection. Alugel S (Serva, 69 Heidelberg, FRG) is a stabilized aluminum hydroxide preparation. *Bordetella pertussis* antigen is purchased from Schweizerisches Seruminstitut, 3012 Bern, Switzerland, or from Difco.

Occasionally hybridoma lines producing monoclonal antibodies against the contact site A glycoprotein have been obtained by injecting twice 0.5 mg protein of a plasma membrane fraction with Freund's adjuvant, or by injecting 100–500 μg protein of that fraction following the aluminum hydroxide–*B. pertussis* schedule.

Hybridoma supernatants are screened either in solid-phase ELISA or radioimmunoassays with butanol–water extracted or further purified glycoproteins. Depending on the purity of the antigens, 20–50 μg/ml is used for coating the plastic wells. Mini-immunoblots are used for screening in cases where not enough purified protein is available, the protein is difficult to extract, or where only antibodies of high specificity are of interest. SDS–polyacrylamide gel electrophoresis of membrane fractions is stopped when the front has moved not more than 3 cm. The proteins are blotted onto nitrocellulose and the pieces are cut into strips of 2 mm width to fit into 3 × 40 mm polyethylene tubes. The tubes are

filled with hybridoma supernatants and labeled as usual for immunoblots. Biotinylated second antibody in combination with streptavidin–peroxidase labeling
should not be used, since an 80-kDa protein may be labeled unspecifically at the
same position where the contact site A glycoprotein would appear.

B. Deglycosylation of Glycoproteins with Anhydrous Hydrogen Fluoride (HF)

First note that all operations should be done under a well-ventilated hood with
face mask and acid-resistant plastic gloves. In case of contact with HF the skin
should be rinsed with diluted ammonia and a physician should be consulted.

The following procedure has been applied to *D. discoideum* and *P. pallidum*
cell surface glycoproteins (J. Stadler, personal communication). The procedure
is based on the method of Mort and Lamport (1977) and its modifications by
Edge *et al.* (1981) and Manjunath *et al.* (1982).

Preparations of membranes or of purified glycoproteins containing at least 1
mg protein/ml are precipitated in a plastic vial, e.g., an Eppendorf microfuge
tube, by adding nine volumes of acetone to one volume of the sample. The pellet
is dried carefully under vacuum. Oxygen is removed with an argon stream, and
20 μl of anisol is added while the samples are kept on ice.

About 0.5 ml of HF is condensed from a Baker (Phillipsburg, NJ) no. 4297
"lecture bottle" through a Baker no. 4472 pressure reducer into a plastic vial
kept on ice. Only connections made from stainless steel or Teflon should be
used. Warming of the lecture bottle by hand suffices to distill the HF. Then, 200
μl of the freshly condensed HF is added to the sample using a dry, precooled
plastic pipet. The tube is kept closed for 1 hour on ice. The specimens should be
completely dissolved. (Pipetman, e.g., Eppendorf or Gilson, should not be used,
since the rubber O-rings become corroded.)

Most of the HF is removed in a stream of N_2 or argon until foam is no longer
formed; the material is resuspended in ≥ 200 μl of 1 *M* ammonium carbonate
buffer, pH 7.0, with phenol red as indicator. It is important that remnants of HF
are neutralized. After drying under vacuum, enough HF would be left to acidify a
sample when it is resuspended in a buffer of low capacity, and proteins may be
hydrolyzed. If necessary, the material is precipitated in acetone–water 9:1 and
dried as above with N_2 or argon to remove the volatile buffer.

The HF treatment leads to the removal of most but not all of the carbohydrate
residues of the contact site A glycoprotein in *D. discoideum* membrane fractions
(Bertholdt *et al.*, 1985). The method has proved its practical value in raising
monoclonal antibodies against the protein portion of the contact site 1 glycoprotein of *P. pallidum* in which the immunogenicity of carbohydrate moieties is
strongly prevailing (Toda, *et al.*, 1987).

C. Characterization of the Epitopes Recognized by Antibodies

The contact sites A of *D. discoideum* and the contact sites 1 of *P. pallidum* are both glycoproteins. The contact site A protein is modified by two types of carbohydrate chains. Type 1 carbohydrate is recognized by concanavalin A and is sulfated. Type 2 carbohydrate is recognized by wheat germ agglutinin, and is highly immunogenic. Furthermore, the protein is phosphorylated and acylated with palmitic acid (for review see Gerisch, 1986). The contact site 1 protein carries carbohydrate residues which are modified during development by an L-fucose-containing epitope that dominates the immune response against the entire glycoprotein (Toda *et al.*, 1984b, 1987).

The following criteria can be applied to the identification of anticarbohydrate antibodies:

1. Carbohydrate residues can be cleaved off from the glycoproteins by hydrazinolysis, purified, and used after reacetylation for the neutralization of antibodies (Toda *et al.*, 1984a). Competition of the oligosaccharide for binding of the antibodies to the intact glycoprotein can be assayed using immunoblots or solid-phase radioimmunoassays (Toda *et al.*, 1984a).

2. Some anticarbohydrate antibodies are blocked by sugars. Anti-contact site A antibodies reacting with type 2 carbohydrate have been obtained which are blocked by mannose, maltose, and/or *N*-acetylglucosamine (Bozzaro and Merkl, 1985). Anti-contact site 1 antibodies are often blocked by L-fucose. The binding of simple sugars to antibodies can be revealed by using 100 m*M* sugar for competing with binding of the antibodies to glycoprotein in immunoblots or solid-phase radioimmunoassays (Toda *et al.*, 1984b). Alternatively, the antibodies are bound to agarose beads derivatized with specific sugars (E. Y. Laboratories, San Mateo, CA) and eluted from the beads by the corresponding sugar (Bozzaro and Merkl, 1985).

3. Nonbinding to partially or completely unglycosylated forms of the glycoproteins. Development of *D. discoideum* in the presence of tunicamycin results in the production of a 66-kDa form of the contact site A glycoprotein which carries type 2 and lacks type 1 carbohydrate, and of a 53-kDa form which lacks both types of carbohydrate (Bertholdt *et al.*, 1985). Mod B mutants produce a 68-kDa form which resembles a precursor of the contact site A glycoprotein in the wild type carrying only type 1 carbohydrate (Hohmann *et al.*, 1985). Binding of antibodies to these forms is tested by immunoblotting. To produce cells that contain the 66-kDa and 53-kDa forms, 0.5 µg tunicamycin is added at 2 hours of starvation to 1×10^7 AX2 cells in 1 ml of phosphate buffer, pH 6.0. Development is enhanced by stimulation with 20-n*M* pulses of cAMP applied every 6 minutes, and the cells are harvested at 6–8 hours of starvation. The 68-kDa form is obtained from the mod B mutant HL220 or from HG220, an axenically

growing derivative of that mutant, by starving cells for 8–14 hours under pulsing with cAMP.

For identification of antibodies against the protein portion of glycoproteins, their usually high specificity is not unequivocal. On one hand, cross-reactivity between a lectin, discoidin I, and the contact site A glycoprotein has been observed (Stadler *et al.*, 1984), on the other hand, some antibodies against type 2 carbohydrate are also highly specific (Bertholdt *et al.*, 1985). Reactivity of an antibody with the 53-kDa protein of tunicamycin-treated cells appears to be a reliable criterion for an antiprotein antibody, although reactivity with a modification other than carbohydrate cannot be excluded. Proof for recognition of a polypeptide epitope is provided by reactivity of an antibody with the product of an expression vector carrying the cloned contact site A gene (Noegel *et al.*, 1985, 1986).

IX. Application of Monoclonal Antibodies to Affinity Purification and Precipitation of the Contact Site A Glycoprotein

A. Binding to an Anticarbohydrate Antibody Column and Elution with Sugar

The contact site A glycoprotein can be purified under mild conditions by binding to a monoclonal antibody which is blocked by a simple sugar, and by dissociation from the antibody by that sugar. To guarantee efficient purification, the antibody should cross-react with as few other glycoproteins as possible. An adequate antibody is mAb 40-178-3 directed against type 2 carbohydrate (Bozzaro and Merkl, 1985).

First, 0.5 ml of swollen protein A–Sepharose beads (Pharmacia, Uppsala, Sweden) are filled into a small column, and a solution of 2 mg antibody IgG per milliliter of 150 mM NaCl buffered with 100 mM sodium phosphate, pH 8.0, is circulated through the column until no more antibodies are bound as revealed by monitoring of the flowthrough at 280 nm. For stabilizing the antibody linkage (Schneider *et al.*, 1982), the column is washed with 200 mM triethanolamine-HCl pH 8.2, and the beads are suspended in 10 ml of a freshly made solution of 30 mM dimethylpimelimidate dihydrochloride (Pierce, Rockford, IL) in the triethanolamine-HCl buffer and gently shaken for 45 minutes at room temperature. Subsequently the beads are pelleted and the reaction stopped by resuspending them for 5 minutes in 0.5 ml of 30 mM ethanolamine-HCl buffer, pH 8.2. The beads are washed with 150 mM NaCl buffered with 100 mM sodium phosphate, pH 7.2, containing 0.02% sodium azide, and stored in the refrigerator.

Contact sites A are extracted from membranes with butanol–water or with 1% octyl-oligooxyethylene, loaded in the NaCl–phosphate buffer, pH 7.2, on the column, and recycled three times. The column is washed with the buffer and successively eluted with 100, 250, and 500 mM N-acetylglucosamine in the same buffer. No detergent is required for elution. Fractions containing the contact site A glycoprotein are identified by dot-blotting using a specific antibody against its protein portion (see Section VII, C).

B. Isolation by Precipitation with an Antiprotein Antibody

Plasma membrane-enriched fractions equivalent to 2×10^7 aggregation-competent cells are extracted for 15 minutes on ice in 100 μl of 10 mM HEPES buffer, pH 7.4, containing 1 mM DTT, 150 mM NaCl, and 1% octyl-oligooxyethylene (see Section VII,A), and the insoluble material removed by centrifugation for 5 minutes in a microfuge (Eppendorf). The supernatant is incubated for 30 minutes on ice with 5 μg of a monoclonal antibody specific for the protein moiety of the contact site A glycoprotein, ҏ.g., mAb 41-71-21 or mAb 41-448-9 (Table I), dissolved in a minimum of buffer. One volume of swollen protein A–Sepharose beads (Pharmacia, Uppsala, Sweden) is suspended in four volumes of the HEPES buffer; 50 μl of the suspension is added to the reaction mixture and shaken for 30 minutes in the cold. The beads are pelleted in a microfuge and washed with 1 ml of buffer. For immunoblotting of the precipitated material, the beads are suspended in 30 μl of sample buffer containing 2% SDS, heated, and subjected to SDS–polyacrylamide gel electrophoresis.

X. Isolation of Contact Site Mutants

A. Distinction between Various Classes of Mutants

Several categories of *D. discoideum* mutants can be selected by cell surface labeling with anti-csA antibodies followed by selection of unlabeled cells using a cell sorter: (1) pleiotropic mutants with an early block in development which do not express a number of developmentally regulated proteins, including the contact site A glycoprotein, (2) developing mutants specifically defective in the synthesis of the contact site protein moiety, (3) mutants defective in glycosylation or other modifications of the protein, and (4) mutants defective in transport of the glycoprotein to the cell surface.

Mutants of the first category are characterized by absence of other developmentally regulated RNAs or proteins, and are totally deficient in aggregation. Mutants of the second category are rare. Their yield relative to pleiotropic

TABLE I

CHOICE OF MONOCLONAL ANTIBODIES AND *Dictyostelium discoideum* STRAINS FOR THE SELECTION AND CHARACTERIZATION OF CONTACT SITE A MUTANTS[a]

Contact site A defect	Specificity of antibody	Antibody number[b]	D. discoideum strain	Characterization of mutant
Developmental regulation	Antiprotein	mAb 41-71-21 mAb 41-448-9	AX2	Nonexpression of other developmentally regulated gene products
Synthesis of the protein	Specific anti-type 2 carbohydrate Antiprotein	mAb 12-120-94 mAb 40-178-3 mAb 41-71-21 mAb 41-448-9	HG592	Expression of other developmentally regulated gene products; type 2 glycosylation of other proteins
Type 1 glycosylation	Anti-type 1 carbohydrate	None identified	HG592	Production of 66-kDa glycoprotein recognized by antiprotein and by anti-type 2 carbohydrate antibodies
Type 2 glycosylation (mod B⁻)	Specific anti-type 2 carbohydrate Cross-reacting anti-type 2 carbohydrate	mAb 12-120-94 mAb 40-178-3 mAb 20-121-1 mAb 40-62-5	AX2; HG592	Production of 68-kDa glycoprotein recognized by antiprotein antibodies, but not labeled with anti-type 2 carbohydrate antibodies
Transport to the cell surface	Antiprotein	mAb 41-71-21 mAb 41-448-9	AX2; HG592	No or reduced fluorescence labeling of intact cells but labeling in immunoblots with antiprotein antibody

[a] Antibody numbers and *D. discoideum* strains are those used in our laboratory (Bertholdt *et al.*, 1985; Bozzaro and Merkl, 1985; Noegel *et al.*, 1985). The AX2 strain can be replaced by AX3 provided that development of starved cells is stimulated in suspension cultures by pulses of 20 nM cAMP applied every 6 minutes. Another cross-reacting anti-type 2 antibody is E28D8 (Murray *et al.*, 1984). Only antiprotein antibodies that recognize an epitope accessible on the cell surface are suited for mutant selection using a cell sorter. Anti-type 2 carbohydrate antibodies designated as "specific" show nevertheless some cross-reaction with other glycoproteins. The *D. discoideum* strains should be cultivated axenically, since bacteria-grown cells produce more proteins that react with these antibodies. For recognizing the contact site A protein after cloning in the λgt11 expression vector (Noegel *et al.*, 1986) or after SDS–polyacrylamide gel electrophoresis (Hohmann *et al.*, 1985), mAG 33-294-17 (Bertholdt *et al.*, 1985) is best suited.

[b] mAb, Monoclonal antibody.

mutants can be increased by mutagenizing a strain such as HG592 in which the requirements for several developmental signals have been bypassed (Gerisch *et al.*, 1985b). Mutants of the third category can be selected by antibodies recognizing the modification of interest. Mutants defective in the mod B modification (Murray *et al.*, 1984) lack type 2 carbohydrate (Yoshida *et al.*, 1984; Gerisch *et al.*, 1985a). Such mutants can be selected together with mutants defective in synthesis of the contact site A protein moiety or in transport of the contact site A glycoprotein to the cell surface if an anti-type 2 carbohydrate antibody that is relatively specific for the contact site A glycoprotein is used for cell surface labeling. These classes of mutants can be distinguished by immunoblotting after SDS–polyacrylamide gel electrophoresis (Table I).

B.   Selection of Mutants by Cell Sorting

For the selection of mutants defective in the contact site A glycoprotein or in other cell surface antigens, a cell sorter can be used in combination with fluorescent-antibody labeling of the cells. The following procedure has been worked out for a FACS IV cell sorter equipped with a single-cell deposition system, but this combination is not necessary.

Cells mutagenized with nitrosoguanidine (see Section II,D) or by UV irradiation are divided into aliquots which are grown up separately for about three generations. For the selection of contact site A mutants, AX2 cells are starved for 6 hours in phosphate buffer, pH 6.0 (see Section IV,A). Cells of the AX3 strain are stimulated under the same conditions at intervals of 6 minutes with 20-nM pulses of cAMP for full expression of the contact site A glycoprotein. The cells are washed in the buffer and labeled for 15 minutes strictly on ice with monoclonal IgG in 150 mM NaCl buffered with phosphate, pH 7.2 (PBS). Normally 100 μg IgG/ml is used. After washing three times in ice-cold PBS, the cells are labeled with FITC-conjugated anti-mouse IgG in PBS, washed three times in PBS, and resuspended in phosphate buffer, pH 6.0. Cells should be kept on ice during labeling and kept cool during sorting, since the antibody label is removed by cell surface capping at temperatures >5°C. Anti-mouse IgG sera use to react primarily with IgG$_1$. If the monoclonal antibody is of another subclass, fluorescent antibodies against this particular subclass should be used as second antibodies. For cell sorting the cells must not be clumped. PBS suppresses clumping but diminishes the plating efficiency when present during sorting.

All tubings and connections of the cell sorter which come into contact with cells are sterilized with 5% hydrogen peroxide overnight and rinsed with sterile phosphate buffer before use. The windows of the cell sorter are adjusted such that particles within the light-scattering and the autofluorescence ranges of single cells are selected. The cells are regrown for several generations, starved as described, and sorted again. With a single-cell deposition system the cells are directly cloned after the second selection into a lawn of *E. coli* B/2 or other food bacteria on plates with nutrient agar containing 0.1% glucose, 0.1% peptone, and 2% agar in phosphate buffer, pH 6.0. If 8 × 12 cm microtiter plates without wells (Dynatech cat. no. MA1701) are used, 96 droplets containing single cells can be deposited per plate. If no single-cell deposition system is available, the cells are plated out by conventional procedures.

The method has been worked out for the selection of *P. pallidum* mutants defective in the expression of an L-fucose-containing cell surface epitope (Francis *et al.*, 1985) and has been applied to *D. discoideum* mutants defective in

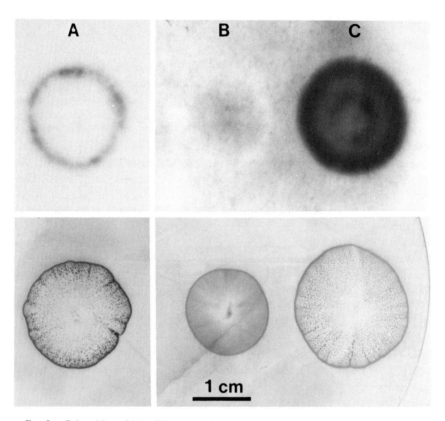

Fig. 5. Colony blots of (A) wild-type and (B, C) mutants of *D. discoideum*. *Top:* autoradiogram of colony immunoblots labeled with ^{125}I-IgG 41-71-21 specific for the contact site A protein. *Bottom:* Cellular proteins of the same blots stained with Ponceau S. The wild type forms aggregates; as reflected in the patched staining with Ponceau S. The contact site A glycoprotein is maximally expressed during the aggregation stage, as indicated by the ringlike antibody labeling. The mutant in (B) does not express the contact site A and does not appear to aggregate properly. The mutant in (C) forms aggregates but does not develop further. The contact site A glycoprotein is not downregulated, as shown by the homogeneous labeling with the antibody.

contact site A expression (Noegel *et al.*, 1985). The proportion of mutants after two runs of selection has been ~5% of the total number of colonies, and the plating efficiency in the order of 50%.

XI. Colony Immunoblotting for Identifying and Characterizing Mutants

Colonies derived from the cloned cells are blotted onto nitrocellulose filters (e.g., BA 85, Schleicher and Schuell, D-3354 Dassel, FRG, and Keene, NH). Most of the cells adhere to the filter, but enough cells will be left on the agar to start new cultures from the interesting clones after storage of the plates in the cold. The filters are dried, washed with Tween–NaCl buffer as described under Section VII,C, and incubated either with ^{125}I-labeled antibody for autoradiography or with biotinylated antibody for peroxidase staining. Afterwards, proteins can be stained with 0.2% Ponceau S (Serva, Heidelberg, FRG) in 3% TCA, followed by washing in the TCA to visualize the colonies.

Colony immunoblots provide not only information on the presence or absence of specific proteins or carbohydrate epitopes but also on their developmental regulation (Fig. 5). The blotting technique can be adapted to the labeling of intracellular proteins by placing the freshly made blot, instead of drying it, onto a metal plate which is cooled from below with dry ice. After thawing, the cells are broken and intracellular proteins are bound to the filter. Without allowing the filters to dry, they are washed with Tween–NaCl buffer and the above procedure is followed up. This version of colony immunoblotting has been used for isolating mutants defective in cytoskeletal proteins (Wallraff *et al.*, 1986).

REFERENCES

Bertholdt, G., Stadler, J., Bozzaro, S., Fichtner, B., and Gerisch, G. (1985). *Cell Differ.* **16,** 187–202.

Beug, H., and Gerisch, G. (1972). *J. Immunol. Methods* **2,** 49–57.

Beug, H., Gerisch, G., Kempf, S., Riedel, V., and Cremer, G. (1970). *Exp. Cell Res.* **63,** 147–158.

Beug, H., Katz, F. E., and Gerisch, G. (1973). *J. Cell Biol.* **56,** 647–658.

Born, G. V. R., and Garrod, D. (1968). *Nature (London)* **220,** 616–618.

Bozzaro, S., and Gerisch, G. (1978). *J. Mol. Biol.* **120,** 265–280.

Bozzaro, S., and Merkl, R. (1985). *Cell Differ.* **17,** 83–94.

Bozzaro, S., and Roseman, S. (1982). *In* "Embryonic Development" (M. M. Burger and R. Weber, eds.), pp. 183–192. Liss, New York.

Bozzaro, S., and Roseman, S. (1983a). *J. Biol. Chem.* **258,** 13882–13889.

Bozzaro, S., and Roseman, S. (1983b). *J. Biol. Chem.* **258,** 13890–13897.

Bozzaro, S., Tsugita, A., Janku, M., Monok, G., Opatz, K., and Gerisch, G. (1981). *Exp. Cell Res.* **134,** 181–191.

Bozzaro, S., Bernstein, R., and Roseman. S. (1983). *Cell Differ.* **12,** 109–114.

Bozzaro, S., Perlo, C., Ceccarelli, A., and Mangiarotti, G. (1984). *EMBO J.* **3,** 193–200.

Brunette, D. M., and Till, J. E. (1971). *J. Membr. Biol.* **5,** 215–224.

Chadwick, C. M., Ellison, J. E., and Garrod, D. R. (1984). *Nature (London)* **307,** 646–647.

Curtis, A. S. G. (1969). *J. Embryol. Exp. Morphol.* **22,** 305–325.

Edge, A. S. B., Faltinek, C. R., Hof, L., Reichert, L. E., and Weber, P. (1981). *Anal. Biochem.* **118,** 131–137.

Firtel, R. A., and Brackenbury, R. W. (1972). *Dev. Biol.* **27,** 307–321.

Francis, D., Toda, K., Merkl, R., Hatfield, T., and Gerisch, G. (1985). *EMBO J.* **4,** 2525–2532.

Gerisch, G. (1961). *Exp. Cell Res.* **25,** 535–554.

Gerisch, G. (1980). *In* ''Current Topics in Developmental Biology'' (A. Moscona and A. Monroy, eds.), pp. 243–270. Academic Press, New York.

Gerisch, G. (1986). *J. Cell Sci.* (Suppl.) **4,** 201–219.

Gerisch, G., and Hess, B. (1974). *Proc. Natl. Acad. Sci. U.S.A.* **71,** 2118–2122.

Gerisch, G., Malchow, D., Wilhelms, H., and Lüderitz, O. (1969). *Eur. J. Biochem.* **9,** 229–236.

Gerisch, G., Fromm, H., Huesgen, A., and Wick, U. (1975). *Nature (London)* **255,** 547–549.

Gerisch, G., Krelle, H., Bozzaro, S., Eitle, E., and Guggenheim, R. (1980). *In* ''Cell Adhesion and Motility'' (A. S. G. Curtis and J. D. Pitts, eds.), pp. 293–307. Cambridge Univ. Press, London and New York.

Gerisch, G., Weinhart, U., Bertholdt, G., Claviez, M., and Stadler, J. (1985a). *J. Cell Sci.* **73,** 49–68.

Gerisch, G., Hagmann, J., Hirth, P., Rossier, C., Weinhart, N., and Westphal, M. (1985b). *Cold Spring Harbor Symp. Quant. Biol.* **50,** 813–822.

Hohmann, H.-P., Gerisch, G., Lee, R. W. H., and Huttner, W. B. (1985). *J. Biol. Chem.* **260,** 13869–13878.

Huesgen, A., and Gerisch, G. (1975). *FEBS Lett.* **56,** 46–49.

Manjunath, P., Sairam, M. R., and Schiller, P. W. (1982). *Biochem. J.* **207,** 11–19.

Mort, A. J., and Lamport, D. T. A. (1977). *Anal. Biochem.* **82,** 289–309.

Müller, K., and Gerisch, G. (1978). *Nature (London)* **274,** 445–449.

Müller, K., Gerisch, G., Fromme, I., Mayer, H., and Tsugita, A. (1979). *Eur. J. Biochem.* **99,** 419–426.

Murray, B. A., Wheeler, S., Jongens, T., and Loomis, W. F. (1984). *Mol. Cell. Biol.* **4,** 514–519.

Noegel, A., Harloff, C., Hirth, P., Merkl, R., Modersitzki, M., Stadler, J., Weinhart, U., Westphal, M., and Gerisch, G. (1985). *EMBO J.* **4,** 3805–3810.

Noegel, A., Gerisch, G., Stadler, J., and Westphal, M. (1986). *EMBO J.* **5,** 1473–1476.

Ochiai, H., Stadler, J., Westphal, M., Wagle, G., Merkl, R., and Gerisch, G. (1982). *EMBO J.* **1,** 1011–1016.

Orr, C. W., and Roseman, S. (1969). *J. Membr. Biol.* **1,** 109–124.

Raper, K. B., and Thom, C. (1941). *Am. J. Bot.* **28,** 69–78.

Rosenbusch, J. P., Garavito, R. M., Dorset, D. L., and Engel, A. (1982). *In* ''Protides of the Biological Fluids'' 29th Colloquium 1981 (Peeters, H., ed.), pp. 171–174. Pergamon, New York.

Schnaar, R. L., Weigel, P. H., Kuhlenschmidt, M. S., Lee, Y. C., and Roseman, S. (1978). *J. Biol. Chem.* **253,** 7940–7951.

Schneider, C., Newman, R. A., Sutherland, D. R., Asser, U., and Greaves, M. F. (1982). *J. Biol. Chem.* **257,** 10766–10769.

Springer, W. R., and Barondes, S. H. (1978). *J. Cell Biol.* **78,** 937–942.

Springer, W. R., and Barondes, S. H. (1983). *J. Biol. Chem.* **258,** 4698–4701.

Stadler, J., Bordier, C., Lottspeich, F., Henschen, A., and Gerisch, G. (1982). *Hoppe Seyler's Z. Physiol. Chem.* **363,** 771–776.

Stadler, J., Bauer, G., Westphal, M., and Gerisch, G. (1984). *Hoppe Seyler's Z. Physiol. Chem.* **365,** 283–288.

Tetsch, F. (1970). Thesis, University of Freiburg im Breisgau.

Toda, K., Bozzaro, S., Lottspeich, F., Merkl, R., and Gerisch, G. (1984a). *Eur. J. Biochem.* **140,** 73–81.

Toda, K., Tharanathan, R. N., Bozzaro, S., and Gerisch, G. (1984b). *Eur. J. Biochem.* **143,** 477–481.

Toda, K., Francis, D., and Gerisch, G. (1987). *J. Cell Sci.* In press.

Vogel, G. (1983). *In* "Methods in Enzymology" (S. Fleischer and B. Fleischer, eds.), Vol. 98, pp. 421–430. Academic Press, New York.

Vogel, G., Thilo, L., Schwarz, H., and Steinhart, R. (1980). *J. Cell Biol.* **86,** 456–465.

Wallraff, E., Schleicher, M., Modersitzki, M., Rieger, D., Isenberg, G., and Gerisch, G. (1986). *EMBO J.* **5,** 61–67.

Walther, B. T., Öhman, R., and Roseman, S. (1973). *Proc. Natl. Acad. Sci. U.S.A.* **70,** 1569–1573.

Weigel, P. H., Naoi, M., Roseman, S., and Lee, Y. C. (1979). *Carbohydr. Res.* **70,** 83–91.

Yoshida, M., Stadler, J., Bertholdt, G., and Gerisch, G. (1984). *EMBO J.* **3,** 2663–2670.

Chapter 21

Discoidins I and II: Endogenous Lectins Involved in Cell–Substratum Adhesion and Spore Coat Formation

S. H. BARONDES,[1] D. N. W. COOPER, AND W. R. SPRINGER

Department of Psychiatry
University of California, San Diego
La Jolla, California 92093
and
Veterans Administration Medical Center
San Diego, California 92161

I. Introduction

As *Dictyostelium discoideum* differentiates it synthesizes two lectins, discoidin I and discoidin II (Rosen *et al.*, 1973; Simpson *et al.*, 1974; Frazier *et al.*, 1975). These carbohydrate-binding proteins are absent in vegetative cells but become abundant with development. Discoidin I reaches maximal levels (~1% of soluble cell protein) during aggregation, whereas discoidin II peaks during spore differentiation (Cooper and Barondes, 1984).

The slime mold lectins are examples of a large class of proteins of this kind found in many organisms (Barondes, 1984). Like discoidins I and II, many other lectins are both abundant and developmentally regulated (Barondes, 1984). Although it seems obvious that such proteins must function by interacting with complementary glycoconjugates in the tissues that make them, identification of their cellular roles has been difficult. Because of its favorable experimental properties, *D. discoideum* has been useful for such functional studies. We here

[1]Present address: Department of Psychiatry & Langley Porter Institute, University of California, San Francisco, San Francisco, California, 94143.

METHODS IN CELL BIOLOGY, VOL. 28

describe the basic biochemical and biological techniques used to study discoidins I and II.

II.   Purification of Discoidins I and II

Both lectins bind to Sepharose beads, which are made of polymers rich in galactose. For large-scale preparations we use a 1-liter column of Sepharose 4B. It is possible to recover as much as 100 mg of discoidin I and 10 mg of discoidin II by affinity chromatography on this column using an extract derived from ~5 $\times$ 10^{10} cells.

Initial studies monitored purification of discoidins I and II with a hemagglutination assay (Rosen *et al.*, 1973). We do not recommend setting up this assay, since purification, essentially a one-step procedure, can be monitored by polyacrylamide gel electrophoresis. Testing for retention of carbohydrate-binding activity is easily done by rebinding an aliquot to a small affinity column.

The lectins can be purified from either strain Ax-3 grown axenically or from NC-4 grown on a lawn of *Klebsiella pneumoniae*. We prefer the latter, since we find it easier to grow large numbers of cells in that way. To extract both lectins from bacterially grown cells, it is critical that 0.3 M galactose be present in the extraction medium. This helps dissociate discoidin I from bacterial polysaccharides with which it is associated in multilamellar bodies (Cooper *et al.*, 1986). In axenically grown cells, which do not form multilamellar bodies, discoidin I can be solubilized in media without galactose.

Since added Ca^{2+} augments the carbohydrate-binding activity of discoidin I (Alexander *et al.*, 1983a; Cooper *et al.*, 1983), we routinely include it in all buffers. Ethylene glycol (25%, v/v) is also used in purification and storage, since it stabilizes discoidin I (Cooper *et al.*, 1983), presumably by reducing the favorability of hydrophobic interactions. Discoidin I tends to aggregate and come out of solution at high concentrations. This is inhibited by addition of 25% ethylene glycol and by storage at concentrations that do not exceed ~0.1 mg/ml. Discoidin II is not influenced by either Ca^{2+} or ethylene glycol.

A.   Culture of *D. discoideum*

To obtain *D. discoideum* with high concentrations of lectin, large numbers of vegetative cells are first grown on a solid medium containing *K. pneumoniae* and then differentiated in an aqueous medium in the absence of bacteria.

1.   GROWTH OF VEGETATIVE *D. discoideum*

Klebsiella pneumoniae is grown at room temperature with shaking in an autoclaved medium containing 10 g of dextrose, 10 g of proteose peptone no. 2, and

5 g of yeast extract (the latter two from Difco Laboratories, Detroit, MI) per liter of 2 mM potassium phosphate buffer, pH 6.5. When the bacteria have reached stationary phase, NC-4 spores are added (to a concentration of ~4 × 10^{5}ml) from a stock agar plate containing differentiated colonies, preferably no more than 2 weeks old. Then, 3 ml of this mixture is spread over the surface of a solid medium which had been poured to a thickness of 1 cm in a 20 × 25 cm sterilized aluminum pan. The solid medium contains, per liter: 20 g of Bacto-agar, 10 g of Bacto-peptone, and 1 g of yeast extract (all from Difco Laboratories); 10 g of dextrose; 1.5 g of $MgSO_4 \cdot 7H_2O$; 1.9 g of KH_2PO_4; and 0.6 g of K_2HPO_4. It is autoclaved for 20 minutes. After inoculation, the pans are incubated at room temperature for 50 hours. At this time, *D. discoideum* cells are still largely in vegetative phase and have not formed aggregates. Each pan contains ~10^9 slime mold cells. We have used up to 50 pans for a preparation.

2. DIFFERENTIATION OF *D. discoideum* CELLS

The cells grown on pans will eventually differentiate and synthesize abundant discoidins I and II. Such cells can be used for the purification of the lectins. However, differentiation is asynchronous. It is more convenient to harvest the cells when still vegetative and induce them to differentiate by starving than while they are being gyrated in suspension.

To harvest the cells, 30 ml of cold distilled water is added to each pan, and the cells are scraped off with a glass slide. They are separated from the bacteria by centrifugation at 300 g for 5 minutes in a Sorvall GSA rotor. The cells are washed three times by centrifugation under the same conditions in cold distilled water. The washed cells are then counted in an electronic particle counter or a hemocytometer and resuspended to a concentration of ~3 × 10^7/ml in 16.7 mM Na$_2$PO$_4$–KH$_2$PO$_4$, pH 6.2. Aliquots (500 ml) of this suspension are then poured into 2-liter Erlenmeyer flasks and shaken on a gyratory shaker at room temperature for 16 hours. Under these starvation conditions the cells begin to differentiate and synthesize large amounts of lectin. The differentiated cells are centrifuged at 300 g for 5 minutes and resuspended to a concentration of ~2 × 10^8 cells/ml in a solution consisting of 15 mM Tris-HCl, 75 mM NaCl, 75 mM KCl, made to pH 7.3 (this solution is hereinafter called TBS), containing 1 mM CaCl$_2$ (this solution is called TBS-Ca) and 0.3 M galactose. The cells are then lysed by freezing in liquid nitrogen and thawing under running cold water. The lysate is centrifuged at 90,000 g for 75 minutes in a Spinco type 35 rotor; the supernatant is saved.

B. Affinity Chromatography on Sepharose 4B

Sepharose 4B, which is composed of a linear polymer of alternating units of D-galactose and 3,6-anhydro-L-galactose, specifically binds the lectins in extracts of *D. discoideum* (Simpson *et al.*, 1974). Lectin activity is quantitatively and reversibly bound to this material, and can be eluted with D-galactose. To bind the lectins in crude extracts to the Sepharose, it is necessary to remove the galactose. To this end, 1 volume of extract is mixed with 0.5 volume of Sepharose 4B that has been equilibrated with TBS-Ca. This suspension is then placed in a dialysis bag and dialyzed at 4°C for 24 hours against two baths, each containing 25 volumes of TBS-Ca. Dialysis in the presence of Sepharose facilitates binding and also helps preserve lectin activity, which is more stable in the presence of D-galactose and its derivatives.

The contents of the dialysis bag is then layered on top of a 1-liter (bed volume) Sepharose 4B column equilibrated with TBS-Ca at 4°C. The column is fitted with a flow adapter, and TBS-Ca is pumped through at ~60 ml/hr until all unbound material is eluted as indicated by monitoring the optical density of the eluate at 280 nm. The elution buffer is then changed to TBS-Ca containing 0.3 M galactose and 25% (by volume) ethylene glycol, and fractions are collected until all the activity is eluted. Appearance of the eluted activity is signaled by a rise in the optical density; but this does not return to baseline after all lectin activity has been eluted, since high concentrations of commercial galactose solutions contain materials that also absorb somewhat at 280 nm. All the lectin activity applied to the column can be quantitatively eluted with D-galactose along with ~1% of the protein added to the column. Carbohydrate-binding activity (as judged by either hemagglutination assays or binding to an affinity column) is stable when frozen in liquid nitrogen, then kept at −20°C.

In a typical preparation 90–95% of the lectin eluted from the column is discoidin I. It migrates at ~26 kDa on polyacrylamide gel electrophoresis in sodium dodecyl sulfate under reducing conditions, whereas discoidin II migrates at ~24 kDa. Both proteins are tetramers (Simpson *et al.*, 1974; Frazier *et al.*, 1975).

Should a preparation be desired that is more highly enriched in discoidin I or that is somewhat enriched in discoidin II, the initial extracts should be prepared by a preferential solubilization procedure. This is readily achieved, since discoidin II is preferentially solubilized from cells grown on *K. pneumoniae* if the extraction buffer does not contain galactose. If such cells are frozen and thawed in TBS-Ca, most of the discoidin I remains particulate, whereas discoidin II is in the supernatant. An extract that is rich in discoidin I and virtually devoid of discoidin II can then be made by freezing and thawing the pellet in TBS-Ca containing 0.3 M galactose and 25% ethylene glycol. The two extracts can be separately chromatographed on Sepharose 4B, giving preparations already highly enriched in one or the other lectin.

C.   Separation of Discoidins I and II

Discoidin I and discoidin II can be separated by either preparative isoelectric focusing (Simpson *et al.*, 1974), ion exchange chromatography (Frazier *et al.*,

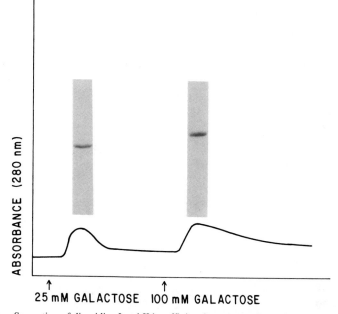

FIG. 1.   Separation of discoidins I and II by affinity chromatography on *N*-acetylgalactosamine-conjugated agarose. Approximately 1 mg of a mixture of discoidins I and II in TBS-Ca containing 25% ethylene glycol and 0.3 *M* galactose was added to 1 ml of *N*-acetylgalactosamine-conjugated agarose beads in a dialysis bag. The galactose was removed by extensive dialysis against TBS-Ca, 25% ethylene glycol. The beads were then poured into a column and washed with this buffer. Discoidin II was eluted by addition of 25 m*M* galactose, and discoidin I was subsequently eluted by raising the galactose concentration to 100 m*M*. Fractions were monitored for absorbance at 280 nm and the protein peaks were analyzed by polyacrylamide gel electrophoresis in sodium dodecyl sulfate. The first peak contains pure discoidin II (apparent subunit MW 24,000) and the second peak pure discoidin I (apparent subunit MW 26,000). For details see Cooper *et al.* (1983).

1975), or a second affinity chromatography step (Cooper *et al.*, 1983) on a column of *N*-acetylgalactosamine-derivatized agarose beads (Selectin 6, Pierce Chemical Co., Rockford, IL). The latter method is easiest and gives the highest yield.

Selective affinity chromatography is based on the relatively greater affinity of discoidin I for *N*-acetylgalactosamine. After binding both lectins to the affinity column, discoidin II is eluted with 25 m*M* galactose, while 100 m*M* galactose is required to elute discoidin I (Fig. 1). Although this procedure is very efficient in separating the two lectins, traces of cross-contamination can occur. This can be reduced to undetectable levels by doing the initial affinity chromatography (on Sepharose 4B) with starting materials already enriched in discoidin I or discoidin II, as indicated above.

III.   Immunoassay for Discoidins I and II

The highly purified lectins have been used to raise antisera for quantitative immunoassays. A solid-phase immunoassay has proved very useful, since specificity of the antigen–antibody reaction is assured both by binding the antigen to immobilized antibody and by quantitation of the bound antigen by reaction with a second, biotin-conjugated antibody (Wilchek, 1980).

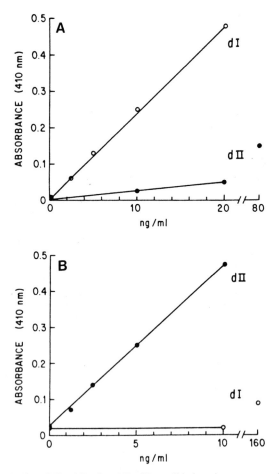

FIG. 2.   Quantitation of discoidins I and II with a solid-phase immunoassay. Polystyrene wells were saturated with anti-discoidin I IgG (A) or anti-discoidin II IgG (B). After washing, the wells were incubated with serial dilutions of either purified discoidin I or discoidin II, as shown, or with extracts containing these lectins (not shown). Wells were washed and incubated with biotinylated anti-discoidin I (A) or biotinylated anti-discoidin II (B). Binding of a biotinylated antibody was measured with avidin-conjugated peroxidase. For other details, see Cooper and Barondes (1984).

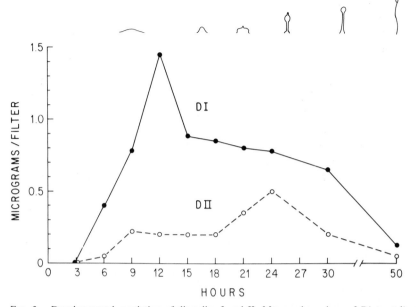

FIG. 3. Developmental regulation of discodins I and II. Measured numbers of *Dictyostelium discoideum* cells were differentiated on small circular filters. Sketches of the appearance of aggregates at the indicated times are shown at the top of the figure. Extracts prepared at the indicated times after initiation of differentiation and the amounts of discoidin I (DI) and discoidin II (DII) in each extract were determined by the immunoassay described in Fig. 2. Results are expressed in micrograms per filter, reflecting the total amount of lectin in and around the constant number of cells that had been applied to each filter. Protein determinations were also made. The peak level of discoidin I is ~9 μg/mg cell protein; the peak level of discoidin II is ~3 μg/mg cell protein. For details, see Cooper and Barondes (1984).

In this assay, polystyrene wells (Immunolon I Removawells, Dynatech Laboratories, Alexandria, VA) are saturated by incubation with 10 μg of IgG prepared from rabbit antisera against discoidin I or discoidin II in 0.2 ml TBS. The IgG can be prepared by chromatography on DEAE Affi-Gel Blue (Bio-Rad Laboratories, Richmond, CA). After washing the wells four times with TBS containing 0.2% bovine serum albumin (BSA) and 0.3 M galactose, wells are incubated for 1 hour with serial dilutions of a sonicated extract of *D. discoideum* in TBS containing 0.1% Emulphogen, 0.2% BSA, and 0.3 M galactose. Wells are then washed four times in this buffer and biotin-conjugated rabbit IgGs specific for discoidin I or discoidin II [which have been prepared with the *N*-hydroxysuccinimide ester of biotin obtained from Calbiochem-Behring Corp. (La Jolla, CA), using the procedures described by Heggeness and Ash (1977)] are then added for 1 hour at a concentration of 10 μg in 0.2 ml of the above buffer.

After washing the wells four times, they are incubated for 30 minutes with 150 μl of avidin-conjugated peroxidase (ABC Vectastain standard kit, Vector Laboratories, Burlingame, CA) prepared according to the manufacturer's instructions and then diluted with four volumes of TBS containing 0.2% BSA and 0.3 M galactose. Wells are then washed six times with this diluent before addition of 200 μl of *o*-phenylenediamine (Bionetics Laboratory products, Kensington, MD) in 0.1 M sodium citrate, pH 4.5, with 0.02% H_2O_2. The colorimetric reaction is allowed to proceed until standard wells containing high discoidin concentration reach saturation, and is stopped by addition of 50 μl of saturated NaF. Results are measured spectrophotometrically at 410 nm (Minireader, Dynatech Laboratories, Alexandria, VA).

This assay discriminates well, but not perfectly, between discoidin I and discoidin II (Fig. 2), since there is often slight cross-reaction between the antisera (Berger and Armant, 1982; Cooper and Barondes, 1984). For most practical purposes a small amount of cross-reaction is acceptable but, where necessary, it is possible to adsorb out cross-reacting antibodies by passing the antiserum raised against one of the lectins through a column to which the other lectin has been coupled. Levels of discoidin I and discoidin II during development have been measured with this assay (Fig. 3).

IV. Assays of Carbohydrate-Binding Activity and Glycoconjugate Ligands

A. Hemagglutination Assay

Discoidins I and II were first identified as agglutinins of erythrocytes. Agglutination activity could be inhibited by specific saccharides such as galactose and *N*-acetylgalactosamine, indicating that agglutination was due to interaction between the carbohydrate-binding sites of these lectins and glycoconjugates on the erythrocyte surfaces (Rosen *et al.*, 1973). However, for studies of protein–carbohydrate interactions hemagglutination has been supplanted by a quantitative solid-phase binding assay. Neither assay of carbohydrate-binding activity is needed to monitor lectin purification, which can be done by gel electrophoresis.

B. Solid-Phase Binding Assay

This assay is advantageous not only because it gives a quantitative measure (as opposed to an "end point" in agglutination), but also because it can be conducted in the presence of certain detergents which have proved useful for solubilizing glycoconjugate ligands. It is based on the finding that discoidin I or discoidin II can be adsorbed on polystyrene microtiter wells (Removawells, Dynatech Laboratories, Alexandria, VA) with retention of carbohydrate-binding activity. The immobilized lectins can bind a radioiodinated neoglycoprotein, [^{125}I]lactosyl-BSA; and binding of other glycoconjugates can be estimated from their potencies as competitive inhibitors.

The lactosyl-BSA we use contains 37 moles of lactose per mole of BSA, and was synthesized by the method of Smith and Ginsburg (1980). Radioiodination

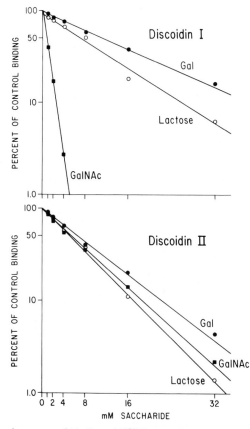

FIG. 4. Inhibition by sugars of binding of [^{125}I]lactosyl-BSA to immobilized discoidin I or discoidin II. Plastic wells were saturated with discoidin I or discoidin II; then, after washing, they were incubated for 1 hour with 4×10^4 counts/min of [^{125}I]lactosyl-BSA in TBS-Ca containing 0.2% BSA and a range of concentrations of galactose (Gal), N-acetylgalactosamine (GalNAc), or lactose. Wells were then washed in 0.2% BSA in TBS-Ca. Bound [^{125}I]lactosyl-BSA is expressed as percentage of control binding in the absence of sugar. About 10% of the added radioactivity bound in the absence of sugar, whereas binding was negligible to wells that had not been preincubated with lectin. For details, see Cooper et al. (1983).

was done by the chloramine-T procedure (McConahey and Dixon, 1966) by incubating 0.5 mg of the neoglycoprotein with 5 mCi [125]I. Incorporation is very efficient, and iodine is removed by exclusion chromatography on Sephadex G-25.

To immobilize either lectin, polystyrene microtiter wells are incubated with 100 μl of a 40-μg/ml lectin solution in TBS-Ca containing 25% ethylene glycol and 0.3 M galactose for 1 hour, then washed four times with 0.4 ml of 0.2% BSA (radioimmunoassay grade) in TBS-Ca. To assay binding, the coated wells are incubated for 1 hour with 4×10^4 counts/min [125I]lactosyl-BSA ($\sim 5 \times 10^9$ counts/min/mg) in 150 ml of the same buffer with or without samples to be tested for inhibitory activity. Wells are again washed four times with 0.4 ml of 0.2% BSA in TBS-Ca, and the bound radioactivity is assayed with a γ-scintillation counter.

Under the conditions that we use, $\sim 10\%$ of the radioactive neoglycoprotein is bound to the immobilized lectin in the absence of competitive inhibitors. Binding to discoidin II is inhibited to about the same extent by galactose, N-acetylgalactosamine, and lactose (Fig. 4), whereas N-acetylgalactosamine is much more potent in inhibiting binding to discoidin I (Fig. 4).

V. Purification of a Discoidin-Binding Polysaccharide from *D. discoideum*

The carbohydrate-binding site of discoidin I and discoidin II must function by interacting with glycoconjugates. The search for endogenous glycoconjugates in *D. discoideum* that bind these lectins has been confounded by the fact that the bacteria they eat contain polysaccharides which can interact with these ligands (Madley and Hames, 1981; Cooper *et al.*, 1983). Indeed, recent evidence indicates that the natural ligands for the carbohydrate-binding site of discoidin I are bacterial polysaccharides, which remain packaged in multilamellar bodies in *D. discoideum* early in differentiation (Barondes *et al.*, 1985), and which direct compartmentalization of discoidin I into this structure (Cooper *et al.*, 1986).

A. Axenic Growth in Macromolecule-Depleted Medium

To search for endogenous ligands for discoidin I or discoidin II it is therefore necessary to grow the cells in the absence of bacteria or other exogenous ligands, including those found in the axenic media that are normally used (Bartles *et al.*, 1981). This can be accomplished by raising the cells on macromolecule-depleted axenic medium. This is prepared by mixing 40 g protease peptone, 20 g yeast extract, and 40 g dextrose per liter of H_2O, pouring this concentrated medium into large dialysis bags (taking care to allow for expansion of contents) and dialyzing against 3 liters of H_2O. The material that passes through the dialysis tubing contains no ligands for discoidin I or II and, after autoclaving, will support axenic growth of strain Ax-3.

Cells grown in this way and then allowed to differentiate on filters contain no detectable ligands for the carbohydrate-binding site of either discoidin I or discoidin II until the tight-aggregate stage (Cooper et al., 1983). With spore maturation, ligand activity becomes prominent. It can be solubilized either by sonication, pronase digestion, or addition of 1% Emulphogen, and blocks [^{125}I]lactosyl-BSA binding to either discoidin I or discoidin II (Cooper et al., 1983). The properties of the ligand found late in development suggest that it is the spore polysaccharide; and it may be partially purified from fruiting bodies by a procedure based on that used by White and Sussman (1963) to isolate spore polysaccharide.

B. Extraction and Initial Purification of Polysaccharide

To obtain an extract for purification, we use fruiting bodies that developed from Ax-3 cells that had been raised in macromolecule-depleted medium. They are differentiated on filters, harvested in cold water, extensively sonicated, boiled for 10 minutes, and then centrifuged at 10,000 g for 30 minutes at 4°C. The pellet is reextracted by sonication and boiling in distilled water and recentrifuged. The supernatants are pooled, adjusted to 35% ethanol, and the precipitate that forms overnight at 4°C is centrifuged at 10,000 g for 30 minutes. It contains ~25% of the glycoconjugate in the initial extract and essentially no discoidin I- or discoidin II-binding activity (Cooper et al., 1983).

The supernatant is adjusted to 80% ethanol, and the precipitate formed at 4°C overnight is collected by centrifugation, dissolved in distilled water, and adjusted to 10% trichloroacetic acid. After storage at 4°C overnight, the precipitate is centrifuged and the supernatant is dialyzed extensively against 75 mM NaCl, 75 mM Na/KPO$_4$, pH 7.2. The dialyzed material is again precipitated overnight with 80% ethanol, collected by centrifugation, dialyzed extensively against distilled water, and lyophilized. At this stage of purification, specific activity [i.e., inhibition of [^{125}I]lactosyl-BSA binding to immobilized discoidin I or discoidin II per microgram glycoconjugate, the latter estimated by the anthrone method (Morris, 1948)] is increased about sixfold compared to the starting material, with complete recovery of activity (Cooper et al., 1983).

C. Final Purification by Precipitation with Discoidin

For final purification, the lyophilized material is dissolved in distilled water at 2 mg of anthrone-reactive material per milliliter and, after centrifugation at 100,000 g for 60 minutes, to remove any undissolved material, 0.4 ml of the supernatant is added to 2.5 mg of discoidin (a mixture of ~90% discoidin I and 10% discoidin II) in 10 ml TBS-Ca containing 25% ethylene glycol and 0.1 M galactose. This mixture is then dialyzed extensively against the same buffer without galactose. The precipitate that forms in the dialysis tube, which is a discoidin–polysaccharide complex, is collected by centrifugation at 100,000 g for 60 minutes. To separate the glycoconjugate from the lectin, the precipitate is suspended in 8 M urea, boiled for 15 minutes, dialyzed extensively against distilled water, and adjusted to 10% trichloroacetic acid. With this treatment the glycoconjugate is solubilized and the discoidin is denatured and precipitated. After storage overnight at 4°C the precipitate is removed by centrifugation at 100,000 g for 60 minutes and the supernatant, which contains the glycoconjugate, is extensively dialyzed against distilled water and lyophilized.

The final material (Cooper *et al.*, 1983) still contains all the discoidin-binding activity and ~5% of the carbohydrate of the starting extract, indicating that the purified polysaccharide represents ~5% of the total glycoconjugate of the fruiting bodies. The purified material contains 77 mol of galactose and 15 mol of *N*-acetylgalactosamine per 100 mol of carbohydrate and ~3% amino acids by weight. It is a more potent inhibitor of discoidin II than of discoidin I (Cooper *et al.*, 1983). Based on this and on the fact that it apparently colocalizes with discoidin II in intracellular vesicles and in the spore coat (Cooper and Barondes, 1984), it seems very likely that it is the endogenous glycoconjugate ligand for discoidin II. It remains possible, however, that it is heterogeneous and contains some ligand with preferential affinity for discoidin I.

VI.   Methods for Studying the Role of Discoidin I in Cell–Substratum Adhesion and Migration into Aggregates

A number of lines of evidence indicate that discoidin I is involved in cell–substratum adhesion and ordered cell migration during aggregation. This conclusion has come from a series of studies using synthetic peptides derived from the primary structure of discoidin I (Springer *et al.*, 1984), antibodies directed against either the lectin (Springer *et al.*, 1984; Cano and Pestaña, 1984) or a receptor for its cell-binding site (Gabius *et al.*, 1985), and mutants markedly deficient in discoidin I (Springer *et al.*, 1984; Alexander *et al.*, 1983b).

For these studies, it has been necessary to use methods for measuring adhesion of *D. discoideum* to a substratum and for observing the aggregation process. We describe here some methods which we have found useful for this work. It should be emphasized, however, that these laboratory techniques, although they offer certain experimental advantages, are not typical of environments that *D. discoideum* encounters in nature. For example, tissue culture plastic may be fundamentally different from many substrata which slime molds normally encounter. Nevertheless, the effects of specific reagents on the assays described below increase our confidence that what is being measured is relevant to cellular behavior in more natural environments.

A.   Adhesion of Cells to Tissue Culture Plastic

Adhesion to tissue culture plates can be measured (Springer *et al.*, 1984) by allowing the cells to attach and then detaching relatively poorly adhering cells by centrifugation of inverted plates. This method is based on a procedure developed by McClay *et al.* (1981) for other purposes. With this procedure, differences in binding between vegetative cells (which have no discoidin) and aggregating cells (with abundant discoidin I) can be observed. Furthermore, the binding of aggregating cells to tissue culture plastic can also be disrupted by specific peptides whose sequence is based on that of a component of the cell-binding site of discoidin I (Springer *et al.*, 1984).

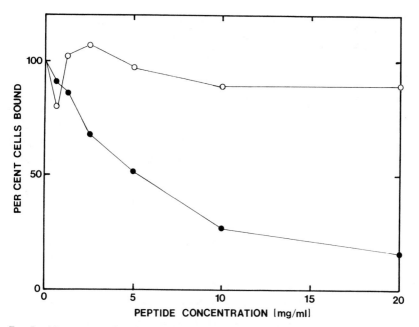

Fig. 5.   Measurement of binding of vegetative or aggregating *Dictyostelium discoideum* to tissue culture plastic as a function of varying concentrations of a synthetic peptide. Aggregating (●) or vegetative (○) cells were mixed with the indicated concentration of a synthetic decapeptide whose structure was based on the primary amino acid sequence of discoidin I. Binding was quantitatively measured 30 minutes later, as described in the text. For details see Springer *et al.* (1984).

In the assay, 5 µl of vegetative or aggregating cells at a concentration of 2×10^7/ml is added to 50 µl of PDF (20 mM KCl, 2 mM MgSO$_4$, 300 µM strep-tomycin sulfate, 40 mM potassium phosphate, pH 6.5) in a well of a 96-well tissue culture plate and allowed to settle and attach at room temperature for 30 minutes. The wells are then completely filled with PDF without disturbing the attached cells, sealed with an adhesive plastic sheet, inverted, and centrifuged on swinging platforms for 5 minutes at 300 g in a Sorvall GLC-1 centrifuge at room temperature. After draining unbound cells, 50 µl of undiluted Bio-Rad Protein Assay Dye Reagent Concentrate (Bio-Rad, Richmond, CA) is added to each well followed by 0.2 ml water. The number of cells remaining in the well is estimated by determining the amount of protein present in the well. This colorimetric measurement is calibrated using known concentrations of cells.

Virtually all vegetative or aggregating cells remain attached to the plastic substratum under these conditions (Fig. 5). However, the mechanism of adhe-sion at these two stages appears to be different, since a synthetic peptide, whose sequence is based on that of a component of the cell-binding site of discoidin I, inhibits adhesion of aggregating but not vegetative cells (Fig. 5).

B.   Observations of Cell Migration into Aggregates

Dictyostelium discoideum cells normally differentiate at an air–water inter-face. The process of aggregation can be readily observed either with the naked eye or with a dissecting microscope; but it is difficult to observe individual cellular behavior during aggregation by looking down from above with a high-power microscope, because the cells requirements for humidity are incompatible with standard microscopic techniques. However, we and others (Brodie *et al.*, 1983) have found that the process of aggregate formation can be studied by keeping the cells in submerged culture in small tissue culture wells, and using an inverted microscope.

For this purpose, we plate cells in a 96-well tissue culture plate under exactly the same conditions described above for the adhesion assay (Springer *et al.*, 1984). The plate is mounted on the stage of an inverted microscope. After varying periods of time, each well is observed for formation of aggregates. We usually find it most convenient to begin the study by using cells that have already begun to aggregate.

After about 12–16 hours in the wells, very prominent streaming into aggre-gates is observed (Fig. 6A). This can be completely blocked by a synthetic peptide whose sequence is based on a part of discoidin I implicated in cell binding (Springer *et al.*, 1984), or by univalent antibodies specific for discoidin I (Fig. 6B) or for the receptor for the cell-binding site of this lectin (Gabius *et al.*, 1985). Many wells can be rapidly examined, so that concentration curves for various reagents are easy to perform. The time of examination is critical and

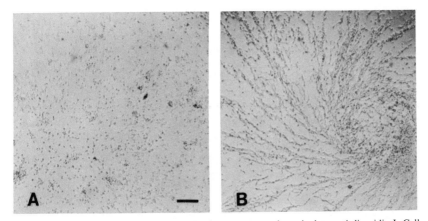

Fɪɢ. 6. Inhibition of ordered cell migration into aggregates by univalent anti-discoidin I. Cells were incubated in the wells of tissue culture plates with 2 mg/ml univalent antibody fragments of anti-discoidin I (A) or an identical concentration of nonimmune univalent antibody fragments (B). The wells were examined 20 hours later. Bar = 100 μm. The cells in the nonimmune antibody were streaming into aggregates and were indistinguishable from other samples in which no antibody had been added. No ordered streams were found in anti-discoidin I. For details see Springer *et al.* (1984).

variable. With a given population of cells, streams may appear within ~8 hours or not until ~20 hours. Over longer periods of time large aggregates are formed, but differentiation beyond this point is not possible under submerged conditions.

C. Time-Lapse Video Microscopy

The assays described above permit observations of aggregation at various times, but the movement of individual cells and aggregates is impossible to discern. Time-lapse video microscopy can be used for this purpose. Behavior of cells used in the above assays can be recorded using an inverted microscope, a video camera, and a time-lapse recorder such as a Panasonic NV40 reel recorder set to record at 18–76 times slower than normal. Using various objectives, it is possible to follow individual cells or whole aggregates over periods as long as 20–30 hours. When the tape is played back at normal speed, cell movement is easily observed and abnormalities in this behavior can be studied. Using such a system we have been able to ascertain that mutants very deficient in discoidin I (Alexander *et al.*, 1983b) are capable of aggregating in submerged culture, but that they do not show normal streaming into aggregates (Springer *et al.*, 1984). Instead they use an alternative, and less effective, mechanism by which small aggregates move around accumulating additional cells (Springer *et al.*, 1984).

VII.   Histochemical Localization of Lectins and Glycoconjugate Ligands

Much of our knowledge about the functions of discoidin I and II comes from histochemical studies of the cellular distribution of the lectins and the glycoconjugate ligands that bind them. A general discussion of such techniques is not appropriate here. However, a few points of special interest will be mentioned, especially because they are also applicable to other work with *D. discoideum*.

A.   Fixation and Sectioning

A major problem in doing immunohistochemical studies is to preserve morphology without destroying the antigenicity of the proteins to be localized and without making sections impermeable to the antibodies used for the localization studies. With *D. discoideum*, preservation of individual cellular morphology is possible with a wide variety of fixatives, but permeation of aggregates and their preservation in intact form is difficult. With standard fixatives containing paraformaldehyde and glutaraldehyde the cells at the periphery of an aggregate may be well preserved, but those deeper within it are not; and aggregates tend to fall apart with handling. As observed by Gregg (1965), Carnoy's fixative (glacial acetic acid, chloroform, absolute ethyl alcohol, 1:3:6 by volume) produces adequate fixation of complete aggregates. Fixed aggregates can be frozen, and reasonably thin sections (1 μm) can be prepared.

In a typical experiment (Barondes *et al.*, 1983), aggregates are allowed to form by differentiation on a layer of 2% Noble agar (Difco Laboratories) using glass-distilled water in a 100 $\times$ 200 mM Petri dish and, at the desired stage of differentiation, selected areas of agar containing aggregates are cut out and floated in the fixative in culture dishes. After fixation at room temperature for 48–96 hours, the fixed aggregates are scraped from the agar, allowed to settle in 1.5-ml microfuge tubes containing 30% sucrose in PBS for 1 hour at room temperature and the supernatant is aspirated. One drop of O.C.T. compound (Miles Laboratories, Napeville, IL) is added to the tube and mixed, and the pellet is then quick-frozen and mounted on a microtome chuck using O.C.T. compound. Frozen sections of 1-2 μm are cut at $-35°C$ in an H/I Bright Cryostat (Hacker Instruments, Inc., Fairfield, NJ) fitted for sectioning with glass knives with an LKB "Ralph" knife adapter (LKB Instruments, Bromma, Sweden). Sections are picked up on slides that had been cleaned, then dipped while warm in a mixture of 0.5% gelatin and 0.05% chromic potassium sulfate heated to 50°C. Sections on slides may be stored at $-20°C$ before staining.

B. Fluorescence Microscopy

Sections prepared in this way can be stained with antiserum against either discoidin I or discoidin II, and bound antibody can be localized with fluorescent second antibody. A double-label technique can also be used. For example, labeling the section of a late aggregate with a mixture of rabbit anti-discoidin I and mouse anti-discoidin II, followed by washing and reaction with rhodamine-conjugated goat anti-rabbit IgG and fluorescein-conjugated goat anti-mouse IgG permits visualization of both lectins in the same section. In this instance the discoidin I is exclusively extracellular around the aggregate (Fig. 7A), whereas the discoidin II remains intracellular in a punctate distribution, probably representing prespore vesicles (Fig. 7B). This fixation and sectioning technique clearly preserves the integrity of aggregates and sufficient morphology and antigenicity to allow such localization studies. However, preservation is not adequate for ultrastructural studies.

For localization of endogenous ligands for these lectins, we have used fixation with a mixture of 3% paraformaldehyde, 1% glutaraldehyde, in PBS for 30 minutes at room temperature. Such prolonged exposure to a relatively high glutaraldehyde concentration markedly reduces the immunological reactivity of both discoidin I and discoidin II. Whereas this precludes localization of endoge-

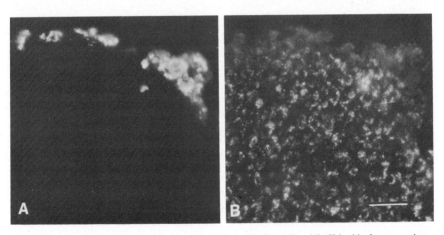

FIG. 7.   Immunohistochemical localization of discoidin I and discoidin II in thin frozen sections of a maturing aggregate at the Mexican hat stage. The section was reacted with a mixture of rabbit anti-discoidin I and mouse anti-discoidin II followed by rhodamine-conjugated goat anti-rabbit IgG and fluorescein-conjugated goat anti-mouse IgG. Fluorescence attributed to fluorescein (A) indicates the localization of discoidin I around the aggregate; fluorescence attributed to rhodamine (B) shows the punctate intracellular distribution of discoidin II, which is apparently concentrated in prespore vesicles. $\times$ 1250; bar = 8 μm. For details see Barondes *et al.* (1983).

nous lectins by fluorescence immunohistochemistry, it is actually advantageous for studying the localization of endogenous ligands for these lectins, by an indirect procedure (without direct derivatization of the lectin which might impair its carbohydrate-binding activity). Since immunoreactivity of the endogenous lectins is destroyed, whereas the structure of glycoconjugates is unaffected by this fixation, addition of exogenous lectin to the sections followed by specific antiserum and fluorescent second antibody can be used to localize glycoconjugates that bind the lectins. In this way we found that the endogenous ligand for

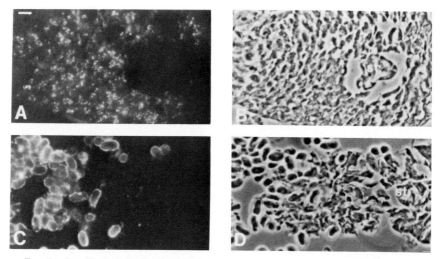

Fig. 8.   Localization of glycoconjugate-binding sites for discoidin II late in differentiation. Frozen sections of a maturing aggregate at the standing slug stage (A, B) and of a partially differentiated fruiting body (C, D) were prepared after glutaraldehyde fixation to block immunological reactivity of endogenous discoidin II in the tissue. The sections were incubated with 50 μg/ml of discoidin II, then with affinity-purified cross-adsorbed rabbit antibody to discoidin II followed by rhodamine-conjugated goat anti-rabbit IgG. They were visualized by fluorescence (A, C) or phase-contrast (B, D) microscopy. At the standing slug stage, the discoidin II ligand is found in a punctate intracellular distribution like that of prespore vesicles. In mature spores it is found in the spore coat. The stalk (st) does not stain. If these sections are stained only with anti-discoidin II without first adding exogenous discoidin II, no immunofluorescence is found because the immunoreactivity of the endogenous lectin has been obliterated by the fixation procedure. Likewise, addition of 0.3 *M* galactose along with the exogenous discoidin II blocks binding (not shown). × 250; bar = 2 μm. For details see Cooper and Barondes (1984).

Fig. 9.   Immunohistochemical localization of discoidin I in differentiating *Dictyostelium discoideum* cells. Cells were differentiated, fixed, embedded in Epon, then stained with affinity-purified rabbit anti-discoidin I and goat anti-rabbit IgG complexed with 4-nm colloidal gold. The colloidal gold, which indicates the localization of endogenous discoidin I, is concentrated in multilamellar bodies (mb) and vesicles (v). × 105,000; bar = 0.1 μm.

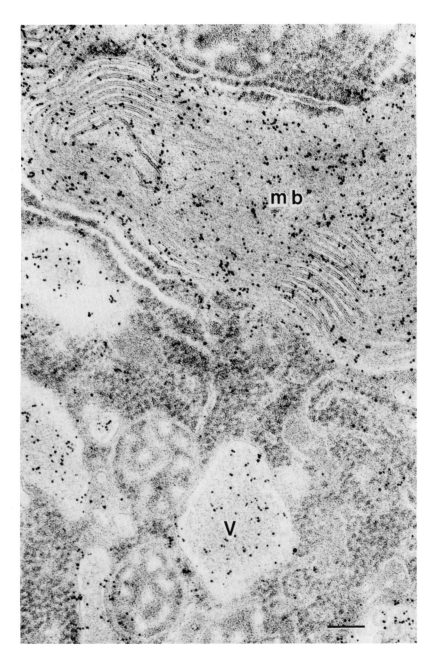

discoidin II is located in a punctate distribution within prespore cells, probably representing prespore vesicles (Fig. 8A). Later in development, the glycoconjugate that binds discoidin II is localized in the spore coat (Fig. 8C).

C. Electron Microscopy

The ultrastructural localization of discoidin I has been determined by electron microscopy after staining of thin Epon sections with either colloidal gold-conjugated antibodies or colloidal gold-conjugated discoidin I (Barondes *et al.*, 1985). The lectin in the sections apparently maintains sufficient reactivity with the antibodies for its location to be determined. These studies showed high concentrations of discoidin I in multilamellar bodies within differentiating cells (Fig. 9). The lectin is externalized for its extracellular function within these organelles (Barondes *et al.*, 1985).

To localize glycoconjugate ligands for discoidin I, complexes of discoidin I–gold can be used. These are prepared using highly purified discoidin I, free of discoidin II, by a modification of the procedure of Roth (1984). To remove the salts in which it is normally stored, which interfere with conjugation, 2 ml of discoidin I (25 μg/ml) is dialyzed overnight against 1 liter of water that contains 50 mM galactose and 1% ethylene glycol. The latter ingredients preserve lectin activity and keep it from forming aggregates. The pH of 9-nm colloidal gold sol (obtained from Janssen Pharmaceutical Inc., Piscataway, NJ) is adjusted to 7.0 with 2% K_2CO_3, and 1 ml is added dropwise with stirring to the dialyzed lectin solution. After 5 minutes, 1.0 ml of 1% aqueous polyethylene glycol (MW 20,000) is added, and the suspension is diluted to 9 ml with 50 mM Tris-HCl (pH 7.6), 150 mM NaCl that contains 1% BSA (RIA grade; Sigma Chemical Co., St. Louis, MO), and 1 mM $CaCl_2$. The discoidin I–gold complex is then sedimented by centrifugation at 60,000 g for 20 minutes. The supernatant is carefully removed, and the sedimented lectin–gold complex is collected and resuspended in 1.0 ml 50 mM Tris-HCl (pH 7.6) containing 1% BSA and 1 mM $CaCl_2$, then used for staining.

When discoidin I–gold complexes are added to Epon sections of differentiating *D. discoideum*, glycoconjugate ligands for the lectin are found in multilamellar bodies (Fig. 10). Binding of the discoidin I–gold complex at these sites can be completely blocked with *N*-acetylgalactosamine but not *N*-acetylglucosamine, confirming the specificity of the reaction (Barondes *et al.*, 1985). It is notable that cells that had been grown on bacteria depleted of glycoconjugate ligands for discoidin I contain no discoidin I-binding glycoconjugate in multilamellar bodies, and discoidin I is found in the cytoplasm rather than multilamellar bodies (Cooper *et al.*, 1986).

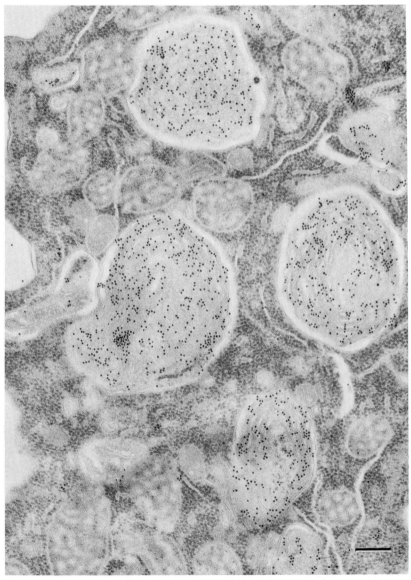

FIG. 10.   Discoidin I–colloidal gold binding to a section of differentiating *Dictyostelium dis-coideum*. A discoidin I–colloidal gold complex was applied to thin sections of differentiating *D. discoideum* cells that had been fixed, then embedded in Epon. The discoidin I–colloidal gold complex binds specifically to glycoconjugates in multilamellar bodies. Addition of *N*-acetylgalactosamine completely blocks this binding (not shown). × 45,000; bar = 0.5 μm. For details see Barondes *et al.* (1985).

VIII.   Concluding Remarks

The methods we describe have been very useful in identifying the functions of discoidins I and II. Like soluble lectins from vertebrates (Barondes, 1984), they are apparently designed to play extracellular roles—discoidin I in cell–substratum adhesion and discoidin II in spore coat formation.

A number of lines of evidence point to a role for discoidin I in a specific form of cell–substratum adhesion. Antibodies against the lectin (Springer *et al.*, 1984; Cano and Pestaña, 1984) and against a 67-kDa glycoprotein with properties of a discoidin I receptor (Gabius *et al.*, 1985) both block ordered cell migration into aggregates. In addition, a mutant deficient in discoidin I (Alexander *et al.*, 1983b) does not aggregate normally (Springer *et al.*, 1984). Nor does an antisense transformant in which there was a >90% reduction in discoidin I (Crowley *et al.*, 1985).

One interesting conclusion from studies of the cellular role of discoidin I is that this lectin shares several properties with fibronectin, a vertebrate molecule implicated in the adhesion of fibroblasts to substrata. Like Fibronectin, discoidin I contains the sequence Arg-Gly-Asp-x, which has been implicated in the cell-binding properties of both molecules (Pierschbacher and Ruoslahti, 1984; Springer *et al.*, 1984). Both molecules also have carbohydrate-binding sites. Firbronectin's binding sites for glycosaminoglycans may participate in anchoring it to the substratum. Presumably the carbohydrate-binding site of discoidin I plays a similar role, by association with extracellular glycoconjugates; but this is only a speculation, since the only known role for the carbohydrate-binding site of this lectin is in packaging it into multilamellar bodies by association with bacterial polysaccharides (Cooper *et al.*, 1985).

Little is presently known about the regulation of discoidin I synthesis. There is evidence that cyclic AMP (Williams *et al.*, 1980) and cell contact (Berger and Clark, 1983) play a part. The significance of the distinct subunits of discoidin I encoded by different genes (Tsang *et al.*, 1981; Poole *et al.*, 1982) is also unknown.

A role for discoidin II in spore coat formation is indicated by its localization in vesicles which also contain a polysaccharide which binds well to this lectin, and by the secretion of lectin and the polysaccharide into the spore coat (Barondes *et al.*, 1983; Cooper and Barondes, 1984). However, the lectin is only associated transiently with the spore coat and does not behave like a permanent structural component. It may be needed to help organize the polysaccharides, which then come to associate by polysaccharide–polysaccharide interactions; but the mechanism of spore coat formation requires further examination.

Although discoidins I and II are among the most intensively studied of all the lectins, much remains to be learned. Combining the findings and approaches described here with recombinant DNA techniques described elsewhere in this volume, and already successfully employed (Crowley *et al.*, 1985), should lead to a much deeper understanding in the near future.

ACKNOWLEDGMENTS

The work described here was supported by grants from the NIH, the NSF, and the McKnight Foundation, and by the Veterans Administration Medical Center.

REFERENCES

Alexander, S., Cibulsky, A. M., and Lerner, R. A. (1983a). *Differentiation* **24**, 209–212.
Alexander, S., Shinnick, T. M., and Lerner, R. A. (1983b). *Cell* **34**, 467–475.
Barondes, S. H. (1984). *Science* **223**, 1259–1264.
Barondes, S. H., Cooper, D. N., and Haywood-Reid, P. L. (1983). *J. Cell Biol.* **96**, 291–296.
Barondes, S. H., Haywood-Reid, P. L., and Cooper, D. N. W. (1985). *J. Cell Biol.* **100**, 1825–1833.
Bartles, J. R., Santoro, B. C., and Frazier, W. A. (1981). *Biochim. Biophys. Acta* **674**, 372–382.
Berger, E. A., and Armant, D. R. (1982). *Proc. Natl. Acad. Sci. U.S.A.* **79**, 2162–2166.
Berger, E. A., and Clark, J. M. (1983). *Proc. Natl. Acad. Sci. U.S.A.* **80**, 4983–4987.
Brodie, C., Klein, C., and Swierkosz, J. (1983). *Cell* **32**, 1115–1123.
Cano, A., and Pestaña, A. (1984). *J. Cell. Biochem.* **25**, 31–43.
Cooper, D. N. W., and Barondes, S. H. (1984). *Dev. Biol.* **105**, 59–70.
Cooper, D. N., Lee, S.-C., and Barondes, S. H. (1983). *J. Biol. Chem.* **258**, 4698–4701.
Cooper, D. N., Haywood-Reid, P. L., Springer, W. R., and Barondes, S. H. (1986). *Dev. Biol.* **114**, 416–425.
Crowley, T. E., Nellen, W., Gomer, W. H., and Firtel, R. A. (1985). *Cell* **43**, 633–641.
Frazier, W. A., Rosen, S. D., Reitherman, R. W., and Barondes, S. H. (1975). *J. Biol. Chem.* **250**, 7714–7721.
Gabius, H. J., Springer, W. R., and Barondes, S. H. (1985). *Cell* **42**, 449–456.
Gregg, J. (1965). *Dev. Biol.* **12**, 377–393.
Heggeness, M. H., and Ash, J. F. (1977). *J. Cell Biol.* **73**, 783–788.
McClay, D. R., Wessel, G. M., and Marchage, R. B. (1981). *Proc. Natl. Acad. Sci. U.S.A.* **78**, 4975–4979.
McConahey, P. J., and Dixon, F. J. (1966). *Int. Arch. Allergy* **29**, 185–189.
Madley, I. C., and Hames, B. D. (1981). *Biochem. J.* **200**, 83–91.
Morris, D. L. (1948). *Science* **107**, 254–255.
Pierschbacher, M. D., and Ruoslahti, E. (1984). *Nature (London)* **309**, 30–33.
Poole, S., Firtel, R. A., Lamar, E., and Rowekamp, W. (1982). *J. Mol. Biol.* **153**, 273–289.
Rosen, S. D., Kafka, J. A., Simpson, D. L., and Barondes, S. H. (1973). *Proc. Natl. Acad. Sci. U.S.A.* **70**, 2554–2557.
Roth, J. (1984). *J. Cell Biol.* **98**, 399–406.
Simpson, D. L., Rosen, S. D., and Barondes, S. H. (1974). *Biochemistry* **13**, 3287–3493.
Smith, D. F., and Ginsburg, V. (1980). *J. Biol. Chem.* **255**, 55–59.
Springer, W. R., Cooper, D. N. W., and Barondes, S. H. (1984). *Cell* **39**, 557–564.
Tsang, A., Devine, J. M., and Williams, J. G. (1981). *Dev. Biol.* **84**, 212–217.
White, G. J., and Sussman, M. (1963). *Biochim. Biophys. Acta* **74**, 179–187.
Wilchek, M. (1980). *J. Solid-Phase Biochem.* **3**, 193–195.
Williams, J., Tsang, A., and Mahbubani, H. (1980). *Proc. Natl. Acad. Sci. U.S.A.* **77**, 7171–7175.

Part VI. Development and Pattern Formation

Once *Dictyostelium* cells have formed aggregates, they differentiate into a fruiting body consisting of stalk cells and spore cells. This developmental process has been the subject of a great deal of study by many investigators and represents the most complicated cell biological problem to try to understand at the molecular level. In this last part, the authors describe some of the approaches being used to provide insights into this area of biology using *D. discoideum*. The first chapter of this section, by Soll, begins by detailing methods for obtaining reproducible timing of the *D. discoideum* developmental program. Soll also describes methods for dissecting and characterizing the program of temporal regulation and for obtaining timer mutants. Additional information on cell differentiation in monolayers is given in the next chapter by Kay, who also describes approaches to assay for and to purify morphogens involved in the morphogenetic changes that occur. More on developmental mutants is also described by Kay. In order to study pattern formation at the molecular level, methods must be available to distinguish readily between the two different cell types in the developing aggregate. Such methods are described by Weijer *et al.* in the third chapter of this section. Vital staining procedures are described, including double-label staining in individual aggregates. Assays for studying cell sorting are presented using both fluorescently labeled cells and genetic markers.

In the final three chapters of the book, considerations using molecular genetic approaches to study developmental genes are discussed. In Chapter 25, Chisholm discusses various screening techniques that can be used to isolate developmentally regulated genes. He also comments on the choice of cDNA or genomic clones and on construction of libraries with special regard for *Dictyostelium*. In Chapter 26, Gomer describes a cloning strategy to study development and pattern formation which uses antibodies against the products of cloned developmental genes. Detailed information, with special reference to peculiarities in working with *Dictyostelium,* is given on immunological techniques, including immunostaining procedures, and on a microassay for cell type-specific gene expression. In the final chapter, De Lozanne describes the use of homologous recombination to shut down the expression of native myosin and to express instead the first half of the myosin molecule. This directed gene disruption results in a complete block in development.

Chapter 22

Methods for Manipulating and Investigating Developmental Timing in Dictyostelium discoideum

DAVID R. SOLL

Department of Biology
The University of Iowa
Iowa City, Iowa 52242

I. Introduction

A major task of developmental biologists is to elucidate the mechanisms which dictate when things happen in a developmental program. It has been suggested that particular rate-limiting processes, or "developmental timers," have evolved as cues to coordinate the complex interactions leading to the genesis of each developmental stage (Soll, 1979, 1983), and the isolation of putative "heterochrony" mutants in the nematode suggests that alterations in timing have the dramatic effect of actually changing the developmental program, in the case of nematodes by changing cell lineages (Ambros and Horvitz, 1984). It is therefore surprising that so little attention has been paid to the question of developmental timing. One major reason for this may be the tacit assumption that because developmental stages occur in a temporal sequence, the developmental timers which dictate when they occur must also be in sequence. Alternatively, clock mechanisms have been suggested for simpler programs like the cell cycle (Lloyd and Edwards, 1982; Klevecz, 1976), in which a single rate-limiting process has been hypothesized to cue sequential events. Unfortunately, few methods have been available in the past to test unambiguously these assumptions or hypotheses, and few systems have availed themselves for such studies.

As for many other unsolved problems in cellular, molecular, and developmental biology, *Dictyostelium discoideum* again provides an excellent system, per-

METHODS IN CELL BIOLOGY, VOL. 28

haps in this case an extraordinary system, for investigating timing regulation (Soll, 1979, 1983; Varnum et al., 1983). First, because the developmental program includes an ordered, easily monitored, and highly reproducible sequence of morphogenetic stages (Soll, 1979), it lends itself to detailed investigations of the number, complexity, and dependencies of those processes which are rate-limiting for developmental stages (Soll, 1979; Varnum et al., 1983). These processes have been referred to as "developmental timers" (Soll, 1979, 1983). Second, because of the cellularity of Dictyostelium morphogenesis, developing cultures can be disaggregated and challenged to reinitiate development without an intermittent period of growth. Disaggregated cultures will rapidly recapitulate the stages they had progressed through during initial development, but in roughly one-tenth the original time (Loomis and Sussman, 1966; Newell et al., 1971; Soll and Waddell, 1975). This characteristic allows the investigator to monitor temporal progress through the developmental program under a variety of conditions (e.g., see Finney et al., 1985a). Third, developmental progress can be "erased" by disaggregating and feeding developing cultures (Soll and Waddell, 1975; Waddell and Soll, 1977; Finney et al., 1979, 1981). "Erasure" occurs in a synchronous, rapid step resulting in the complete transition from rapid recapitulation timing to the slow timing of naive log-phase cells. This erasure event sets in motion a program of dedifferentiation during which cells lose developmentally acquired characteristics during a period in excess of 400 minutes (Finney et al., 1979; Varnum and Soll, 1981; Soll et al., 1984). Finally, cells can be tricked into progressing simultaneously through the opposing programs of differentiation and dedifferentiation (Soll and Mitchell, 1982; Finney et al., 1983).

In this chapter, methods will be described for obtaining reproducible developmental timing, dissecting and characterizing the program of temporal regulation, obtaining timer mutants, initiating rapid recapitulation, initiating erasure and the subsequent program of dedifferentiation, and stimulating simultaneous programs of dedifferentiation and redifferentiation in the same cells. Since several aspects of these methods are based upon concepts which are not normally dealt with by molecular or cellular biologists, I will touch upon them only to the extent that it allows the methods to be readily employed, and I will refer the interested reader to more extensive treatments of these subjects in the literature.

II. A Standardized Set of Conditions for Reproducible Developmental Timing

Virtually every environmental condition as well as the growth history of the cells affect progress through the developmental program and therefore developmental timing. For this reason, it is necessary to follow the same regime for each

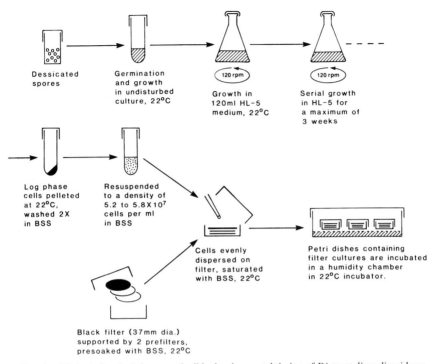

FIG. 1. Methods for obtaining reproducible developmental timing of *Dictyostelium discoideum*. BSS, Buffered salts solution (recipe in text).

experiment in order to obtain reproducible timing. The following method (diagrammed in Fig. 1) results in reproducible timing as well as synchronous morphogenesis, and bypasses the migratory phase of the pseudoplasmodium. This method was adapted from one developed by Sussman and co-workers (Sussman, 1966), and employs (1) an axenic strain of *D. discoideum*, Ax-3, clone RC-3, which is readily obtainable from our laboratory, (2) the axenic medium HL5 (Cocucci and Sussman, 1970), (3) amoebae from the mid-log phase of growth (Yarger *et al.*, 1974), (4) development pads saturated with a buffered salts solution (BSS) of high ionic strength (Newell *et al.*, 1969), and (5) black paper filters for visualizing the stages of morphogenesis. In this method, log-phase amoebae are inoculated into a 1-liter Erlenmeyer flask containing 120 ml of the nutrient medium HL5. The initial cell density is $\sim2 \times 10^5$. Flasks are rotated at 100 rpm at 22°C (±0.5°C). When cells reach a mid- to late log-phase density of 2×10^6/ml (Yarger *et al.*, 1974), they are harvested for development. The generation time in axenic medium will vary from 8 to 12 hours, depending on the origin and batch of the protease peptone and yeast extract employed. Generation times in excess of 13 hours sometimes lead to irreproducible developmental timing.

It should be noted that cells begin to enter stationary phase at a concentration of 4–6×10^6/ml, stabilizing at $\sim 2 \times 10^7$/ml (Yarger *et al.*, 1974). Since stationary-phase cells aggregate roughly 2.5 hours faster than log-phase cells, and cells entering stationary phase exhibit intermediate timing (Ferguson and Soll, 1977), it is imperative that cells are harvested prior to a cutoff density of 3×10^6 cells/ml.

To initiate development, cells are pelleted at room temperature, resuspended in BSS (20 mM KCl, 0.24 mM MgCl$_2$, 40 mM Phosphate buffer, pH 6.4, and 0.34 mM streptomycin sulfate; previously referred to as LPS. Newell *et al.*, 1969), and washed twice in BSS. The final pellet is resuspended in BSS (22°C) to a final concentration of 5.2–5.8×10^7 cells/ml. Then, 1 ml of the cell slurry is gently dispersed on the surface of a black filter pad (Eaton-Dikenson no. 4740 C20), 37 mm in diameter, which is supported by two Millipore prefilters (AP1003700 MF) in a 60-mm plastic Petri dish. Prior to cell addition, the black filter and supporting pads are saturated with BSS. Any excess BSS which is emitted as a ring around the pads should be aspirated off. The final cell density is $\sim 5 \times 10^6$ cells/cm^2, and results in a carpet several cells deep. The top of the Petri dish is applied and the Petri dish placed in a humidity chamber (a simple plastic box with a support screen and shallow water reservoir), which in turn is placed in an incubator set at 22°C, with overhead incandescent lighting. Black filters and supporting prefilters with far smaller or far larger diameters can be employed depending on experimental demands. Proportionate changes in cell inoculum and saturating BSS are easily calculated. In cases in which cultures are frequently monitored, parallel plates should be inoculated at the same time. Cultures are monitored at time intervals under a dissection microscope, no single plate being scored more than once in a 1-hour period in order to safeguard against dehydration and heat effects caused by removal from the humidity chamber and the microscope light source, respectively.

For reproducibility, synchrony, and complete development, amoebae should never be mass-cultured in HL5 for more than a 3-week period. Fresh stocks should be clonally generated from desiccated stocks of spores and tested before use. This procedure ensures a high level of genotypic and, therefore, phenotypic homogeneity.

This general procedure results in a preaggregative period of ~ 7 hours. The onset of aggregation under these conditions is reflected by a rippling of the thick cell carpet. Because of the high ionic concentration of the BSS employed, no period of slug migration occurs and formation of fruiting bodies is complete after ~ 26 hours (Soll, 1979). A diagram of the developmental stages and timing is presented in Fig. 2.

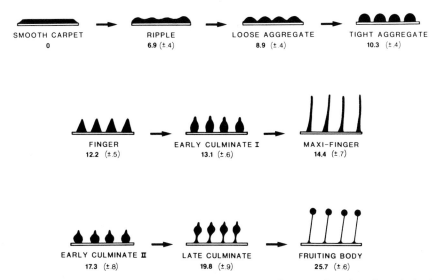

SMOOTH CARPET   RIPPLE   LOOSE AGGREGATE   TIGHT AGGREGATE
0   6.9 (± .4)   8.9 (± .4)   10.3 (± .4)

FINGER   EARLY CULMINATE I   MAXI-FINGER
12.2 (± .5)   13.1 (± .6)   14.4 (± .7)

EARLY CULMINATE II   LATE CULMINATE   FRUITING BODY
17.3 (± .8)   19.8 (± .9)   25.7 (± .6)

FIG. 2.   Diagram of the developmental stages and timing of *Dictyostelium discoideum* obtained when employing the methods for development described in Fig. 1. The mean time (hours) and standard deviation (in parentheses) for six independent experiments are presented below each stage.

A.   The Effects on Timing of Growth History

Two aspects of growth affect the length of the preaggregative period of the developmental program. The first is the nutrient composition of the growth medium. If cells are grown in association with bacteria rather than in axenic medium, harvested in mid-log phase, and development initiated as described, the preaggregative period varies between 8 and 12 hours, 1–5 hours longer than the highly reproducible preaggregative period of axenically grown log-phase cells. This increase appears to be due to a rate-limiting component added to the beginning of the preaggregative period (Varnum *et al.*, 1986; Soll *et al.*, 1986), and probably represents the time it takes for the complete consumption of ingested bacteria.

The second aspect of growth which affects developmental timing is the phase of the growth culture. As we have reiterated, stationary-phase cells aggregate faster than log-phase cells and the 2.5-hour decrease in the preaggregative period is due to a contraction of the first component of the preaggregative period (Varnum *et al.*, 1983; Soll *et al.*, 1986). Timing beginning with the second component and continuing through the subsequent stages of development is identical to log-phase cells (Varnum *et al.*, 1983).

B. The Effects on Timing of the Ionic Composition and Temperature of the Development Medium

Timing as well as the morphogenetic program are dramatically affected by the ionic composition of the nonnutrient medium which saturates the development pads (Newell *et al.*, 1969; Mercer and Soll, 1980). When development filters are saturated with a medium of low ionic strength or with distilled water, the preaggregative period is reduced by 2 hours, and the cells form migratory slugs rather than progressing through fruiting body formation. It has been demonstrated that the reduction in time of the preaggregative period again reflects a selective decrease in the first component of the preaggregative period (Mercer and Soll, 1980).

Temperature has a dramatic and selective effect on developmental timing (Soll, 1979). Within the limited range of 18°–24°C, there is virtually no effect on the extent of the preaggregative period, but the total time for fruiting body formation differs by 7.7 hours (31.5 hours at 18°C and 23.8 hours at 24°C; see Soll, 1979). As we will see, the selective effects of temperature on the stages of morphogenesis provide a method for dissecting timer complexity (Soll, 1979, 1983).

C. The Effects on Timing of "Development" in Suspension

With increasing frequency, researchers have been employing suspension cultures to obtain high yields of starved cells which have been referred to in many cases as "developing cells." There is no doubt that starvation triggers the developmental program (Marin, 1976, 1977), but there is also accumulating evidence that suspension cultures do not progress through the second rate-limiting component of the preaggregative period, and that some aggregation-associated changes (e.g., the appearance of cAMP-binding sites) occur while others (e.g., the appearance of EDTA-resistant cohesion) do not (Finney *et al.*, 1985a,b; Chisholm *et al.*, 1984), or do so belatedly and less synchronously. Of course, the conditions of suspension are very important, especially those dealing with shear force and the stability of transient cell interactions. Treating cell suspensions with pulses of cAMP will drive the system through the entire preaggregative period (Finney *et al.*, 1985a), but it is not clear how selective or complete even this treatment is. Rather than belabor this point in any more detail, it should just be emphasized that researchers should be acutely aware of the lack of normal cell–cell interactions and signals in suspension conditions. This may prove to be critical to researchers investigating the developmental regulation of the cytoskeleton and other elements involved in cell motility in light of the recent observation that the basal level of single-cell motility increases transiently at the onset of aggregation on filters (Varnum *et al.*, 1986). I therefore strongly recommend that when it is at all possible and even if it creates added experimental complexity, real development on filters or even nonnutrient agar should be selected over suspension cultures so that progress through morphogenesis can be accurately assessed.

III. Elucidating Minimum Number and Relationships of Developmental Timers: The "Reciprocal-Shift Experiment"

With reproducible timing, one can begin to dissect the program of timing regulation by the "reciprocal-shift experiment" (Soll, 1983; Varnum *et al.*, 1983). Since this method has been described in detail elsewhere for general application (Soll, 1983), only those features relevant to its application to *Dictyostelium* development will be described in this section.

A "developmental timer" is simply defined as the rate-limiting pathway to a specific developmental stage under a single set of experimental conditions. A developmental timer may be complex (i.e., it may be composed of more than one component), may not be rate-limiting under other sets of environmental conditions, and may regulate more than one developmental stage in a program. A developmental timer probably represents one of many essential pathways involved in the genesis of a particular developmental stage, but it is distinct in its role as a timing cue. There are four basic models for the regulation of timing of sequential developmental stages: the single-timer model, the sequential-timer model, the parallel-timer model, and a branching model which is a composite of sequential and parallel models (Fig. 3).

The reciprocal-shift experiment allows one not only to assess timer complexity, but also to distinguish between timer models (Soll, 1983). Two sets of conditions must first be decided on, which result in short and long timing to one or more developmental stages. In the case of *D. discoideum*, both ionic strength (Mercer and Soll, 1980) and temperature (Varnum *et al.*, 1983) have been employed to obtain short and long conditions, but temperature has proved to be a more pervasive and useful condition. In this case, 18°C provides long timing and 24°C short timing. Shifts are performed at time intervals from high to low temperature (short to long shift) and from low to high temperature (long to short shift) by gently transferring the black filter supporting the cell culture from the underpads at the initial temperature to fresh underpads prewarmed or prechilled to the second temperature. The total time to each stage (time at first temperature plus time at second temperature) is measured and plotted as a function of the timing of the shift, resulting in two plots, one for shifts from high to low temperature and the other for shifts from low to high temperature. Each plot is then analyzed for the number of components, slope of each component, absolute

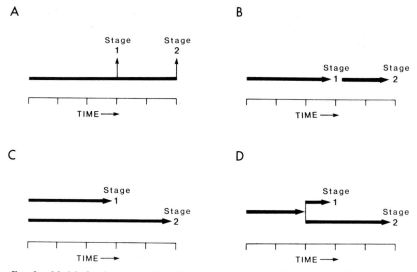

FIG. 3.   Models for the relationships of developmental timers for the sequential stages of *Dictyostelium discoideum* morphogenesis. The horizontal axis of each arrow reflects the progress of each timer. (A) Single; (B) sequential; (C) parallel; (D) branching.

times of origin and termination of each component, and discontinuities between components (Fig. 4). The information obtained from the plots of a reciprocal-shift experiment can be used (1) to distinguish between single- and multiple-component timers, (2) to define accurately the time of transition from one component to another, (3) to test for the addition of a new component under long conditions or elimination of a component under short conditions, (4) to assess the sensitivity of each timer component to the change in temperature, including reversibility, (5) to test for an identity change in a timer component, and (6) to distinguish between timer models for sequential developmental stages (Soll, 1983).

When the reciprocal-shift experiment employing temperature as the variable was applied to the first seven stages of *Dictyostelium* morphogenesis (Varnum *et al.*, 1983), a relatively simple picture of minimum timer complexity emerged (Fig. 5) which included (1) between four and eight individual timer components, (2) parallel timer relationships, (3) sequential timer relationships, (4) a reversible timer component (the first component in the developmental program), (5) discrete transition points between rate-limiting components including a major branch point at the onset of aggregation for a number of parallel timers, and (6) the different temperature sensitivities of the individual timer components.

It should be emphasized that the reciprocal-shift experiment gives us a picture of minimum timer complexity based on sensitivity to one set of environmental

I. Two conditions are developed for short and long timing to a developmental stage (D.S.). In this case, 24°C and 18°C respectively.

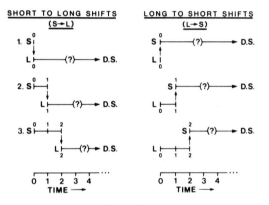

II. Shifts are performed at time intervals from short (24°C) to long (18°C) and long (18°C) to short (24°C) conditions. The total time to the stage (time under first condition plus time under second condition) is then scored.

III. Total developmental time is plotted as a function of time of shift for shifts from short (24°C) to long (18°C) conditions (S→L) and long (18°C) to short (24°C) conditions (L→S).

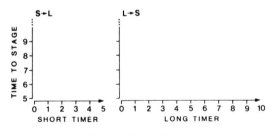

TIME OF SHIFT

IV. Each plot is then analysed for number of components, slopes of components, absolute times of origins and termini of components, and discontinuities between components. Timer interpretations are then made according to the combinations of these characteristics for the two plots

FIG. 4. The "reciprocal-shift experiment" for analyzing timer complexity and timer relationships in *Dictyostelium discoideum*.

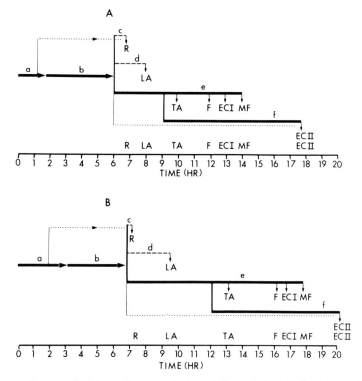

FIG. 5. A model of minimum timer complexity and timer relationships for *Dictyostelium discoideum* which has emerged from the reciprocal-shift experiment. (A) Short condition (24°C); (B) long condition (18°C). The horizontal, solid arrows represent clearly distinguished components. Dashed arrows represent tentative parallel timers. Dotted arrows represent timers which become rate-limiting after shifts from high to low temperature. The details of this model and the data used to generate it are presented in Varnum *et al.* (1983). R, Ripple; LA, loose aggregate; TA, tight aggregate; F, finger; ECI, early-culminate I; MF, maxifinger; ECII, early-culminate II.

conditions. A more detailed picture of complexity should emerge if the reciprocal-shift experiment is repeated, but with different environmental conditions used as variables. Besides temperature and ionic strength, the use of pH, selective inhibitors of metabolism and macromolecular synthesis, and the inhibition of cell cohesion can be employed in reciprocal-shift experiments. Interestingly, at least for the preaggregative period, two major components continue to be delineated by shifts between high and low temperature (Varnum *et al.*, 1983), between high and low ionic strength (Mercer and Soll, 1980), between solutions with and without cycloheximide (Finney *et al.*, 1985a), and between suspension and filter culture (Finney *et al.*, 1985a). Conditional experiments of these types delineate the major components of the regulatory program which sets the stage for molecular analyses of regulation.

IV. A Method for Isolating and Characterizing Timer Mutants

Mutants with aberrant timing have been isolated throughout the years of *Dictyostelium* research since the early 1960s, and many of these mutants (e.g., FR17) have served as very effective tools for understanding normal morphogenesis and differentiation (Sonneborn *et al.*, 1963; Sussman and Sussman, 1969; Brackenbury and Sussman, 1975). With the elucidation of developmental timers, it becomes possible to isolate timer mutants in which particular developmental timer components are selectively expanded, contracted, or even eliminated (Soll *et al.*, 1986). The method for isolating such mutants is relatively simple and is based upon colony morphology. Cells are mutagenized with nitrosoguanidine according to the methods outlined by Dr. William Loomis in "Genetic Tools for *Dictyostelium discoideum*", Chapter 3, this volume. After roughly a 90% kill, cells are grown in HL5 medium, mixed with *Klebsiella aerogenes,* and plated at very low density on nutrient agar. After several days, each amoeba forms a "plaque" in the bacterial lawn as it feeds on the bacteria and divides. In each clonal colony, amoebae continue to feed at the periphery, but amoebae in the area cleared of bacteria begin to starve and therefore enter morphogenesis. When a colony is roughly 15 mm in diameter, the colony center contains fruiting bodies, and between this core of fruiting bodies and the outer clear ring of amoebae, there is a zone of intermediate stages. Moving centripedally, there is first a ring of aggregating amoebae with visible streams, then a ring of tight aggregates and early-finger morphologies, a ring of slugs and maxifingers, a ring of culmination stages, and finally the core of fruiting bodies. The relative proportions of these zones are constant for most colonies of wild-type cells if the cloning conditions are rigorously controlled and if the bacterial lawns, agar surface, and amoebic density are uniform. Putative timer mutants are easily identified by modified zonal proportions (see diagram in Fig. 6). For instance, timer mutants with expanded preaggregative periods exhibit an expanded outer ring of unaggregated amoebae and timer mutants with protracted preaggregative periods exhibit a narrower ring. Timer mutants with selective changes in the intervals between stages exhibit differences in ring width and the proportions of intermediate morphologies in these rings. Spores from the colony center of putative mutants are transferred to axenic medium and grown to high cell densities, serially transferred to fresh growth medium, grown to mid-log phase, and development on filters initiated according to the method described in a previous section. If a clone exhibits an aberration in timing, it is carefully assessed for (1)

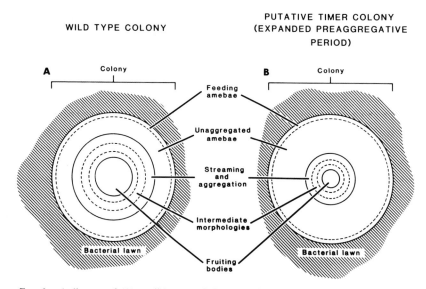

FIG. 6. A diagram of (A) a wild-type and (B) a putative timer mutant colony after 7 days at 22°C. Colony morphologies can be used to isolate putative timer mutants (see text).

the generation time (Yarger *et al.*, 1974; Soll *et al.*, 1976), (2) the density at stationary phase (Yarger *et al.*, 1974), (3) the sequence of morphological stages during development (Soll, 1979), (4) the formation of normal spores (Raper and Fennell, 1952), (5) the capacity to dedifferentiate (Soll and Waddell, 1975; Waddell and Soll, 1977; Finney *et al.*, 1979, 1981), and (6) the capacity to recapitulate rapidly (Loomis and Sussman, 1966, Newell *et al.*, 1971; Soll and Waddell, 1975). The most elegant developmental-timer mutants isolated by this method are normal for all of the above parameters, but exhibit altered timing intervals for one or more stages. For instance, variant FM-1 lacks the first component of the preaggregative period, FM-2 possesses a contracted first component, and SM-1 possesses an expanded first component (Soll *et al.*, 1986). All other timing components appear to be relatively normal in these mutants. To assess which timer component or components are altered in a timing mutant, the standard reciprocal-shift experiment is performed and the plots for the short to long and long to short shifts are compared to wild type (Soll, 1983). Timer mutants are useful not only in dissecting timer complexity but also in correlating molecular changes which occur during development with morphogenetic events and in uncoupling changes which occur simultaneously during normal development (Alexander *et al.*, 1985).

Other methods have also been employed to isolate timer mutants systematically. One which seems to be effective takes advantage of the sensitivity of amoebae and the resistance of spores to heat (Cotter and Raper, 1968). Cells are mutagenized, allowed to grow, spread on nutrient agar plates in association with *K. aerogenes*, allowed to clear the plate of bacteria, harvested, and dispersed on a development filter. After 17 hours, the cultures are treated for two 30-minute intervals at 45°C. This procedure selects for precocious spore formers and results in strains either phenotypically similar to the intensively studied strain FR17 or to other fast-developing phenotypes (Kessin, 1977). This method should be cautiously applied, since it is more likely to select for individual cells which differentiate rapidly to spores rather than for mutations in the developmental timers associated with morphogenesis. Mutant FR17 exhibits completely abnormal morphogenesis (Sonneborn *et al.*, 1963).

V.   A Method for Stimulating Rapid Recapitulation

Besides lending itself so well to analyses of developmental timers, *D. discoideum* exhibits several unique timing features which make it an exceptional system for monitoring temporal progress through the developmental program. The first of these characteristics is the capacity to recapitulate morphogenesis rapidly (Loomis and Sussman, 1966; Newell *et al.*, 1971; Soll and Waddell, 1975). If developing cultures are washed from development filters, vortexed into a single-cell suspension, then redispersed on a fresh development pad saturated with BSS (see Fig. 7), the amoebae will progress once more through the stages they had progressed through during the original morphogenetic program, only the second time through they do so in a fraction of the original time. For instance, if log-phase cells are washed free of nutrient medium, dispersed on development filters, and allowed to develop to the early-culminate I stage, they progress through the ripple, loose-aggregate, tight-aggregate, finger, and early-culminate stages after 414, 537, 618, 732, and 786 minutes, respectively (Table I). If the cultures are disaggregated into a single-cell suspension at the early-culminate I stage (14 hours) and dispersed on a fresh development filter, the cultures then progress through the same stages after 20, 40, 55, 90, and 130 minutes, respectively (Table I). Minimum recapitulation time to a stage is achieved roughly at the time that stage is generated during the original developmental program.

Based on a comparison of sensitivities of original timing and rapid recapitulation timing to inhibitors and environmental changes, and an analysis of the stability during rapid recapitulation of developmental changes acquired during the preaggregative and aggregative periods of initial development, we have concluded that rapid recapitulation represents for the most part the reutilization of developmental machinery acquired during initial development (Soll and co-

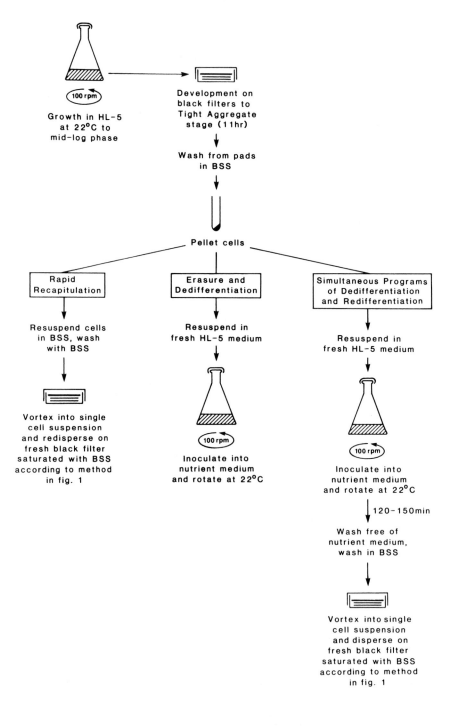

Growth in HL-5
at 22°C to
mid-log phase

Development on
black filters to
Tight Aggregate
stage (11hr)

Wash from pads
in BSS

Pellet cells

Rapid
Recapitulation

Erasure and
Dedifferentiation

Simultaneous Programs
of Dedifferentiation
and Redifferentiation

Resuspend cells
in BSS, wash
with BSS

Resuspend in
fresh HL-5 medium

Resuspend in
fresh HL-5 medium

Vortex into single
cell suspension
and redisperse on
fresh black filter
saturated with BSS
according to method
in fig. 1

Inoculate into
nutrient medium
and rotate at 22°C

Inoculate into
nutrient medium
and rotate at 22°C

120–150min

Wash free of
nutrient medium,
wash in BSS

Vortex into single
cell suspension
and disperse on
fresh black filter
saturated with BSS
according to method
in fig. 1

TABLE I

TOTAL DEVELOPMENTAL TIME (A) AND
INTERVAL TIME (B) FOR THE EARLY STAGES
OF INITIAL SLIME MOLD MORPHOGENESIS (1°)
AND RAPID RECAPITULATION (2°)[a]

	1°	2°
A. Total time to each stage (minutes)		
0 → R	414 ± 22	20 ± 8.6
0 → LA	537 ± 24	40 ± 8.6
0 → TA	618 ± 26	55 ± 8.6
0 → F	732 ± 29	90 ± 0
0 → ECI	786 ± 37	130 ± 8.6
B. Interval time between stages (minutes)		
0 → R	414	20
R → LA	120	20
LA → TA	84	15
TA → F	114	35
F → ECI	54	40

[a]Rapid recapitulation was stimulated in 14-hour cultures, at the early-culminate I stage. The results represent the means and standard deviations for three independent experiments. The interval times were calculated by subtracting the means. 0, Time at which development was initiated according to the method in Fig. 1; R, ripple stage; LA, loose-aggregate stage; TA, tight-aggregate stage; F, finger stage; ECI, early-culminate I stage.

workers, unpublished observations). One can stimulate rapid recapitulation up to five times in sequence without interim growth periods, but each time a percentage of cells drop out of the actively developing population (Soll and Mitchell, unpublished observations).

Rapid recapitulation has been effectively employed to study the role of cell–cell interactions in enzyme regulation (Loomis and Sussman, 1966; Newell *et al.*, 1971; Sussman and Newell, 1972). It also serves as a method for assessing progress through the temporal program of development (Soll and Waddell, 1975; Waddell and Soll, 1977; Finney *et al.*, 1985a).

FIG. 7. A diagram of methods for stimulating rapid recapitulation, erasure and the subsequent program of dedifferentiation, and simultaneous programs of dedifferentiation and redifferentiation.

VI. A Method for Stimulating Erasure and Subsequent Dedifferentiation

The second unique timing feature of *D. discoideum* is "erasure" and the subsequent program of dedifferentiation. When developing cultures are disaggregated and resuspended in nutrient medium, they synchronously and completely lose the capacity to recapitulate rapidly in a single step which has been referred to as the "erasure event" (Finney *et al.*, 1981). The method for initiating erasure is quite simple. Developing cultures are gently washed from development pads in BSS, pelleted, and resuspended in a small volume of fresh nutrient medium at a density of roughly 3×10^6/ml. The erasure culture is then rotated at 100 rpm at 22°C, and, for all intents and purposes the methods and conditions are similar to those employed for standard axenic growth. At time intervals, cells are removed from the "erasure flask," washed free of nutrient medium, and stimulated to undergo development on fresh black filters according to the method described in an earlier section. The timing to developmental stages is assessed, and the timing to each stage plotted as a function of the time of removal from the erasure flask.

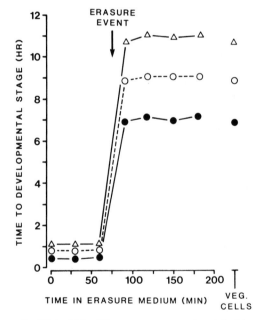

Fig. 8.   An example of data obtained from an erasure experiment. In this case, 11-hour developing cultures were disaggregated, resuspended in fresh nutrient medium (see diagram in Fig. 7), and at time intervals washed free of nutrient medium and initiated to undergo development. Note the discreteness of the erasure event. (●) Ripple; (○) loose aggregate; (△) tight aggregate.

An example of such an experiment is presented in Fig. 8. Eleven-hour developing cultures, at the tight-aggregate stage, were disaggregated, inoculated into
fresh growth medium, and recapitulation timing monitored. For the initial 70
minutes, cells progressed through the ripple, loose-aggregate, and tight-aggregate stages in minimum recapitulation timing. However, between 70 and 90
minutes, the timing reverted completely to that of mid-log-phase cells, defining
precisely the time at which the ''erasure event'' occurred. It has been demonstrated that the erasure event does not represent the complete and rapid loss of all
developmentally acquired components involved in morphogenesis, but rather
sets in motion a program of dedifferentiation during which different developmentally acquired characteristics are lost at different times (Finney *et al.,* 1979,
1981, 1983; Varnum and Soll, 1981; Soll *et al.,* 1984; Soll and Mitchell, 1982;
Hedberg and Soll, 1984). Therefore, it provides a method for studying the loss of
developmentally acquired components in a precise, reproducible, and programmed fashion, and should be included in future studies of developmental
characteristics. It also appears to play a role in the regulation of developmental
and vegetative gene expression (Finney *et al.,* 1986), and should be included in
the repertoire of developmental manipulations by molecular biologists employing *D. discoideum.*

VII.   A Method for Stimulating Parallel Programs of Dedifferentiation and Redifferentiation

It has been demonstrated that the erasure event completely reverses the progress which cells have made in the temporal program of development. If cells are
stimulated to reenter development immediately after the erasure event, they must
once again proceed through the first and second rate-limiting components of the
preaggregative period, even though they still possess a number of developmentally acquired characteristics (Soll and Mitchell, 1982; Finney *et al.,* 1983). It
has been demonstrated that these redeveloping erased cells still lose their developmental characteristics at the prescribed times in the program of dedifferentiation while reacquiring these characteristics at the prescribed times according to
the program of development (Soll and Mitchell, 1982; Finney *et al.,* 1983).
These results have been interpreted to mean that programs of dedifferentiation
and redifferentiation can function independently and simultaneously in the same
cell, leading to a remarkable situation for studying regulation. To initiate parallel
programs of dedifferentiation and redifferentiation (see diagram in Fig. 7), a
developing culture, for example at the tight-aggregate stage, is disaggregated,
resuspended in nutrient medium, and rotated at 100 rpm at 22°C. These cells
normally erase at 90 minutes. At 120–150 minutes, cells are simply washed free

of erasure medium and dispersed on fresh development filters according to the methods described in a previous section. In performing such an experiment, the researcher should carefully assess the time of the erasure event in the original erasure flask to be sure that it has occurred.

VIII. Concluding Remarks

In the preceding sections, I have reviewed the methods which have been developed for (1) investigating the complexity and relationships of those processes which serve as developmental timers, (2) isolating timer mutants, (3) initiating rapid recapitulation, (4) initiating erasure and the subsequent program of dedifferentiation, and (5) stimulating simultaneous progress through dedifferentiation and redifferentiation. Most scientists employing D. discoideum as a system for studying the developmental regulation of gene expression, chemotaxis, cell interactions, cohesion, cytoskeleton, and a long list of other cell functions and components have not taken advantage of these extraordinary temporal manipulations in their pursuits. Hopefully, this chapter will underline how simple these manipulations really are so that they will become part of the experimental repertoire of less biologically oriented investigators.

REFERENCES

Alexander, S., Cibulsky, A. M., Mitchell, L., and Soll, D. R. (1985). *Differentiation,* **30,** 1–6.
Ambros, V., and Horvitz, H. R. (1984). *Science* **226,** 409–416.
Brackenbury, R., and Sussman, M. (1975). *Cell* **4,** 347–352.
Chisholm, R. L., Barklis, E., and Lodish, H. F. (1984). *Nature (London)* **310,** 67–69.
Cocucci, S., and Sussman, M. (1970). *J. Cell Biol.* **45,** 399–407.
Cotter, D. A., and Raper, K. B. (1968). *J. Bacteriol.* **96,** 86–92.
Ferguson, R., and Soll, D. R. (1977). *Exp. Cell Res.* **106,** 159–165.
Finney, R., Varnum, B., and Soll, D. R. (1979). *Dev. Biol.* **73,** 290–303.
Finney, R., Slutsky, B., and Soll, D. R. (1981). *Dev. Biol.* **84,** 313–321.
Finney, R., Mitchell, L. H., Soll, D. R., Murray, B. A., and Loomis, W. F. (1983). *Dev. Biol.* **98,** 502–509.
Finney, R., Langtimm, C., and Soll, D. R. (1985a). *Dev. Biol.* **110,** 157–170.
Finney, R., Langtimm, C., and Soll, D. R. (1985b). *Dev. Biol.* **110,** 171–191.
Finney, R., Langtimm, C., Ellis, M., Rosen, E., Firtel, R., and Soll, D. R. (1986). *Dev. Biol.,* in press.
Hedberg, C., and Soll, D. R. (1984). *Differentiation* **27.** 168–174.
Kessin, R. H. (1977). *Cell* **10,** 703–708.
Klevecz, R. R. (1976). *Proc. Natl. Acad. Sci. U.S.A.* **73,** 4012–4016.
Lloyd, D., and Edwards, S. W. (1982). *In* "Synergetics: Phénomènes Non-Linéaires de la Dynamique". Springer-Verlag, Berlin.
Loomis, W. F., and Sussman, M. (1966). *J. Mol. Biol.* **22,** 401–404.
Marin, F. T. (1976). *Dev. Biol.* **48,** 110–117.

Marin, F. T. (1977). *Dev. Biol.* **60**, 389–395.

Mercer, J., and Soll, D. R. (1980). *Differentiation* **16**, 117–124.

Newell, P. C., Telser, A., and Sussman, M. (1969). *J. Bacteriol.* **100**, 763–768.

Newell, P. C., Longlands, M., and Sussman, M. (1971). *J. Mol. Biol.* **58**, 541–554.

Raper, K., and Fennell, D. (1952). *Bull. Torrey Bot. Club* **79**, 25–51.

Soll, D. R. (1979). *Science* **203**, 841–849.

Soll, D. R. (1983). *Dev. Biol.* **95**, 73–91.

Soll, D. R., and Mitchell, L. H. (1982). *Dev. Biol.* **91**, 183–190.

Soll, D. R., and Waddell, D. R. (1975). *Dev. Biol.* **47**, 292–302.

Soll, D. R., Yarger, J., and Mirick, M. (1976). *J. Cell Sci.* **20**, 513–523.

Soll, D. R., Mitchell, L. H., Hedberg, C., and Varnum, B. (1984). *Dev. Genet.* **4**, 167–184.

Soll, D. R., Mitchell, L., Kraft, B., Alexander, S., Finney, R., and Varnum-Finney, B. (1986). *Dev. Biol.,* in press.

Sonneborn, D. R., White, G. J., and Sussman, M. (1963). *Dev. Biol.* **7**, 79–93.

Sussman, M. (1966). *Methods Cell Physiol.* **2**, 397–410.

Sussman, M., and Sussman, R. R. (1969). *Symp. Soc. Gen. Microbiol.* **19**, 403–435.

Sussman, M., and Newell, P. C. (1972). *In* "Molecular Genetics and Developmental Biology," pp. 275–302. Prentice-Hall, New York.

Varnum, B., and Soll, D. R. (1981). *Differentiation* **18**, 151–160.

Varnum, B., Mitchell, L., and Soll, D. R. (1983). *Dev. Biol.* **95**, 92–107.

Varnum, B., Edwards, K., and Soll, D. R. (1986). *Dev. Biol.* **113**, 218–227.

Waddell, D., and Soll, D. R. (1977). *Dev. Biol.* **60**, 83–92.

Yarger, J., Stults, K., and Soll, D. R. (1974). *J. Cell Sci.* **10**, 315–333.

Chapter 23

Cell Differentiation in Monolayers and the Investigation of Slime Mold Morphogens

ROBERT R. KAY

Laboratory of Molecular Biology
Medical Research Council
Cambridge CB2 2QH, England

I. Introduction

The cells in a slime mold aggregate exist within a distinct microenvironment, isolated from the outside world by the slime sheath and by the tight apposition of their plasma membranes. Within this environment extracellular metabolites, proteins, and signal molecules accumulate and play out their roles in controlling cell differentiation, movement, and the formation of the prestalk–prespore pattern. It is hard to imagine how these processes will be understood, without first taking the aggregate apart and discovering the nature of the important actors in the microenvironment and of the cells' responses to them. This chapter describes a methodology to do just this. Accordingly in the "monolayer techniques," isolated cells or monolayers of cells are induced to differentiate into stalk or spore cells, while submerged under a simple salts solution in tissue culture dishes. Under these conditions the degree of interaction between the cells, whether mediated by cell contact or by diffusible factors, can be varied at will, merely by varying the cell density. Conversely, the cells remain accessible to added inducers, inhibitors, and so on, and often their responses to these agents can be seen directly by scoring stalk and spore cell differentiation by phase-contrast microscopy (see Fig. 1).

The initial impetus to developing the monolayer techniques came from the discovery of Bonner (1970) that a small proportion of NC-4 cells plated on agar, containing 1 mM cyclic AMP (cAMP), differentiate into stalk cells without ever undergoing normal morphogenesis. Here matters rested for several years until

METHODS IN CELL BIOLOGY, VOL. 28

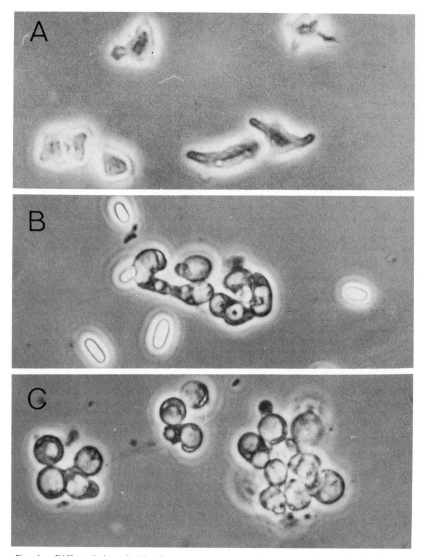

Fɪɢ. 1.   Differentiation of cells of a sporogenous mutant in submerged monolayers. The phase-contrast micrographs show: (A) amoebae, shortly after plating, (B) vacuolated stalk cells and refractile spores that differentiate by 30 hours of incubation in the presence of 2 mM cAMP, and (C) a pure population of stalk cells that differentiated when 2000 units/ml of DIF was included together with cAMP. HM29 cells were freed of bacteria and plated at a density of 10^5/cm^2 in tissue culture dishes containing a medium consisting of 10 mM MES, 20 mM KCl, 20 mM NaCl, 1 mM CaCl$_2$, 1 mM MgCl$_2$, 100 μg/ml streptomycin sulfate, 15 μg/ml tetracycline (pH 6.2), plus 2 mM cAMP and DIF as indicated. See Section II for further details.

the serendipitous observation by Town (Town *et al.*, 1976) that cells of strain V12M2 were capable of far more efficient stalk cell differentiation than NC-4 in the same conditions. Shortly thereafter it was found that stalk cell induction requires not only cAMP but also a low-MW factor termed DIF (Town *et al.*, 1976). DIF can be assayed by its ability to induce the differentiation of isolated amoebae into stalk cells in the presence of cAMP (Town and Stanford, 1979). Meanwhile, cAMP turned out to have an additional role, namely, of inducing prespore cell differentiation (Kay *et al.*, 1978). The prespore cells that are induced in this way do not usually mature into spore cells, and at high cell density, as DIF accumulates, they are eventually diverted to stalk formation. However, strains in which prespore cells can complete spore formation have been isolated and monolayers of these sporogenous mutants form stalk and spore cells when incubated with cAMP (Town *et al.*, 1976; Kay *et al.*, 1978). Although the underlying alteration in the sporogenous mutants is still obscure, they have proved extremely useful for examining the factors that regulate spore cell differentiation. As a result of efforts to induce isolated sporogenous amoebae to differentiate into spores, it was found that the agar substratum usually employed contained inhibitors of differentiation, and that cells differentiated much better when submerged (Kay and Trevan, 1981; Kay, 1982). These improvements, incorporated into the DIF assay, increased its reliability and sensitivity, facilitating the eventual purification of DIF (Kay *et al.*, 1983). Recently DIF-1 has been identified and is a novel type of effector molecule, containing chlorine (G. Taylor, M. Masento, H. Morris, K. Jermyn, and R. Kay, unpublished).

Most work with the monolayer techniques has used strain V12M2 and its derivatives because, as mentioned above, the more common laboratory strain of *Dictyostelium*, NC4, does not form stalk cells efficiently when simply provided with cAMP and DIF at the start of development. However recently we have finally succeeded in inducing NC4 amoebae to form stalk cells efficiently (M. Berks and R. Kay, unpublished). cAMP and DIF can act sequentially to induce stalk cells in V12M2 (Sobolewski *et al.*, 1983) and we have found, surprisingly, that in the period when DIF is acting, cAMP is actually inhibitory (M. Berks and R. Kay, unpublished). Strain NC4 is much more sensitive to this inhibition than V12M2 and will only produce stalk cells if the cAMP is completely removed when DIF is added. There is also an early period of cAMP sensitivity, so an optimum induction regime for NC4 involves an initial period of about 4 hours without cAMP, followed by 14 hours with cAMP alone and then a period with DIF alone, until stalk cells form. However, as this regime is rather inconvenient and we do not yet understand all the variables involved, the methods are all described in terms of strain V12M2.

The major uses of the monolayer techniques are as follows:

1. Revealing the signals controlling stalk and spore cell differentiation (see Town *et al.*, 1976; Kay and Trevan, 1981; Kay, 1982; Gross *et al.*, 1983).

Of these, cAMP, DIF and ammonia are firmly established. Other conditioned medium factors exist, and these might also be assayed in a suitable monolayer system (Kay, 1982; Wilkinson *et al.*, 1985; Mehdy *et al.*, 1983). The bioassay for DIF is described, as are methods for its production and purification (Section II,B).

2. The isolation and characterization of developmental mutants, such as the sporogenous and DIF-less ones (Section II,C).
3. Directed cell differentiation, in which the objective is to obtain pure populations of cells differentiating into either stalk or spore cells (Section II,D).

Before proceeding further, it may be worthwhile to counter certain past objections to the monolayer techniques. First, as already pointed out, NC4 amoebae can be induced to form stalk cells in a monolayer and are fundamentally similar to V12M2 amoebae in their requirements for cAMP and DIF. Second, although high cAMP concentrations are routinely used in the medium (1–5 mM), essentially physiological concentrations will suffice, provided these are not destroyed by cAMP phosphodiesterase (Rossier *et al.*, 1978; Sobolewski *et al.*, 1983). Finally, the stalk and spore cells formed in the monolayers are very similar to those formed in normal development, as judged by (1) light and electron microscopy (Kay *et al.*, 1979), (2) the viability of the spores after detergent treatment (Town *et al.*, 1976), and (3) the spectrum of polypeptides synthesized by the stalk cells (Kopachik *et al.*, 1985). Furthermore, the timing and levels of a number of developmentally regulated enzymes are very similar in the monolayers and *in vivo* (Sobolewski *et al.*, 1983; Town, 1984). These similarities between the differentiation of cells in the monolayer and in normal development indicate that many of the fundamental processes of development can be analyzed using the monolayer methodology.

The principle alternative to the monolayer technique is to allow cells to develop in shaken suspension (Klein, 1975; Town and Gross, 1978). Typically cells at 10^7/ml are used, shaken at 160 rpm. This technique is very convenient where applicable: the cells are synchronous, and inducers can readily be added and samples withdrawn. However, almost irrespective of cell density, aggregates form after a few hours which produce a slime sheath and so resemble a submerged slug. Thus the degree of cell contact cannot be controlled, nor can the cells readily be prevented from accumulating their own inducers of differentiation within the aggregates. Conversely, the cells within the aggregates may not be accessible to exogenous factors and inducers. Some attempt has been made to reduce the degree of cell contact by shaking the cells rapidly (Town and Gross, 1978), sometimes with added EDTA. When this is done, the expression of a whole range of developmentally regulated genes is blocked. As many of the fast-shaken cells remain isolated, it has been concluded that the expression of the blocked genes is dependent on cell contact (Chisholm *et al.*, 1984; Mehdy *et al.*,

1983). This conclusion is not warranted. First, the fast-shaken cells are still colliding and therefore making transient contacts, but more importantly, the high shearing forces (and EDTA) to which the cells are exposed are likely to be deleterious and inhibit processes which are not contact-dependent. One sign of this is that cAMP signaling—which, of course, is mediated by a diffusible molecule—is inhibited by fast shaking (Town and Gross, 1978).

II.   Methods

A.   General Methods and Notes

1.   SOLUTIONS AND MEDIA

SM agar: (per liter) Difco Bactopeptone, 10 g; Difco yeast extract, 1 g; glucose, 10 g; KH_2PO_4 2·2 g; Na_2HPO_4, 1 g; $MgSO_4·7H_2O$, 1 g; Bactoagar, 18 g.

NS, stalk, and spore salts:

	KCl	NaCl	$CaCl_2$	$MgCl_2$
NS	20 mM	20 mM	1 mM	—
Stalk salts	10 mM	2 mM	1 mM	—
Spore salts	20 mM	20 mM	1 mM	1 mM

Stalk and spore salts were devised to optimize stalk and spore cell differentiation, respectively, in the monolayer incubation system (Brookman *et al.*, 1982; Kay, 1982). Stocks are normally made up as sterile $100\times$ concentrates.

KK_2: 20 mM K_1K_2 phosphate, 2 mM $MgSO_4$, pH 6.2. The $MgSO_4$ is added from a sterile 1 M stock after autoclaving the phosphate.

MES: 2(N-morpholino)ethane sulfonic acid (K^+), used at a final concentration of 10 mM, pH 6.2. Kept as 1 M sterile stock.

Antibiotics: A mixture of 200 μg/ml streptomycin sulfate and 15 μg/ml tetracycline. These are kept as 10 mg/ml stocks in water and ethanol at 4°C and −20°C, respectively.

cAMP: Used as the K^+ salt, kept as 200 mM sterile stock.

2.   GROWTH AND HANDLING OF CELLS

Cell biological techniques such as those that follow are often sensitive to the details of cell growth and handling. To allow comparison with practice in other laboratories, the methods used in this laboratory will be briefly described. Points which are known to be critical are indicated; it is not known which, if any, of the other procedures are also critical.

Working stocks of cells are maintained as growth zones on lawns of *Klebsiella aerogenes* on SM agar held at 8°C. In these conditions it takes about a month for a growth zone to traverse the plate. As the phenotype of many strains alters after prolonged growth at 8°C, stocks must be renewed every month from spores stored on silica gel or amoebae frozen in liquid nitrogen. Cells for experiments are grown up at 22°C, from an inoculum of the stock-plate growth zone, on SM agar in association with *K. aerogenes*. Plates are harvested after 40–50 hours, when they are half-cleared (half the bacterial lawn has turned translucent, but before there is any significant aggregation). Amoebae are separated from residual bacteria by five centrifugal washes at 200 g × 3 min, the first four being in KK_2 and the final one normally in NS. The cell pellet is resuspended at $1-3 \times 10^8$/ml and counted using a Coulter counter (100-μm orifice) before diluting as necessary.

Ax-2 for large-scale DIF production (see Section II,B) is grown in 20-liter Pyrex aspirators containing 16 liters of medium (Watts and Ashworth, 1970), to which antifoam A (Sigma) is added at 0.1 ml/liter. The medium is stirred with a magnetic follower and aerated by compressed air passed through a sterile filter (miniature line filter from Microflow Ltd., Fleet, Hampshire, UK) at a rate of ~8 liters/min. Such cells grow slowly (doubling time 12–20 hours) and are harvested as they approach stationary phase ($5-10 \times 10^6$/ml). The cells are used after pelleting and a single wash in NS.

3. IMPORTANT NOTES

1. Tissue culture plastic dishes are used for all the monolayer techniques. These dishes are specially treated so as to have a hydrophilic surface, suitable for cells. Sterilin, NUNC, or Falcon dishes are all suitable, with Sterilin being marginally the best. Hydrophobic bacteriological dishes are *not* suitable. For 5-cm-diameter dishes, 2 ml of medium is used, and this volume is adjusted in proportion to the surface area for other dish sizes.

2. Cells at low density, as in the DIF assay, are particularly sensitive to traces of detergent from the glassware. All glassware should be rinsed with glass-distilled water before use.

3. At cell densities $>5 \times 10^5$/ml, it is necessary to increase the cAMP concentration to >1 mM, to compensate for hydrolysis during the incubation. If the cAMP becomes exhausted during the incubation, large aggregates with streams will often form.

4. The greatest source of variability in the monolayer techniques is undoubtedly due to the initial physiology of the cells. This variability can be minimized

by renewing stock plates regularly and using cells from midclearing growth plates, as already described.

4. GETTING STARTED

Perhaps the easiest way to get started with the monolayer techniques is to plate washed cells of strain V12M2 at a density of 10^6/ml in the same conditions as for the DIF assay (Section II,B) but omitting DIF and with the cAMP concentration increased to 2 mM. After 2 days of incubation, large numbers of convincing stalk cells should have differentiated. This result should be sufficient encouragement to try the other techniques!

B. DIF: Assay, Production, and Purification

DIF is a low-MW, nonpolar molecule, released during development, and which in combination with cAMP, induces isolated cells of strain V12M2 to differentiate into stalk cells. This fact is the basis of the bioassay. DIF is soluble in organic solvents (ethanol, methanol, acetonitrile, and hexane are commonly used) and is stable for months or years at $-20°C$. It is also stable in aqueous solution, but here it tends to bind to glass or plastic.

1. BIOASSAY FOR DIF

The assay is as described by Brookman et al. (1982), except that the antioxidant butylated hydroxytoluene is omitted. It supersedes the earlier method using agar (Town and Stanford, 1979). The assay mix contains 10 mM MES (K$^+$), 10 mM KCl, 2 mM NaCl, 1 mM CaCl$_2$, pH 6.2, plus 1 mM cAMP, 200 µg/ml streptomycin sulfate, and 15 µg/ml tetracycline. It is stored as a sterile, 20× concentrate at $-20°C$ and diluted into sterile distilled water for use. Washed V12M2 cells are added to a density of 7.5×10^3/ml and 2-ml aliquots pipeted into 5-cm-diameter tissue culture dishes (Sterilin 302V). DIF is then mixed into the medium, normally dissolved in an organic solvent. Up to 2 µl/ml of most solvents can be added without adverse effect. The dishes are incubated at 22°C in the dark and scored after 40–50 hours by phase-contrast microscopy. A 20× objective and 15× eyepiece are ideal. Cells in which >50% of the cross-sectional area is vacuolated are considered to be stalk cells: typical examples are shown in Fig. 2. The distinction between a cell that just is and one that just is not a stalk cell is necessarily arbitrary, but these doubtful cells do not normally constitute a large proportion of the total. The assay is roughly linear, with amounts of DIF giving between 10 and 50% stalk cell formation; outside these limits the amount of DIF is underestimated. One unit of DIF is defined as that

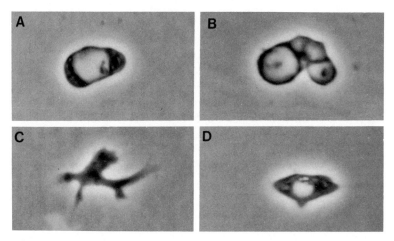

FIG. 2.   Scoring the DIF assay. Typical cell types seen by phase-contrast microscopy after 40 hours incubation. The fully vacuolated cells (A, B) are scored as stalk cells, and the amoeba (C) and partially vacuolated cell (D) are scored as nonstalk cells. The DIF assay was carried out exactly as described in Section II,B. Normally 100–200 cells are scored per plate.

amount which induces 1% stalk cells in the above assay. The assay may vary about twofold in sensitivity from day to day.

Cell-associated DIF can be measured after extraction with chloroform–methanol by the method of Bligh and Dyer (1959) or better after hexane extraction (Sobolewski *et al.*, 1983). However, the cellular lipids also extracted will kill the tester cells when added at too high a concentration, so these extracts must be titrated with care.

2.   DIF PRODUCTION AND PURIFICATION (see Kay *et al.*, 1983)

DIF is collected across a cellophane membrane from developing cells. Washed Ax-2 cells are spread at a density of up to $1.5 \times 10^7/cm^2$ on sheets of thin cellophane (325P, Canning Parry Packaging, Avonmouth Way, Avonmouth, Bristol BS11 9DZ, UK) which were previously boiled in EDTA–NaHCO$_3$ and then water. Each cellophane sheet is in turn supported by a stainless-steel sieve on low legs (0.5 cm), the whole fitting inside a surgical tray (35 × 30 cm). The underside of the mesh is bathed in a shallow layer of stalk salts (~200 ml) containing 5–10 g/liter Amberlite XAD-2 resin (previously washed extensively in ethanol then water). It is essential that the cells should not be submerged during development; otherwise, at the high densities used, they become anaerobic and die. The trays are shaken slowly (16 rpm) on a gyratory shaker to aid leaching of the DIF. Development is slow and DIF is collected over 3–4 days. At the end of this period the XAD-2 beads, now yellow with adsorbed material, are collected, washed extensively with water, and the DIF eluted with three to five batches of ethanol. The typical yield is 50–100 units/10^6 cells.

The ethanol eluate is concentrated by rotary evaporation, dispersed in water, and the process repeated so that an aqueous suspension results. An aliquot is dried down and weighed, then the rest is made to 5 mg/ml in 0.1 M KOH and extracted three times with an equal volume of hexane. It may be necessary to centrifuge at 200 g × 5 min to aid separation of the phases. This extraction removes neutral and basic nonpolar substances but leaves DIF in the aqueous phase. The DIF is then extracted from the aqueous phase, after neutralization with HCl, by three more extractions with hexane.

The hexane-extracted DIF is concentrated first by rotary evaporation and then using a Speedvac vacuum concentrator (Savant), and the resulting brown oil taken up in ethanol. This DIF can be used for many experiments, but it is far from pure. Further purification is achieved by reverse-phase high-pressure liquid chromatography (HPLC). In the first step DIF is eluted from a 50-cm Whatman Partisil M9 ODS-3 column with a methanol–water gradient containing acetic acid (solvent A = 2% acetic acid; solvent B = 98% methanol, 2% acetic acid; the gradient goes from 40 to 80% B in 80 minutes, then from 80 to 100% in 20 minutes, and finally, 100% B for 20 minutes, at a flow rate of 3 ml/min). The major activity is DIF-1 and accounts for >95% of the recovered activity. It elutes at ~95 minutes and is preceded by several minor activities. The peak fractions are concentrated as before and rerun on a 25-cm Whatman Partisil 5 ODS-3 analytical column, eluting with an acetonitrile–water gradient again containing acetic acid (solvent A = 5% acetic acid; solvent B = 95% acetonitrile, 5% acetic acid; the gradient goes from 50 to 80% B in 60 minutes at a flow rate of 1 ml/min). DIF-1 elutes after ~35 minutes associated with a distinct A_{254} peak. The recovery of DIF from a large-scale purification is quite low (2–5%), although some of the losses appear to be due to the separation of DIF away from factors that stimulate its activity in the bioassay.

C. Developmental Mutants

The monolayer incubation techniques can be used as a way of obtaining mutants with altered cell differentiation. For example, the "DIF-less" mutants accumulate very little DIF but remain responsive to it and are blocked in development as mounds (Kopachik et al., 1983), and the sporogenous mutants make both stalk and spore cells in the monolayer and form bizarre fruit (see Table II). In the future one can envisage recognizing or selecting additional classes such as DIF overproducers or DIF-resistant mutants, by variants of the procedures to be described.

1. Mutagenesis

Cells are mutagenized as described in Chapter 3 by Loomis, this volume, with the following modifications: (1) filter-sterilized N-methyl-N'-nitro-N-nitrosoguanidine (NMG) is used, since yeast and fungal contaminants from the mutagen can cause problems for the selection procedures; (2) the cells (at 10^7/ml) are mutagenized more lightly (150 µg/ml NMG for 30 minutes, giving 5–10% survival) to reduce the number of secondary mutations introduced; and (3) since the plaques from V12M2-derived strains are more diffuse than those from NC-4, fewer viable cells can be plated, when clonal growth is desired (~20 per 9-cm-diameter plate). After mutagenesis, the amoebae can be either plated directly on nutrient agar and the resulting plaques tested for altered differentiation (Section II,C,2), or they can be subjected to a selection procedure designed to enrich for a particular type of developmental mutant (Section II,C,3). In the latter case it is advisable to grow up the mutagenized cells before performing a sporogenous selection; otherwise, the cells killed by the mutagen inhibit cell differentiation. To do this, spread 10^7 cells on a pregrown lawn of bacteria on a 9-cm SM plate, and harvest when the plate is half-cleared (2–3 days).

2. Spot Tests for Altered Cell Differentiation

Two procedures have been used, in which cells are taken directly from the growth zone of an interesting plaque and their ability to differentiate tested in a monolayer. In the first, a small square of washed cellophane (~1 × 1 cm) is touched onto the growth zone and then deposited cells-down onto thin non-nutrient agar containing stalk salts, antibiotics, and the appropriate test substances (2 mM cAMP, with or without 4000 units/ml DIF). With practice a monolayer of cells can be made in this way. After 24–48 hours, cell differentiation, under the cellophane square, is scored by phase-contrast microscopy. In a slightly more refined version of this test, a loopful (1–2 × 10^6) of cells from the growth zone is dispersed in 5 ml of stalk salts and partially freed of bacteria by centrifugation. The cell pellet is resuspended in 2 ml of buffered salts and antibiotics medium, in a 5-cm-diameter tissue culture dish supplemented with the appropriate test substances. This method is more quantitative than the cellophane square method, and in addition, the supernatant can be taken after 36 hours, freed of cells, and its DIF content measured by adding aliquots to V12M2 cells in a standard DIF assay. Using these spot tests, DIF-less mutants can easily be recognized, since at high cell density they do not make stalk cells in the presence of cAMP alone (whereas the wild type does), but they do when DIF is supplied. Such a phenotype is confirmed, if, as expected, the mutant accumulates little if any DIF in the medium (Kopachik *et al.*, 1983).

3. SELECTION OF SPOROGENOUS MUTANTS

Mutagenized amoebae are grown up, harvested, and plated at 2.5×10^6/ml in tissue culture dishes containing spore salts, 10 mM MES, pH 6.2, antibiotics, and 10 mM cAMP. After 2 days incubation at 22°C, sterile 3% cemulsol NP12 (Rhone-Polenc) is added to 0.3% to kill all nonspores. After 2 hours the cell residues are scraped off the plate with a silicone rubber scraper and centrifuged at 500 $g \times 10$ min. The pellet is resuspended and plated out on SM agar in association with bacteria (10^4–10^6 cell equivalents per plate). Clones should appear from the surviving spores after 3–6 days and their sporogenous phenotype can be confirmed using the spot test. This selection is extremely powerful (10^5- to 10^6-fold) and works equally with NC-4 as with V12-derived strains (e.g., Kay and Jermyn, 1983). At least four phenotypic classes can be distinguished based on the morphology of the final fruiting structure formed (mound, tipped mound, first finger, and culminate with spores at the base). Most of the sporogenous mutants develop more rapidly than their parents, and, conversely, the rapidly developing (*rde*) mutants described previously (Sonneborn *et al.*, 1963; Abe and Yanagisawa, 1983) all turn out to be sporogenous (Kay, unpublished). The sporogenous and *rde* mutants may all be affected in cAMP metabolism or the cAMP effector pathway, (Kessin, 1977; Coukell and Chan, 1980).

TABLE I

STRAINS AND MARKERS SUITABLE FOR GENETIC ANALYSIS OF V12

Strain	Markers[a]	Origin[b]	References[c]
HM2	*tsg-900, acr-900, cob-900*	V12M2	1
HM3	*whiA900, cycA900, tsg901, acr-900*	V12M2	1
HM27	*whiA900, cycA900, tsg901* and/or *tsg903*	V12M2	1
NP157	*acrA354, sprB2, tsgN350, axe-352*	V12	2
HU571	*whiA1, acrA1, tsgD12, bsgA5, bwnA1*	NC-4/V12	3
HU864	*whiA1, tsgD12, bsgA5, manA2*	NC-4/V12	2

[a]tsg, Not grown at 27.0°C; acr, resistant to 100 μg/ml acriflavine; cob, resistant to 200 μg/ml CoCl$_2$; cyc, resistant to 500 μg/ml cycloheximide; whi, white sori. The *cycA900* and *whiA900* alleles were assigned to their complementation groups by Professor K. L. Williams.

[b]All strains are of the V12 mating type and therefore can be crossed with each other by parasexual methods (Robson and Williams, 1979). However, as indicated in this column, some strains are pure V12 whereas others contain both V12 and NC-4 genetic material and are derived from (rare) V12 × NC-4 crosses.

[c]References: 1. This work; 2. K. L. Williams (personal communication), who should be contacted for these strains at School of Biological Sciences, Macquarie University, North Ryde, New South Wales 2113, Australia; 3. Williams *et al.* (1980).

TABLE II

DEVELOPMENTAL MUTANTS[a]

Strain	Parent	Developmental mutation	Phenotype
HM15	HM2	sci-904	Sporogenous, rapid development, "fruit" is a mound
HM18	HM2	sci-907	Sporogenous, rapid development, "fruit" is stalk with spores at the base
HM28	HM27	sci-909	Similar to HM15
HM29	HM27	sci-910	Similar to HM18
HM43	V12M2	—	Reduced DIF accumulation, responsive to DIF, arrests as tipless mound
HM44	HM27	—	As HM43

[a]The four sporogenous mutations are all recessive (Kay, unpublished). Nothing is yet known of the genetics of the "DIF-less" mutants HM43 and HM44.

4. GENETIC ANALYSIS OF V12

V12-Derived strains are very difficult to cross by parasexual methods with NC-4-derived strains (Robson and Williams, 1979), so that many of the common genetic markers are not available in V12. However, it can be seen from Table I that there are good markers available for linkage groups I (*cycA*), II (*acrA, whiA, sprB*), III (*bsgA, tsgN*), IV (*bwnA*), and VI (*manA*). Strains HM2 and HM27 are fully stalk cell-inducible and are recommended as parental strains for isolating developmental mutants (HM3 is weedy). It is not known whether any of the other strains listed in Table I are stalk cell-inducible or not, and their main use should be for mapping mutants. The developmental mutants mentioned elsewhere in this chapter are listed in Table II. It is to be hoped that the relative genetic isolation of the V12 strains will be broken in the future by improvements in macrocyst genetics, since V12 and NC-4 are of opposite mating types.

D. Directed Cell Differentiation

The principle of the two main methods for directed cell differentiation is to have cells developing in conditions where little DIF accumulates—either at low density or using a DIF-less mutant—so that only differentiation along the spore pathway is permitted. Then DIF or the nonphysiological inducer, diethylstilbestrol (DES; Gross *et al.*, 1983), can be added to induce stalk cells and suppress all spore differentiation.

1. STALK AND SPORE CELLS AT LOW DENSITY
(see Kay and Jermyn, 1983)

Cells of a sporogenous mutant (preferably HM29) are plated at a density of $10^3/cm^2$ (10^4/ml of medium) in tissue culture dishes in 10 mM MES, spore salts, pH 6.2, 0.1 mM cAMP, and antibiotics. These cells differentiate to produce 60–80% spores, the rest remaining as amoebae. If DIF (2000 units/ml) is added between t_0 and t_8, spore differentiation is almost completely abolished and >90% of the amoebae differentiate into stalk cells. This technique is very useful when individual cells can be studied, as with fluorescent antibodies, but requires heroic efforts to obtain sufficient material for conventional molecular analysis.

2. STALK AND PRESPORE CELL DIFFERENTIATION AT HIGH CELL DENSITY

Washed amoebae of the DIF-less mutant HM44 are plated at a density of $10^5/cm^2$ (10^6/ml of medium) in tissue culture dishes under 10 mM MES, stalk salts, pH 6.2, plus 2 mM cAMP and antibiotics. After 8–10 hours the medium is removed and replaced with fresh medium containing 2 mM cAMP, and 4000 units/ml of DIF or 12 μM DES added as necessary. In the absence of DIF or DES many prespore cells differentiate, but they remain amoeboid and few if any stalk cells form. On the other hand, DIF or DES induces >90% stalk cells differentiation, with the first stalk cells appearing ~6 hours after addition of the inducer. The concentration of DES is extremely critical as too much will kill the cells; DIF has no such adverse effects. It should be noted that it is necessary to change the medium to suppress a background at stalk cells that otherwise may form, even in the absence of added inducer. The reasons for this are not fully understood at the moment.

This method is now in routine use for studying DIF-induced gene expression. However, further improvements may be possible; for instance, a DIF-less, sporogenous double mutant should make spores and not just prespores at high cell density in the absence of added DIF but stalk cells in its presence.

III.   Discussion

The nature of the transactions between the cells in an aggregate is the key mystery if we are to understand the later events in slime mold morphogenesis. Much of what we do know comes from observations of cell differentiation in monolayers, and the conclusions from this work—especially on the role of cAMP—are increasingly being confirmed using cloned genes and other molecular probes for cell differentiation (Landfear and Lodish, 1980; Barklis and Lodish, 1983; Mehdy *et al.*, 1983; but see below for cell contact).

There is now much evidence that cAMP induces the differentiation of amoebae to the postaggregative (tight mound) stage of development (see Williams *et al.*, 1986) and that beyond this stage, it also induces prespore and spore cell differentiation (Kay *et al.*, 1978; Kay, 1979, 1982). However cAMP is actually inhibitory to DIF-induced stalk cell formation (M. Berks and R. Kay, unpublished), while DIF is inhibitory to cAMP-induced spore formation (Kay and Jermyn, 1983). Thus the *in vitro* work suggests that the decision between stalk and spore cell differentiation in *Dictyostelium* depends on the balance between DIF and cAMP. Probably other morphogens also weigh in this balance. Ammonia can favor spore over stalk differentiation (Gross *et al.*, 1983; Town, 1984), possibly by raising intracellular pH (pH_i). However recent measurements of pH_i do not support the original idea of Gross *et al.* (1983) that cytoplasmic pH is the central switch between stalk and spore cell differentiation (Jentoft and Town, 1985; Ratner, 1986; Kay *et al.*, 1986). Adenosine inhibits prespore cell differentiation in aggregates (Weijer and Durston, 1985; Schaap and Wang, 1986) and this can be readily understood in terms of its known antagonism to cAMP signaling (Newell and Ross, 1982), since cAMP induces prespore differentiation. Other soluble signals may also be involved in controlling stalk and spore cell differentiation but these are not yet as well characterized as those already mentioned (Sternfeld and David, 1979; Kay, 1982; Weeks, 1984; Mehdy *et al.*, 1985; Wilkinson *et al.*, 1985). However, it is not unreasonable to hope that in the near future we shall know all the signals necessary to direct an individual amoebae to either stalk or spore differentiation.

One area that is still very confused is the role, if any, of cell contact in inducing cell differentiation. All the inducers mentioned above are soluble molecules. For a while spore cell induction appeared to be contact dependent (Kay *et al.*, 1978; Weeks, 1982), but this impression was subsequently corrected when ways were found of inducing the differentiation of isolated prespore and spore cells (Kay and Trevan, 1981; Kay, 1982; Weeks, 1984). It is likely that the conflict about the existence of contact-dependent interactions reflects differences in the ways in which the cells are handled to prevent contact, and it has already been noted in the introduction that the fast-shaking–EDTA regimes used to minimize cell contact are almost certain to have additional deleterious effects. Given this, it is not surprising that one of the claimed instances of a contact-dependent interaction has been withdrawn following reexamination of the question in a monolayer system (Mehdy and Firtel, 1985). It is important that the other claims should also be reexamined in this light (Landfear and Lodish, 1980; Chisholm *et al.*, 1984).

The monolayer techniques for cell differentiation have been in use for almost a decade in the Mill Hill laboratories of the ICRF and more recently at Cambridge, and have evolved to the stage where they are both straightforward and reliable. Recently they have been used successfully in several other laboratories (e.g., Sobolewski *et al.*, 1983; Town, 1984; Inouye, 1985). The time is now ripe for their more widespread application.

ACKNOWLEDGMENTS

I am grateful to many people at the Mill Hill laboratories of the ICRF for collaboration over a number of years but most especially to Keith Jermyn and Julian Gross. This manuscript was improved thanks to the help of Mary Berks, Jenny Brookman, Ralph Pogge, and Stuart McRobbie.

REFERENCES

Abe, K., and Yanagisawa, K. (1983). *Dev. Biol.* **95,** 200–210.
Barklis, E., and Lodish, H. F. (1983). *Cell* **32,** 1139–1148.
Bligh, E. G., and Dyer, W. J. (1959). *Can. J. Biochem. Physiol.* **37,** 911–912.
Bonner, J. T. (1970). *Proc. Natl. Acad. Sci. U.S.A.* **65,** 110–113.
Brookman, J. J., Town, C. D., Jermyn, K. A., and Kay, R. R. (1982). *Dev. Biol.* **91,** 191–196.
Chisholm, R. L., Barklis, E., and Lodish, H. F. (1984). *Nature (London)* **310,** 67–71.
Coukell, M. B., and Chan, F. K. (1980). *FEBS Lett.* **110,** 39–42.
Gross, J. G., Bradbury, J., Kay, R. R., and Peacey, M. J. (1983). *Nature (London)* **303,** 244–245.
Inouye, K. (1985). *J. Cell. Sci.,* **76,** 235–245.
Ishida, S. (1980). *Dev. Growth Differ.* **22,** 781–788.
Jentoft, J. E., and Town, C. D. (1985). *J. Cell Biol.* **101,** 778–784.
Kay, R. R. (1979). *J. Embryol. Exp. Morphol.* **52,** 171–182.
Kay, R. R. (1982). *Proc. Natl. Acad. Sci. U.S.A.* **79,** 3228–3231.
Kay, R. R., and Jermyn, K. A. (1983). *Nature (London)* **303,** 242–244.
Kay, R. R., and Trevan, D. (1981). *J. Embryol. Exp. Morphol.* **62,** 369–378.
Kay, R. R., and Garrod, D., and Tilly, R. (1978). *Nature (London)* **271,** 58–60.
Kay, R. R., Town, C. D., and Gross, J. D. (1979). *Differentiation* **13,** 7–14.
Kay, R. R., Gadian, D. G., and Williams, S. R. (1986). *J. Cell Sci.* **83,** 165–169.
Kay, R. R., Dhokia, B., and Jermyn, K. A. (1983). *Eur. J. Biochem.* **136,** 51–56.
Kessin, R. (1977). *Cell* **10,** 703–708.
Klein, C. (1975). *J. Biol. Chem.* **250,** 7134–7138.
Kopachik, W., Oohata, A., Dhokia, B., Brookman, J. J., and Kay, R. R. (1983). *Cell* **33,** 397–403.
Kopachik, W., Dhokia, B., and Kay, R. R. (1985). *Differentiation* **28,** 209–216.
Landfear, S. M., and Lodish, H. F. (1980). *Proc. Natl. Acad. Sci. U.S.A.* **77,** 1044–1048.
Mehdy, M. C., and Firtel, R. A. (1985). *Mol. Cell. Biol.* **5,** 705–713.
Mehdy, M. C., Ratner, D., and Firtel, R. A. (1983). *Cell* **32,** 763–771.
Newell, P., and Ross, F. M. (1982). *J. Gen. Microbiol.* **128,** 2715–2724.
Ratner, D. (1986). *Nature (London)* **321,** 180–182.
Robson, G. E., and Williams, K. L. (1979). *Genetics* **93,** 861–875.
Rossier, C., Gerisch, G., Malchow, D., and Eckstein, F. (1978). *J. Cell Sci.* **35,** 321–338.
Schaap, P., and Wang, M. (1986). *Cell* **45,** 137–144.
Sobolewski, A., Neave, N., and Weeks, G. (1983). *Differentiation* **25,** 93–100.
Sonneborn, D. R., White, G. J., and Sussman, M. (1963). *Dev. Biol.* **7,** 79–93.
Sternfeld, J., and David, C. N. (1979). *J. Cell Sci.* **38,** 181–191.
Town, C. D. (1984). *Differentiation* **27,** 29–35.

Town, C., and Gross, J. (1978). *Dev. Biol.* **63,** 412–420.

Town, C., and Stanford, E. (1979). *Proc. Natl. Acad. Sci. U.S.A.* **76,** 308–312.

Town, C. D., Gross, J. D., and Kay, R. R. (1976). *Nature (London)* **262,** 717–719.

Watts, D. J., and Ashworth, J. M. (1970). *Biochem. J.* **119,** 171–174.

Weijer, C. J., and Durston, A. J. (1985). *J. Embryol. Exp. Morphol.* **86,** 19–37.

Weeks, G. (1982). *Exp. Cell Res.* **137,** 301–308.

Weeks, G. (1984). *Exp. Cell Res.* **153,** 81–90.

Wilkinson, D. G., Wilson, J., and Hames, B. D. (1985). *Dev. Biol.* **107,** 38–46.

Williams, K. L., Robson, G. E., and Welker, D. L. (1980). *Genetics* **95,** 289–304.

Williams, J. G., Pears, C. J., Jermyn, K. A., Driscoll, D. M., Mahbubani, H., and Kay, R. R. (1986). *In* "Symposium of the Society for General Microbiology: Regulation of Gene Expression" (I. Booth and C. Higgins, eds.), pp. 277–298. Cambridge University Press, Cambridge.

Chapter 24

Vital Staining Methods Used in the Analysis of Cell Sorting in Dictyostelium discoideum

CORNELIS J. WEIJER AND CHARLES N. DAVID

Zoologisches Institut der Universität München
D-8000 München 2, Federal Republic of Germany

JOHN STERNFELD

State University College at Cortland
Cortland, New York 13045

I. Introduction

There has been a long history of experiments investigating the role of cell sorting during development of the slime mold *Dictyostelium discoideum*. Early experiments of Bonner (1952, 1959) indicated that slugs contain cells in the anterior part which can be stained with the vital dye neutral red. Bonner observed that when neutral red-stained slugs were mixed, new slugs formed that again showed a stained anterior part (Bonner, 1952). He proposed that the pattern was reformed by cell sorting (Bonner, 1959). Since there was some concern that transfer of the dye might be partly responsible for the observed result, similar experiments using [³H]thymidine labeling were done by Takeuchi (1969). These experiments confirmed that slug reformation is brought about by cell sorting.

Subsequent work has led to the idea that cell sorting is also responsible for pattern formation in normal development. It was shown that cells separated on density gradients (Takeuchi, 1969; Bonner *et al.*, 1971; Maeda and Maeda, 1974; Feinberg *et al.*, 1979; Weijer *et al.*, 1984b) and cells grown under different conditions (Leach *et al.*, 1973) would sort out from each other and preferen-

METHODS IN CELL BIOLOGY, VOL. 28

tially form prespore or prestalk cells. In these experiments a variety of markers was used to follow cell fate: neutral red staining, [³H]thymidine labeling, and genetic markers. More recently there have been a number of reports assaying cell sorting by labeling cells with the fluorescent dyes fluorescein isothiocyanate (FITC) and rhodamine XR isothiocyanote (XRITC) (Springer and Barondes, 1978; Feinberg *et al.*, 1979; Weijer *et al.*, 1984a,b; Siu *et al.*, 1983; McDonald and Durston, 1984; Noce and Takeuchi, 1985).

The vital dyes neutral red and nile blue have also been used extensively in the investigation of the mechanism involved in cell sorting (Durston and Vork, 1978; Matsukuma and Durston, 1979; Sternfeld and David, 1981) and in studies of cell type proportioning. The proportion of prestalk and anteriorlike cells and their formation under regulation conditions has been followed after restaining with neutral red (Sternfeld and David, 1982; Weijer and Durston, 1985).

II. Vital Staining with Neutral Red and Nile Blue

The vital dyes neutral red and nile blue provide a rapid and convenient method of marking certain cell types in *Dictyostelium* (Fig. 1). These dyes are accumulated in large granules in the prestalk, anteriorlike, and prebasal disk cells; prespore cells are stained only weakly (Fig. 2, and Yamamoto and Takeuchi, 1983). This difference accounts for the darkly staining anterior tissue and lightly staining posterior tissue in slugs. Neutral red and nile blue are the vital dyes most commonly used to stain *Dictyostelium* cells. Their toxicity is low and they cause cells to be highly visible both in transmitted and reflected light.

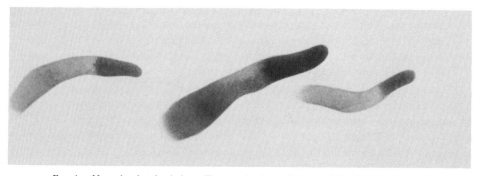

Fig. 1.   Neutral red-stained slugs. The anterior (prestalk) part of the slugs is stained.

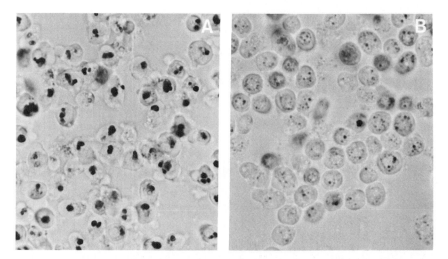

FIG. 2.   Neutral red-stained, dissociated slug cells. (A) Cells from a dissociated anterior region. Note the large stained autophagic vacuoles in the cells. (B) Cells from a dissociated posterior region. The few cells that contain stained vacuoles are anteriorlike cells.

A.   Staining Procedure

Staining is routinely accomplished by adding the dye during the cell harvesting process. The cells are stained rapidly by exposing them to a relatively high concentration of dye for a brief time.

1. Cells are harvested and washed by centrifugation (200 g, 3 minutes) and resuspension in cold 17 mM NaK-phosphate buffer, pH 6.8.
2. Fresh buffer is added to the pellet to bring the total volume to 2 ml.
3. Four or five drops of a stock solution of 0.2 mg/ml neutral red or 0.1 mg/ml nile blue in distilled water are added with a Pasteur pipet.
4. The cell pellet is resuspended in the concentrated stain by pipeting and then immediately diluted by the addition of more buffer to fill the centrifuge tube. The contents of the tube are quickly mixed and then centrifuged.
5. The cells are washed once or twice more in buffer until no more stain remains in the supernatant.

The procedure above is designed to stain $1-2 \times 10^8$ cells ($\sim$0.1 ml cell pellet). If the cell pellet is larger or smaller than this, the amount of dye should be appropriately adjusted. The duration of staining (step 4) should be only 10–15 seconds prior to dilution; longer periods cause overstaining of the cells.

Although these dyes are toxic at very high doses, this is not a concern in the procedure described here. Moderate overstaining, however, does delay development. Six to eight drops of the stock solution will cause a delay of about 2–3 hours in slug formation compared with unstained cells. Ten drops causes a delay of 4–6 hours and results in an observable increase in the number of cells that are left on the agar and do not participate in development.

The toxic effects of the stains vary with cell "age." Toxicity is greatest if vegetative cells are stained. It is best to stain cells which have been starved for 4–6 hours.

B.   Stability of the Dyes in Stained Cells

Use of these dyes in sorting experiments requires that they be stable cell markers. Neutral red and nile blue have been shown to be stable in slug cells by mixing neutral red-stained cells with nile blue-stained cells and permitting slugs to re-form (Sternfeld and David, 1981). The slugs appeared purple in color, but even after migrating for 30 hours individual cells contained either red or blue granules, indicating the dyes are not exchanged between cells.

At other developmental stages (vegetative cells and aggregation-competent cells), however, the dyes are not so stably bound, and cell mixing experiments demonstrate exchange of dye between cells. Also, prestalk cells, as they differentiate into stalk cells during culmination, appear to lose some of their stain to surrounding tissue.

III.   Vital Staining with XRITC

The rhodamine derivative XRITC (Serva no. 34282) has been used extensively to stain cells vitally. XRITC is covalently bound to cell surface proteins when live cells are incubated with the dye. After binding, the dye is internalized and appears to be located in lysosomes (Fig. 3). All cell types are labeled with XRITC. The label does not appear to be toxic and does not interfere with cell differentiation. Vegetative-stage cells labeled with XRITC can be followed through development to spore and stalk formation.

Staining Procedure

1. Prepare a solution of XRITC (50 mg/ml) in dimethyl sulfoxide (DMSO) and freeze ($-80°C$) in 10- or 20-μl aliquots.
2. Dissolve 10 μl dye/ml of phosphate buffer (potassium phosphate, pH 7.0, KK2). A precipitate forms which can be partly dissolved by sonication (1–

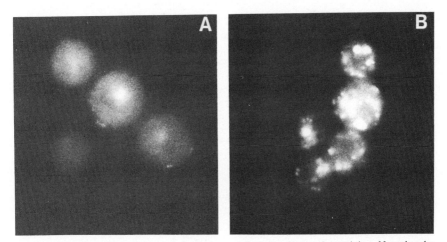

FIG. 3. XRITC-Stained cells. (A) Cells photographed 15 minutes after staining. Note that the staining is still mostly on the cell surface. (B) Stained cells after 30-minute internalization of the dye. Note that most of the dye is in vacuoles.

5 minutes). Pellet the bigger pieces by low-speed centrifugation (200 g, 3 minutes).

3. Wash cells in KK2, prepare a cell pellet, and add the dye solution to the pellet. Optimal staining occurs when the cell concentration does not exceed 5×10^7–ml of dye solution.

4. Incubate the cells under continuous shaking for at least 30 minutes at room temperature in the dark.

5. Wash cells three times by centrifugation. Resuspend the cells at 10^7/ml and shake for 30 minutes at room temperature in the dark. During this time the cells lose a substantial amount of dye.

6. Cells are washed two to three times by centrifugation and are ready for use.

It is important to treat control cells with 1% DMSO for comparable periods of time to avoid possible complications caused by slight differences in rate of development due to the exposure of stained cells to 1% DMSO during staining.

IV. Vital Staining with FITC

FITC is a less suitable stain for long-term experiments, since it is no longer visible after internalization in acidic lysosomes, due to the strong pH dependence of its fluorescence. However, in short-term experiments, where not much FITC is internalized, it may be a suitable label. Another alternative is to fix the FITC-labeled cells before observation and thereby raise the lysosomal pH (Noce and Takeuchi, 1985). The dye is then visible again.

Staining Procedure

The protocol for staining cells with FITC is the same as that described for XRITC. It is important to avoid bright light, since FITC is very sensitive to photobleaching.

V. Double-Label Staining of Cells with XRITC and a Cell Type-Specific Antibody

In many experiments it is desirable to follow the differentiation state of the rhodamine-labeled cells. We have therefore developed a procedure which allows the simultaneous detection of the XRITC label and anti-cell surface antibodies, using a sandwich technique with an FITC-labeled second antibody.

Procedure

1. Slugs containing cells labeled with XRITC are dissociated to a single-cell suspension. (We have found a 10-minute treatment of slugs with Cellulase Onozaku R-10 (1%) in KK2 buffer, pH 6.0, containing 5 mM EDTA, to be very effective.) Afterwards the cells are washed once or twice (10 seconds in an Eppendorf centrifuge) with KK2 and resuspended as a concentrated suspension in KK2.
2. Cells are fixed by adding one volume of neutralized formaldehyde (8%, pH 7.0).
3. A drop of the cell suspension is put on a microscope slide and air-dried. It is important not to use chemically cleaned slides, since the cells do not stick very well to those slides.
4. Slides are stained with a mixture of an anti-cell surface monoclonal antibody (MUD1, an anti-prespore cell surface antibody described by Krefft *et al.*, 1984) and an FITC-labeled goat anti-mouse antibody (Medac 1:40 dilution; 20 μl/drop of cells) and incubated for at least 30 minutes in the dark.
5. The slides are rinsed once carefully in KK2 and a coverslip mounted with a 50% glycerin–Tris mixture, pH 8.0. The slides are now ready for examination in a fluorescence microscope equipped with suitable rhodamine and fluorescein filter combinations.
6. Slides can be kept in the dark at 4°C for prolonged periods of time.

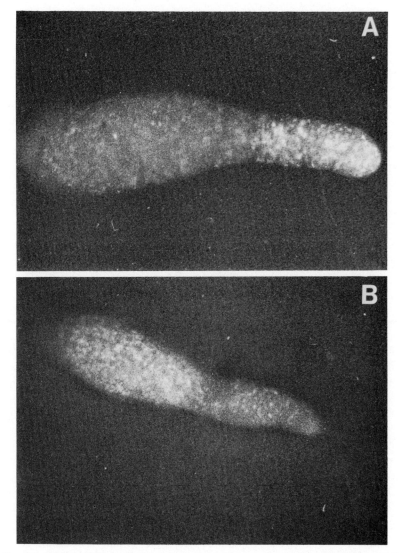

FIG. 4.   XRITC-Labeled cells that have sorted in slugs made of 10% labeled cells and 90% unlabeled cells. (A) Labeled cells have sorted to the prestalk zone. (B) Labeled cells have sorted to the prespore zone.

VI. Sorting Assay

Sorting experiments are performed by mixing a small amount of XRITC-labeled cells (1–10%) with unlabeled cells. The mixtures are allowed to develop to slugs, and the sorting pattern can be observed under a fluorescence microscope (Fig. 4). Individual cells can easily be detected. The best observations (avoiding reflections and lens effects) are made when slugs are gently squashed under a coverslip on an agar surface.

In order to quantitate cell sorting, slugs can be dissociated, the cells fixed in neutralized formaldehyde, and the percentage of XRITC-labeled cells determined. The label is conserved due to the virtual absence of cell division during development; there is at most one doubling of cell number during development (Zada Hames and Ashworth, 1978).

VII. Genetic Markers

It also is possible to use genetically marked strains in cell sorting assays. Sorting experiments using strains marked with certain mutations like spore size (Bonner *et al.*, 1971) and temperature-sensitive growth have been described (Leach *et al.*, 1973). In principle, any easily assayable mutation that does not affect sorting behavior, like drug resistance, can be used. In their classical study of the cell sorting properties of cells grown under different conditions, Leach *et al.* (1973) used both normal strains and strains genetically marked with a temperature-sensitive mutation for growth. Wild-type and mutant cells grown under different conditions were mixed in different ratios at the vegetative stage and allowed to develop to fruiting bodies. The spores of the fruiting bodies were collected, replica-plated at both the permissive and the nonpermissive temperatures, and the percentage of mutant spores determined. The drawback of this study was that it did not determine whether both strains had always efficiently coaggregated, and what role cell division played under these circumstances.

It is, however, possible to assay cell sorting at the slug stage using genetic markers. Slugs can be cut into pieces using vital staining to estimate the location of the prespore–prestalk boundary and the percentage of mutant cells determined in each region.

An interesting variation on the use of genetically marked strains is to use mutants of known antigens, whose mutation can be detected with an antibody. This mutation can then be used directly in conjunction with a cell type-specific antibody to follow cell differentiation.

VIII.   Formation of Slugs

For cell sorting experiments it is desirable to have large, clean slugs. This can be accomplished as follows:

1. After the last washing step of the staining procedure, the cell pellet is centrifuged at high speed in a clinical centrifuge for 30 seconds and the supernatant completely removed.
2. The cell pellet is taken up in a Pasteur pipet and a streak of cells ~2 mm wide is made across the middle of a Petri dish containing fresh 2% nonnutrient agar.
3. After 15–20 minutes the preparation should be examined for any liquid surrounding the streak. If present, it should be removed with a pipet or by absorption with filter paper.
4. Incubate the agar dish in a moist environment (wet paper towels in a plastic box) in the dark to permit slug formation.

IX.   Manipulating Slug Tissue

When mixing tissue from a number of slugs, slugs can be picked up with a small-bore (150–200 μm) Pasteur pipet attached to a mouth pipet. Several slugs can be transferred at once in this manner. But note that the tissue is damaged if held in the pipet for more than a few seconds. This is especially true of anterior tissue which should be held in the pipet no longer than it takes quickly to pick up two or three anteriors. Individual slugs or portions of slugs can also be conveniently picked up on the end of a steel needle and transferred to a mass of slug tissue on agar for mixing.

Gentle mixing of slug tissue is best done by stirring it with an eyelash attached to a Pasteur pipet or with the steel needle used for transferring slug tissue. Mixing can also be achieved by gentle pipeting of the tissue mass. If tissue is vitally stained with neutral red or nile blue it is essential to mix gently, since vigorous mixing lyses some cells and releases dye which is then taken up by other cells. When mixing XRITC-labeled tissue this consideration is less important.

To graft slug tissue, a tool such as an aluminum foil microknife is needed. It is made by attaching a small piece of foil to the tip of a Pasteur pipet. The microknife is easy to make and difficult to damage. It can be trimmed into different shapes, bent at various angles, and is strong enough to slice through agar.

The major problem in grafting is desiccation of the tissues as they are moved around. Thus it is best to perform all manipulations in a single agar dish. The slugs or pieces of slugs can be picked up with the microknife and transferred to the graft site, on clean agar. The slugs should be placed side by side on the agar in the correct orientation. The slugs are cut at the point desired and the unwanted tissue scraped away. The portions for grafting are then butted together and the tip of the microknife is used to mix the cells at the graft region.

X. Concluding Remarks

Vital dyes and XRITC labeling of cells provide simple methods to visualize single cells in living slugs. Thus the behavior of these cells can be followed. Such methods have provided the basic experimental evidence for cell sorting in *Dictyostelium*. While vital dyes only stain prestalk and anteriorlike cells, XRITC stains all cell types and thus permits investigation of the behavior of prespore cells as well as prestalk cells.

Genetic and radioactive methods can also be used to label cells. However, these methods have the drawback that the behavior of cells cannot be followed directly *in vivo*. The fate of marked cells can only be determined following clonal analysis of the genetic phenotype or autoradiography in the case of radioactive labeling. Hence these methods have been little used in recent years.

References

Bonner, J. T. (1952). *Am. Nat.* **86**, 79–89.
Bonner, J. T. (1959). *Proc. Natl. Acad. Sci. U.S.A.* **45**, 379–384.
Bonner, J. T., Sieja, J. W., and Hall, E. M. (1971). *J. Embryol. Exp. Morphol.* **25**, 457–565.
Durston, A. J., and Vork, F. (1979). *J. Cell Sci.* **36**, 261–279.
Feinberg, A. P., Springer, W. R., and Barondes, S. H. (1979). *Proc. Natl. Acad. Sci. U.S.A.* **76**, 3977–3981.
Krefft, M., Voet, L., Gregg, J., Mairhofer, H., and Williams, K. L. (1984). *EMBO J.* **3**, 201–206.
Leach, C. K., Ashworth, J. M., and Garrod, D. R. (1973). *J. Embryol Exp. Morphol.* **29**, 647–661.
McDonald, S. A., and Durston, A. J. (1984). *J. Cell Sci.* **66**, 195–204.
Maeda, Y., and Maeda, M. (1974). *Exp. Cell Res.* **84**, 88–94.
Matsukuma, S., and Durston, A. J. (1979). *J. Embryol. Exp. Morphol.* **50**, 243–252.
Noce, T., and Takeuchi, I. (1985). *Dev. Biol.* **109**, 157–164.
Siu, C. H., Des Roches, B., and Lam, T. Y. (1983). *Proc. Natl. Acad. Sci. U.S.A.* **80**, 6596–6600.
Springer, W. R., and Barondes, S. H. (1978). *J. Cell Biol.* **78**, 937–942.
Sternfeld, J., and David, C. N. (1981). *Differentiation* **20**, 10–21.
Sternfeld, J., and David, C. N. (1982). *Dev. Biol.* **93**, 111–118.

Takeuchi, I. (1969). *In* "Nucleic Acid Metabolism, Cell Differentiation and Cancer Growth" (E. V. Cowdry and S. Seno, eds.), pp. 297–304. Pergamon, Oxford.

Weijer, C. J., and Durston, A. J. (1985). *J. Embryol. Exp. Morphol.* **86,** 19–37.

Weijer, C. J., Duschl, G., and David, C. N. (1984a). *J. Cell Sci.* **70,** 133–145.

Weijer, C. J., McDonald, S. A., and Durston, A. J. (1984b). *Differentiation* **28,** 13–23.

Yamamoto, A., and Takeuchi, I. (1983). *Differentiation* **24,** 83–87.

Zada-Hames, I., and Ashworth, J. M. (1978). *Dev. Biol.* **63,** 307–320.

Chapter 25

Isolation of Developmentally Regulated Genes

REX L. CHISHOLM

Department of Cell Biology and Anatomy
Northwestern University Medical School
Chicago, Illinois 60611

I. Introduction

Dictyostelium discoideum has proved a popular and extremely useful system for the investigation of cellular differentiation and the mechanisms by which a developmental program is regulated (see Loomis 1975, 1982, for extensive reviews). In large part these studies have been facilitated by the ability to characterize the expression of developmentally regulated genes and the mRNAs they encode. A relatively large number of developmentally regulated *Dictyostelium* genes have been cloned, including both cDNAs and genomic DNAs. Some of these cloned genes encode known proteins, including actin (Kindle and Firtel 1978), myosin (De Lozanne *et al.*, 1985), discoidin (Williams *et al.*, 1979; Rowekamp *et al.*, 1980), and a *ras* oncogene homolog (Reymond *et al.*, 1984). An even larger number of developmentally regulated genes which encode unknown polypeptides, but serve as useful markers of development, have been reported (Rowekamp and Firtel, 1980; Barklis and Lodish, 1983; Mehdy *et al.*, 1983; Mangiarotti *et al.*, 1981). The goal of this chapter is to consider the approaches available for the isolation of developmentally regulated genes from *Dictyostelium,* and some of the advantages and drawbacks of each.

II.   cDNA or Genomic Clones?

One of the first considerations is the type of library to use to isolate regulated clones: genomic DNA or cDNA. Each type of library has its advantages and its drawbacks, and each has generated useful developmentally regulated clones. Genomic clones have traditionally been long, usually >15 kb in length, and this can complicate the identification of developmentally regulated clones. Although detailed structural information concerning *Dictyostelium* genes is limited, what is known suggests that their introns are relatively small, as is the distance between adjacent genes (Firtel and Cockburn, 1976; Firtel and Kindle, 1975; Kimmel and Firtel, 1980, 1982). The *Dictyostelium* genome consists of ~35,000 kb of single-copy DNA, and solution hybridization analysis has led to the conclusion that between 10,000 and 12,000 genes are expressed during growth and development (Firtel, 1972; Blumberg and Lodish, 1980; Jaquet *et al.,* 1981). This is most likely an underestimate of the total number of genes, as these experiments have not addressed gene expression during macrocyst development, which presumably requires additional gene expression. Thus, there is a maximum of roughly 2.5 kb of DNA per gene. Consequently, a typical 15-kb *Dictyostelium* genomic DNA clone generally contains more than one gene, perhaps several. Because the evidence suggests that there is no clustering of developmentally regulated genes, many 15- to 20-kb genomic clones will carry both regulated and unregulated genes. This represents a potential problem for the application of ''plus-minus'' screening techniques (see Section V). This technique distinguishes clones carrying regulated genes from those carrying unregulated genes by their ability to hybridize to a probe specific for development, but not to one specific for growing cells. If a cloned DNA fragment carries both a regulated and an unregulated gene, both probes will react with that clone, despite the fact that it contains a regulated gene. That regulated gene would be overlooked. Use of a subtracted probe, described below, should avoid this potential problem. It may also be minimized by the use of genomic fragments 1–2 kb in length when preparing the library. Unfortunately this will dramatically increase the number of clones which must be screened to find a particular gene. For example, if 15,000 clones containing 20-kb inserts are required for a 99% chance of finding a sequence of interest, reducing the size of inserts to 1.5 kb increases the number of clones which it would be necessary to screen to nearly 200,000! However, if the goal is isolation of genes with a given pattern of expression, and not a particular gene, it is usually not necessary to screen that many clones.

A good genomic library has the advantage of having (at least in principle) one copy of every gene in the cell. It would contain, for example, genes induced at 15 hours of differentiation as well as those expressed in growing cells. In contrast, a cDNA library has the advantage that preparation of RNA from the appropriate cells can result in a significant enrichment for a message of interest due to the fact that a cDNA library will contain clones for a given mRNA in proportion to the mass representation of that message in the mRNA population. For example, if one were interested in isolating a clone for an abundant mRNA amounting to 1% of the mRNA in the cell, ~1% of the clones in the cDNA library should contain sequences for that message. Thus instead of 1 in 15,000–20,000 clones containing the sequences of interest, as would be the case for the genomic library, the cDNA library would contain 1 clone for the desired sequence per 100 cDNA clones. Clearly, this enrichment can be a significant argument in favor of the use of cDNA clones for certain projects.

However, depending on the screening technique used, it may be almost as easy to screen 30,000 clones as it is to screen 500. In such cases the enrichment for sequences of interest provided by cDNA clones may be of little advantage. The amount of time saved by not having to construct a new library each time a new gene is to be isolated can be a persuasive argument in favor of genomic libraries.

Once a gene of interest has been identified, cDNA clones have the advantage that they represent the mRNA sequences, while genomic clones carry intergenic spacer and introns in addition to the message sequences. cDNA clones make determining the structure of the message, which can often be difficult to deduce directly from the genomic sequences, much easier.

Based on this discussion it should be clear that careful consideration of the type of clone bank to be used for a gene isolation project, and of the configuration that bank will take, can greatly improve chances of success in finding the desired genes, as well as minimizing effort required for that success.

III. Construction of a Genomic Library

The construction of genomic libraries has been well covered in "Molecular Cloning" (Maniatis *et al.*, 1982). However, a couple of recent developments are worth noting. Several new cloning vectors have become available. Among them, the λEMBL phages represent a significant advance (Frischauf *et al.*, 1983). Because they carry a polylinker region surrounding the *Bam*HI cloning site, the removal of the cloned fragment of DNA is facilitated. The structure of these vectors also simplifies library construction. Most vectors used in genomic library construction require removal of the "stuffer" fragment—the DNA which is replaced by the DNA to be cloned—to achieve high efficiencies. The EMBL vectors eliminate the need to remove the insert physically with a neat trick. By digesting the vector DNA with two different enzymes (e.g., *Bam*HI and *Sal*I), small eight-nucleotide fragments are clipped from the ends of the stuffer fragment. Due to their small size, these small double-digestion fragments fail to precipitate in isopropanol at 0°C. The removal of these small fragments prevents the reinsertion of the vector "stuffer" fragments during ligation, because the remaining fragments of the "stuffer" have different termini than the functional vector arms. Consequently, the stuffer fragment can no longer compete with added genomic DNA for insertion into the vector.

It has been reported that certain DNA sequences are unstable when propagated through *Escherichia coli* (Wyman et al., 1985; Nader et al., 1985). This appears to be particularly true of repeats and palindromes. The consensus among laboratories involved in construction of *Dictyostelium* gene banks is that this instability is a serious problem with *Dictyostelium*. A number of laboratories have experienced difficulties obtaining genomic libraries with reasonable gene representation. Despite its relatively small genome size, "complete" gene libraries have been difficult to generate. In addition, several investigators have noted instability of sequences during the propagation of specific clones. The presence of regions with extremely high A-T base compositions (Firtel et al., 1979; Kimmel and Firtel, 1983) may contribute to this instability. A partial solution may be the use of recombination-deficient E. coli strains, in particular, those carrying recBC and sbc mutations (Leach and Stahl, 1983). Such an approach has recently solved similar problems of gene representation and instability for *Physarum* clones.

As mentioned in the previous section, there may be advantages to construction of genomic libraries with smaller inserts. The use of smaller genomic inserts may also help minimize problems of sequence instability. The vector of choice for such projects is probably λgt10 (Hyunh et al., 1984). This vector will accept fragments in the 1- to 7-kb range. Unlike the EMBL phage, this phage is an "insertion" rather than a "replacement" vector. Religation of the vector without an insert produces a viable phage. This means that precautions must be taken to ensure that a reasonable proportion of the clones in the library actually contain *Dictyostelium* DNA. Fortunately, alkaline phosphatase treatment of the vector prior to the ligation of insert DNA will accomplish this. One note of caution needs to be sounded here. It is important to avoid the use of excess enzyme when removing terminal phosphates. Low levels of contaminating exonuclease activities present in most phosphatase preparations will remove terminal nucleotides in addition to terminal phosphates, resulting in difficulties removing cloned fragments. With most "molecular biology"-grade calf intestine alkaline phosphatases, 0.1 units of enzyme is sufficient to treat 1 μg of λ-vector arms.

Alternatively, plasmid vectors can be used for "small-insert" libraries. Plasmid vectors offer a significant advantage in that the insert represents a fairly significant proportion of the sequences present in DNA preparations of the clones of interest. This is in contrast to the relatively small 2–5% of the total DNA, which represents the sequences of interest when λ-phage vectors are used; e.g., compare 42 kb of vector DNA (with λ) to 3 kb with typical plasmids. Because of this, it is often necessary to subclone phage inserts into plasmids. The use of a plasmid vector in the original library construction avoids this step. However, the efficiency that results from λ *in vitro*-packaging systems, and the ease with which λ libraries can be stored and screened, probably gives them the advantage.

IV. Construction of cDNA Libraries

Construction of cDNA libraries has been well covered in a variety of recent publications, and the reader is referred to these (Maniatis *et al.*, 1982; Huynh *et al.*, 1984). Again however, some consideration of vector choice may be useful. The development of really effective phage vectors for the construction of cDNA libraries has dramatically improved the efficiency of cDNA cloning. Efficiencies of 10^7 clones/μg of double-stranded cDNA have been reported with the λgt10 and λgt11 vectors (Young and Davis, 1983; Huynh *et al.*, 1984). Despite reports of *E. coli* transformation efficiencies, of 10^7 or 10^8/μg, many laboratories have experienced difficulty achieving these levels, thus the efficiency inherent in λ *in vitro* packaging is advantageous.

The small genome size of *Dictyostelium* does not help much when working with cDNA clones, and efficiencies approaching those required for mammalian cells are necessary when generating cDNA libraries if a reasonable representation of the cell messages is desired. An average mammalian cell contains roughly 350,000 mRNA molecules, while a *Dictyostelium* cell contains ~200,000 mRNA molecules. For this reason one needs *Dictyostelium* cDNA libraries nearly the size of those needed for mammalian systems.

V. Screening Techniques

Once one has constructed the desired library, the question becomes how to find the desired developmentally regulated genes. If something is known about the gene of interest, such as the protein it encodes, standard approaches like the use of synthetic oligonucleotide probes or screening of an expression library with a specific antibody can be employed to identify clones containing the coding sequences for that protein. It is also possible to identify and isolate genes which have a particular pattern of expression. The following sections discuss some of the approaches that can be employed to isolate developmentally regulated genes.

A. Differential Hybridization

Differential hybridization is perhaps the most widely used technique for the identification of differentially expressed genes. In the case of *Dictyostelium* it has been used not only to isolate genes expressed at a particular time during development (Rowekamp and Firtel, 1980; Mangiarotti *et al.*, 1981), but also to identify genes expressed in a cell type-specific manner (Medhy *et al.*, 1983; Barklis and Lodish, 1983). Duplicate filters containing colonies (if plasmid clones are being screened) or duplicate plaque-lifts (if a phage vector was used) are made. One of the filters is incubated with a "plus" probe—one which reacts with the desired clones—and the duplicate filter is hybridized with a "minus" probe, which should fail to react with the desired clones. This type of screening has also been called "plus–minus" screening for the obvious reasons. One example of the application of differential hybridization is the isolation of *Dictyostelium* genes which are expressed in a cell type-specific fashion. Cells were allowed to develop to a stage at which the cells have begun the process of biochemical differentiation along either the spore or stalk cell pathways [either migrating slugs (Mehdy *et al.*, 1983) or 16-hour filter-developed cells (Barklis and Lodish, 1983)]. The aggregates were disaggregated and the two cell types, "prespore" and "prestalk," were separated by centrifugation in Percoll (Ratner and Borth, 1983; Tsang and Bradbury, 1981) and mRNA prepared from the purified fractions. Probes were made by kinase-labeling the two mRNA populations. If the goal were isolation of a "prespore"-specific gene, the probe made from prespore cell RNA becomes the "plus" probe, and the probe made prestalk cell RNA the "minus" probe. Obviously, genes specifically expressed in prespore cells will be absent from the prestalk cells. Hybridization of duplicate filters with these two probes allows identification of clones which reacted with the prespore probe, but not the prestalk probe. It can be used in any case where a "plus" and "minus" state can be identified.

The probes can take two forms: kinase-labeled RNA or single-stranded cDNA. Each type of probe has its advantages and disadvantages. Kinase-labeled RNA actually gives the most complete sequence representation. Because the mRNA is fragmented (usually by treating it with 0.1 M NaOH for 30 minutes) to produce a distribution of 5' ends available for labeling, the label is more or less randomly introduced over the entire length of the mRNA sequence. In contrast, cDNA is usually synthesized using an oligo(dT) primer, which tends to bias the label incorporation toward the 3' ends of the mRNA. Alternatively, random primers (often prepared from limit DNase I digestion of calf thymus DNA) can be employed to help eliminate this bias. An advantage of cDNA is that label is distributed throughout the length of whatever the resulting fragments are, while kinased RNA fragments contain their label only at the 5' termini, resulting in significantly higher specific activities for the cDNA probes.

Possible sources of difficulty with differential hybridization stem from different colony sizes on the duplicate filters. Clearly, if there are more cells containing the clone being tested on one of the two duplicate filters, a stronger signal is likely to be seen for that particular clone, even if the mRNA represented by that cDNA is unregulated. Plaque lifts seem somewhat less susceptible to this problem. In both cases retesting of the clones is crucial.

B. Competitive Hybridization

This screening technique represents a variation on the differential screening discussed above. In this case the minus probe is unlabeled and is included in the same hybridization as the labeled plus probe (Mangiarotti et al., 1981). For example, if the goal were the isolation of genes expressed at 15 hours of differentiation, but not in growing cells, the hybridization would contain labeled 15-hour RNA, and a vast excess of growing-cell RNA. The idea is that the growing-cell RNA, being in excess, would compete for hybridization to any clone representing a message expressed in growing cells, dramatically reducing the hybridization of the labeled 15-hour RNA. Any clone which bound labeled RNA should represent a clone with the desired characteristics, avoiding the comparison of two different filters. In practice, however, it is necessary to compare hybridization of duplicate filters, one which has been hybridized in the presence of the competitor with one hybridized in the absence of competitor. The clones of interest show hybridization that is unaffected by the presence of competitor mRNA. A potential source of trouble is the loss of clones encoding sequences which are transcribed at low levels in the minus preparation, but are induced to much higher levels in the plus RNA sample. A good example of these is the prestalk class I genes we have described (Chisholm et al., 1984). They are present at low levels in growing cells and induced to higher levels at ~10 hours of development. Clones of this class would probably be missed in competitive hybridization screening.

C. Subtractive Hybridization

This is perhaps the most glamorous of the techniques, and it has received a lot of attention because it contributed to the cloning of the elusive T-cell receptor (Zimmerman et al., 1980; Sargent and Dawid, 1983; Davis et al., 1984). It is really a variation on the competitive hybridization theme. Labeled cDNA is made on the "plus" mRNA template, and purified away from the mRNA following alkaline hydrolysis of the template. The cDNA is then incubated with excess "minus" mRNA. Those cDNA molecules which were templated on a "minus" message will hybridize, while those templated on an mRNA with the desired regulation will have no mRNA complementary to them, and will remain single-stranded. Hydroxyapatite chromatography is then used to separate the double-stranded ("minus") cDNA from the unreacted single-stranded cDNAs. The labeled, single-stranded cDNA is then used to probe filters containing the clones to be screened. As with competitive hybridization, care must be taken to avoid the possible loss of sequences which might be present at low levels in the minus mRNA preparation. One advantage to this procedure is that the cDNA/mRNA "subtraction" is done in solution where it is possible to control the level of completion of the reaction (by controlling the c_0t), as well as the level of subtraction. It is (at least in theory) possible to ask for subtraction of all sequences which are differentially expressed in the two samples by less than a given amount, e.g., fivefold. Thus the probe might still contain a cDNA for a mRNA which is present in the minus mRNA, but at a level $<10\%$ of what is seen in the plus sample. One real advantage of this system is that a relatively large number of clones can be screened, since a positive signal is obtained only for clones of potential interest.

D.   Use of Antibodies and Expression Libraries

The development of efficient expression vectors and techniques for screening for clones by virtue of the peptide sequences they express has opened up new possibilities for the identification of developmentally regulated genes (Young and Davis, 1983). The same principles described above for nucleic acid probes can be applied to screening with antibodies. For example, differential screening of an expression library with antibodies which react with overlapping but not identical antigens has been employed in the isolation of clones for malarial antigens using patient sera (Stahl et al., 1984). This approach could be beneficially applied to Dictyostelium to isolate, among other things, developmentally regulated cell surface antigens. Antibodies made against growing cells and against aggregated cells should be used to screen duplicate plaque lifts of a λgt11 library. Clones reacting with the antibody against aggregated cells, but not the growing cells, would be selected for further study. These clones could then be tested by a differential hybridization using the appropriate cDNA or mRNA probes to confirm their pattern of expression. Alternatively, it may be possible to use an antibody-screening procedure analogous to the subtractive hybridization approach described above. In the case of the example just mentioned, the antibody prepared against aggregated cells could be absorbed with growing cells to prepare a "subtracted" antibody specific for antigens present in the aggregated cells, and then used to probe an expression library for clones which react.

VI.   Conclusions and Future Prospects

The power of the techniques discussed above for the isolation of genes having a particular pattern of regulation is limited only by imagination. Using these techniques it would be possible to identify genes induced by heat shock, by starvation, expressed at a particular time in the cell cycle, in a particular cell type, or any other fashion which allows the investigator to identify a set of conditions where the genes of interest are expressed and another where they are not. The use of developmental mutants in such gene selection schemes may increase the kinds of schemes which can be devised. For example, it may be possible to use null mutants and subtractive hybridization to isolate a wild-type copy of the mutant gene.

The recent development of successful protocols and vectors for introduction of DNA into *Dictyostelium* (see Nellen *et al.*, Chap. 4, this volume) opens the very exciting possibility of isolation of developmentally important genes by direct complementation of developmental mutants. In this scheme, a developmental mutant would be transformed with a library of *Dictyostelium* genes cloned into a transformation vector, preferably one capable of extrachromosomal replication. Cells which take up and express DNA which encodes the wild-type function which is defective in the mutant should be able to develop normally. Recovery of the exogenous DNA from resulting developmentally competent cells enables identification of the gene responsible for the developmental defect in the mutant used in the initial complementation.

A great deal has already been learned from the developmentally regulated genes which have been isolated to date. The general scheme of the developmental program has become clearer as a consequence of many of these studies. The availability of cloned developmentally regulated genes has facilitated investigation of the mechanisms by which the developmental program is regulated, and has given us significant insight into the molecular details of that regulation. As the techniques for isolation and expression of genes become more sophisticated, and the reintroduction of those genes facilitates functional analysis (see, for example, De Lozanne, this volume, Chapter 27), even more is likely to be learned from a molecular approach to the study of *Dictyostelium* development.

REFERENCES

Barklis, E., and Lodish, H. F. (1983). *Cell* **32**, 1139–1148.
Blumberg, D. D., and Lodish, H. F. (1980). *Dev. Biol.* **78**, 285–301.
Chisholm, R. L., Barklis, E., and Lodish, H. F. (1984). *Nature (London)* **310**, 67–69.
Davis, M. M., Cohen, D. I., Nielsen, E. A., Steinmetz, M., Paul, W. E., and Hood, L. (1984). *Proc. Natl. Acad. Sci. U.S.A.* **81**, 2194–2198.
De Lozanne, A., Lewis, M., Spudich, J. A., and Leinwand, L. (1985). *Proc. Natl. Acad. Sci. U.S.A.* **82**, 6807–6810.
Firtel, R. A. (1972). *Dev. Biol.* **66**, 363–377.
Firtel, R. A., and Cockburn, A. F. (1976). *J. Mol. Biol.* **102**, 831.
Firtel, R. A., and Kindle, K. (1975). *Cell* **5**, 401–411.
Firtel, R. A., Timm, R., Kimmel, A. R., and McKeown, M. (1979). *Proc. Natl. Acad. Sci. U.S.A.* **76**, 6206–6210.
Frischauf, A. M., Lehrach, H., Poustka, A., and Murray, N. (1983). *J. Mol. Biol.* **170**, 827–842.

Huynh, T. V., R. A. Young, and Davis, R. W. (1984). In "DNA Cloning Techniques: A Practical Approach" (D. A. Glover, ed.). IRL Press.

Jacquet, M., Part, D., and Felenbok, B. (1981). Dev. Biol. 81, 155–166.

Kimmel, A. R., and Firtel, R. A. (1980). Nucleic Acids Res. 8, 5599–5610.

Kimmel, A. R., and Firtel, R. A. (1982). In "The Development of Dictyostelium" (W. L. Loomis, ed.). Academic Press, New York.

Kimmel, A. R., and Firtel, R. A. (1983). Nucleic Acids Res. 11, 541–550.

Kindle, K., and Firtel, R. A. (1978). Cell 15, 763–778.

Leach, D. R. F., and Stahl, F. W. (1983). Nature (London) 305, 448–451.

Loomis, W. L. (1975). "Dictyostelium discoideum: A Developmental System." Academic Press, New York.

Loomis, W. L. (1982). "The Development of Dictyostelium." Academic Press, New York.

Mangiarotti, G. et al. (1981). Nucleic Acids Res. 9, 947–963.

Maniatis, T., Fritsch, E., and Sambrook, J. (1982). "Molecular Cloning: A Laboratory Manual." Cold Spring Harbor Laboratory, Cold Spring Harbor, New York.

Mehdy, M. C., Ratner, D., and Firtel, R. A. (1983). Cell 32, 763–771.

Nader, W. F., Edlind, T. D., Huetermann, A., and Sauer, H. W. (1985). Proc. Natl. Acad. Sci. U.S.A. 82, 2698.

Ratner, D., and Borth, W. (1983). Exp. Cell Res. 143, 1–12.

Reymond, C. D. Gomer, R. H., Mehdy, M. C., and Firtel, R. A. (1984). Cell 39, 141–148.

Rowekamp, W., and Firtel, R. A. (1980). Dev. Biol. 79, 409–418.

Rowekamp, W., Poole, S., and Firtel, R. A. (1980). Cell 20, 495–505.

Sargent, T. D., and Dawid, I. B. (1983). Science 222, 135–139.

Stahl, H. D., Coppel, R. L., Brown, G. V., Baint, R., Lingelbach, K., Cowman, A. F., Anders, R. F., and Kempt, D. S. (1984). Proc. Natl. Acad. Sci. U.S.A. 81, 2456–2460.

Tsang, A. S., and Bradbury, J. M. (1981). Exp. Cell Res. 132, 433–441.

Williams, J. G., Lloyd, M., and Devine, J. M. (1979). Cell 17, 903–913.

Wyman, A. R., Wolfe, L. B., and Botstein, D. (1985). Proc. Natl. Acad. Sci. U.S.A. 82, 2880.

Young, R. A., and Davis, R. W. (1983). Proc. Natl. Acad. Sci. U.S.A. 80, 1194–1198.

Young, R. A., and Davis, R. W. (1985). In "Genetic Engineering," (J. Setlow and A. Hollaender, eds.). Vol. 7. Plenum, New York.

Zimmerman, C. R., Orr, W. C., Lellerc, R. F., Barnard, E. C., and Timberlake, W. E. (1980). Cell 21, 709–715.

Chapter 26

A Strategy to Study Development and Pattern Formation: Use of Antibodies against Products of Cloned Genes

RICHARD H. GOMER

Department of Biology
Center for Molecular Genetics
University of California, San Diego
La Jolla, California 92093

I. Introduction

In order to study pattern formation and cellular development, one needs to be able to localize within a multicellular structure those cells expressing developmentally regulated genes. A general strategy for studying pattern formation and development at the molecular level would thus be to localize the products of specific cloned developmentally regulated genes. Because of its simplicity, *Dictyostelium discoideum* is an excellent system for studying multicellular developments and pattern formation (see Loomis, 1975, for review). During development, *Dictyostelium* cells differentiate into two isolatable cell types: prespore and prestalk cells (Ratner and Borth, 1983). Several groups have isolated developmentally regulated genes that are specifically expressed in one of the two cell types (Mehdy *et al.*, 1983; Barklis and Lodish, 1983). In order to study the ontogeny of prestalk and prespore cells, and the pattern of regulation of cell type-specific genes, we have made antibodies against the products of developmentally regulated genes.

Antibodies against the product of a cloned gene can be used in a variety of studies to complement the examination of the gene itself. Western blots allow a determination of the protein's molecular weight, spectrum of variants, and distribution in cell fractions. Using immunofluorescence, antibodies can be used to determine both the subcellular location of a protein and its distribution within a

471

tissue. Because of the very small number of cells needed for immunofluorescence (and concomitant small scale of the assay), this technique is ideally suited for assaying extracellular substances or physiological conditions that affect the spatial and temporal pattern of expression of the gene in question.

Since the antibody is against the product of a cloned gene, the antibody can be used to assay for proper expression of a protein in cells transformed with vectors encoding an antisense RNA (see Nellen et al., Chap. 4, this volume). The effect of changes in a putative promoter region can be studied for proper temporal, subcellular, and cell type expression. The effect of mutations in heterologous genes on development and pattern formation can also be studied with such antibodies.

To make antibodies, we have chosen to use the λgt11 expression vector of Young and Davis (1983) to produce antigen. This is then used to make polyclonal antibodies in rabbits. The λgt11 system produces in bacteria large amounts of a fusion protein consisting of β-galactosidase and a gene of interest. The fusion protein has a molecular weight greater than the vast majority of the other bacterial proteins. This allows a simple purification based on excision from SDS–polyacrylamide gels. The antiserum against the fusion protein will contain antibodies directed against the *Escherichia coli* β-galactosidase portion of the fusion protein. We have found that antibodies against *E. coli* β-galactosidase do not react with antigens in any developmental stage of *Dictyostelium*, as assayed by Western blots and immunofluorescence. Other types of expression vectors could also be used, although some of the methods will be different from those described here.

Using a protein synthesized in bacteria as an antigen has many benefits. The foremost is that one can easily produce vast quantities of antigen. Another advantage is that the antibodies are directed against a selected region of a protein. For instance, if two proteins have both homologous and heterologous domains, appropriate selection of the coding region used for antigen production can yield an antibody that recognizes both proteins or an antibody that recognizes only one. A third advantage is that the bacteria will not be making most posttranslational modifications found in *Dictyostelium* proteins, reducing the possibility that the resulting antiserum will contain antibodies against a posttranslational modification such as a glycosyl group that might be found on products of more than one gene or that might be regulated in a manner other than that of the gene transcription and translation itself (Grant and Williams, 1983; Loomis et al., 1983; Knecht et al., 1984). Finally, the antiserum will be polyclonal and contain antibodies directed against (theoretically) a large number of epitopes on a relatively large segment of the protein (or the entire protein). This is an advantage over monoclonal antibodies, which can give incorrect results due to modification or inaccessibility of the binding site (Blose et al., 1982; Franke et al., 1983; Danto and Fischman, 1984).

λgt11 DNA has a unique *Eco*RI site located near the 3′ end of the *E. coli lac Z* gene, which codes for β-galactosidase. In the next section we discuss briefly the molecular biological technqiues for constructing the gene fusion with the *Dictyostelium* ORF (open reading frame) in the proper reading frame and orientation to produce DNA coding for a fusion protein.

II. Cloning Strategy

A. General Method

The insertion of a DNA fragment containing an ORF into the λgt11 expression vector is relatively straightforward. The detailed methods used can be obtained from Maniatis *et al.* (1982). After sequencing the gene or cDNA clone, a typically 500-bp restriction fragment containing an uninterrupted ORF (or one with a stop near the 3' end) is chosen. Unless the fragment can be readily cloned in proper reading frame into existing phage M13 polylinker restriction sites, the fragment is blunt-ended, and, depending on the reading frame, ligated into the *Hinc*II or *Sma*I site of the polylinker of an appropriate M13 vector. Since the M13 polylinker has an *Eco*RI site, all of these operations put an *Eco*RI site near one end of the *Dictyostelium* coding region fragment. In all these cases, the coding region becomes ligated in the M13 polylinker between the polylinker *Eco*RI site and *Hin*dIII site. A synthetic *Eco*RI linker is then ligated into the *Hin*dIII site of the M13 polylinker after the *Hin*dIII site has been blunted with the large (Klenow) fragment of DNA polymerase I (see Maniatis *et al.*, 1982).

The *Dictyostelium* coding region fragment is then closely flanked on either side by *Eco*RI sites. After sequencing the resultant M13 DNA to verify that the correct reading frame will be obtained (Sanger *et al.*, 1977; Gomer *et al.*, 1985), the *Eco*RI fragment is excised, purified by gel electrophoresis, and ligated into the λgt11 *Eco*RI site. Orientation of the fragment in the λgt11 is determined by restriction mapping and by the size of the resultant fusion protein. Fusion protein size can be predicted for both orientations by location of in-frame stop codons. The sequence for the λgt11 β-galactosidase stop codons is shown in Galas *et al.* (1980). Detailed cloning methods can be found in Maniatis *et al.* (1982) and Nellen *et al.* (Chap. 4, this volume). The only major modification to generally used cloning methods is to eliminate the use of isopropyl thiogalactoside (IPTG) and 5-bromo-4-chloro-3-indoly-β-D-galactoside (X-gal), a chromogenic indicator of the absence or presence of inserts within a β-galactosidase gene. Since IPTG induces expression of the β-galactosidase gene, use of these compounds could cause expression of a possibly deleterious fusion protein which conceivably could be neutralized in the bacteria by a missense mutation within the fusion gene.

B.   Specific Example

As a specific example, we describe here the steps used to produce a fusion protein using a fragment from a *Dictyostelium* prestalk-specific gene (see Gomer *et al.*, 1986). A cDNA clone was isolated by Mehdy *et al.* (1983) and found to hybridize to mRNA from prestalk cells but not mRNA from vegetative or pre-spore cells. This cDNA was used to isolate a larger fragment of the gene from a genomic library (Datta *et al.*, 1986; Datta and Firtel, 1986). Portions of the genomic and cDNA clones were sequenced. From a region where the sequences overlapped and matched, we chose a 273-bp *Dde*I–*Bgl*II fragment of coding region (Fig. 1). From the sequence and M13 hybridization to Northern blots, S. Datta determined the reading frame and polarity of this fragment (Datta and Firtel, 1986). We saw that by filling in the fragment ends using the Klenow fragment of DNA polymerase I and deoxynucleotides, a correct reading frame would be obtained by ligating the resulting fragment into the *Hinc*II site of M13mp11. Before ligating in the *Dictyostelium* sequence, the blunt ends of the cut *Hinc*II site were dephosphorylated with alkaline phosphatase to prevent re-closure. After obtaining the appropriate construct, the next step was to cut the M13mp11 at the unique *Hind*III site in the polylinker, fill in the ends with Klenow fragment and deoxynucleotides, dephosphorylate the ends with alkaline phosphatase, and ligate in a phosphorylated *Eco*RI linker fragment. The resulting population of DNA contained, among other things, some circular DNA contain-

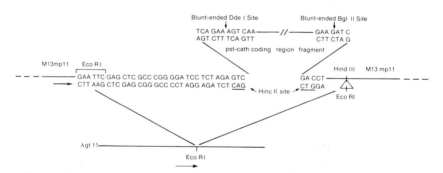

FIG. 1.   Diagram showing a construct used to make fusion protein antigen. A blunt-ended *Dictyostelium* coding region fragment from a gene designated *pst-cath* is shown at top. The fragment was ligated into the *Hinc*II site of M13mp 11. The M13mp11 *Hind*III site was converted to an *Eco*RI site (represented at right in the diagram). The resulting *Eco*RI fragment was purified and ligated into the unique *Eco*RI site of λgt11. Arrow at bottom represents the coding region for β-galactosidase, the tip represents termination. Codon triplets shown represent the reading frame used in the β-galac-tosidase–M13mp11–*Dictyostelium* fusion protein gene. Arrow at left represents the M13 dideoxy sequencing primer-binding site and sequencing direction (see Gomer *et al.*, 1986, for other details).

ing unbroken *Hin*dIII sites. To eliminate these, the M13 DNA was digested with *Hin*dIII and then used to transform *E. coli* (strain 71-18). Selected clones were sequenced to verify insert polarity and reading frame by the Sanger *et al.* (1977) dideoxy method.

The resulting *Eco*RI fragment containing the *Dictyostelium* prestalk gene coding region was excised and purified by gel electrophoresis. This was then ligated into λgt11. The λgt11 DNA had been previously ligated into a circle, cut with *Eco*RI and treated with alkaline phosphatase. The two possible orientations of the *Eco*RI fragment in λgt11 can be distinguished by a *Bam*HI–*Kpn*I digest. Fragment orientation can also be determined by making fusion protein from different recombinant λgt11–*Dictyostelium* clones. From the DNA sequence of the two *Eco*RI insert orientations in λgt11, we located termination codons and were able to predict accurately the size of the fusion proteins produced by the two different classes of clones (see next section for fusion protein production). This was then used to verify that the insert polarities determined by *Bam*HI–*Kpn*I digests were correct.

III.   Antigen Preparation

A.   Fusion Protein Preparation

Preparation of fusion protein follows Young and Davis (1983) with two modifications. First, fusion protein yields are greatly increased by using a very high multiplicity of infection (100:1, virus to bacteria) while infecting BNN103 bacteria. Second, much cleaner preparations are obtained when using more SDS sample buffer to solubilize fusion protein-containing bacteria than described in Young and Davis (1983). Appropriate quantities can be determined by titration.

We have found that fusion proteins are often insoluble and thus can be easily partially purified from other bacterial proteins. The frozen bacterial pellets are thawed in a room-temperature water bath and resuspended in ice-cold 50 mM NaCl–0.1 mM phenylmethylsulfonyl fluoride to the original volume of LB. To break open the bacteria, the sample is then refrozen in dry ice, thawed, refrozen, and, finally, thawed. The thawed mixture is then centrifuged at 10,000 g at 4°C for 15 minutes to separate soluble from insoluble proteins. A majority of the fusion proteins we have constructed have been found to partition in the insoluble fraction, whereas most bacterial proteins are soluble. This then allows rapid partial purification of the fusion protein. The pellet should be resuspended by vortexing in a small amount of ice-cold distilled water and then boiled in SDS sample buffer.

B.   Purification of Fusion Protein by SDS–Polyacrylamide Gel Electrophoresis

Using a very pure antigen to make antibodies for cell biology work will save an enormous amount of time trying to clean up a "dirty" antibody. The fusion proteins have molecular weights of 118,000–130,000. There are very few *E. coli* proteins that have molecular weights this high (Fig. 2). Thus separation on the basis of molecular weight can be a very efficient one-step purification. We have found purification of fusion protein by excision of stained bands from SDS–polyacrylamide gels to be one of the best and simplest methods, yielding a very pure antigen.

We purify the fusion protein on 10% polyacrylamide Laemmli (1970) gels. Bands are stained with Coomassie and excised. Fusion protein is then electroeluted out of the polyacrylamide matrix following Hunkapiller *et al.* (1983), and concentrated by lyophilization. Rabbits are immunized following standard protocols (Garvey *et al.*, 1977). Lymph node inoculation is best, especially for the first immunization (Sigel *et al.*, 1983).

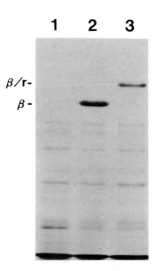

Fɪɢ. 2.   A 10% polyacrylamide–SDS gel stained for total protein with Coomassie dye. Lane 1, Uninfected BNN103 bacteria; lane 2, BNN103 bacteria infected with λgt11 and induced to express β-galactosidase; the bacteria in lane 1 were treated in an identical manner; lane 3, similar to lane 2 except the cells have been infected with a recombinant λgt11 containing a *Dictyostelium ras* gene coding sequence in the λgt11 *Eco*RI site. The resulting fusion protein (β/r) has a molecular weight greater than that of β-galactosidase (β).

IV. Antiserum Preparation

A. Preparation of Immunoglobulins from Serum

Purification of IgG is often not necessary for procedures such as Western blots. When whole serum is used for immunofluorescence, however, a strong diffuse background staining can be seen in *Dictyostelium* cells. Purifying γ-globulins from immune and preimmune rabbit serum eliminates most of the background staining. This is done by ammonium sulfate precipitation; further purification can be obtained by chromatography over DEAE–Sephacel (Garvey *et al.*, 1977).

B. Affinity Purification of Antisera

To obtain the lowest possible backgrounds with Western blots and immunofluorescent staining, we affinity-purify sera. The total protein from bacteria producing fusion protein can be coupled to a solid support. Passage of immune serum over one resin will allow binding of antibodies to *E. coli* proteins, β-galactosidase, and the *Dictyostelium*-derived moiety of the fusion protein. Other antibodies and serum components can be washed out and the bound antibodies eluted. These can then be passed over resin coupled to total protein from bacteria infected with "pure" λgt11. Antibodies to *E. coli* proteins and β-galactosidase will bind to the resin, while antibodies to the *Dictyostelium* protein can be collected in the flowthrough.

We prepare affinity resin following Marsh *et al.* (1974), using bacterial lysates made by repeated freezing and thawing of bacteria. Antibody binding and elution are performed as described in Gomer and Lazarides (1981). Antibodies can also be affinity-purified by using as the solid-phase immunoabsorbent the appropriate band cut from a Western blot (see Section VI,B) of a preparative gel (Talian *et al.*, 1983; Smith and Fisher, 1984). These methods, however, yield relatively small amounts of affinity-purified antibody.

V. Immunostaining of Electrophoresed Proteins

A. Gel Electrophoresis of *Dictyostelium* Proteins

Whole vegetative and preculmination *Dictyostelium* cells can be solubilized by boiling in Laemmli (1970) SDS sample buffer. A good rule of thumb is to use 1 ml sample buffer per 10^7 cells and to load $1 \times 10^5 – 5 \times 10^5$ cells equivalent of lysate in a 6-mm-wide well of a 1.5-mm-thick gel. Cells from fruiting bodies, especially stalks, do not solubilize readily and require mechanical disruption such as sonication.

For two-dimensional electrophoresis, we take 1×10^6 cells boiled in 20 μl 1% SDS sample buffer and add five volumes of 2.5% NP-40 saturated with urea at room temperature. This is loaded on an electrofocusing tube gel (O'Farrell, 1975). Second-dimension electrophoresis is on Laemmli (1970) gels as described above. A somewhat different method of two-dimensional gel electrophoresis of *Dictyostelium* proteins is presented in Morrissey *et al.* (1984).

B.   Western Blots

Western blotting allows a rapid determination of the molecular weight and variant distribution of a protein antigen (Towbin *et al.*, 1979; reviewed in Gershoni and Palade, 1983; and in Towbin and Gordon, 1984). When one-dimensional gels are used, samples from many cell types, cell fractions, or developmental stages can be assayed in parallel. In addition, this method furnishes crude relative quantification of antigens. We use a Bio-Rad Transphor apparatus to transfer proteins electrophoretically from Laemmli (1970) gels to nitrocellulose. The details of how to use the apparatus are well presented in the user's manual. We have found that different transfer buffers must be used for different antigens. Some proteins transfer well in Laemmli gel running buffer/ 20% methanol. Other *Dictyostelium* proteins show no reactivity on Western blots when transferred in this buffer, but do show reactivity when transferred in the above buffer without the SDS. In this case the gel must be equilibrated in this buffer with several changes over 1 hour before transfer, presumably to remove SDS.

Many methods can be used to immunostain Western blots (Barinaga *et al.*, 1981). We have found two general rules of thumb for *Dictyostelium* proteins: First, 0.05% Tween-20 works better than gelatin, bovine serum albumin (BSA), or milk for blocking nonspecific binding, and second, ^{125}I-labeled protein A (labeled following Greenwood *et al.*, 1963) is more sensitive and gives less background than enzyme-based colorimetric stains for detection of bound antibody. As with immunofluorescence, background and nonspecific staining can be reduced by increasing salt and/or detergent concentrations (see Section VI,C). However, excessive concentrations can block the specific staining.

VI.   Immunofluorescence

A.   Introduction

This technique allows one to determine the location of an antigen within a multicellular aggregate or within a cell. It is extremely rapid and can be used as an assay for gene expression in a very small population of cells. In a population of cells, immunofluorescence will show the percentage of the cells expressing the gene. If the antibody recognizes a cell surface antigen, live cells expressing the antigen can be separated in a fluorescence-activated cell sorter (Krefft *et al.*, 1983). By using double-label immunofluorescence (see Fig. 3), two different antibodies can be used (Brandtzaeg, 1973). This can show the distribution of two different antigens within the same cell or field of cells. This is often the only way to determine whether two different proteins are ever present within the same cell. For higher resolution localization of an antigen within the cell, one can use

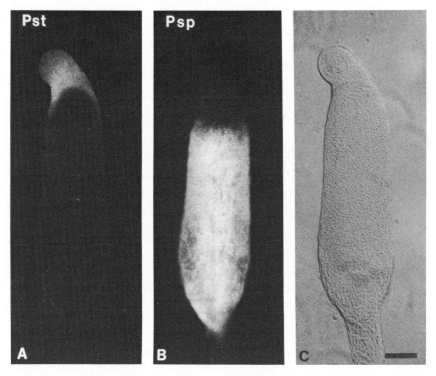

FIG. 3. Double-label indirect-immunofluorescence staining of a *Dictyostelium* slug. A Kessin Ax-3 slug was prepared and fixed as described in Section VI,F. After fixation, the slug was incubated with a rabbit polyclonal antiserum against a *Dictyostelium* prestalk gene product–β-galactosidase fusion protein, washed, then incubated with FITC-conjugated goat anti-rabbit antibody, and washed again. After blocking with a nonimmune rabbit antibody, the slug was incubated with a biotin-conjugated antiserum made against a prespore gene product–β-galactosidase fusion protein. After washing, biotin-conjugated bound antibody was stained with Texas Red (T.M.)-conjugated streptavidin. Fluorescence of the two fluorochromes could be visualized separately by appropriate choice of excitation, emission, and dichroic filters in the epifluorescence microscope. (A) Staining of prestalk (Pst) cells; (B) staining of prespore (Psp) cells; (C) A phase-contrast image of the slug. All images are at the same magnification; bar in (C) is 100 μm.

immunoelectron microscopy (Devine *et al.*, 1983). The first step in immunofluorescence involves fixation of disaggregated cells or, as described in the
following sections (no pun intended), preparation and fixation of cryosections.

B.   Attachment of Cells to Coverslips

The easiest way to prepare *Dictyostelium* cells for immunofluorescence is to
allow them to adhere to a glass coverslip. This procedure works for axenic
vegetative cells growing in shaking culture, vegetative wild-type cells dislodged
from an agar plate with water or buffer squirted from a Pasteur pipet, developing
cells disaggregated as above in water or buffer, or separated prespore and prestalk cells (Ratner and Borth, 1983). Until culmination, all cell types will adhere
fairly well to uncoated glass coverslips. Improved adhesion of vegetative and
developing cells and adhesion of spores and stalks (stalk cells cannot be readily
disaggregated) can be obtained by coating glass coverslips with rat tail collagen
(Bornstein, 1958). Coverslips are coated with collagen solution and then baked
at 70°C in a vacuum for ~3 hours. Coverslips can also be coated with polylysine,
gelatin, or BSA to improve adhesion of cells (Clarke *et al.*, 1975).

C.   Fixation, Permeabilization, and Staining of Cells

Cells should be allowed to adhere to collagen-coated or plain coverslips for
approximately 5–10 minutes at room temperature. Cells can be in water, buffer,
or growth medium; typically 100 µl of cells at 2×10^6 cells/ml are placed as a
drop in the center of a 18×18 mm no. 1 glass coverslip. Coverslips are then
placed in a rack in 100% methanol at room temperature for 10 minutes, and then
dried at room temperature in air. Some researchers prefer to have the methanol at
0° or −20°C, or to use methanol–water solutions. The coverslips will be dry in
~1 hour and can then be immersed in staining buffer (see below) at room
temperature.

Fixation conditions can be varied. An agarose overlay has been reported for
improved fixation (Yumura *et al.*, 1984; Fukui *et al.*, Chap. 19, this volume).
Fixation solutions containing detergent and glutaraldehyde have been used
(Rubino *et al.*, 1983). Cells adhering to collagen-coated coverslips can be fixed
in 3.7% formaldehyde in water or low-salt buffer for 10 minutes at room temperature. Cells on uncoated coverslips do not seem to survive formaldehyde
fixation well. Cells are then rinsed for 10 minutes in staining buffer (see below).
If this buffer does not contain an appreciable amount of detergent, cells should
first be permeabilized for 10 minutes at room temperature in 100% methanol or
buffer containing 0.5–1.0% NP-40, and then rinsed in staining buffer.

The choice of staining buffer depends on the antigen and antibody used. The
basic buffer is 130 mM NaCl buffered with 20 mM Tris-HCl or K or NaPO$_4$, pH

7.0–7.5. To reduce background staining, the salt concentration can be raised, up to a practical limit of ~1.0 M. To improve wettability of the coverslip and also reduce nonspecific staining, one can include detergent. Most antigen–antibody combinations can tolerate 0.005% Tween-20 or NP-40. For harsher conditions one can raise the detergent concentration or use an ionic detergent such as 0.1% SDS with 0.5% NP-40 to reduce the critical micelle concentration. Immunofluorescent staining should be tested under as wide a variety of conditions as possible. Indirect and double-label immunofluorescence is performed following generally used procedures (Garvey *et al.*, 1977; Gomer and Lazarides, 1981). Antibody solutions should always be clarified by centrifugation for 3–5 minutes in an Eppendorf centrifuge immediately before use. To prevent fading of fluorescein, I use the mounting medium described by Johnson and Araujo (1981).

D. Embedding *Dictyostelium* Slugs and Aggregates

Migrating slugs are developed on 1.5% agar plates made with distilled water. Before adding cells, let the plates age 1–4 days, 1 day being preferable. Wash the vegetative cells in water once or twice by centrifugation at ~1500 g for 5 minutes in a Sorvall HB4 at room temperature. Resuspend the cells to ~5 × 10^7/ml—a moderately thick solution. Put ~200 μl on a line on the agar plate, seal the plate with parafilm, and put alternating layers of black paper disks and plates in a slit tube with the light coming in perpendicular to the line of cells. The slit should be ~1 mm wide. The container for Schleichler and Schuell BA85 makes an ideal slit tube—just add the slit. Make a handle out of tape to be able to lower the plate gently into the slit tube. Leave the plates to form slugs ~1 day and migrate ~1 day (2 days total) in a lighted room.

Unfixed *Dictyostelium* slugs and fruiting bodies are extremely fragile and cannot be manipulated or embedded in viscous media (such as OCT) without prior fixation. Slugs can also be embedded in paraffin (Tasaka *et al.*, 1983). To fix the slugs, gently pour room-temperature 3.7% formaldehyde in distilled water at one edge onto the plate. The slugs and their trails will float up on top of the fixative. Put the lid back on and let the plate sit for 12–48 hours at room temperature.

To embed the slugs, take a piece of exposed (blackened) X-ray film ~7 × 5 cm and put on it a piece of Saran Wrap, ~5 × 5 cm, with a 6-mm-diameter circle drawn on its underside in the middle. Put a drop of Tissue-Tek OCT compound on the Saran Wrap over the circle. Take one finger of a toothless straight forceps and lay it parallel to a fixed slug. We normally do this under a dissecting microscope. Bring the forcep finger underneath the slug and pick up the slug so that the slug is lying along the flat side of the finger. Since this is the inside of the forceps finger, we place a small plug made from a Kimwipe in the forceps to keep the fingers spread apart.

The slug is now transferred to the drop of OCT by putting the forceps finger in the OCT at a point up the finger from the slug, then pulling the forceps finger through the drop of OCT, allowing the slug to enter the OCT and hopefully detach from the forceps. Occasionally the slug will be completely immersed in OCT, but generally it remains on the surface of the OCT. To immerse it, suspend a second, smaller drop of OCT between the tips of a forceps and gently lower this drop onto the slug. Move the slugs into the circle by sweeping OCT away from the slugs in the direction you want to move the slugs. The slugs will be drawn into the resulting "vacuum." Slugs should not be pushed or touched with the forceps once in OCT. Sweeping away OCT also serves to thin the block. Making the block as thin as possible at this stage will save much time when cryosectioning. A mark can be made on the Saran Wrap and later on the frozen block to designate the position of the tip of the slug.

To freeze the slugs, place on the benchtop a large piece of dry ice with a flat, horizontal upper surface. The exposed X-ray film can then be placed on the ice and slid out from under the Saran Wrap to freeze the OCT and slugs. After drawing a second circle directly on the frozen OCT, the piece inside the circle can be cut out with a prechilled razor blade and placed in a chilled Eppendorf tube with chilled forceps. The Saran Wrap can also be removed from the piece of OCT. Use of room-temperature utensils will cause the block to freeze to the utensil or even completely thaw. The 1- to 2-mm-thick chips of OCT and slug can be kept in parafilm-sealed Eppendorf tubes at $-20°C$ for up to 1 week. The same procedure can be used to fix and embed aggregates from stages other than the slug stage. For example, by appropriately orienting "Mexican hat"-stage aggregates we have been able to obtain cryosections oriented parallel and others oriented perpendicular to the axis of radial symmetry.

E.   Cryosectioning Slugs and Aggregates

The OCT chips containing slugs (see previous section) can be mounted on cryostat chucks and sectioned at $-20°C$. Sections do not stick to glass coverslips and must be picked up on collagen-coated coverslips. It has been reported that gelatin-subbed coverslips also work for this procedure (Krefft et al., 1984). Cryosections, typically 10 μm thick, are thawed onto the coverslips and then immediately immersed in 100% methanol at room temperature for 10 minutes, and then air-dried. This method appears to be marginally better than first drying the sections and then permeabilizing them, or permeabilizing with detergent instead of methanol. After air-drying for 2 hours at room temperature, coverslips with sections are rinsed in staining buffer at room temperature for 10 minutes and can then be stained by immunofluorescence (see Section VI,C).

F.   Preparation of Whole-Mount Slugs and Fruiting Bodies

Cryosectioning slugs is quite difficult and time-consuming. I have found that staining whole-mount aggregates by immunofluorescence to be a convenient way to examine cell type localization in, for instance, aggregates of transformed cells. For whole mounts, vegetative cells are pelleted, resuspended in water, and repelleted as described in Section VI,D. Cells are resuspended in PDF (20 mM KCl, 1.2 mM MgSO$_4$, 20 mM potassium phosphate, pH 6.4) to 10^7 cells/ml. A 200-μl drop of cell suspension is placed, without spreading out the drop, on a plastic coverslip (Fisher type 12-547, Fisher Scientific Co., Pittsburgh, PA) supported in a humid chamber. Aggregates will form inside the drop, and then slugs will crawl radially outward from the drop. The slugs develop into fruiting bodies. The developmental time course is somewhat slower than that for development on filter pads. To fix the aggregates at any stage, the coverslips are immersed in room-temperature 100% methanol for 20 minutes and then allowed to air-dry. Aggregates can then be stained by immunofluorescence following Section VI,C.

VII.   Microassay for Cell Type-Specific Gene Expression

Isolated starved cells can differentiate into prespore and prestalk cells under certain conditions (Town *et al.*, 1976; Kay, 1982). Mehdy and Firtel (1985) showed that starved cells plated at a low density will express prespore and prestalk mRNAs when incubated in a medium previously conditioned by high-density cells that contains no cAMP. The medium is removed from the high-density cells, and then the cAMP is added. The assay for the developmentally regulated mRNAs required cells grown in a total of 150 ml of medium in 10 × 14 cm diameter Petri plates for each assay point. We have found that cells grown in the above conditions synthesize the proteins encoded by the developmentally regulated mRNAs. This allows detection of the expression of the mRNAs by immunofluorescence staining of the proteins encoded by the mRNAs. Since this technique involves the direct examination of individual cells, the assay can be performed on a relatively small number of cells in a correspondingly small volume. Starved cells are plated at low density in the wells of a multiwell slide (Lab Tek type 4818, Miles Scientific, Naperville, IL) or microtiter plate (Falcon type 3040, Becton, Dickinson and Co., Oxnard, CA). After incubations in fresh or conditioned medium and some or no cAMP (following the large-scale conditions described by Mehdy and Firtel, 1985), the cells are fixed on the slides or

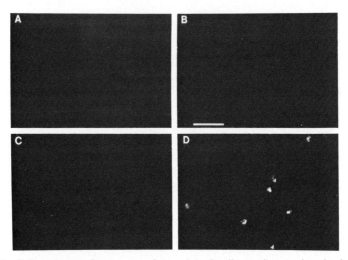

Fig. 4. Indirect immunofluorescence of starved Ax-3 cells growing at a low density in submerged culture. In (A) and (C) cells were grown in fresh medium (FM); in (B) and (D) cells were in conditioned medium (CM) (Mehdy and Firtel, 1985). Five hours after starvation, cAMP was added to cells in (C) and (D) to a final concentration of 200 μM. At 24 hours after starvation cells were fixed *in situ* and stained for both prespore and prestalk proteins using a 1:1 mixture of the two antibodies used in Fig. 3. Cells in (A), (B), and (C) do not show appreciable staining; only cells treated with both conditioned medium and cAMP (D) show positive staining. Photos are all at the same magnification; bar in (B) is 50 μm.

plates and stained for prespore or prestalk proteins by immunofluorescence (Fig. 4). This assay uses cells grown in 100–200 μl of medium for each assay point, and takes ~6 hours as opposed to ~1 week for the mRNA–Northern blot method. Using the antibodies against the products of cloned developmentally regulated *Dictyostelium* genes will allow us to isolate the factor from conditioned medium that, in conjunction with cAMP, causes the expression of developmentally regulated genes. This assay is also being used to examine the effect of cAMP analogs and other substances on prespore and prestalk gene expression.

VIII. Immunoprecipitation

The synthesis and posttranslational modification of proteins can often best be studied by immunoprecipitation (Kessler, 1975, 1976). This involves labeling the protein, solubilizing it, binding antibodies to it in solution, and precipitating and washing the immune complex. Generally, the antigen is then dissociated from the antibody by boiling in SDS sample buffer and purified by electrophoresis on a polyacrylamide–SDS gel. The first step of labeling proteins will depend on the phenomenon being studied. Labeling can include [^{35}S]methionine, ^{32}P$_i$, [^{35}S]sulfate, or ^{3}H- or ^{14}C-labeled sugars (Wallace and Frazier, 1979; Wang et al., 1981; Devine et al., 1982; Schmidt and Loomis, 1982; Das and Henderson, 1983a,b; Stadler et al., 1983; Ivatt et al., 1984). When solubilizing the protein, one should be careful that the conditions used will actually solubilize all the protein. This can be ensured by boiling the cells or in vitro translation mixture in a buffer containing 1% SDS, then diluting the solubilized cells in a buffer containing NP-40 to obtain a final concentration of 1% NP-40 and 0.1–0.5% SDS (Gomer and Lazarides, 1983a,b). The final concentration of SDS and NP-40 will depend on the antigen–antibody combination and must be determined empirically. Occasionally, antibody–antigen interaction is improved by other detergents such as sodium deoxycholate; in other instances a particular detergent can inhibit immunoprecipitation. The pH and salt concentration of the immunoprecipitation buffer can also be varied (see Section VI,C). The remaining procedures for immunoprecipitation will follow those given in the above references and must be optimized empirically.

IX. Summary

The λgt11 system is very useful for studying products of cloned *Dictyostelium* genes. We have successfully used this system to make antibodies against four different *Dictyostelium* cell type-specific proteins. One antibody was against *Dictyostelium ras*. We were able to use this antibody to show that *ras* protein is present in vegetative cells and developing cells. After ~15 hours of development, levels of *ras* begin to decrease (Reymond et al., 1984). We have recently used this antibody to examine the distribution of *ras* in prespore and prestalk cells and in cell fractions at various times in development (Gomer and Firtel, unpublished). Another antibody is against a prespore-specific protein and a third is against a prestalk protein. These two antibodies each stain a single band on Western blots. The fourth antibody is at present uncharacterized. As shown in Fig. 3, the antibodies stain different parts of slugs. We are currently using these antibodies to examine the distribution of prestalk and prespore cells in aggregates at different stages of development. Other studies are using these antibodies for immunofluorescent assays of cell differentiation. We feel that these antibodies will become the central assay for many cell biological, molecular, and physiological studies to be done in this laboratory.

ACKNOWLEDGMENTS

I thank Richard A. Firtel for support and helpful suggestions. Many of the techniques used specifically for *Dictyostelium* were developed or improved in his laboratory. I would like to thank J.

Roth for assisting in the preparation of this chapter. Work on this chapter was supported by USPHS grants GM24279 and GM30693 to Richard A. Firtel. R.H.G. was supported by an USPHS postdoctoral fellowship.

References

Barinaga, M., Franco, R., Meinkoth, J., Ong, E., and Wahl, G. M. (1981). *In* "Methods for the Transfer of DNA, RNA and Protein to Nitrocellulose and Diazotized Paper Solid Supports." Scheicher & Schuell, NH.

Barklis, E., and Lodish, H. F. (1983). *Cell,* **32,** 1139–1148.

Blose, S. H., Matsumara, F., and Lin, J. C. (1982). *Cold Spring Harbor Symp. Quant. Biol.* **46,** 455–463.

Bonner, J. T. (1952). *Am. Nat.* **86,** 79–89.

Bornstein, M. B. (1958). *Lab. Invest.* **7,** 134–137.

Bradford, M. M. (1976). *Anal. Biochem.* **72,** 248–254.

Brandtzaeg, P. (1973). *Scand. J. Immunol.* **2,** 273–290.

Clarke, M., Schatten, G., Maria, D., and Spudich, J. A. (1975). *Proc. Natl. Acad. Sci. U.S.A.* **72,** 1758–1762.

Danto, S. I., and Fischman, D. A. (1984). *J. Cell Biol.* **98,** 2179–2191.

Das, O. P., and Henderson, E. J. (1983a). *J. Cell Biol.* **97,** 1544–1558.

Das, O. P., and Henderson, E. J. (1983b). *Biochim. Biophys. Acta* **736,** 45–56.

Datta, S., and Firtel, R. A. (1987). *J. Mol. Cell Biol.* in press.

Datta, S., Gomer, R. H., and Firtel, R. A. (1986). *Mol. Cell. Biol.* **6,** 811–820.

Devine, K. M., Morrissey, J. H., and Loomis, W. F. (1982). *Proc. Natl. Acad. Sci. U.S.A.* **79,** 7361–7365.

Devine, K. M., Bergmann, J. E., and Loomis, W. F. (1983). *Dev. Biol.* **99,** 437–446.

Franke, W. W., Schmid, E., Wellsteed, J., Grund, C., Gigi, O., and Geiger, B. (1983). *J. Cell Biol.* **97,** 1255–1260.

Galas, D. J., Calos, M. P., and Miller, J. H. (1980). *J. Mol. Biol.* **144,** 19–41.

Garvey, J. S., Cremer, N. E., and Sussdorf, D. H. (1977). *In* "Methods in Immunnology," 3rd Ed. Benjamin, Reading, MA.

Gershoni, J. M., and Palade, G. E. (1983). *Anal. Biochem.* **131,** 1–15.

Gomer, R. H., and Lazarides, E. (1981). *Cell* **23,** 524–532.

Gomer, R. H., and Lazarides, E. (1983a). *J. Cell Biol.* **96,** 321–329.

Gomer, R. H., and Lazarides, E. (1983b). *J. Cell Biol.* **97,** 818–823.

Gomer, R. H., Datta, S., and Firtel, R. A. (1985). *Focus* **7,** 6–7.

Gomer, R. H., Datta, S., and Firtel, R. A. (1986). *J. Cell Biol.* **103,** 1999–2015.

Grant, W. N., and Williams, K. L. (1983). *EMBO J.* **2,** 935–940.

Greenwood, F. C., Hunter, W. M., and Glover, J. S. (1963). *Biochem. J.* **89,** 114–123.

Hancock, K., and Tsang, V. (1983). *Anal. Biochem.* **133,** 157–162.

Hunkapiller, M. W., Lujon, E., Ostrander, F., and Hood, L. E. (1983). *In* "Methods in Enzymology" (L. H. W. Hirs and S. N. Timasheff, eds.), Vol. 91, pp. 227–236. Academic Press, New York.

Ivatt, R. L., Das, O. P., Henderson, E. J., and Robbins, P. W. (1984). *Cell* **38,** 561–567.

Johnson, G. D., and Araujo, G. M., de C. N. (1981). *J. Immunol. Methods* **43,** 349–350.

Kay, R. R. (1982). *Proc. Natl. Acad. Sci. U.S.A.* **79,** 3228–3231.

Kessler, S. W. (1975). *J. Immunol.* **115,** 1617–1624.

Kessler, S. W. (1976). *J. Immunol.* **117,** 1482–1490.

Knecht, D. A., Dimond, R. L., Wheeler, S., and Loomis, W. F. (1984). *J. Biol. Chem.* **259,** 10633–10640.

Krefft, M., Voet, L., Mairhofer, H., and Williams, K. L. (1983). *Exp. Cell Res.* **147**, 235–239.

Krefft, M., Voet, L., Gregg, J. H., Mairhofer, H., and Williams, K. L. (1984). *EMBO J.* **3**, 201–206.

Laemmli, U. K. (1970). *Nature (London)* **227**, 680–685.

Laskey, R. A., and Mills, A. D. (1977). *FEBS Lett.* **82**, 314–316.

Loomis, W. F., Jr., ed. (1975). "*Dictyostelium discoideum*, A Developmental System." Academic Press, New York.

Loomis, W. F., Murray, B. A., Yee, L., and Jongens, T. (1983). *Exp. Cell Res.* **147**, 231–234.

Maniatis, T., Fritsch, E. F., and Sambrook, J. (1982). "Molecular Cloning: A Laboratory Manual." Cold Spring Harbor Laboratory, Cold Spring Harbor, New York.

Marsh, S. C., Parikh, I., and Cuatrecasas, P. (1974). *Anal. Biochem.* **60**, 149–152.

Mehdy, M. C., and Firtel, R. A. (1985). *Mol. Cell. Biol.* **5**, 705–713.

Mehdy, M. C., Ratner, D., and Firtel, R. A. (1983). *Cell* **32**, 763–771.

Morrissey, J. H., Devine, K. M., and Loomis, W. F. (1984). *Dev. Biol.* **30**, 414–424.

Nellen, W., Silan, C., and Firtel, R. A. (1984). *Mol. Cell. Biol.* **4**, 2890–2898.

O'Farrell, P. (1975). *J. Biol. Chem.* **250**, 4007–4021.

Palfree, R. G. E., and Elliot, B. E. (1982). *J. Immunol. Methods* **52**, 395–408.

Ratner, D., and Borth, W. (1983). *Exp. Cell Res.* **143**, 1–13.

Reymond, C., Gomer, R. H., Mehdy, M. C., and Firtel, R. A. (1984). *Cell* **39**, 141–148.

Rubino, S., Small, J. V., Claviez, M., Sellitto, C., and Cappuccinelli, P. (1983). *In* "Contractile Proteins in Muscle and Non-Muscle Cell Systems" (E. E. Aliz, N. Arena, and M. A. Russo, eds.). Praeger, New York.

Sanger, F., Nicklen, S., and Coulson, A. (1977). *Proc. Natl. Acad. Sci. U.S.A.* **74**, 5463–5467.

Schmidt, J. A., and Loomis, W. F. (1982). *Dev. Biol.* **91**, 296–304.

Sigel, M. B., Sinha, Y. N., and Vanderlaan, W. P. (1983). *In* "Methods in Enzymology" (J. J. Langone and H. Van Vunakis, eds.), Vol. 93, pp. 3–12. Academic Press, New York.

Smith, D. E., and Fisher, P. A. (1984). *J. Cell Biol.* **99**, 20–28.

Stadler, J., Gerisch, G., Suchanek, C., and Huttner, W. B. (1983). *EMBO J.* **2**, 1137–1143.

Talian, J. C., Olmstead, J. B., and Goldman, R. D. (1983). *J. Cell Biol.* **97**, 1277–1282.

Tasaka, M., Noce, T., and Takeuchi, I. (1983). *Proc. Natl. Acad. Sci. U.S.A.* **80**, 5340–5344.

Towbin, H., and Gordon, J. (1984). *J. Immunol. Methods* **72**, 313–340.

Towbin, H., Staehelin, T., and Gordon, J. (1979). *Proc. Natl. Acad. Sci. U.S.A.* **76**, 4350–4354.

Town, C. D., Gross, J. D., and Kay, R. R. (1976). *Nature (London)* **262**, 717–719.

Wallace, L. J., and Frazier, W. A. (1979). *Proc. Natl. Acad. Sci. U.S.A.* **76**, 4250.

Wang, C., Gomer, R. H., and Lazarides, E. (1981). *Proc. Natl. Acad. Sci. U.S.A.* **78**, 3531–3535.

Young, R. A., and Davis, R. W. (1983). *Proc. Natl. Acad. Sci. U.S.A.* **80**, 1194–1198.

Yumura, S., Mori, H., and Fukui, Y. (1984). *J. Cell Biol.* **99**, 894–899.

Chapter 27

Homologous Recombination in Dictyostelium as a Tool for the Study of Developmental Genes

ARTURO DE LOZANNE

Department of Cell Biology
Stanford University School of Medicine
Stanford, California 94305

I.   Introduction

The development of reliable vectors for *Dictyostelium* transformation (Nellen *et al.*, this volume, Chapter 4; Knecht *et al.*, 1986) makes it possible to study in more detail the biological questions discussed throughout this book. As described in the previous two chapters, the ability to introduce cloned genes into *Dictyostelium* provides new approaches to the study of developmentally regulated genes, including the use of antisense RNA to repress the expression of many cloned genes (Knecht and Loomis, 1987). Another exciting possibility for the use of *Dictyostelium* transformation is the targeted disruption or alteration of genes by homologous recombination. In the course of exploring the use of *Dictyostelium* as an expression system for myosin and its subfragments, we found the first evidence that gene disruption by homologous recombination can indeed occur in *Dictyostelium* (De Lozanne and Spudich, 1987). We are now using this approach to study the role of myosin in cell motility and during development.

II.   Methods

A.   General Considerations

Since our work on the myosin gene is the first example of integration of a transformation plasmid by homologous recombination in *Dictyostelium,* there is much to be learned about the extent to which this event can occur with other genes. A few general considerations might be useful for those interested in exploring this approach in their own research. It is likely that when *Dictyostelium* is transformed with a plasmid that is not maintained extrachromosomally, integration can occur either randomly or by homologous recombination. Thus, one should grow cultures from individual transformants and analyze the restriction pattern of the genomic copy of the gene in each one to determine whether it has been altered by plasmid integration. Since there is often integration of multiple copies of the transformation plasmid, it is advisable to use a DNA probe against sequences that are not present in the plasmid to generate the restriction map. This

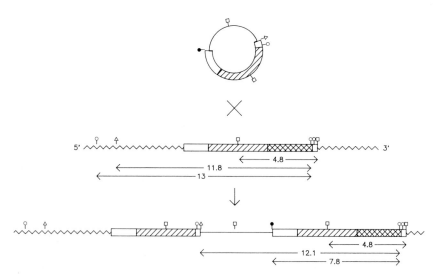

FIG. 1.   Representation of the insertion of the transformation plasmid into the *mhcA* gene. The circle represents the plasmid pNEO-HMM140 that codes for a portion of the *Dictyostelium* myosin heavy chain equivalent to muscle heavy meromyosin (HMM). Below the plasmid is the map of the native *mhcA* gene locus. Integration of the plasmid can occur at any point along the HMM-coding region or 5′-flanking sequence to give rise to the map below. The sites for the restriction enzymes utilized in Fig. 2 are indicated as follows: (●) *Bam*HI, (○) *Bcl*I, (□) *Bgl*II, and (△) *Sal*I. The horizontal arrows indicate the size in kilobases of the restriction fragments expected to hybridize with a probe directed against the LMM (light meromyosin) portion of the gene (see cross-hatched portion). (▭) 5′- and 3′-flanking sequences, (▨) HMM-coding portion, (▩) LMM-coding portion, (〰) chromosomal DNA, and (—) plasmid sequences.

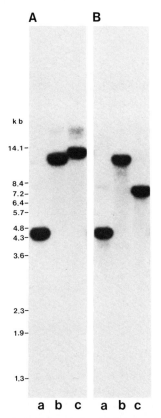

FIG. 2.   Southern blot analysis of the *mhcA* restriction map in wild-type and transformed cells. Total DNA from AX4 wild-type (lanes a–c in A) or transformed cells with the *mhcA* gene disruption (lanes a–c in B) was digested with *Bgl*II (lanes a in both A and B), *Bcl*I–*Sal*I (lanes b in both A and B), or *Bcl*I–*Bam*HI (lanes c in both A and B) and was subjected to Southern blot analysis. The probe was a 1.5-kb *Eco*RI fragment from the LMM portion of the *mhcA* gene.

will ensure that any changes in the locus of the gene can be identified. Figures 1 and 2 illustrate this type of analysis with the myosin heavy chain gene (*mhcA*). We constructed a transformation plasmid containing a 5.5-kb fragment of the *mhcA* gene, using the transformation vector A15TX (Cohen *et al.*, 1986). Integration of this plasmid into the single *mhcA* locus results in the alteration of the restriction pattern of the gene as indicated in Fig. 1. Southern blot analysis was carried out using a probe against a 1.5-kb fragment (cross-hatched segment shown in Fig. 1) of the *mhcA* gene that is not present in the transformation plasmid. The restriction patterns of DNA from wild-type and recombinant cells (Fig. 2) are those predicted from the schematic maps shown in Fig. 1.

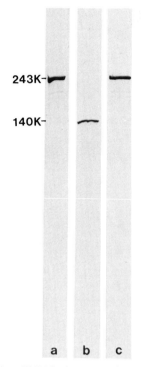

FIG. 3. Expression of myosin or HMM fragment. Axenically growing AX4 wild-type cells (lane a), transformed cells with homologous integration in the *mhcA* gene (lane b), and transformed cells with nonhomologous integration into the genome (lane c) were collected, lysed, and loaded onto a 4.5% stacking–7.5% running SDS–PAGE. Cells (1×10^5) were loaded per lane. The gel was transferred onto nitrocellulose paper and was stained with an antibody directed against the *Dictyostelium* myosin.

Given that the relative frequency of homologous versus nonhomologous integration events may depend on the specific gene and size of insert in the transformation plasmid, it is useful to devise a screening method for the selection of the transformants that have incorporated the plasmid into the desired locus. Several screening techniques are discussed in detail by Loomis (this volume, Chapter 3). In the particular case of the *mhcA* gene disruption described above, the insertion of the plasmid causes the interruption of native myosin heavy chain expression and the induction of expression of the myosin fragment encoded by the plasmid. Figure 3 shows a Western blot analysis of the expression of myosin or its fragment in wild-type cells (lane a), transformed cells with homologous integration in the *mhcA* gene (lane b), and transformed cells with nonhomologous integration into the genome (lane c). As a result of the lack of expression of native myosin heavy chain, the transformed cells fail in cytokinesis and become large and multinucleate (Knecht and Loomis, 1987; De Lozanne and Spudich, 1987). Thus, either of these two phenotypes can be used to screen for homologous recombination in the *mhcA* gene that results in the elimination of a functional myosin.

B. Transformation Protocol

Detailed protocols for the transformation of *Dictyostelium* have been described (Nellen *et al.*, this volume, Chapter 4; Knecht *et al.*, 1986). It is important to revise the transformation protocol for each particular case in order to avoid steps that would select against the predicted phenotype of the cells where insertion by homologous recombination had occurred. For example, in the case of the *mhcA* gene disruption, the transformed cells become large and multinucleate and do not survive in a shaking flask culture. This makes it necessary to carry out all steps during transformation and growth with the cells attached to a solid substratum. The transformation protocol of Knecht *et al.* (1986) was modified as follows: after the first round of G418 selection on plastic Petri dishes, the population of transformants are grown without selection, such that those cells that have not incorporated the plasmid into the genome will lose the plasmid and become G418 sensitive. This step can be carried out very effectively by plating the transformants on a lawn of live *Klebsiella aerogenes* on an SM plate (Sussman, this volume, Chapter 2). Under these conditions, the transformed cells grow much faster than when feeding on autoclaved bacteria on a filter paper or in a shaking flask. When the cultures start to clear the bacterial lawn, the cells are collected by scraping them off the agar plate. They are then washed free of remaining bacteria by several cycles of low-speed centrifugation and resuspension into HL-5 media supplemented with penicillin, streptomycin, and G418. The cells are then allowed to attach to Petri dishes and grown under selection for about 6–8 days with a change of media after 3 days. At this point, colonies of stable transformed cells are visible under the microscope. The cells in the population of transformants are then cloned by dilution into microtiter wells with 60 μl of HL-5 medium with penicillin, streptomycin, and G418 and are allowed to grow until they form colonies. These clonal populations are then screened for the occurrence of homologous integration events in the manner devised for the particular gene studied.

III. Perspectives

If gene targeting by homologous recombination proves to be a more general phenomenon in *Dictyostelium*, it will be a very powerful technique for the study

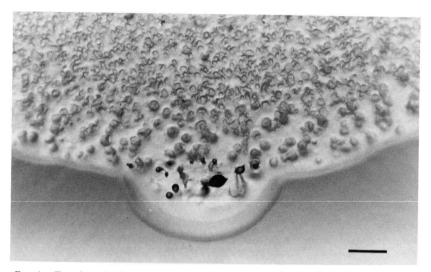

FIG. 4.   Transformed cells growing on a lawn of bacteria. The transformed cells with homologous integration in the *mhcA* gene were inoculated onto lawns of live *K. aerogenes* on SM plates. The
cultures were allowed to grow for several days. The leading edge of one such culture is shown above.
The upper region is where the cells have depleted the plate of bacteria. The cells aggregate and then
fail to develop further. In the center of the photograph, there is a reversion event. The revertant cells
grow faster than the transformed cells and are capable of developing into fruiting bodies. The bar
represents 1 mm.

of any aspect of *Dictyostelium* biology. In the case of the *mhcA* gene, it is now
possible to do fine mapping of the portions of the myosin molecule that are
necessary for its function. It is clear that a complete myosin molecule is required
for proper cytokinesis, but is not essential for growth. The giant cells survive as
long as they are attached to a substratum. These cells can also ingest bacteria by
phagocytosis, as is shown in Fig. 4. When the cells are placed in starvation
conditions, they start development, although at a very slow rate compared to that
of wild-type cells. They are capable of chemotaxis and of forming streams that
lead to aggregates like those seen in Fig. 4. These aggregates cannot develop
further.

The integration of the plasmid into the *mhcA* gene is reversible, with a frequency that we have not determined. The reversion event leads to cells that
recover the expression of native myosin and become G418 sensitive. Figure 4
shows one reversion event that is clearly seen by the faster rate of growth on the
bacterial lawn and the appearance of fully developed fruiting bodies. This result
suggests the possibility of creating deletions of specific genes in a way similar to
gene deletion in yeast.

ACKNOWLEDGMENTS

The author thanks Drs. David Knecht and William Loomis for numerous discussions in the course of this work and for providing the transformation vector A15TX. This work was supported by NIH grant GM33289 to James A. Spudich.

REFERENCES

Cohen, S. M., Knecht, D., Lodish, H. F., and Loomis, W. F. (1986). *EMBO J.*, **5,** 3361–3366.
De Lozanne, A., and Spudich, J. A. (1987). Submitted.
Knecht, D. A., and Loomis, W. F. (1987). Submitted.
Knecht, D. A., Cohen, S. M., Loomis, W. F., and Lodish, H. F. (1986). *Mol. Cell. Biol.* **6,** 3973–3983.

Appendix

Codon Frequency in Dictyostelium discoideum

HANS M. WARRICK

Department of Cell Biology
Stanford University School of Medicine
Stanford, California 94305

A compilation of codon utilization in *Dictyostelium discoideum* is presented in Table I. Several large genes have recently been sequenced, contributing statistically to the trends since the compilation was last published (Kimmel and Firtel, 1983). Table I contains information on coding regions from 12 genes or gene fragments. Coding regions which belong to gene families are only represented once (i.e., discoidin, actin, and cysteine proteases) in order not to bias the count with numbers of proteins which differ only slightly. The results indicate that the codon choice in *Dictyostelium* is strongly biased. This bias can be applied to help identify open-reading frames in *Dictyostelium* DNA and has implications for heterologous gene expression. To date, one of the three possible termination codons (TAA) appears to be used exclusively.

REFERENCES

Kimmel, A. R., and Firtel, R. A. (1983). *Nucleic Acid Res.* **11**, 541–552.

METHODS IN CELL BIOLOGY, VOL. 28

TABLE I

Codon Frequency in *Dictyostelium discoideum*[a]

Amino acid	Codon	Total	MHC	preD11	discIA	lowM4	RAS	cysPRO1	DHDH	csitA	actin8	calmo	actinin	eb4
G	GGG	1	0	0	0	0	0	1	0	0	0	0	0	0
G	GGA	27	0	0	0	0	1	8	5	12	1	0	0	0
G	GGT	226	61	15	15	2	10	18	22	24	29	10	19	1
G	GGC	9	1	1	1	0	1	0	2	2	0	1	0	0
Q	GAG	36	15	3	0	1	3	5	6	1	0	0	2	0
Q	GAA	450	282	12	8	7	13	20	12	12	28	18	37	1
D	GAT	258	117	9	14	4	14	15	20	13	17	12	21	2
D	GAC	47	23	0	0	0	1	2	3	5	5	4	4	0
V	GTG	5	0	1	0	0	0	0	0	3	0	0	0	1
V	GTA	41	3	3	3	1	3	7	7	8	2	0	1	3
V	GTT	171	53	11	15	2	9	13	9	25	11	8	15	0
V	GTC	81	50	2	4	0	0	0	4	6	10	5	5	0
A	GCG	0	0	0	0	0	0	0	0	0	0	0	0	0
A	GCA	46	6	2	2	0	7	7	13	4	0	2	3	0
A	GCT	200	100	3	13	3	2	11	11	16	12	6	19	4
A	GCC	109	67	1	3	0	0	2	1	5	18	0	10	2
R	AGG	1	1	0	0	0	0	0	0	0	0	0	0	0
R	AGA	77	32	4	5	0	8	6	3	1	6	4	6	2
S	AGT	57	14	3	0	2	6	7	15	5	0	1	3	1
S	AGC	15	7	0	1	2	0	1	0	2	0	0	1	1
K	AAG	133	105	2	1	1	2	6	4	2	2	0	8	0
K	AAA	308	164	20	6	1	13	14	23	11	17	8	30	1
N	AAT	167	50	4	14	8	4	24	20	27	3	3	8	2
N	AAC	110	49	8	7	4	0	5	1	7	7	5	10	7
M	ATG	94	20	8	2	1	3	5	13	5	18	9	7	3
I	ATA	19	0	2	0	1	0	4	3	8	0	0	0	1

	Codon													
I	ATT	173	44	7	9	1	14	20	19	30	10	4	14	1
I	ATC	116	56	1	4	7	0	3	7	8	17	3	8	2
T	ACG	1	0	0	0	1	0	0	0	0	0	0	0	0
T	ACA	68	3	9	1	0	4	7	8	33	1	0	2	0
T	ACT	143	42	8	12	3	3	11	5	36	10	4	9	0
T	ACC	120	49	6	13	0	0	0	6	16	14	5	11	0
W	TGG	37	9	1	5	0	0	6	6	0	4	0	6	0
End	TGA	0	0	0	0	0	0	0	0	0	0	0	0	2
C	TGT	76	5	36	7	0	3	9	4	4	3	0	3	0
C	TGC	14	2	7	1	1	0	0	1	1	1	0	0	0
End	TAG	0	0	0	0	0	0	0	0	0	0	0	0	0
End	TAA	8	1	1	1	0	0	1	1	0	0	1	0	0
Y	TAT	76	20	7	2	0	1	12	11	9	4	1	3	0
Y	TAC	64	23	1	9	3	7	4	1	2	11	1	8	0
L	TTG	56	26	2	1	0	1	2	8	2	0	5	10	0
L	TTA	213	100	8	8	1	4	10	19	16	19	3	17	0
F	TTT	92	14	1	4	8	10	16	8	16	3	5	10	0
F	TTC	90	46	2	0	2	5	6	3	4	11	0	4	3
S	TCG	6	0	0	11	3	0	0	1	2	0	3	0	0
S	TCA	147	48	3	2	0	0	11	11	23	15	1	12	1
S	TCT	103	48	3	1	8	9	4	6	14	5	0	9	1
S	TCC	35	16	3	0	1	2	2	0	3	4	0	4	0
R	CGG	2	0	1	0	0	1	0	1	0	0	0	0	0
R	CGA	2	0	1	0	0	0	0	0	0	0	2	0	0
R	CGT	132	90	3	7	0	0	1	2	1	12	0	0	1
R	CGC	0	0	0	0	0	3	0	0	0	0	0	10	1
E	CAG	7	0	3	14	0	0	1	0	1	0	0	0	0
E	CAA	229	115	13	2	1	12	12	11	13	10	4	22	1
H	CAT	46	10	2	3	2	3	4	8	4	3	0	8	1
H	CAC	24	10	0	0	2	0	1	1	0	6	1	0	0
L	CTG	0	0	0	0	0	0	0	0	1	0	0	0	0
L	CTA	10	0	6	0	0	2	1	0	0	0	0	0	0

(continued)

499

TABLE I (*Continued*)

Amino acid	Codon	Total	MHC	preD11	discIA	lowM4	RAS	cysPRO1	DHDH	csitA	actin8	calmo	actinin	eb4
L	CTT	60	22	6	1	1	1	6	2	8	1	3	9	0
L	CTC	112	71	5	5	0	0	1	3	3	7	0	15	2
P	CCG	1	0	0	0	0	0	0	0	1	0	0	0	0
P	CCA	153	27	18	7	2	2	10	18	36	19	2	9	3
P	CCT	14	0	4	2	0	0	2	2	3	0	0	1	0
P	CCC	1	0	1	0	0	0	0	0	0	0	0	0	0
Total amino acids		5119	2117	283	254	87	187	344	370	495	377	140	414	51
References[b]			1	2	3	4	5	6	7	8	9	10	11	12

[a]Abbreviations: large myosin heavy chain gene (MHC), prestalk D11 gene (preD11), discoidin I A gene (discIA), low abundance class M4 gene fragment (lowM4), ras gene (RAS), cysteine proteinase 1 gene (cysPRO1), dihydroorotate dehydrogenase gene (DHDH), contact site A gene fragment (csitA), actin-8 gene (actin8), calmodulin gene (calmo), α-actinin gene fragment (actinin), prespore EB4 gene fragment (eb4).

[b]References: (1) Warrick, H. M., De Lozanne, A., Leinwand, L., and Spudich, J. A. (1986). *Proc. Natl. Acad. Sci. U.S.A.* **83**, 9433–9437. (2) Barklis, E., Pontius, B., and Lodish, H. F. (1985). *Mol. Cell. Biol.* **5**, 1473–1479. (3) Poole, S., Firtel, R. A., Lamar, E., and Rowekamp, W. (1981). *J. Mol. Biol.* **153**, 273–289. (4) Kimmel, A. R., and Firtel, R. A. (1980). *Nucleic Acids Res.* **8**, 5599–5610. (5) Reymond, C. D., Gomer, R. H., Mehdy, M. C., and Firtel, R. A. (1984). *Cell* **39**, 141–148. (6) Williams, J. G., North, M. J., and Mahbubani, H. (1985). *EMBO J.* **4**, 999–1006. (7) Jacquet, M., Kalekine, M., and Boy-Marcotte, E. (1985). *Biochimie* **67**, 583–588. (8) Nogel, A., Gerisch, G., Stadler, J., and Westphal, M. (1986). *EMBO J.* **5**, 1473–1476. (9) Romans, P., and Firtel, R. A. (1985). *J. Mol. Biol.* **186**, 321–335. (10) Goldhagen, H., and Clarke, M. (1986). *Mol. Cell. Biol.* **6**, 1851–1854. (11) Witke, W., Schleicher, M., Lottspeich, F., and Noegel, A. (1986). *J. Cell Biol.* **103**, 969–975. (12) Barklis, E., Pontius, B., Barfield, K., and Lodish, H. F. (1985). *Mol. Cell. Biol.* **5**, 1465–1472.

Index

A

Acetylation, actin, 234
Actin
 acetylation, inhibition, 234
 cytoskeleton, colloidal gold probes, 196–197
 deuterated
  characterization, 226–227
  purification
   ammonium sulfate backwash, 225
   cell lysis and centrifugation, 223–224
   control of proteolysis, 222–223
   gel filtration chromatography, 224–226
   ion exchange chromatography, 225
   sedimentation and polymerization, 225
 filaments, ultrastructural localization with
  myosin fragments, 203–205
 [^{35}S]methionine-labeled containing NH$_2$-ter-
  minal Ac-Met
   Dictyostelium RNA containing translatable
    actin mRNA, 233
   preactin synthesis, 234
 NH$_2$ terminus, processing specificity, 241–
  242
 processing enzymes
  preparation, 235–236
  rabbit reticulocyte lysate, *Dictyostelium*
   extract, and rat liver enzyme, 236
 removal from plasma membranes, 112
 removal of Ac-Met from actin precursor NH$_2$
  terminus
   inhibition of actin acetylation, 234
   preparation of actin-processing enzymes,
    235–236
   procedure, 237–241
   synthesis of *S*-acetonyl CoA, 235
Actin-binding protein, localization in sonicated
 cells, 184–185
Actin filaments, ultrastructural localization with
 myosin fragments
  artifacts, 205
  methods, 204–205
Actin matrix, cortical

cytoskeleton isolation, 212–213
cytoskeleton preparation
 by detergent lysis, 210–212
 purification of actin from, 213
 quantitation of actin in, 213
Adenylate cyclase
 activation
  cAMP stimulation of intact cells, 315
  GMPPNP stimulation, 315–316
  mutants and reconstitution assays, 317
  soluble activator, 317–318
 assay, 313–314
 basal activity, 314–315
  crude extracts, 314
  membranes, purification, and solubilization, 314–315
Adherence, sonication-loaded cells, 183
Adhesion
 assay, 133
 to bacteria, *see* Cell–bacteria adhesion
 cell–substratum, *see* Cell–substratum
  adhesion
 homotypic cell–cell, *see* Cell–cell adhesion,
  homotypic
Adhesive revertants, isolation, 135
Affinity chromatography, discoidins I and II,
 391
 purification on Sepharose 4B, 389–390
 separation on *N*-acetylgalactosamine-deri-
  vatized agarose beads, 391
Affinity purification
 antiserum, 477
 contact site A glycoprotein with monoclonal
  antibodies, 378–379
 immune serum, 196
Agar-overlay technique for immunofluorescence, 348–351
 extraction of soluble substance, 349
 fixation protocols, 349
 preparation
  agarose sheet, 348
  cells, 348–349